The Atmospheric Chemist's Companion

Peter Warneck • Jonathan Williams

The Atmospheric Chemist's Companion

Numerical Data for Use in the Atmospheric Sciences

 Springer

Dr. Peter Warneck
Max Planck Institute for Chemistry
Mainz
Germany

Dr. Jonathan Williams
Max Planck Institute for Chemistry
Mainz
Germany

ISBN 978-94-007-2274-3   e-ISBN 978-94-007-2275-0
DOI 10.1007/978-94-007-2275-0
Springer Dordrecht Heidelberg London New York

Library of Congress Control Number: 2012930992

Printed on acid-free paper

Springer is part of Springer Science+Business Media (www.springer.com)

Preface

A survey conducted several years ago within the local atmospheric science community had indicated the need for a comprehensive reference book of atmospheric chemistry and physics. The present compilation of data has been prepared in an attempt to fill this need.

While the subject, as a whole or in parts, has received an adequate treatment in textbooks, encyclopedias, and in other overviews, these publications do not generally present numerical data but discuss important observations by way of illustrations. Even the current scientific literature tends to display observations and results graphically and often relegates numerical data, on which the results are based, to unpublished supplements made available only upon request. Because of this practice the data are prone to sink into oblivion sooner or later. Yet, numerical data are required whenever a quantitative assessment is to be made of the processes at work in the atmosphere.

The aim of the present data collection is to assemble, in one handy volume, frequently needed fundamental data as well as observational data on the structure and the chemical composition of Earth and its atmosphere. Thereby, we hope to assist atmospheric scientists in their daily work and to provide detailed information as a reference to other persons interested in the subject. The data are mostly arranged in the form of annotated tables; any explanatory text is kept to a minimum. The sources of the material presented are indicated at the bottom of each table (or figure) with literature citations being given at the end of each section. Because of the range of the subject we have added a Glossary to provide explanations and definitions of technical terms.

In a compilation of this type the occurrence of unwanted errors will be inevitable. We shall be grateful to all users that notify us of such errors so that corrections can be made in a future edition. We also recognize that some of the data might have to be replaced by more recent or more accurate material not yet known to us, and that additional data, currently not included, would be of interest. We encourage all users to alert us of the existence of such material so that we may consider it for inclusion in a future edition. Any suggestions should be directed to the e-mail address atmoschemcompanion@lists.mpic.de.

We are pleased to acknowledge the unswerving support by our editors at Springer: Gert-Jan Geraeds, who initiated this project, and Robert Doe, whose help was essential in bringing the project to completion. We are particularly grateful to Andreas Zimmer at the Max-Planck-Institute's library for supplying us with much of the material from which data were selected.

We thank the AGU, Elsevier Ltd and IUPAC for permissions to reproduce material from other sources. We are also grateful for the help and advice from our colleagues, in particular Janeen Auld, Steven S. Brown, Stephanie Brückner, Tim Butler, John Crowley, Frank Drewnick, Günter Helas, Peter Hoor, Marloes Penning de Vries, Rolf Sander, Lothar Schütz, Claudia Stubenrauch, Thomas Wagner.

Peter Warneck and Jonathan Williams

Contents

IUPAC Periodic Table of the Elements

Key:

atomic number
Symbol
name
standard atomic weight

1																	18
1 **H** hydrogen [1.007, 1.009]	2											13	14	15	16	17	2 **He** helium 4.003
3 **Li** lithium [6.938, 6.997]	4 **Be** beryllium 9.012											5 **B** boron [10.80, 10.83]	6 **C** carbon [12.00, 12.02]	7 **N** nitrogen [14.00, 14.01]	8 **O** oxygen [15.99, 16.00]	9 **F** fluorine 19.00	10 **Ne** neon 20.18
11 **Na** sodium 22.99	12 **Mg** magnesium 24.31	3	4	5	6	7	8	9	10	11	12	13 **Al** aluminium 26.98	14 **Si** silicon [28.08, 28.09]	15 **P** phosphorus 30.97	16 **S** sulfur [32.05, 32.08]	17 **Cl** chlorine [35.44, 35.46]	18 **Ar** argon 39.95
19 **K** potassium 39.10	20 **Ca** calcium 40.08	21 **Sc** scandium 44.96	22 **Ti** titanium 47.87	23 **V** vanadium 50.94	24 **Cr** chromium 52.00	25 **Mn** manganese 54.94	26 **Fe** iron 55.85	27 **Co** cobalt 58.93	28 **Ni** nickel 58.69	29 **Cu** copper 63.55	30 **Zn** zinc 65.38(2)	31 **Ga** gallium 69.72	32 **Ge** germanium 72.63	33 **As** arsenic 74.92	34 **Se** selenium 78.96(3)	35 **Br** bromine 79.90	36 **Kr** krypton 83.80
37 **Rb** rubidium 85.47	38 **Sr** strontium 87.62	39 **Y** yttrium 88.91	40 **Zr** zirconium 91.22	41 **Nb** niobium 92.91	42 **Mo** molybdenum 95.96(2)	43 **Tc** technetium	44 **Ru** ruthenium 101.1	45 **Rh** rhodium 102.9	46 **Pd** palladium 106.4	47 **Ag** silver 107.9	48 **Cd** cadmium 112.4	49 **In** indium 114.8	50 **Sn** tin 118.7	51 **Sb** antimony 121.8	52 **Te** tellurium 127.6	53 **I** iodine 126.9	54 **Xe** xenon 131.3
55 **Cs** caesium 132.9	56 **Ba** barium 137.3	57-71 lanthanoids	72 **Hf** hafnium 178.5	73 **Ta** tantalum 180.9	74 **W** tungsten 183.8	75 **Re** rhenium 186.2	76 **Os** osmium 190.2	77 **Ir** iridium 192.2	78 **Pt** platinum 195.1	79 **Au** gold 197.0	80 **Hg** mercury 200.6	81 **Tl** thallium [204.3, 204.4]	82 **Pb** lead 207.2	83 **Bi** bismuth 209.0	84 **Po** polonium	85 **At** astatine	86 **Rn** radon
87 **Fr** francium	88 **Ra** radium	89-103 actinoids	104 **Rf** rutherfordium	105 **Db** dubnium	106 **Sg** seaborgium	107 **Bh** bohrium	108 **Hs** hassium	109 **Mt** meitnerium	110 **Ds** darmstadtium	111 **Rg** roentgenium	112 **Cn** copernicium						

57 **La** lanthanum 138.9	58 **Ce** cerium 140.1	59 **Pr** praseodymium 140.9	60 **Nd** neodymium 144.2	61 **Pm** promethium	62 **Sm** samarium 150.4	63 **Eu** europium 152.0	64 **Gd** gadolinium 157.3	65 **Tb** terbium 158.9	66 **Dy** dysprosium 162.5	67 **Ho** holmium 164.9	68 **Er** erbium 167.3	69 **Tm** thulium 168.9	70 **Yb** ytterbium 173.1	71 **Lu** lutetium 175.0
89 **Ac** actinium	90 **Th** thorium 232.0	91 **Pa** protactinium 231.0	92 **U** uranium 238.0	93 **Np** neptunium	94 **Pu** plutonium	95 **Am** americium	96 **Cm** curium	97 **Bk** berkelium	98 **Cf** californium	99 **Es** einsteinium	100 **Fm** fermium	101 **Md** mendelevium	102 **No** nobelium	103 **Lr** lawrencium

International Year of **CHEMISTRY** 2011

Notes

- IUPAC 2009 Standard atomic weights abridged to four significant digits (Table 4 published in Pure Appl. Chem. 83, 359-396 (2011);
doi:10.1351/PAC-REP-1009-14). The uncertainty in the last digit of the standard atomic weight value is listed in parentheses following the value.
In the absence of parentheses, the uncertainty is one in that last digit. An interval in square brackets provides the lower and upper bounds of the
standard atomic weight for that element. No values are listed for elements which have no stable isotopes. See PAC for more details.

- "Aluminum" and "cesium" are commonly used alternative spellings for "aluminium" and "caesium."

For updates to this table, see iupac.org/reports/periodic_table/. This version is dated 21 January 2011.
Copyright © 2011 IUPAC, the International Union of Pure and Applied Chemistry.

Chapter 1
Fundamental Quantities and Units

1.1 Fundamental Constants

Table 1.1 Fundamental and frequently used constants

Quantity	Symbol	Value[a]
Universal constants		
Gravitational constant	G	$6.6742(10) \times 10^{-11}$ m^3 kg^{-1} s^{-2}
Speed of light in vacuum	c_0	299 792 458 m s^{-1} (defined)
Permeability of vacuum	μ_0	$4\pi \times 10^{-7}$ N A^{-2} (defined)
Permittivity of vacuum	$\varepsilon_0 = 1/(\mu_0 c_0^2)$	$8.854\ 187\ 816... \times 10^{-12}$ F m^{-1}
Planck constant	h	$6.626\ 069\ 3(11) \times 10^{-34}$ J s
Atomic constants		
Elementary charge	e	$1.602\ 176\ 53(14) \times 10^{-19}$ C
Electron rest mass	m_e	$9.109\ 3826(16) \times 10^{-31}$ kg
Electron specific charge	$-e/m_e$	$1.758\ 820\ 12(15) \times 10^{11}$ C kg^{-1}
Proton rest mass	m_p	$1.672\ 621\ 71(29) \times 10^{-27}$ kg
Neutron rest mass	m_n	$1.674\ 927\ 28(29) \times 10^{-27}$ kg
Unified atomic mass unit	$(m^{12}C/12) = 1$ u	$1.660\ 538\ 86(28) \times 10^{-27}$ kg
Bohr radius	$a_0 = \varepsilon_0 h^2/\pi m_e e^2$	$5.291\ 772\ 108(18) \times 10^{-11}$ m
Hartree energy	$E_h = h^2/4\pi^2 m_e a_0^2$	$4.359\ 744\ 17(75) \times 10^{-18}$ J
Rydberg constant	$R_\infty = E_h/2hc_0$	10 973 731.568 525(73) m^{-1}
Fine structure constant	$\alpha = \mu_0 e^2 c_0/2hc_0$	$7.297\ 352\ 568(24) \times 10^{-3}$
	α^{-1}	137.035 999 11(46)
Bohr magneton	$\mu_B = eh/4\pi m_e$	$9.274\ 000\ 949(80) \times 10^{-24}$ J T^{-1}
Electron magnetic moment	μ_e	$9.284\ 764\ 12(80) \times 10^{-24}$ J T^{-1}
Nuclear magneton	$\mu_N = (m_e/m_p)\,\mu_B$	$5.050\ 783\ 43(43) \times 10^{-27}$ J T^{-1}
Proton magnetic moment	μ_p	$1.410\ 606\ 71(12) \times 10^{-26}$ J T^{-1}
Physicochemical constants		
Avogadro constant	N_A	$6.022\ 14179(30) \times 10^{23}$ mol^{-1}
Molar gas constant	R	$8.314\ 472(15)$ J mol^{-1} K^{-1}
Boltzmann constant	$k = R/N_A$	$1.380\ 6504(24) \times 10^{-23}$ J K^{-1}
Zero of Celsius scale		273.15 K (defined)

(continued)

P. Warneck and J. Williams, *The Atmospheric Chemist's Companion: Numerical Data for Use in the Atmospheric Sciences*, DOI 10.1007/978-94-007-2275-0_1,
© Springer Science+Business Media B.V. 2012

Table 1.1 (continued)

Quantity	Symbol	Value[a]
Molar volume (ideal gas)	V_m	$2.241\ 3996(39) \times 10^{-2}$ m³ mol⁻¹
Loschmidt constant	$n_0 = N_A/V_m$	$2.686\ 7774(47) \times 10^{25}$ m⁻³
Standard atmosphere	atm	101 325 Pa (defined)
Faraday constant	F	$9.648\ 533\ 83(83) \times 10^4$ C mol⁻¹
Radiation constants		
Stefan-Boltzmann constant	$\sigma = 2\pi^5 k^4/15h^3c_0^2$	$5.670\ 400(40) \times 10^{-8}$ W m⁻² K⁻⁴
Radiation density constant	$a = 4\sigma/c_0$	$7.565\ 767(53) \times 10^{-16}$ J m⁻³ K⁻⁴
First radiation constant	$c_1 = 2\pi hc_0^2$	$3.741\ 771\ 38(64) \times 10^{-16}$ W m²
Second radiation constant	$c_2 = hc_0/k$	$1.438\ 7752(25) \times 10^{-2}$ m K
Wien displacement law constant	$b = \lambda_{max} T$	$2.897\ 7685(51) \times 10^{-3}$ m K
Astronomical constants		
Standard acceleration of free fall	g_a	9.806 65 m s⁻¹ (defined)
Sidereal second	s	0.9972696 s
Mean sidereal year	yr	365.256 36 d = 315 581 49.5 s
Tropical year (equinox to equinox)	yr	365.242 189 7 d
Anomalistic year (perihelion to perihelion)	yr	365.259 64 d
Gregorian calendar year	a, yr	365.242 5 d
Astronomical unit	AU	$1.495\ 978\ 061 \times 10^{11}$ m
Light travel time for 1 AU		499.004 783 5 s
Solar constant (at 1 AU)		1.3676 kW m⁻²
Earth mass	M_e	5.9736×10^{24} kg
Mean Earth radius	R_e	6371.01 km
Solar mass	M_s	1.98910×10^{30} kg
Solar radius	R_s	695950 km
Solar effective surface temperature	T_{eff}	5778 K
Solar absolute luminosity	L	3.8268×10^{26} W
Mathematical constants		
Ratio circumference/diameter of a circle	π	3.141 592 653 59
Base of natural logarithm	e	2.718 281 828 46
Natural logarithm of 10	ln 10	2.302 585 092 99

For a continual updating of values compare with www.physics.nist.gov/constants

[a]The standard deviation uncertainty in the least significant digits is shown in parentheses

1.2   Units for Use in Atmospheric Chemistry

Introductory Comments:

The international system of units (système international d'unités, SI) has now largely replaced all earlier systems of units used to describe physical quantities. SI is built upon seven base quantities each having its own dimension: length, mass, time, thermodynamic temperature, amount of substance (chemical amount), electric current, and luminous intensity (Bureau International des Poids et Mesures 1991). All other quantities are derived quantities that acquire dimensions derived algebraically from the seven base quantities by multiplication and division. The possibility of combining SI units with prefixes designating decimal multiples or submultiples of units provides additional flexibility when dealing with quantities that range over many orders of magnitude, a feature which makes SI especially suitable for use in the atmospheric sciences. Accordingly, SI units are strongly recommended for use in atmospheric chemistry. However, some non-SI units are still required, and they will remain in use along with SI. Examples are time periods such as minute, hour, day and year, which can be expressed in terms of the second, but which defy decimalization. The tables that follow provide an overview on SI, on units that are not part of SI but are used along with it, and on a variety of non-SI units still in use together with the conversion factors to the corresponding SI units. For the correct treatment of symbols and units the reader should consult the manual *Quantities, Units and Symbols in Physical Chemistry* by Mills et al. (1993).

The suitability of SI for use in atmospheric chemistry has been examined by Schwartz and Warneck (1995). In the atmospheric sciences, some non-SI units have remained in favor mainly for historic reasons. In most cases, these units can be replaced by SI units without difficulty. Where non-SI units are preferred, the appropriate conversion rule to SI units should be indicated (example: 1 atm = 101325 Pa). There are a number of other aspects regarding units used in atmospheric chemistry that require attention. The most important aspects will be briefly summarized below.

Time: The length of minute, hour and day are defined exactly in terms of the second (see Table 1.4). For the length of the year, which is somewhat variable because of the occurrence of leap years, the mean Gregorian year (365.2425 d) is recommended, or the actual number of days (365 or 366) multiplied by 86 400 s d^{-1} for any specific year. Because of variability in the length of the calendar month, its use as a unit of time should be avoided. Seasonal variations of atmospheric quantities should be reported on the basis of actual calendar days.

Concentration is the amount of a substance of concern in a given volume divided by that volume. It may be expressed in terms of amount of substance per volume (mol m^{-3}), number of molecules (atoms, particles, or other entities) per volume (m^{-3}), mass per volume (kg m^{-3}), or volume per volume (m^3 m^{-3}). Chemical amount concentration is appropriate for a substance of known chemical formula; number concentration is appropriate for aerosol particles, and it is frequently used also for gaseous substances (see further below); mass concentration is required for a substance whose composition or chemical formula is unknown or indeterminate, such as particulate matter suspended in air; volume concentration is appropriate for the volume of condensed phase matter per volume of air, for example, cloud water.

Units of concentration also are involved when dealing with rates of chemical reactions. The most appropriate SI unit in this application would be mol m^{-3}. This unit is, in fact, used for concentrations of solutes in the aqueous phase of clouds and rain water in the form mol dm^{-3}. Regarding gas phase reactions, however, atmospheric chemists have for some time favored number concentration as a unit for concentration, with a widespread use of the unit molecule cm^{-3} (atom cm^{-3}, etc.). This unit violates the rule that a qualifying name should not be part of a unit. On the other hand, it designates the entity involved but not the name of the molecule, atom, etc., which must be specified separately. Thus, it distinguishes between several possibilities and helps to remove potential ambiguities. For this reason, the use of molecule cm^{-3} (atom cm^{-3}, etc.) as a unit for number concentration is currently tolerated.

Mixing ratio in atmospheric chemistry is defined as the ratio of the amount (mass, volume) of the substance of concern in a given volume to the amount (mass, volume) of all constituents of air in that volume. Here, air denoted gaseous substances, including water vapor, but not condensed phase water or particulate matter. Mixing ratio is frequently employed to quantify the abundance of a trace gas in air. The specific advantage of mixing ratio over concentration in this context is that mixing ratio is unchanged by differences in pressure or temperature associated with altitude or with meteorological variability, whereas concentration depends on pressure and temperature in accordance with the equation of state. Since mixing ratio refers to the total gas mixture, the presence of water causes the mixing ratio to vary somewhat with humidity. For this reason it is preferable to refer to dry air when reporting mixing ratios of trace gases in the atmosphere whenever possible.

The above definition of mixing ratio is identical to the fraction that the amount (mass, volume) contributes to the total amount (mass, volume) of the whole mixture. For gaseous species chemical amount fraction and volume fraction are practically identical because air at atmospheric pressure is essentially an ideal gas. Chemical amount fraction is preferable, however, because it is applicable also to condensed phase species present in the same volume. Thus, one can express abundances of gaseous SO_2, NH_3, and HNO_3 as well as aerosol sulfate, ammonium, and nitrate all as chemical amount fraction and thereby immediately infer chemically meaningful relationships among these quantities.

Chemical amount fraction, like any other fraction, is a quantity of dimension one and does not require the application of units. A notation widely used in atmospheric chemistry to quantify mixing ratios includes ppm (parts per million = 10^{-6}), ppb (parts per billion = 10^{-9}), and ppt (parts per trillion = 10^{-12}). Because of inherent ambiguities (for example, whether one is dealing with the mass fraction or the chemical amount fraction), there is an increasing tendency to apply appropriate SI units to specify gas phase mixing ratios. Thus, it is strongly recommended that the above notations be replaced by, respectively, $\mu mol\ mol^{-1}$, $nmol\ mol^{-1}$, and $pmol\ mol^{-1}$. If the mixing ratio is expressed as a mass fraction, the corresponding units would be $mg\ kg^{-1}$, $\mu g\ kg^{-1}$, and $ng\ kg^{-1}$, respectively.

Light intensity: The SI base unit candela is not used in atmospheric chemistry. The intensity of solar radiation is expressed in terms of other base units of SI, depending on the application. With regard to the energy flux, the appropriate unit for irradiance is $W\ m^{-2}$. In photochemical considerations, it is generally desirable to express the radiation flux as a flux of photons, because in the act of optical absorption a single photon is taken up by the absorbing molecule. The process resembles that of a chemical reaction between molecules and photons. In the photochemically important ultraviolet region of the solar spectrum, the spectral irradiance of the sun is presented in terms of photon $m^{-2}\ s^{-1}$ for specific wavelength intervals. Although this usage violates the convention that qualifying names should not be included in the unit, the practice is tolerated for the same reasons as in the case of units for concentration.

References

Bureau International des Poids et Mesures, *Le Système International d'Unités (SI)*, 6th French and English edition (BIPM, Sèvres, 1991)

Mills, I., T. Cvitaš, K. Homann, N. Kallay, K. Kuchitsu, *Quantities, Units and Symbols in Physical Chemistry*, 2nd edn. (International Union of Pure and Applied Chemistry, Blackwell Scientific Publications, Oxford, 1993)

Schwartz, S.E., P. Warneck, Pure and Appl. Chem. **67**, 1377–1406 (1995)

Table 1.2 SI base units

Quantity	Frequently used formula symbol	Name of SI unit	Symbol for SI unit
Length	l	metre	m
Mass	m	kilogram	kg
Time	t	second	s
Amount of substance	n	mole	mol
Thermodynamic temperature	T	kelvin	K
Electric current	I	ampere	A
Luminous intensity	I_v	candela	cd

Definitions of SI Base Units (Bureau International des Poids et Mesures 1991):

metre: The metre is the length of path traveled by light in vacuum during a time interval of 1/299 792 458 of a second.

kilogram: The kilogram (unit of mass) is equal to the mass of the international prototype of the kilogram.

second: The second is the duration of 9 192 631 770 periods of the radiation corresponding to the transition between the two hyperfine levels of the ground state of the cesium-133 atom.

mole: The mole is the amount of substance of a system, which contains as many elementary entities as there are atoms in 0.012 kg of carbon-12. When the mole is used, the elementary entities must be specified (atoms, molecules, ions, particles, etc.).

ampere: The ampere is that constant current which, if maintained in two straight parallel conductors of infinite length, of negligible cross section, and placed 1 m apart in vacuum, would produce between them force equal to 2×10^{-7} N/m of length.

candela: The candela is the luminous intensity in a given direction of a source that emits monochromatic radiation of frequency 540×10^{12} Hz and has a radiant intensity in that direction of 1/683 W/per steradian.

Table 1.3 SI prefixes[a] to indicate decimal multiples and sub-multiples of SI units

Prefix	Submultiple	Symbol	Prefix	Multiple	Symbol
deci	10^{-1}	d	deca	10	da
centi	10^{-2}	c	hecto	10^2	h
milli	10^{-3}	m	kilo	10^3	k
micro	10^{-6}	μ	mega	10^6	M
nano	10^{-9}	n	giga	10^9	G
pico	10^{-12}	p	tera	10^{12}	T
femto	10^{-15}	f	peta	10^{15}	P
atto	10^{-18}	a	exa	10^{18}	E
zepto	10^{-21}	z	zetta	10^{21}	Z
yocto	10^{-24}	y	yotta	10^{24}	Y

[a]Prefix symbols should be printed in roman (upright) type without space between prefix and unit symbol (example: kilometre, km). The combination of prefix and SI symbol is taken to represent a new symbol that can be raised to any power without the use of parentheses (example: $cm^3 = 10^{-6}\ m^3$)

Table 1.4 Time periods used in atmospheric chemistry

Name of unit	Symbol	Relation to SI
second	s	SI unit
minute	min	60 s
hour	h	3600 s
day	d	86400 s
year (mean Gregorian)[a]	a	31 556 952 s (exactly)

[a]Mean Gregorian year = 365.2425 d. The symbol a (abbreviation for annum) is a recommendation taken from the ISO Standards Handbook 2 (1993) *Quantities and Units*, International Organization for Standardization, Central Secretariat, Geneva, Switzerland

Table 1.5 Definitions of SI derived units with special names and symbols

Physical quantity	Name of SI unit	Symbol for SI unit	Expression in terms of SI base units
Frequency[a]	hertz	Hz	s^{-1}
Force	newton	N	$m\ kg\ s^{-2}$
Pressure, stress	pascal	Pa	$N\ m^{-2} = m^{-1}\ kg\ s^{-2}$
Energy, work, heat	joule	J	$N\ m = m^2\ kg\ s^{-2}$
Power, radiant flux	watt	W	$J\ s^{-1} = m^2\ kg\ s^{-3}$
Electric charge	coulomb	C	$A\ s$
Electric potential	volt	V	$J\ C^{-1} = m^2\ kg\ s^{-3}\ A^{-1}$
Electric resistance	ohm	Ω	$V\ A^{-1} = m^2\ kg\ s^{-3}\ A^{-2}$
Electric conductance	siemens	S	$\Omega^{-1} = m^{-2}\ kg^1\ s^3\ A^2$
Electric capacitance	farad	F	$C\ V^{-1} = m^{-2}\ kg^{-1}\ s^4\ A^2$
Magnetic flux density	tesla	T	$V\ s\ m^{-2} = kg\ s^{-2}\ A^{-1}$
Magnetic flux	weber	Wb	$V\ s = m^2\ kg\ s^{-2}\ A^{-1}$
Inductance	henry	H	$V\ A^{-1}\ s = m^2\ kg\ s^{-2}\ A^{-2}$
Luminous Flux	lumen	lm	$cd\ sr$
Illuminance	lux	lx	$cd\ sr\ m^{-2}$
Radioactive decay rate[b]	becquerel	Bq	s^{-1}

(continued)

Table 1.5 (continued)

Physical quantity	Name of SI unit	Symbol for SI unit	Expression in terms of SI base units
Absorbed radiation dose[b]	gray	Gy	$J\ kg^{-1} = m^2\ s^{-2}$
Dose equivalent[b]	sievert	Sv	$J\ kg^{-1} = m^2\ s^{-2}$
Plane angle[c]	radian	rad	$1 = m\ m^{-1}$
Solid angle[c]	steradian	sr	$1 = m^2\ m^{-2}$

[a]The unit Hz should be used only for frequency in the sense of cycles per second
[b]The units becquerel, gray and sievert are admitted for reasons of safeguarding human health
[c]The units radian (defined as the angle subtended by an arc equal to the radius) and steradian (the solid angle which encloses a surface on the sphere equivalent to the square of the radius) are treated as SI supplementary units. Since they are of dimension 1 they may as well be omitted from the list of SI derived units. In practice, rad and sr may be used when appropriate and may be omitted if this does not lead to loss of clarity

Table 1.6 Conversion factors for selected units

length, l	metre (SI unit), m	
	Ångström: $1\ \text{Å} = 10^{-10}$ m	foot: 1 ft = 12 in = 0.3048 m
	x unit: $1\ X = 1.00206 \times 10^{-13}$ m	yard: 1 yd = 3 ft = 0.9144 m
	fermi: $1\ fm = 10^{-15}$ m	mile: 1 mi = 1760 yd = 1609.344 m
	inch: $1\ in = 2.54 \times 10^{-2}$ m	nautical mile: = 1852 m
plane angle, α	radian (SI unit), rad	
	degree: $1° = 17.4532925$ mrad	minute: $1' = 1°/60 = 0.290888209$ mrad
	second: $1'' = 1°/3600 = 4.8481368$ μrad	
area, A	square metre (SI unit), m^2	
	barn: $1\ b = 10^{-28}\ m^2$	are: $1\ a = 100\ m^2$
	acre: $1\ acre \approx 4046.856\ m^2$	hectare: $1\ ha = 10^4\ m^2$
	square mile: $1\ mi^2 = 2.589988\ km^2$	
volume, V	cubic metre (SI unit), m^3	
	litre: $1\ L = 1\ dm^3$	bushel (US, dry): 1 bu = 35.238 dm^3
	gallon (US): 1 gal = 3.78541 dm^3	bushel (UK, dry): 1 bu = 36.3687 dm^3
	gallon (UK): 1 gal = 4.54609 dm^3	barrel (US, oil): 1 bbl $\approx$ 158.987 dm^3
mass, m	kilogram (SI unit), kg	
	gram (cgs unit): $1\ g = 10^{-3}$ kg	pound (avoirdupois): 1 lb = 453.59237 g
	tonne: $1\ t = 10^3$ kg = 1 Mg	ounce (avoirdupois): 1 oz $\approx$ 28.34952 g
	long ton: = 1016.047 kg	ounce (troy): 1 oz (troy) $\approx$ 31.1035 g
	short ton (US): = 907.185 kg	grain: 1 gr = 64.79891 mg
force, F	newton (SI unit), N	
	dyne (cgs unit): $1\ dyn = 10^{-5}$ N	kilogram force: 1 kgf = 9.80665 N
pressure, p	pascal (SI unit), Pa	
	bar: $1\ bar = 10^5$ Pa, 1 mbar = 1 hPa	
	atmosphere: 1 atm = 101325 Pa	torr: 1 torr = 1/760 atm $\approx$ 133.3224 Pa
	millimetre of mercury: 1 mmHg = 13.5951×9.80665 Pa $\approx$ 133.3224 Pa	
	pounds per square inch: 1 psi = $6.894\ 757 \times 10^3$ Pa	

(continued)

Table 1.6 (continued)

energy, E, U	joule (SI unit), J
	erg (cgs unit): $= 10^{-7}$ J; electronvolt: $1\ eV = 1.602\ 1765 \times 10^{-19}$ J;
	calorie, thermochemical: $1\ cal_{th} = 4.184$ J;
	calorie, international: $1\ cal_{IT} = 4.1868$ J; 15 °C calorie: $1\ cal_{15} = 4.1855$ J;
	British thermal unit: $1\ Btu = 1055.06$ J.
power, P	watt (SI unit), W
	horse power: $1\ hp = 745.7$ W; kilocalorie per hour: $1\ kcal/h = 1.1628$ W
dynamic viscosity, η	SI unit: Pa s $=$ kg m^{-1} s^{-1}
	poise: $1\ P = 10^{-1}$ Pa s; centipoise: $1\ cP = 1$ mPa s
kinematic viscosity, ν	SI unit: m^2 s^{-1}
	stokes: $1\ St = 10^{-4}$ m^2 s^{-1}
temperature, T	kelvin (SI unit), K
	Celsius: $\theta/°C = T/K - 273.15$; Fahrenheit: $\theta_F/°F = (9/5)(\theta/°C) + 32$
molar volume, Vm	SI unit: m^3 mol^{-1}
	amagat[a]: 1 amagat $= 2.24141 \times 10^{-2}$ m^3 mol^{-1}(ideal gas);
	1 amagat $\approx 2.24 \times 10^{-2}$ m^3 mol^{-1}(real gas)
Molar column density	SI unit: mol m^{-2}
	Dobson unit: 1 DU ≈ 446.149 µmol m^{-2} (for atmospheric ozone)
Radioactivity, A	becquerel (SI unit), Bq
	curie: 1 Ci $= 3.7 \times 10^{10}$ Bq
Absorbed radiation dose	gray (SI unit), Gy
	roentgen[b]: 1 R $= 2.58 \times 10^{-4}$ C kg^{-1}; rad: 1 rd $= 10^{-2}$ Gy
magnetic flux density, B	tesla (SI unit), T
	gauss: 1 G $= 10^{-4}$ T
magnetic flux, ϕ	weber (SI unit), Wb
	maxwell: 1 Mx $= 1$ G cm^{-2} $= 10^{-8}$ Wb
magnetic field, H	SI unit: A m^{-1}
	oersted[c]: 1 Oe $= 10^3/4\pi$ A m^{-1}

[a]The amagat is defined as the mole volume of a real gas at 1 atm and 273.15 K
[b]The amount of radiation that will produce one electrostatic unit of ions per cm^3
[c]In practice, the oersted is only used as a unit for $4\pi H$

Table 1.7 Conversion factors for pressure units

Unit	Pa	bar	atm	torr	psi
1 Pa	1	10^{-5}	9.869233×10^{-6}	7.50062×10^{-3}	1.45038×10^{-4}
1 bar	10^5	1	0.986923	750.0615	14.50377
1 atm	101325	1.01325	1	760	14.69595
1 torr	133.3224	1.33322×10^{-3}	1.31579×10^{-3}	1	1.933678×10^{-2}
1 psi	6894.757	6.89476×10^{-2}	6.804596×10^{-2}	51.71491	1

Table 1.8 Conversion factors for concentrations and rate coefficients of chemical reactions

Quantity[a]	Unit	mol m⁻³, s	mol dm⁻³, s	molecule cm⁻³, s
c	1 mol m⁻³	1	10^{-3}	6.022014×10^{17}
	1 mol dm⁻³	10^3	1	6.022014×10^{20}
	1 molecule cm⁻³	1.66054×10^{-18}	1.66054×10^{-21}	1
k_{bim}	1 m³ mol⁻¹ s⁻¹	1	10^3	1.66054×10^{-18}
	1 dm³ mol⁻¹ s⁻¹	10^{-3}	1	1.66054×10^{-21}
	1 cm³ molecule⁻¹ s⁻¹	6.022014×10^{17}	6.022014×10^{20}	1
k_{term}	1 m³ mol⁻¹ s⁻¹	1	10^6	2.757389×10^{-36}
	1 dm⁶ mol⁻² s⁻¹	10^{-6}	1	2.757389×10^{-42}
	1 cm⁶ molecule⁻² s⁻¹	3.62662×10^{35}	3.62662×10^{41}	1

[a]Symbols: c concentration, k_{bim} rate coefficient for bimolecular reactions, k_{term} rate coefficient for termolecular reactions. To convert chemical amount concentration to mass concentration multiply mol m⁻³ by molar mass

Comments: To convert gas phase concentration to (partial) pressure requires the knowledge of ambient temperature. The quantities are related by the (ideal) gas law $p = nkT = cRT$, where p (Pa) is pressure, n (molecule m⁻³) is number concentration, $k \approx 1.38 \times 10^{-23}$ (J K⁻¹) is the Boltzmann constant, c (mol m⁻³) is molar concentration, $R \approx 8.31$ J mol⁻¹ K⁻¹ is the gas constant, and T (K) is temperature. For the conversion of pressure units use Table 1.7

Table 1.9 Conversion factors[a] for optical absorption cross section σ, absorption coefficient k, and molar absorption coefficient ε

To convert from	Base	To	Base	Multiply by
σ (cm² molecule⁻¹)	e	ε (dm³ mol⁻¹ cm⁻¹)	10	2.615325×10^{20}
σ (cm² molecule⁻¹)	e	ε (m² mol⁻¹)	10	2.615325×10^{19}
σ (cm² molecule⁻¹)	e	ε (m² mol⁻¹)	e	6.022141×10^{19}
ε (dm³ mol⁻¹ cm⁻¹)	10	σ (cm² molecule⁻¹)	e	3.823600×10^{-21}
ε (m² mol⁻¹)	10	σ (cm² molecule⁻¹)	e	3.823600×10^{-20}
k (cm⁻¹)	e	σ (cm² molecule⁻¹)	e	3.721931×10^{-20}
σ (cm² molecule⁻¹)	e	k (cm⁻¹)	e	2.686777×10^{19}

[a]The relations are $\ln(I_0/I) = \sigma n l$, $\ln(I_0/I) = kl = \sigma n_0 l$, and $\log_{10}(I_0/I) = \varepsilon c l$ where I_0/I is the ratio of incident to transmitted radiation along the path length l, n is the number concentration (molecule cm⁻³), $n_0 = 2.686777 \times 10^{19}$ (molecule cm⁻³) is Loschmidt's constant, both used with l (cm), c is molar concentration, (mol dm⁻³) used with l (cm) or (mol m⁻³) with l (m)

Table 1.10 Conversion factors for energy units[a]

Unit	Wave number cm⁻¹ [b]	Molar energy		Energy eV[c]	Temperature K
		J mol⁻¹	cal mol⁻¹		
1 cm⁻¹	1	11.96266	2.859144	1.239842×10^{-4}	1.438769
1 J mol⁻¹	0.08359347	1	0.239006	1.036427×10^{-5}	0.120272
1 cal mol⁻¹	0.3499891	4.184	1	4.339312×10^{-5}	0.503217
1 eV	8065.544	96485.34	23060.54	1	1160.448
1 K	0.695039	8.314472	1.987207	8.61738×10^{-5}	1

[a]Conversion formulae are: $E = h\nu = hc\nu' = kT$; $E_{molar} = N_A E$, where $h = 6.62607 \times 10^{-34}$ is the Planck constant, $c = 2.9979 \times 10^8$ m s⁻¹ is the speed of light, $k = 1.3807 \times 10^{-23}$ is the Boltzmann constant, T is temperature, and $N_A = 6.02214 \times 10^{23}$ is the Avogadro constant
[b]Wave number $\nu' = 1/\lambda$ is a measure of energy used in spectroscopy
[c]One electron volt is equivalent to the kinetic energy of a singly charged particle (electron or ion) after acceleration by a potential difference of 1 V

1.3 Properties of the Elements

Table 1.11 Standard atomic weights of naturally occurring elements, listed alphabetically[a]

Name, Z	Symbol	Atomic weight	Name, Z	Symbol	Atomic weight
Actinium, 89	Ac	227.0278[b]	Magnesium, 12	Mg	24.3050
Aluminum, 13	Al	26.9815386	Manganese, 25	Mn	54.938045
Antimony, 51	Sb	121.760	Mercury, 80	Hg	200.59
Argon, 18	Ar	39.948	Molybdenum, 42	Mo	95.96
Arsenic, 33	As	74.92160	Neodymium, 60	Nd	144.242
Astatine, 85	At	209.9871[b]	Neon, 10	Ne	20.1797
Barium, 56	Ba	137.327	Nickel, 28	Ni	58.6934
Beryllium, 4	Be	9.012182	Niobium, 41	Nb	92.90638
Bismuth, 83	Bi	208.98040	Nitrogen, 7	N	14.0067
Boron, 5	B	10.811	Osmium, 76	Os	190.23
Bromine, 35	Br	79.904	Oxygen, 8	O	15.9994
Cadmium, 48	Cd	112.411	Palladium, 46	Pd	106.42
Calcium, 20	Ca	40.078	Phosphorus, 15	P	30.973762
Carbon, 6	C	12.0107	Platinum, 78	Pt	195.084
Cerium, 58	Ce	140.116	Polonium, 84	Po	208.9824[b]
Cesium, 58	Cs	132.9054519	Potassium, 19	K	39.0983
Chlorine, 17	Cl	35.453	Praseodymium, 59	Pr	140.90765
Chromium, 24	Cr	51.9961	Promethium, 61	Pm	144.9127[b]
Cobalt, 27	Co	58.933195	Protactinium, 91	Pa	231.0359
Copper, 29	Cu	63.546	Radium, 88	Ra	226.0254[b]
Dysprosium, 66	Dy	162.500	Radon, 86	Rn	222.0176[b]
Erbium, 68	Er	167.259	Rhenium, 75	Re	186.207
Europium, 63	Eu	151.964	Rhodium, 45	Rh	102.90550
Fluorine, 9	F	18.9984032	Rubidium, 37	Rb	85.4678
Francium, 87	Fr	223.0197[b]	Ruthenium, 44	Ru	101.07
Gadolinium, 64	Gd	157.25	Samarium, 62	Sm	150.36
Gallium, 31	Ga	69.723	Scandium, 21	Sc	44.955912
Germanium, 32	Ge	72.64	Selenium, 34	Se	78.96
Gold, 79	Au	196.966569	Silicon, 14	Si	28.0855
Hafnium, 72	Hf	178.49	Silver, 47	Ag	107.8682
Helium, 2	He	4.002602	Sodium, 11	Na	22.98976928
Holmium, 67	Ho	164.93032	Strontium, 38	Sr	87.62
Hydrogen, 1	H	1.00794	Sulfur, 16	S	32.065
Indium, 49	In	114.818	Tantalum, 73	Ta	180.94788
Iodine, 53	I	126.90447	Technetium, 43	Tc	97.9072[b]
Iridium, 77	Ir	192.217	Tellurium, 52	Te	127.60
Iron, 26	Fe	55.845	Terbium, 65	Tb	158.92535
Krypton, 36	Kr	83.798	Thallium, 81	Tl	204.3833
Lanthanum, 57	La	138.90547	Thorium, 90	Th	232.03806
Lead, 82	Pb	207.2	Thulium, 69	Tm	168.93421
Lithium, 3	Li	6.941	Tin, 50	Sn	118.710
Lutetium, 71	Lu	174.9668	Titanium, 22	Ti	47.867

(continued)

Table 1.11 (continued)

Name, Z	Symbol	Atomic weight	Name, Z	Symbol	Atomic weight
Tungsten, 74	W	183.84	Ytterbium, 70	Yb	173.054
Uranium, 92	U	238.02891	Yttrium, 39	Y	88.90585
Vanadium, 23	V	50.9415	Zinc, 30	Zn	65.38
Xenon, 54	Xe	131.293	Zirconium, 40	Zr	91.224

[a]Based on the unified atomic mass unit $(1/12)m(^{12}C) = 10^{-3}$ kg mol^{-1}/N$_A$. Data Source: Wieser, M.E., Berglund, M. (2009) Pure Appl. Chem. **81**, 2131–2156
[b]The element has no stable isotope. The value given refers to the isotope with the longest half-life. Thorium, protactinium, and uranium have no stable isotopes but the terrestrial composition is sufficiently uniform for a standard atomic weight to be specified

Table 1.12 Isotopic composition (natural abundance) of the elements[a]

Element Isotope	Abundance %	Atomic Mass	Element Isotope	Abundance %	Atomic Mass
Hydrogen		1.00794	**Magnesium**		24.3050
^{1}H	99.9885	1.007825032	^{24}Mg	78.99	23.98504187
^{2}H	0.0115	2.014101778	^{25}Mg	10.00	24.98583700
Helium		4.002602	^{26}Mg	11.01	25.98259300
^{3}He	0.000134	3.0160293094	**Aluminum**		26.981538
^{4}He	99.999866	4.0026032497	^{27}Al	100	26.98153841
Lithium		6.941	**Silicon**		28.0855
^{6}Li	7.59	6.0151223	^{28}Si	92.223	27.97692649
^{7}Li	92.41	7.0160041	^{29}Si	4.685	28.97649468
Beryllium		9.0121822	^{30}Si	3.092	29.97377018
^{9}Be	100	9.0121822	**Phosphorus**		30.973761
Boron		10.811	^{31}P	100	30.97376149
^{10}B	19.9	10.0129371	**Sulfur**		32.065
^{11}B	80.1	11.0093055	^{32}S	94.99	31.97207073
Carbon		12.0107	^{33}S	0.75	32.97145854
^{12}C	98.93	12 (defined)	^{34}S	4.25	33.96786687
^{13}C	1.07	13.003354838	^{36}S	0.01	35.96708088
Nitrogen		14.0067	**Chlorine**		35.453
^{14}N	99.636	14.003074007	^{35}Cl	75.76	34.96885271
^{15}N	0.364	15.000108973	^{37}Cl	24.24	36.96590260
Oxygen		15.9994	**Argon**		39.948
^{16}O	99.757	15.994914622	^{36}Ar	0.3365	35.96754626
^{17}O	0.038	16.99913150	^{38}Ar	0.0632	37.9627322
^{18}O	0.205	17.9991604	^{40}Ar	99.6003	39.962383124
Fluorine		18.9984032	**Potassium**		39.0983
^{18}F	100	18.9984032	^{39}K	93.2581	38.9637069
Neon		20.1797	^{40}K*	0.0117	39.96399867
^{20}Ne	90.48	19.992440176	^{41}K	6.7302	39.96182597
^{21}Ne	0.27	20.99384674	**Calcium**		40.078
^{22}Ne	9.25	21.99138550	^{40}Ca	96.941	39.9625912
Sodium		22.989770	^{42}Ca	0.647	41.9586183
^{23}Na	100	22.98976966	^{43}Ca	0.135	42.9587668

(continued)

Table 1.12 (continued)

Element Isotope	Abundance %	Atomic Mass	Element Isotope	Abundance %	Atomic Mass
^{44}Ca	2.086	43.9554811	**Germanium**		72.64
^{46}Ca	0.004	45.9536927	^{70}Ge	20.38	69.9242500
^{48}Ca	0.187	47.952533	^{72}Ge	27.31	71.9220763
Scandium		44.955910	^{73}Ge	7.76	72.9234595
^{45}Sc	100	44.9559102	^{74}Ge	36.72	73.9211784
Titanium		47.867	^{76}Ge	7.83	75.9214029
^{46}Ti	8.25	45.9526295	**Arsenic**		74.92160
^{47}Ti	7.44	46.9517637	^{75}As	100	74.9215966
^{48}Ti	73.72	47.9479470	**Selenium**		78.96
^{49}Ti	5.41	48.9478707	^{74}Se	0.89	73.9224767
^{50}Ti	5.18	49.9447920	^{76}Se	9.37	75.9192143
Vanadium		50.9415	^{77}Se	7.63	76.9199148
^{50}V	0.25	49.9471627	^{78}Se	23.77	77.9173097
^{51}V	99.75	50.9439635	^{80}Se	49.61	79.9165221
Chromium		51.9961	^{82}Se	8.73	81.9167003
^{50}Cr	4.345	49.9460495	**Bromine**		79.904
^{52}Cr	83.789	51.9405115	^{79}Br	50.69	78.9183379
^{53}Cr	9.501	52.9406534	^{81}Br	49.31	80.916291
^{54}Cr	2.365	53.9388846	**Krypton**		83.798
Manganese		54.938049	^{78}Kr	0.355	77.920388
^{55}Mn	100	54.9380493	^{80}Kr	2.286	79.916379
Iron		55.845	^{82}Kr	11.593	81.9134850
^{54}Fe	4.345	53.9396147	^{83}Kr	11.500	82.914137
^{56}Fe	91.754	55.9349418	^{84}Kr	56.987	83.911508
^{57}Fe	2.119	56.9353983	^{86}Kr	17.279	85.910615
^{58}Fe	0.282	57.9332801	**Rubidium**		85.4678
Cobalt		58.933200	^{85}Rb	72.17	84.9117924
^{58}Co	100	58.9331999	^{87}Rb	27.83	86.9091858
Nickel		58.6934	**Strontium**		87.62
^{58}Ni	68.0769	57.9353477	^{84}Sr	0.56	83.913426
^{60}Ni	26.2231	59.9307903	^{86}Sr	9.86	84.9092647
^{61}Ni	1.1399	60.9310601	^{87}Sr	7	86.9088816
^{62}Ni	3.6345	61.9283484	^{88}Sr	82.58	87.9056167
^{64}Ni	0.9256	63.9279692	**Ytterbium**		88.90585
Copper		63.546	^{89}Y	100	88.9058485
^{63}Cu	69.15	62.9296007	**Zirconium**		91.224
^{65}Cu	30.85	64.9277938	^{90}Zr	51.45	89.9047022
Zinc		65.409	^{91}Zr	11.22	90.9056434
^{64}Zn	48.268	63.9291461	^{92}Zr	17.15	91.9050386
^{66}Zn	27.975	65.9260364	^{94}Zr	17.38	93.9063144
^{67}Zn	4.102	66.9271305	^{96}Zr	2.80	95.908275
^{68}Zn	19.024	67.9248473	**Niobium**		92.90638
^{70}Zn	0.631	69.925325	^{93}Nb	100	92.9063762
Gallium		69.723	**Molybdenum**		95.94
^{69}Ga	60.108	68.925581	^{92}Mo	14.77	91.906810
^{71}Ga	39.892	70.9247073	^{94}Mo	9.23	93.9050867

(continued)

Table 1.12   (continued)

Element Isotope	Abundance %	Atomic Mass	Element Isotope	Abundance %	Atomic Mass
^{95}Mo	15.9	94.9058406	^{122}Sn	4.63	121.9034411
^{96}Mo	16.68	95.9046780	^{124}Sn	5.79	123.9052745
^{97}Mo	9.56	96.9060201	**Antimony**		121.76
^{98}Mo	24.19	97.9054069	^{121}Sb	57.21	120.9038222
^{100}Mo	9.67	99.907476	^{123}Sb	42.79	122.9042160
Ruthenium		101.07	**Tellurium**		127.60
^{96}Ru	5.54	95.907604	^{120}Te	0.09	119.904026
^{98}Ru	1.87	97.905287	^{122}Te	2.55	121.9030558
^{99}Ru	12.76	98.9059385	^{123}Te	0.89	122.9042711
^{100}Ru	12.60	99.9042189	^{124}Te	4.74	123.9028188
^{101}Ru	17.06	100.9055815	^{125}Te	7.07	124.9044241
^{102}Ru	31.55	101.9043488	^{126}Te	18.84	125.903049
^{104}Ru	18.62	103.905430	^{128}Te	31.74	127.9044615
Rhodium		102.90550	^{130}Te	34.08	129.9062229
^{103}Rh	100	102.905504	**Iodine**		126.90447
Palladium		106.42	^{127}I	100	126.904468
^{102}Pd	1.02	101.905607	**Xenon**		131.293
^{104}Pd	11.14	103.904034	^{124}Xe	0.0952	123.9058954
^{105}Pd	22.33	104.905083	^{126}Xe	0.0890	125.904268
^{106}Pd	27.33	105.903484	^{128}Xe	1.9102	127.9035305
^{108}Pd	26.46	107.903895	^{129}Xe	26.4006	128.9047799
^{110}Pd	11.72	109.905153	^{130}Xe	4.071	129.9035089
Silver		107.8682	^{131}Xe	21.2324	130.9050828
^{107}Ag	51.839	106.905093	^{132}Xe	26.9086	131.9041546
^{109}Ag	48.161	108.904756	^{134}Xe	10.4357	133.9053945
Cadmium		112.411	^{136}Xe	8.8573	135.907220
^{106}Cd	1.25	105.906458	**Cesium**		132.90545
^{108}Cd	0.89	107.904183	^{133}Cs	100	132.905447
^{110}Cd	12.49	109.903006	**Barium**		137.327
^{111}Cd	12.8	110.904182	^{130}Ba	0.106	129.906311
^{112}Cd	24.13	111.9027577	^{132}Ba	0.101	131.90506
^{113}Cd	12.22	112.9044014	^{134}Ba	2.417	133.904504
^{114}Cd	28.73	113.9033586	^{135}Ba	6.592	134.905684
^{116}Cd	7.49	115.904756	^{136}Ba	7.854	135.904571
Indium		114.818	^{137}Ba	11.232	136.905822
^{113}In	4.29	112.904062	^{138}Ba	71.698	137.905242
^{115}In	95.71	114.903879	**Lanthanum**		138.9055
Tin		118.71	^{138}La	0.090	137.907108
^{112}Sn	0.97	111.904822	^{139}La	99.910	138.906349
^{114}Sn	0.66	113.902783	**Cerium**		140.116
^{115}Sn	0.34	114.903347	^{136}Ce	0.185	135.907140
^{116}Sn	14.54	115.901745	^{138}Ce	0.251	137.905986
^{117}Sn	7.68	116.902955	^{140}Ce	88.450	139.905435
^{118}Sn	24.22	117.901608	^{142}Ce	11.114	141.909241
^{119}Sn	8.59	118.903311	**Praseodymium**	140.90765	160Gd
^{120}Sn	32.58	119.9021985	^{141}Pr	100	140.907648

(continued)

Table 1.12 (continued)

Element Isotope	Abundance %	Atomic Mass	Element Isotope	Abundance %	Atomic Mass
Neodymium		144.24	**Thulium**		168.93421
^{142}Nd	27.2	141.907719	^{169}Tm	100	168.934211
^{143}Nd	12.2	142.909810	**Ytterbium**		173.04
^{144}Nd	23.8	143.910083	^{168}Yb	0.13	167.933895
^{145}Nd	8.3	144.912569	^{170}Yb	3.04	169.934759
^{146}Nd	17.2	145.913113	^{171}Yb	14.28	170.936323
^{148}Nd	5.7	147.916889	^{172}Yb	21.83	171.936378
^{150}Nd	5.6	149.920887	^{173}Yb	16.13	172.938207
Samarium		150.36	^{174}Yb	31.83	173.938858
^{144}Sm	3.07	143.911996	^{176}Yb	12.76	175.942569
^{147}Sm	14.99	146.914894	**Lutetium**		174.967
^{148}Sm	11.24	147.914818	^{175}Lu	97.41	174.9407682
^{149}Sm	13.82	148.917180	^{176}Lu	2.59	175.9426827
^{150}Sm	7.38	149.919272	**Hafnium**		178.49
^{152}Sm	26.75	151.919729	^{174}Hf	0.16	173.940042
^{154}Sm	22.75	151.922206	^{176}Hf	5.26	175.941403
Europium		151.964	^{177}Hf	18.6	176.9432204
^{151}Eu	47.81	150.919846	^{178}Hf	27.28	177.9436981
^{153}Eu	52.19	152.921227	^{179}Hf	13.62	178.9458154
Gadolinium		157.25	^{180}Hf	35.08	179.9465488
^{152}Gd	0.20	151.919789	**Tantalum**		180.9479
^{154}Gd	2.18	153.920862	^{180}Ta	0.012	179.947466
^{155}Gd	14.80	154.922619	^{181}Ta	99.988	180.947996
^{156}Gd	20.47	155.922120	**Tungsten**		183.84
^{157}Gd	15.65	156.923957	^{180}W	0.12	179.946706
^{158}Gd	24.84	157.924101	^{182}W	26.50	181.948205
^{160}Gd	21.86	159.927051	^{183}W	14.31	182.9502242
Terbium		158.92534	^{184}W	30.64	183.9509323
^{159}Tb	100	158.925343	^{186}W	28.43	185.954362
Dysprosium		162.50	**Rhenium**		186.207
^{156}Dy	0.056	155.92478	^{185}Re	37.40	184.952955
^{158}Dy	0.095	157.924405	^{187}Re	62.60	186.9557505
^{160}Dy	2.329	159.925194	**Osmium**		190.23
^{161}Dy	18.889	160.926930	^{184}Os	0.02	183.952491
^{162}Dy	25.475	161.926795	^{186}Os	1.59	185.953838
^{163}Dy	24.896	162.928728	^{187}Os	1.96	186.9557476
^{164}Dy	28.26	163.929171	^{188}Os	13.24	187.9558357
Holmium		164.93032	^{189}Os	16.15	188.958145
^{165}Ho	100	164.930319	^{190}Os	26.26	189.958445
Erbium		167.259	^{192}Os	40.78	191.961479
^{162}Er	0.139	161.928778	**Iridium**		192.217
^{164}Er	1.601	163.92920	^{191}Ir	37.3	190.960591
^{166}Er	33.503	165.930293	^{193}Ir	62.7	192.962923
^{167}Er	22.869	166.932048	**Platinum**		195.078
^{168}Er	26.978	167.932368	^{190}Pt	0.014	189.959930
^{170}Er	14.91	169.935461	^{192}Pt	0.782	191.961035

(continued)

Table 1.12 (continued)

Element Isotope	Abundance %	Atomic Mass	Element Isotope	Abundance %	Atomic Mass
^{194}Pt	32.967	193.962663	^{205}Tl	70.48	204.974412
^{195}Pt	33.832	194.964774	**Lead**[b]		207.2
^{196}Pt	25.242	195.964934	^{204}Pb	1.4	203.973028
^{198}Pt	7.163	197.967875	^{206}Pb	24.1	205.974449
Gold		196.966	^{207}Pb	22.1	206.975880
^{197}Au	100	196.966551	^{208}Pb	52.4	207.976636
Mercury		200.59	**Bismuth**		208.98038
^{196}Hg	0.15	195.965814	^{209}Bi	100	208.980384
^{198}Hg	9.97	197.966752	**Thorium**		232.0381
^{199}Hg	16.87	198.968262	^{232}Th*	100	232.0380495
^{200}Hg	23.10	199.968309	**Protactinium**		231.03588
^{201}Hg	13.18	200.970285	^{231}Pa*	100	231.03588
^{202}Hg	29.86	201.970625	**Uranium**		238.02891
^{204}Hg	6.87	203.973475	^{234}U*	0.0054	234.0409447
Thallium		204.3833	^{235}U*	0.72	235.0439222
^{203}Tl	29.52	202.972329	^{238}U*	99.2	238.0507835

Source of data: De Laeter, J.R., Böhlke, J.K., De Bièvre, P., Hidaka, H., Peiser, H.S., Rosman, K.J.R., Taylor, P.D.P. (2003) Pure Appl. Chem. **75**, 683–800
[a]Radioactive isotopes are indicated by an asterisk. For radioactive species that occur only in trace amounts see Table 1.13
[b]The isotope composition varies somewhat with location

Table 1.13 Radioactive isotopes in the environment, origins and half-lifetimes[a]

Element	Isotope	Atomic mass	Half-lifetime	Decay process[b]	Origin[c]
Tritium	^{3}H	3.016049278	12.33 a	β$^-$	cosmic radiat. (N, O), artificial
Beryllium	^{7}Be	7.0169298	53.28 d	ec	cosmic radiation (N, O)
	^{10}Be	10.0135338	1.52×10^6 a	β$^-$	cosmic radiation (N, O)
Carbon	^{14}C	14.003242	5715 a	β$^-$	cosmic radiat. (N, O), artificial
Sodium	^{22}Na	21.9944364	2.605 a	β$^+$	cosmic radiation (Ar)
Aluminum	^{26}Al	25.9868917	7.1×10^5 a	β$^+$	cosmic radiation (Ar)
Silicon	^{32}Si	31.9741481	160 a	β$^-$	cosmic radiation (Ar)
Phosphorus	^{32}P	31.9739073	14.28 d	β$^-$	cosmic radiation (Ar)
	^{33}P	32.971726	25.3 d	β$^-$	cosmic radiation (Ar)
Chlorine	^{36}Cl	35.968307	3.01×10^5 a	β$^-$	cosmic radiation (Ar)
Argon	^{37}Ar	36.9667763	35.0 d	ec	cosmic radiation (Ar)
	^{39}Ar	38.964313	268 a	β$^-$	cosmic radiation (Ar)
Potassium[d]	^{40}K	39.9639985	1.248×10^9 a	β$^+$/ec, β$^-$	primordial
Krypton	^{81}Kr	80.916592	2.1×10^5 a	ec	cosmic radiation (Kr)
Krypton	^{85}Kr	84.912527	10.73 a	β$^-$	artificial
Rubidium	^{87}Rb	86.9091805	4.88×10^{10} a	β$^-$	primordial
Strontium	^{90}Sr	89.907738	29.1 a	β$^-$	artificial
Iodine	^{129}I	128.904988	1.7×10^7 a	β$^-$	cosmic radiat. (Xe), artificial

(continued)

Table 1.13 (continued)

Element	Isotope	Atomic mass	Half-lifetime	Decay process[b]	Origin[c]
Cesium	^{137}Cs	136.907089	30.2 a	β^-	artificial
Polonium	^{210}Po	209.982874	138.4 d	α	Radon decay
Lead	^{210}Pb	209.984189	22.6 a	β^-	Radon decay
Radon	^{222}Rn	222.017578	3.823 d	α	Uranium decay
Radium	^{223}Ra	223.018502	11.43 d	α	Uranium decay
	^{224}Ra	224.020212	3.66 d	α	Uranium decay
	^{226}Ra	226.025410	1599 a	α	Uranium decay
Actinium	^{227}Ac	227.027752	21.77 a	β^-	Uranium decay
Thorium	^{227}Th	227.027704	18.72 d	α	Uranium decay
	^{228}Th	228.028741	1.913 a	α	Uranium decay
	^{230}Th	230.033134	7.54×10^4 a	α	Uranium decay
	^{232}Th	232.038055	1.4×10^{10} a	α	primordial
	^{234}Th	234.043601	24.1 d	β^-	Uranium decay
Protactinium	^{231}Pa	231.035884	3.25×10^4 a	α	Uranium decay
Uranium	^{234}U	234.040952	2.455×10^5 a	α	primordial
Uranium	^{235}U	235.043930	7.04×10^8 a	α	primordial
	^{238}U	238.050788	4.47×10^9 a	α	primordial

[a]Half-lifetimes (>1 d) are taken from Holden, N.E. (2006), in Lide, D.R. (Editor in Chief) *CRC Handbook of Chemistry and Physics*, Taylor & Francis, Boca Raton, pp. 11-51–11-203
[b]Decay modes: α emission of alpha-particles, β^- emission of electrons, β^+ emission of positrons, ec orbital electron capture
[c]Cosmic radiation: the element undergoing disintegration to form the isotope is indicated. Artificial sources include nuclear weapons tests and nuclear industry
[d]The disintegration proceeds to about 90% to form ^{40}Ar and 10% to form ^{40}Ca

Table 1.14 Natural radioactive decay series: Isotopes, decay processes and half-lifetimes[a]

^{92}U$_{238}$	**Uranium**-238	^{92}U$_{235}$	**Uranium**-235				
$\downarrow \alpha$	4.47×10^9 a	$\downarrow \alpha$	7.04×10^8 a				
^{90}**Th**$_{234}$	Thorium-234	^{90}**Th**$_{231}$	Thorium-231	^{90}**Th**$_{232}$		**Thorium**-232	
$\downarrow \beta$	24.1 d	$\downarrow \beta$	1.063 d	$\downarrow \alpha$		1.4×10^{10} a	
^{91}**Pa**$_{234}$	Protactinium-234	^{91}**Pa**$_{231}$	Protactinium-231	^{88}**Ra**$_{228}$		Radium-228	
$\downarrow \beta$	1.17 min	$\downarrow \alpha$	3.25×10^4 a	$\downarrow \beta$		5.76 a	
^{92}**U**$_{234}$	Uranium-234	^{89}**Ac**$_{227}$	Actinium-227	^{89}**Ac**$_{228}$		Actinium-228	
$\downarrow \alpha$	2.46×10^5 a	$\downarrow \beta$	21.77 a	$\downarrow \beta$		6.15 h	
^{90}**Th**$_{230}$	Thorium-230	^{90}**Th**$_{227}$	Thorium-227	^{90}**Th**$_{228}$		Thorium-228	
$\downarrow \alpha$	7.54×10^4 a	$\downarrow \alpha$	18.72 d	$\downarrow \alpha$		1.913 a	
^{88}**Ra**$_{226}$	Radium-226	^{88}**Ra**$_{223}$	Radium-223	^{88}**Ra**$_{224}$		Radium-224	
$\downarrow \alpha$	1600 a	$\downarrow \alpha$	11.43 d	$\downarrow \alpha$		3.66 d	

(continued)

Table 1.14 (continued)

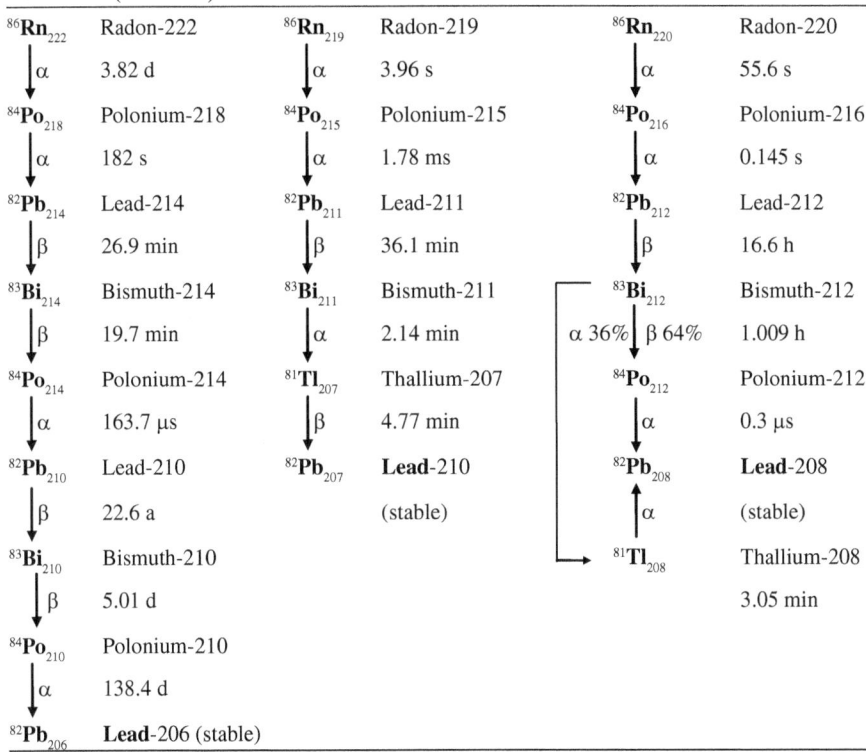

$^{86}Rn_{222}$	Radon-222	$^{86}Rn_{219}$	Radon-219	$^{86}Rn_{220}$	Radon-220
↓α	3.82 d	↓α	3.96 s	↓α	55.6 s
$^{84}Po_{218}$	Polonium-218	$^{84}Po_{215}$	Polonium-215	$^{84}Po_{216}$	Polonium-216
↓α	182 s	↓α	1.78 ms	↓α	0.145 s
$^{82}Pb_{214}$	Lead-214	$^{82}Pb_{211}$	Lead-211	$^{82}Pb_{212}$	Lead-212
↓β	26.9 min	↓β	36.1 min	↓β	16.6 h
$^{83}Bi_{214}$	Bismuth-214	$^{83}Bi_{211}$	Bismuth-211	$^{83}Bi_{212}$	Bismuth-212
↓β	19.7 min	↓α	2.14 min	α 36% ↓ β 64%	1.009 h
$^{84}Po_{214}$	Polonium-214	$^{81}Tl_{207}$	Thallium-207	$^{84}Po_{212}$	Polonium-212
↓α	163.7 µs	↓β	4.77 min	↓α	0.3 µs
$^{82}Pb_{210}$	Lead-210	$^{82}Pb_{207}$	**Lead**-210	$^{82}Pb_{208}$	**Lead**-208
↓β	22.6 a		(stable)	↑α	(stable)
$^{83}Bi_{210}$	Bismuth-210			$^{81}Tl_{208}$	Thallium-208
↓β	5.01 d				3.05 min
$^{84}Po_{210}$	Polonium-210				
↓α	138.4 d				
$^{82}Pb_{206}$	**Lead**-206 (stable)				

[a] Half-life times from Holden, N.E. (2006) in Lide, D.R. (Editor in Chief) *CRC Handbook of Chemistry and Physics*, Taylor & Francis, Boca Raton, pp. 11-51–11-203

Chapter 2
Data Regarding the Earth

2.1 Physical Properties and Interior Structure of the Earth

Table 2.1 Physical properties of the Earth[a]

Mean distance to sun (10^6 km)	149.598	Surface area (10^{12} m²), total	510
Eccentricity of orbit	0.0167	Ocean surface area	361
Inclination of equator to orbit (°)	23.45	Continental surface area	149
Period of sideral revolution (d)	365.256	Mean height of continents (m)	875
Sideral rotation period (h)	23.9345	Mean depth of oceans (m)	3,794
Radius (km), mean	6,371.0	Acceleration due to gravity (m s⁻²)	
Equatorial	6,378.14	Mean for entire surface[b]	9.7978
Polar	6,356.75	Equatorial	9.78036
Mass (10^{21} kg), total	5,973.6	Polar	9.83208
Metallic core	1,941.	Escape velocity (km s⁻¹)	11.18
Silicate mantle	4,007.	Mean surface temperature[c] (K)	288
Crust, total	27.1	Heat flow (mW m⁻²)	
Continental crust[d]	22.5	Global mean	87 ± 2.0
Oceanic crust[d]	4.6	Oceanic mean	101 ± 2.2
Sedimentary rocks	2.24	Continental mean	65 ± 1.6
Hydrosphere[e]	1.664	Total global heat loss (TW)	44 ± 1
Mean density (g cm⁻³)	5.515	Solar constant (kW m⁻²)	1.3676

[a]Sources of data: Garrels and MacKenzie (1971), Lang (1992), Lodders and Fegley (1998), Ronov and Yaroshevsky (1969), Taylor and McLennan (1985), Wedepohl (1995)
[b]The internationally adopted standard value is $g = 9.80665$ m s⁻²
[c]At sea level
[d]Continental crust: 40 km average thickness including continental shelf regions and margins; Oceanic crust: 5 km average thickness and 3.0×10^3 kg m⁻³ average density
[e]Includes pore waters in sediments and water in rocks

P. Warneck and J. Williams, *The Atmospheric Chemist's Companion: Numerical Data for Use in the Atmospheric Sciences*, DOI 10.1007/978-94-007-2275-0_2, © Springer Science+Business Media B.V. 2012

Structure: The solid earth consists of the core, the mantle and the crust. Seismic data show that the (iron-rich) core is divided into a solid inner core, which is denser than iron and is probably an Fe-Ni alloy, and a molten outer core, which is slightly less dense than iron. The silicate mantle is also subdivided into a lower and an upper mantle by seismic discontinuities occurring at depths between 400 and 670 km. The crust is the uppermost layer of the earth. The continental crust (granitic) is enriched in incompatible elements and contains most of the earth alkaline elements as well as a large fraction of uranium and other radioactive elements. Most of the continental crust is 20–50 km thick, with an average thickness of 35–40 km. The oldest known rocks are nearly 4×10^9 years old. The (basaltic) oceanic crust is about 5–10 km thick. It is fairly young with an average age of 60 million years. The oceanic crust is less enriched with incompatible elements than the continental crust. The upper 100–200 km of the earth is made up of about 12 plates that float on the upper mantle. The thickness of oceanic plates is about 60 km, and that of continental plates 100–200 km. The mutual interaction of plates gives rise to divergence zones, such as the mid-ocean ridges where they spread apart, and convergence zones, where the higher density oceanic plates are subducted under the lower density continental plates and mountain ranges are formed. These zones are active regions of volcanism.

Table 2.2   The interior structure of the Earth[a]

Region or boundary	Outer radius (km)	Depth (km)	Density (g cm^{-3})	Pressure (10^8 Pa)	Temperature (K)
Inner core (metallic solid)	1,220	5,150–6,370	12.8–13.1	3,290–3,570	5,000–5,500
Outer core (metallic liquid)	3,485	2,890–5,150	9.9–12.2	1,390–3,290	3,930–5,000
Lower mantle	5,700	670–2,890	4.3–5.7	240–1,390	2,000–3,930
Transition zone	5,970	400–670	3.8–4.0	140–240	1,770–2,000
Upper mantle	6,333	35–400	3.3–3.5	2–140	200–1,770
Continental crust	–	0–35	2.7–2.8	0–13	290–770
Oceanic crust	–	0–10	3.0	0–3.3	290–500

[a]Source of data: Lodders and Fegley (1998)

References

Garrels, R.M., F.T. MacKenzie, *The Evolution of Sedimentary Rocks* (Norton, New York, 1971)

Lang, K.R., *Astrophysical Data: Planets and Stars* (Springer, New York, 1992)

Lodders, K., B. Fegley Jr., *The Planetary Scientist's Companion* (Oxford University Press, New York, 1998)

Ronov, A.B., A.A. Yaroshevsky, in: *The Earth's Crust and Upper Mantle*, American Geophysical Union, Monograph Series 13 (American Geophysical Union, Washington, DC, 1969), pp. 37–57; Geochimiya **12**, 1763–1795 (1976)

Taylor, S.R., S.M. McLennan, *The Continental Crust: Its Composition and Evolution* (Blackwell Scientific Publications, Oxford, 1985)

Wedepohl, K.H., Geochim. Cosmochim. Acta **59**, 1217–1232 (1995)

2.2   Geographic Distribution of Continents and Oceans

Table 2.3 Distribution of continents and oceans by latitude zones, and continental mean annual surface temperature

Latitude belt	Total area 10⁶ km²	Oceans 10⁶ km²	%	Continents 10⁶ km²	%	$T_{surface}$ [a] K
80–90°N	3.873	3.618	93.4	0.256	6.6	249.6
70–80°	11.506	8.203	71.3	3.302	28.7	257.3
60–70°	18.783	5.522	29.4	13.261	70.6	266.0
50–60°	25.497	10.913	42.8	14.585	57.2	273.7
40–50°	31.429	14.928	47.5	16.500	52.5	280.7
30–40°	36.411	20.827	57.2	15.584	42.8	287.2
20–30°	40.285	25.138	62.4	15.147	37.6	293.6
10–20°	42.934	31.600	73.6	11.334	26.4	298.3
0–10°	44.281	34.185	77.2	10.096	22.8	298.7
0–90°N	255	154.934	60.8	100.065	39.2	286.4
0–10°S	44.281	33.830	76.4	10.450	23.6	298.0
10–20°	42.934	33.489	78.0	9.445	22.0	296.5
20–30°	40.285	30.979	76.9	9.306	23.1	292.0
30–40°	36.411	32.333	88.8	4.306	11.2	286.7
40–50°	31.429	30.486	97.0	0.943	3.0	281.9
50–60°	25.497	25.293	99.2	0.204	0.8	274.4
60–70°	18.783	16.830	89.6	1.953	10.4	262.2
70–80°	11.506	2.830	24.6	8.675	75.4	243.7
80–90°	3.873	–	–	3.873	100	225.3
0–90°S	255	206.070	80.8	48.929	19.2	284.6
Globe	510	361	70.8	149	29.2	285.5

[a]As given by Sellers (1965); estimated mean annual surface temperatures over the continents are lower than the global sea level temperature (288 K) often quoted, because of the higher elevation

Table 2.4 Regional distribution of land and ocean areas[a]

Region	Area (10⁶ km²)	Mean elevation or depth (m)	Percent of total region	Percent of globe
Continents[b]				
Antarctica	14.0	2,200	9.4	2.74
Australia and Oceania	8.9	350	6.0	1.74
America	42.0		28.2	8.24
North	24.2	720	16.2	4.95
South	17.8	590	12.0	3.49
Africa	30.1	750	20.2	5.90
Asia	43.5	960	29.2	8.53
Europe	10.5	340	7.0	2.06
Total	149.0	875	100.0	29.20
Oceans[c]				
Atlantic	106.5	3,332	29.3	20.88
Indian	74.9	3,897	20.8	14.69

(continued)

Table 2.4   (continued)

Region	Area (10^6 km^2)	Mean elevation or depth (m)	Percent of total region	Percent of globe
Pacific	179.7	4,028	49.9	35.23
Total	361.1	3,795	100.0	70.80

[a]Source of data: Oceans: Murray (1992); Continents: UNESCO (1978)
[b]Australia includes Tasmania; North America includes Greenland and Central America; Africa includes Madagascar; Asia includes Japan, the Philippines and Indonesia
[c]Includes adjacent seas

References

Murray, J.W., in: *Global Biogeochemical Cycles*, ed. by S.S. Butcher, R.J. Charlson, G.H. Orians, G.V. Wolfe (Academic, London, 1992), pp. 175–212
Sellers, W.D., *Physical Climatology* (University of Chicago Press, Chicago, 1965)
UNESCO, World Water Balance and Water Resources of the Earth. Studies and Report in Hydrology, No. 25, UNESCO, Paris (1978)

2.3   Elemental Composition of the Earth's Crust

The Earth's crust consists almost entirely of oxygen compounds, especially silicates of aluminum, calcium, magnesium, sodium, potassium and iron. These eight elements make up nearly 99% of total crustal material. The next table below shows data for the ten most abundant elements; the remaining elements are treated separately. Whereas data for the upper crust can be derived from direct observation, data for the lower crust and the bulk crust depend on the applied model.

Table 2.5   Average abundance of major elements in Earth's crust (g kg^{-1})

Element		Continental crustal average			Lower	Upper	Oceanic crust[b]
		a	b	c	Continental crust[c]		
Oxygen	O	466.0	–	472.0	–	–	–
Silicon	Si	277.2	267.7	288.0	271.33	303.48	231.4
Aluminum	Al	81.3	84.1	79.6	82.12	77.44	84.7
Iron	Fe	50.0	70.7	43.2	57.06	30.89	81.6
Calcium	Ca	36.3	52.9	38.5	48.60	29.45	80.8
Sodium	Na	28.3	23.0	23.6	21.20	25.67	20.8
Potassium	K	25.9	9.1	21.4	13.14	28.65	1.25
Magnesium	Mg	20.9	32.0	22.0	31.55	13.51	46.4
Titanium	Ti	4.4	5.4	4.01	5.01	3.12	9.0
Phosphorous	P	1.05	–	0.76	0.872	0.665	–
Carbon	C	0.2	–	1.99	0.588	3.24	–

[a]Mason and Moore (1982)
[b]Taylor and McLennan (1985)
[c]Wedepohl (1995)

Table 2.6   Minor elements in Earth's crust (mg kg^{-1})

Element		Continental crustal average			Lower Continental crust[c]	Upper Continental crust[c]	Oceanic crust[b]
		a	b	c			
Hydrogen	H	1,400	–	–	–	–	–
Lithium	Li	20	13	18	13	22	10
Beryllium	Be	2.8	1.5	2.4	1.7	3.1	0.5
Boron	B	10	10	11	5	17	4
Nitrogen	N	20	–	60	34	83	–
Fluorine	F	625	–	525	429	611	–
Sulfur	S	260	–	697	408	953	–
Chlorine	Cl	130	–	472	278	640	174
Scandium	Sc	22	30	16	25.3	–	38
Vanadium	V	135	230	98	149	53	250
Chromium	Cr	100	185	126	228	35	270
Manganese	Mn	950	1,400	716	929	527	1,000
Cobalt	Co	25	29	24	38	11.6	47
Nickel	Ni	75	105	56	99	18.6	135
Copper	Cu	55	75	25	37.4	14.3	86
Zinc	Zn	70	80	65	79	52	85
Gallium	Ga	15	18	15	17	14	17
Germanium	Ge	1.5	1.6	1.4	(1.4)	1.4	1.5
Arsenic	As	1.8	1.0	1.7	1.3	2.0	1.0
Selenium	Se	0.05	0.05	0.12	0.17	0.083	0.16
Bromine	Br	2.5	–	1.0	0.28	1.6	0.4
Rubidium	Rb	90	32	78	41	110	2.2
Strontium	Sr	375	260	333	352	316	130
Yttrium	Y	33	20	24	27.2	20.7	32
Zirconium	Zr	165	100	203	165	237	80
Niobium	Nb	20	11	19	11.3	26	2.2
Molybdenum	Mo	1.5	1.0	1.1	0.6	1.4	1.0
Ruthenium	Ru	0.01	–	0.0001	–	0.001	–
Rhodium	Rh	0.005	–	6×10^{-5}	–	–	0.0002
Palladium	Pd	0.01	0.001	0.0004	–	–	<0.0002
Silver	Ag	0.07	0.08	0.07	0.08	0.055	0.026
Cadmium	Cd	0.2	0.098	0.1	0.101	0.102	0.13
Indium	In	0.1	0.05	0.05	0.052	0.061	0.072
Tin	Sn	2	2.5	2.3	2.1	2.5	1.4
Antimony	Sb	0.2	0.2	0.3	0.3	0.31	0.017
Tellurium	Te	0.01	–	(0.005)	–	–	0.003
Iodine	I	0.5	–	0.8	0.14	1.4	0.008
Cesium	Cs	3	1.0	3.4	0.8	5.8	0.03
Barium	Ba	425	250	584	568	668	25
Lanthanum	La	30	16	30	26.8	32.3	3.7
Cerium	Ce	60	33	60	53.1	65.7	11.5
Praseodymium	Pr	8.2	3.9	6.7	7.4	6.3	1.8
Neodymium	Nd	28	16	27	28.1	25.9	10.0
Samarium	Sm	6.0	3.5	5.3	6.0	4.7	3.3
Europium	Eu	1.2	1.1	1.3	1.6	0.95	1.3

(continued)

Table 2.6 (continued)

Element		Continental crustal average			Lower	Upper	Oceanic crust[b]
		a	b	c	Continental crust[c]		
Gadolinium	Gd	5.4	3.3	4.0	5.4	2.8	4.6
Terbium	Tb	0.9	0.6	0.65	0.81	0.5	0.87
Dysprosium	Dy	3.0	3.7	3.8	4.7	2.9	5.7
Holmium	Ho	1.2	0.78	0.8	0.99	0.62	1.3
Erbium	Er	2.8	2.2	2.1	–	–	3.7
Thulium	Tm	0.5	0.32	0.3	–	–	0.54
Ytterbium	Yb	3.4	2.2	2.0	2.5	1.5	5.1
Lutetium	Lu	0.5	0.3	0.35	0.43	0.27	0.56
Hafnium	Hf	3	3.0	4.9	4.0	5.8	2.5
Tantalum	Ta	2	1.0	1.1	0.84	1.5	0.3
Tungsten	W	1.5	1.0	1.0	0.6	1.4	0.5
Rhenium	Re	0.001	0.0004	0.0004	–	–	0.0009
Osmium	Os	0.005	5×10^{-5}	5×10^{-5}	–	–	$<4 \times 10^{-6}$
Iridium	Ir	0.001	0.0001	5×10^{-5}	–	–	2×10^{-5}
Platinum	Pt	0.01	–	0.0004	–	–	0.0023
Gold	Au	0.004	0.003	0.0025	–	–	0.00023
Mercury	Hg	0.08	–	0.040	0.021	0.056	0.020
Thallium	Tl	0.5	0.36	0.52	0.26	0.75	0.012
Lead	Pb	13	8.0	14.8	12.5	17	0.8
Bismuth	Bi	0.2	0.06	0.085	0.037	0.123	0.007
Thorium	Th	7.2	3.5	8.5	6.6	10.3	0.22
Uranium	U	1.8	0.91	1.7	0.93	2.5	0.1

[a]Mason and Moore (1982)
[b]Taylor and McLennan (1985, 1995); data for chlorine, bromine and iodine in the oceanic crust from Deruelle et al. (1992)
[c]Wedepohl (1995); the data in brackets on Germanium and Tellurium are uncertain

Table 2.7 Important minerals in Earth's crust

Mineral	Chemical formula	Crystal system	Density (g cm^{-3})	Notes
Quartz	SiO_2	Hexagonal	2.65	a
Hematite	Fe_2O_3	Hexagonal	5.25	a
Ilmenite	$FeTiO_3$	Rhombohedral	4.72	
Olivines				b
Forsterite	Mg_2SiO_4	Orthorombic	3.21	b
Fayalite	Fe_2SiO_4	Orthorombic	4.30	b
Pyroxene				
Enstatite	$MgSiO_3$	Monoclinic	3.19	
Wollastonite	$CaSiO_3$	Monoclinic	2.92	
Diopside	$CaMg(SiO_3)_2$	Monoclinic	3.30	
Hedenbergite	$CaFe(SiO_3)_2$	Monoclinic	3.53	
Gibbsite	$Al(OH)_3$	Monoclinic	2.42	c

(continued)

Table 2.7   (continued)

Mineral	Chemical formula	Crystal system	Density (g cm^{-3})	Notes
Feldspars				
Orthoclase	$KAlSi_3O_8$	Monoclinic	2.56	d
Albite	$NaAlSi_3O_2$	Triclinic	2.63	d
Anorthite	$CaAl_2Si_2O_8$	Triclinic	2.76	d
Amphibole				
Tremolite	$Ca_2Mg_5Si_8O_{22}(OH)_2$	Monoclinic	3.0	e
Iron tremolite	$Ca_2Fe_5Si_8O_{22}(OH)_2$	Monoclinic	3.4	e
Glaucophane	$Na_2Mg_3Al_2Si_8O_{22}(OH)_2$	Monoclinic	3.19	e
Chlorite	$Mg_5Al_2Si_3O_{10}(OH)_8$	Monoclinic	2.6–2.8	f
Montmorillonite	$Na_{0.33}Al_{2.33}Si_{3.67}O_{10}(OH)_2$	Monoclinic	2–3	f
	$Ca_{0.17}Al_{12.33}Si_{3.67}O_{10}(OH)_2$	Monoclinic	variable	f
Muscovite (mica)	$KAl_3Si_3O_{10}(OH)_2$	Monoclinic	2.83	f
Kaolinite	$Al_2Si_2O_5(OH)_4$	Triclinic	2.65	f
Zeolite				g
Analcite	$NaAlSi_2O_6 \cdot H_2O$	Cubic	2.26	
Laumontite	$CaAl_2Si_4O_{12} \cdot 4H_2O$	Monoclinic	2.2–2.3	
Phillipsite	$K(Ca_{0.5},Na)_2Al_3Si5O16 \cdot 2H_2O$	Monoclinic	2.2	
Calcite	$CaCO_3$	Hexagonal	2.71	h
Aragonite	$CaCO_3$	Orthorombic	2.83	h
Dolomite	$CaMg(CO_3)_2$	Rhombohedral	2.86	
Magnesite	$MgCO_3$	Hexagonal	3.05	
Gypsum	$CaSO_4 \cdot 2H_2O$	Monoclinic	2.32	i
Anhydrite	$CaSO_4$	Orthorombic	2.96	i
Pyrite	FeS_2	Cubic	5.02	k
Halite	$NaCl$	Cubic	2.16	i

Notes:

(a) Quartz and hematite are practically insoluble in water and resistant to chemical weathering

(b) Olivines take up all compositions from pure forsterite to pure fayalite. Other bivalent cations such as Ni or Mn also substitute

(c) Produced by extreme chemical weathering of aluminosilicates

(d) Feldspars are the most important rock-forming minerals of the crust. Plagioclase exhibits all compositions from albite to anorthite

(e) Especially important in sheared crustal rocks. Composition is highly variable; Al commonly replaces some of the Si

(f) Representative of a larger group of layer silicates that are the most important minerals of sedimentary rocks

(g) Zeolites represent a large group of hydrated aluminosilicates; the cations frequently are almost completely replaceable

(h) Calcite and aragonite are the chief minerals of limestones. Aragonite is common at high pressures, but is preserved at low pressures

(i) These minerals are the major products resulting from the evaporation of sea water

(k) A large amount of sulfur in sedimentary rocks occurs as pyrite, formed in the sediments by bacterial reduction of sulfate

References

Deruelle, B., G. Dreibus, A. Jambon, Earth Planet. Sci. Lett. **108**, 217–227 (1992)

Mason, B., C.B. Moore, *Principles of Geochemistry*, 4th edn. (Wiley, New York, 1982)

Taylor, S.R., S.M. McLennan, *The Continental Crust: Its Composition and Evolution* (Blackwell Scientific Publications, Oxford, 1985)

Taylor, S.R., S.M. McLennan, Rev. Geophys. **33**, 241–265 (1995)

Wedepohl, K.H., Geochim. Cosmochim. Acta **59**, 1217–1232 (1995)

2.4   Sediments and Sedimentary Rocks

Sedimentary rocks comprise about 11% of the total crust of the earth, or about 8% after the deduction of volcanogenics. Marine sediments are formed by the deposition of particles carried to the ocean with the rivers and by winds. These particles originate from the disintegration of continental rocks by weathering processes. In addition, the ocean is a source of carbonate sediments formed from the calcareous skeletons of organisms. Following deposition, sediments undergo alterations (diagenesis) by redox reactions, ion exchange processes and compaction. In the course of time, sedimentary rocks experience uplift by tectonic forces and are eventually incorporated into the continental crust. About 75% of the continental surface is covered by sedimentary rocks.

By appearance and chemical composition the following types of sedimentary rocks are formally distinguished: sandstone, clays and shales, limestones, evaporite, volcanogenics. They are characterized as follows:

Sandstone: Clastic particles, size 0.06–2 mm; their origin is the weathering of crystalline granular rocks (such as granite); mineral components are mainly quartz, some feldspar and primary rock fragments;

Siliceous sediments: The chemical composition of these marine sediments is similar to that of sandstones. They are formed, at least partly, as biogenic deposits of the silica skeletons of diatoms and radiolarians.

Clays and shales: (Lutites) Particles smaller than those of sandstone, commonly in the range of μm; the term shale denotes lamination in addition to a fine-grained constitution. These rocks are the most abundant among all sediments. Their origin is largely the chemical weathering of primary rocks; dominant constituents are clay minerals in addition to some fine-grained quartz and feldspars.

Carbonate rocks: Grain size is variable, up to cm; they are formed as biochemical precipitate of the skeletal remains of marine calcareous organisms. Calcium-rich carbonate rocks are called limestones, dolomite is a rock containing $CaCO_3$ and $MgCO_3$ in nearly equal proportions.

Evaporites: Grain size is variable; these rocks are formed as a precipitate from evaporating salt water; the major minerals are halite, gypsum and anhydrite.

Volcanogenics: These sediments were originally basaltic lava that was fragmented and altered but retained much of the original basaltic imprint; the origin of volcanogenics thus differs from that of other sedimentary rocks, so that this type of sediment should be considered separately.

Table 2.8 Composition of sedimentary rocks (mass percent)[a]

Constituent	Average sandstone	Average limestone	Dolomite	Average shale	Average slate
Normative minerals					
Na-Feldspar	8.0	1.6	–	8.8	11.3
K-Feldspar	10.1	1.9	–	25.0	23.0
Quartz	62.3	3.8	1.6	22.8	24.2
Kaolinite	7.0	–	–	22.8	23.1
Mg-Chlorite	2.5	1.5	–	4.9	6.3
Ca-Carbonate	6.2	74.2	48.7	2.9	3.4
Mg-Carbonate	–	15.5	48.7	0.8	0.9
Hematite	3.9	1.5	1.0	8.1	7.2
Elements (as oxides)[b]					
SiO_2	77.6	5.2	0.73	53.4	60.6
TiO_2	0.4	0.06	–	0.7	0.7
Al_2O_3	7.1	0.81	–	15.5	17.3
Fe_2O_3	1.7	0.54	0.2	4.0	2.3
FeO	1.5	–	1.0	2.5	3.7
MgO	–	7.9	20.5	2.5	2.6
CaO	3.1	42.6	31.0	3.1	1.5
Na_2O	1.2	0.05	–	1.3	1.2
K_2O	1.3	0.33	–	3.3	3.7
H_2O	2.1	0.77	–	5.0	4.1
CO_2	2.5	41.6	47.5	2.6	1.5
SO_2	0.1	0.05	–	0.7	–
FeS_2	–	–	–	–	0.4
P_2O_5	0.1	0.04	0.05	0.2	–

[a]Source: Garrels and MacKenzie (1971)
[b]Approximate, mass balance not necessarily complete

Table 2.9 Distribution of sedimentary rocks on Earth by region and rock type (Vol.%)[a]

Region	Mass (10^{21} kg)	Sandstone	Clays and shales	Carbonate rocks	Evaporites	Siliceous sediments	Volcanogenics
Continents	1.87	21.0	43.1	18.7	0.8	1.8	14.6
	1.56	24.6	50.5	21.9	0.9	2.1	–
Continental margins[b]	0.62	14.0	49.4	17.2	1.5	1.0	16.9
	0.50	16.9	59.5	20.6	1.8	1.2	–
Oceans	0.18	7.0	49.0	33.9	0.7	6.1	3.3
	0.17	7.3	50.7	35.0	0.7	6.3	–
Totals	2.67	18.1	45.1	19.7	1.0	2.0	14.1
	2.23	21.1	52.5	23.0	1.1	2.3	–

[a]Two distributions are given: one including and one excluding volcanogenic sediments, respectively. Source: Budyko et al. (1987); originally Ronov and Yaroshevsky (1976)
[b]Coastal shelves and slopes of the continents

Chemical weathering: Two types of weathering processes contribute to the erosion of rocks at the surface of the continents: Mechanical forces, such as wind abrasion, and rock splitting due to freeze-thaw cycles of water or the growth of plant roots in rock crevices; and chemical reactions of rock minerals with soil waters that include reactions with acids and the oxidation of reduced compounds. This leads to the leaching of some elements and the formation of new minerals. The dominant acid is carbonic acid, which is present in soil waters due to the dissolution of carbon dioxide. Feldspars (aluminosilicates of sodium, potassium and calcium) comprise a major fraction of the primary minerals occurring in the upper crust. They are transformed by chemical weathering to clay minerals such as kaolinite and quartz, releasing the ions of alkali and earth alkaline elements as solutes to soil waters. Secondary minerals as well as sedimentary rocks undergo similar reactions, but the rates at which they react vary, leading to a complex mixture of original and secondary minerals. Ions and HCO_3^- are collected in ground waters and end up in the rivers that carry them to the oceans. Insoluble material is also transported by rivers as suspended particles.

Rate of continental erosion by weathering and sediment formation: The current global rate of weathering and new sediment formation can be estimated from the loss of material from continents to oceans due to transport in rivers, by ground

Table 2.10 Examples of chemical weathering reactions

Reactions of alumino-silicates

$2NaAlSi_3O_8 + 2H_2CO_3 \rightarrow 2Na^+ + 2HCO_3^- + Al_2Si_4O_{10}(OH)_2 + 2SiO_2$
(albite)   (montmorillonite) (quartz)

$3KAlSi_3O_8 + 2H_2CO_3 \rightarrow 2\ K^+ + 2HCO_3^- + KAl_3Si_3O_{10}(OH)_2 + 6SiO_2$
(orthoclase)   (mica)   (quartz)

$2KAl_3Si_3O_{10}(OH)_2 + 2H_2CO_3 + 3H_2O \rightarrow 2\ K^+ + 2HCO_3^- + 3Al_2Si_2O_5(OH)_4$
(mica)   (kaolinite)

$CaAl_2Si_2O_8 + 2H_2CO_3 + H_2O \rightarrow Ca^{2+} + 2HCO_3^- + Al_2Si_2O_5(OH)_4$
(anorthite)   (kaolinite)

$3Al_2Si_2O_5(OH)_4 + 3H_2O \rightarrow 6Al(OH)_3 + 6SiO_2$
(kaolinite)   (gibbsite)   (quartz)

Dissolution of calcium carbonates (calcite and dolomite)

$CaCO_3 + H_2CO_3 \rightarrow Ca^{2+} + 2HCO_3^-$
$CaMg(CO_3)_2 + 2H_2CO_3 \rightarrow Ca^{2+} + Mg^{2+} + 4HCO_3^-$

Oxidation reactions

$FeSiO_3 + 2H_2CO_3 + H_2O \rightarrow Fe^{2+} + 2HCO_3^- + H_4SiO_4$
(siderite)   (silicic acid)

$2Fe^{2+} + 4HCO_3^- + \frac{1}{2}O_2 \rightarrow Fe_2O_3 + 4CO_2 + 2H_2O$
                              (hematite)

$4FeS_2 + 15O_2 + 8H_2O \rightarrow 2Fe_2O_3 + 8HSO_4^- + 8\ H^+$
(pyrite)   (hematite)

water and atmospheric dust, by the wave-driven erosion of coastal areas and the erosion due to glaciers (mainly in Antarctica). A summary of such estimates is shown below. Human activity has a massive influence on erosion and sedimentation rates. Taylor and McLennan (1985) summarize estimates of pre-human global sedimentation rates that indicate a flux of 12.3 Pg a^{-1}. This is about one half of the total flux estimate shown in Table 2.11. If the cycle involving the loss of continental sediments due to weathering, the formation of new sediments in the ocean, and their eventual incorporation into the continental crust is closed, the mass of continental sediments: $(2.1 \times 10^{21}$ kg), the flux of material (12.3 Pg a^{-1}), and the surface coverage of the continents with sedimentary rocks (75%) can be used to calculate the residence time for continental sediments: 200–300 million years. From a quantitative model for the sedimentary rock cycle Garrels and MacKenzie (1972) have estimated a mean residence time of 400 million years, with 150 million years for new rocks and 600 million years for old rocks.

Table 2.11 Estimates for the annual flux of crustal material into the oceans (Pg a^{-1})[a]

Species	Rivers		Ground water	Aerosols	Marine erosion	Glacial	Totals
	Dissolved	Suspended					
SiO$_2$	0.44	12.10	0.05	0.04	0.15	1.30	14.08
Al	–	1.49	–	0.01	0.02	0.18	1.70
Fe	0.02	1.22	–	–	0.01	0.06	1.31
Na	0.21	0.20	0.02	0.07	–	0.05	0.55
K	0.08	0.39	0.01	0.01	0.01	0.06	0.56
Ca	0.50	0.55	0.06	–	0.01	0.06	1.18
Mg	0.14	0.28	0.02	0.01	0.01	0.03	0.49
HCO$_3$	1.90	0.83	0.20	–	0.02	–	2.98
Cl	0.26	–	0.03	0.14	–	–	0.44
SO$_4$	0.38	–	0.04	0.02	–	–	0.44
Total	3.93	17.06	0.43	0.30	0.23	1.74	23.73

[a]Source: Garrels and MacKenzie (1971)

Table 2.12 Estimated global rates of pelagic sedimentation (Pg a^{-1})[a]

Sediment	Atlantic ocean	Pacific ocean	Indian ocean	Total	Percent of total
Terrigenous fraction	0.642	0.784	0.304	1.730	58
Biogenic carbonate	0.543	0.305	0.231	1.079	36
Biogenic siliceous	0.045	0.065	0.062	0.172	6
Total	1.230	1.154	0.597	2.981	100

[a]Source: Lisitsyn et al. (1982)

References

Budyko, M.I., A.B. Ronov, A.L. Yanshin, *History of the Earth's Atmosphere* (Springer, Heidelberg, 1987)

Garrels, R.M., F.T. MacKenzie, *The Evolution of Sedimentary Rocks* (Norton, New York, 1971)

Garrels, R.M., F.T. MacKenzie, Mar. Chem. **1**, 27–41 (1972)

Lisitsyn, A.P., V.N. Lukashin, Y.G. Gurvich, V.V. Gordeyev, L.L. Demina, Geochem. Int. **19**, 102–110 (1982)

Ronov, A.B., A.A. Yaroshevsky, Geochem. Int. **13**, 89–121 (1976)

Taylor, S.R., S.M. McLennan, *The Continental Crust: Its Composition and Evolution* (Blackwell Scientific Publications, Oxford, 1985)

2.5   Composition of Gas Emissions from Volcanoes

The exact chemical composition of volcanic gases is difficult to measure. Data are derived primarily from volcanic vents (natural fissures as well as holes drilled into the cooled roof top above hot magma); residual gases extracted from fresh lava have also been explored. Working at such collection sites generally requires special protection against heat and gases. The collection procedure involves a stainless steel lead-in tube (silica lined, except the immediate intake section), a condenser for the separation of water and acids, a sample bottle for non-condensable gases, and an aspirator. Temperature is measured with a pyrometer and/or a thermocouple. Contamination of samples with air is almost unavoidable. Volcanic gases are reducing in character and contain essentially no oxygen. The amount of exhaled nitrogen is small, and can be determined only after correction for contamination based on the known ratio of nitrogen to argon in air. Oxygen often reacts with reduced gases in the sample, so that the validity of the measurement is compromised. The bulk of data indicates that a chemical equilibrium exists among individual components at the observation temperature, and this feature can be used to make corrections. Table 2.13 shows that water vapor is the principal component, followed by carbon dioxide, with smaller contributions by sulfur dioxide and hydrogen sulfide. Elemental sulfur derives largely from a reaction between SO_2 and H_2S. Hydrogen and carbon monoxide derive from the thermal dissociation of water and carbon dioxide, respectively. All data exhibit larger variations resulting from temperature variations, exhaustion of lava as it cools after having reached the surface, and the aging of magma in volcanic reservoirs below the surface. Whenever possible, averages have been derived for presentation in Table 2.13.

Comments:

Kilauea, Hawaii (1): Summit lava lake gases collected in March 1918 and March 1919 by T. A. Jaggar, averages of seven samples, $T = 1136 \pm 41°C$. Shepherd, E.S.,

Table 2.13 Chemical composition of volcanic gases (Vol.%)[a]

Location	T(°C)	H_2O	H_2	CO_2	CO	N_2	SO_2	H_2S	OCS	HCl	HF
Kilauea, Hawaii (1)	1,136	52.3	0.64	30.9	1.0	–	14.6	0.21	–	0.14	–
Kilauea, Hawaii (2)	917	77.1	1.1	3.98	0.075	–	15.7	1.36	–	0.18	0.20
Mauna Loa	1,130	69.4	0.66	4.57	0.16	–	24.7	0.38	–	0.015	–
Surtsey, Iceland, 1964	1,125	81.6	2.78	9.54	0.70	–	0.46	0.87	–	0.81	–
Surtsey, Iceland, 1965	1,125	87.7	2.75	5.66	0.38	–	2.57	0.51	–	0.41	–
Surtsey, Iceland, 1967	1,125	92.0	1.65	1.94	0.07	–	3.28	1.18	–	0.91	–
Mt. Etna, Sicily	1,075	49.9	0.54	22.9	0.48	–	25.9	0.33	–	–	–
Erta'Ale, Ethopia	1,134	78.2	1.6	11.1	0.5	0.06	7.1	1.0	0.009	0.42	–
Merapi, Indonesia	824	94.1	0.91	4.08	0.044	–	0.44	0.44	–	–	–
St. Helens, USA	656	96.3	0.42	2.65	0.006	–	0.19	0.48	0.0002	–	–
Momotombo, Nicaragua	844	95.0	0.87	2.38	0.025	0.08	0.7	0.49	–	0.35	0.03
St. Augustine, Alaska	870	84.3	0.59	2.34	0.02	–	6.00	0.92	–	5.69	0.086
Klyuchevskoy, Kamchatka	1,050	90.8	1.70	4.47	0.16	–	0.11	0.005	–	2.02	–
Satsuma-Iwojima, Japan	877	97.5	0.53	0.45	0.006	0.003	0.95[b]		–	0.49	0.03
	760	97.5	0.58	0.51	0.001	0.007	1.12[b]		–	0.17	0.006
Kudryavy, Kuril Islands	863	93.8	1.0	2.20	0.013	0.09	1.68	0.43	–	0.57	0.04

[a]N_2 corrected for contribution of air, if argon data are available. If S_2 was reported, it was allotted to SO_2 and H_2S in equal parts
[b]Only total sulfur was reported

Hawaiian Volcano Obs. Bull. **9**, 83–88 (1921); Jaggar, T.A., Am. J. Sci. **238**, 313–353 (1940); reanalyzed by Gerlach, T.M., J. Volcanol. Geotherm. Res. **7**, 295–317 (1980).

Kilauea, Hawaii (2): Episode 1 of Puu Oo eruption along the east rift zone, Vent B, 14–16 Jan. 1983; Type II volcanic gases, from magma stored for a prolonged time. Averages of ten samples, $T = 917 \pm 15°C$; Gerlach, T.M., Geochim. Cosmochim. Acta. **57**, 795–814 (1993)

Mauna Loa, Hawaii: Samples collected in 1984 during the March/April flank eruption; Greenland, L.P., U.S. Geolog. Surv. Prof. Pap. **1350**, 781–790 (1987).

Surtsey, Iceland: Averages of samples collected Oct. 15, 1964, 21. Feb. 1965, 31 Mar. 1967 at fissures fed by lava; Sigvaldason, G.E., G. Elisson, Geochim. Cosmochim. Acta **32**, 797–805 (1968); reanalyzed by Gerlach, T.M. J. Volcanol. Geotherm. Res. **8**, 191–198 (1980).

Mt. Etna, Sicily: Average of nine samples, $T = 1075°C$, taken in 1970 at three vents in the same location by Huntingdon, A.T., Philos. Trans. R. Soc. Lond. A **274**, 119–128 (1973); reanalyzed by Gerlach, T.M., J. Volcanol. Geotherm. Res. **6**, 165–178 (1979).

Erta'Ale, Ethopia: Active subsurface lava lake, sampled from vents in January 1974, averages of 18 samples, $T = 1125 - 1135°C$; Giggenbach, W.F., F. Le Guern, Geochim. Cosmochim. Acta **40**, 25–30 (1976); see also Gerlach, T.M., J. Volcanol. Geotherm. Res. **7**, 415–441 (1980).

Merapi, Indonesia: Averages of five samples collected at high temperature fumarole vents in July 1978, $T = 760–901°C$ (restored composition); Le Guern, F., T.M. Gerlach, T.M. Nohl, J. Volcanol. Geotherm. Res. **14**, 223–245 (1982).

St. Helens, USA: Fumarole on N40°W radial fissure, average of four samples obtained on 27. May and 2. July, 1981, $T = 656 \pm 30$; Gerlach, T.M., T.J. Casadevall, J. Volcanol. Geotherm. Res. **28**, 107–140 (1986).

Momotombo, Nicaragua: Average of 47 samples collected at representative fumaroles at the crater from March to May in 1983, $T = 844 \pm 14$; Menyailov, I.A., L.P. Nikitina, V.N. Shapar, V.P. Pilipenko, J. Geophys. Res. **91**, 12–214 (1986).

St. Augustine, Alaska: Averages of samples collected on 27. Aug. 1987 from fumaroles in a dome that developed subsequent to the 1986 eruption; Symonds, R.B., M.H. Reed, W.I. Rose, Geochim. Cosmochim. Acta **56**, 633–657 (1992).

Klyuchevskoy, Kamchatka: Samples taken on 18. July, 1988 at a place of magma effusion during the 1988 flank eruption; Taran, Y.A., A.M. Rozhkov, E.K. Serafimova, A.D. Esikov, J. Volcanol. Geotherm. Res. **46**, 255–263 (1991).

Satsuma-Iwojima, Japan: Data from fumaroles at opposite flanks of the volcano, averages of samples collected 24/27 Oct. 1990 at Ohachi-Oku ($T = 877°C$) and 24. Oct. 1990 at Arayama, $T = 760°C$; Shinohara, H., W.F. Giggenbach, K. Kazahaya, J.W. Hedenquist, Geochem. J. **27**, 271–285 (1993).

Kudryavy, Kuril Islands: High-temperature fumarolic gases, averages of five samples at the highest temperatures, $T = 863 \pm 68$, obtained in 1990–1992 from the summit crater. Taran, Y.A., J.W. Hedenquist, M.A. Korzhinsky, S.I. Tkachenko, K.I. Shmulovich, Geochim. Cosmochim. Acta, **59**, 1749–1761 (1995).

2.6   The Hydrosphere

The total amount of water on Earth is about 1.664×10^{21} kg. The oceans contain 1.37×10^{21} kg, most of the remainder occurs as pore water in sediments and sedimentary rocks. Ice caps and glaciers harbor about 3×10^{19} kg, ground water contains about 1.5×10^{19} kg, the amount residing in rivers and lakes is trivial in comparison. The atmosphere contains about 1.4×10^{16} kg precipitable water, which is maintained by the hydrological cycle of evaporation from and precipitation back to the earth's surface. Evaporation from the ocean exceeds precipitation (see Table 2.14 below). The surplus undergoes atmospheric transport toward the continents, where the rates are 69×10^{15} and 109×10^{15} kg a^{-1}, respectively. The difference is balanced by river runoff toward the ocean, about 40×10^{15} kg a^{-1} (Hantel, 2005).

Table 2.14   Summary of data for the world ocean[a]

Volume (10^{15} m^3), total	1,370.3	Average temperature (°C), total	3.51
Pacific ocean[b]	723.7	Pacific ocean	3.14
Atlantic ocean[b]	354.7	Atlantic ocean	3.99
Indian ocean[b]	291.9	Indian ocean	3.88
Water cycle total rates (10^{12} m^3 a^{-1})		Average salinity (‰), total	34.72
Evaporation	435	Pacific ocean	34.60
Precipitation	395	Atlantic ocean	34.92
River influx	40	Indian ocean	34.78
Average depth (m)	3,794	Hydrogen ion concentration (pH)	8.1
Mixed surface layer	75–200	Alkalinity[c]	2.4
Mean density of seawater (kg m^{-3})	1,025		

[a]Source: Murray (1992); Water cycle: Hantel (2005)
[b]Includes adjacent seas
[c]Alkalinity is defined as: $[Alk] = [HCO_3^-] + 2[CO_3^{2-}] + [B(OH)_4^-] + [OH^-] - [H^+]$, where the brackets indicate concentrations [mmol kg^{-1}]

Table 2.15 Average concentrations of major ions in seawater[a] and their residence time in the ocean based on the global input by rivers

Ion or DS[b]	Seawater ($mg\,kg^{-1}$)	Seawater ($mmol\,kg^{-1}$)	River water ($mg\,kg^{-1}$)	Residence time[c] (10^6 a)
Cl^-	19,354	545.91	5.75	115
SO_4^{2-}	2,712	28.23	8.25	11
HCO_3^-	142	2.33	52.0	0.1
Br^-	67.3	0.84	0.02	115
F^-	1.3	0.07	0.1	0.54
B^b	4.5	0.42	0.01	16
Na^+	10,781	468.95	5.15	72
Mg^{2+}	1,294	53.24	3.35	13
Ca^{2+}	412	10.28	13.4	1
K^+	399	10.21	1.3	11
Sr^{2+}	7.9	0.09	0.03	9
Si^b	2.5	0.10	4.9	0.02
Fe	0.04–0.25	$(0.7–4.5)\times 10^{-3}$	0.1	~0.05

[a]Salinity S = 35‰. Seawater: Murray (1992), Li (1991); River water (pristine): Meybeck (1979), Livingstone (1963)
[b]DS = dissolved substance: Boron occurs as $B(OH)_3$ and $B(OH)_4^-$; silicon occurs as silicic acid H_4SiO_4. Silicon is most abundant in suspended material in the form of silicates
[c]Based on ocean mass: 1.37×10^{21} kg, annual river input: 4×10^{16} kg Murray (1992)

Table 2.16 Mean concentration of the elements in seawater ($g\,kg^{-1}$)[a]

Element	Species	Abundance[b] (a)	(b)
Helium	He	7.2 (−9)	−
Lithium	Li^+	1.78 (−4)	1.8 (−4)
Beryllium	$BeOH^+$	2.1 (−10)	2.1 (−10)
Boron	$B(OH)_3$, $B(OH)_4^-$	4.4 (−3)	4.5 (−3)
Carbon	CO_2, HCO_3^-, CO_3^{2-}	2.6 (−2)	2.8 (−2)
Nitrogen	N_2	1.64 (−2)	−
	NO_3^-	1.86 (−3)	4.2 (−4)
Oxygen	O_2	4.8 (−3)	−
Fluorine	F^-, MgF	1.3 (−3)	1.3 (−3)
Neon	Ne	1.6 (−7)	−
Sodium	Na^+	10.781	10.8
Magnesium	Mg^{2+}	1.28	1.29
Aluminum	$Al(OH)_4^-$	1 (−6)	3 (−7)
Silicon	$Si(OH)_4$	1.1 (−4)	2.5 (−3)
Phosphorus	Phosphate	6.8 (−5)	6.5 (−5)
Sulfur	SO_4^{2-}	9.05 (−1)	8.98 (−1)
Chlorine	Cl^-	19.353	18.8
Argon	Ar	6.23 (−4)	−
Potassium	K^+	3.99 (−1)	3.9 (−1)
Calcium	Ca^{2+}	4.15 (−1)	4.48 (−1)

(continued)

Table 2.16 (continued)

Element	Species	Abundance[b]	
		(a)	(b)
Scandium	$Sc(OH)_3$	6.7 (−10)	0.86 (−10)
Titanium	$Ti(OH)_4$	1.0 (−8)	1.0 (−8)
Vanadium	Vanadate	~1 (−6)	2.15 (−6)
Chromium	$Cr(OH)_3$	3.3 (−9)	2.6 (−9)
	CrO_4^{2-}	3.5 (−7)	2.5 (−7)
Manganese	Mn^{2+}, $MnCl^+$	1 (−8)	7.2 (−8)
Iron	$Fe(OH)_2^+$, $Fe(OH)_4^-$	4.0 (−8)	2.5 (−7)
Cobalt	Co^{2+}	2 (−9)	1.2 (−9)
Nickel	Ni^{2+}	4.8 (−7)	5.3 (−7)
Copper	$CuCO_3$, $CuOH^+$	1.2 (−7)	2.1 (−7)
Zinc	$ZnOH^+$, Zn^{2+}	3.9 (−7)	3.2 (−7)
Gallium	$Ga(OH)_4^-$	1.0–2.0 (−8)	1.7 (−9)
Germanium	$Ge(OH)_4$	5 (−9)	4.3 (−9)
Arsenic	$HAsO_4^{2-}$, $H_2AsO_4^-$	2 (−6)	1.7 (−6)
Selenium	Se(IV)	1.7 (−7) (total)	1.0 (−7)
	Se(VI)		5.5 (−8)
Bromine	Br^-	6.7 (−2)	6.7 (−2)
Krypton	Kr	3 (−7)	–
Rubidium	Rb^+	1.24 (−4)	1.2 (−4)
Strontium	Sr^{2+}	7.8 (−3)	7.8 (−3)
Yttrium	$Y(OH)_3$	1.30 (−7)	1.3 (−8)
Zirconium	$Zr(OH)_3$	<1 (−6)	1.7 (−8)
Niobium		1 (−6)	1 (−8)
Molybdenum	MoO_4^{2-}	1.1 (−5)	1.0 (−5)
Ruthenium		5 (−10)	≤ 5 (−12)
Palladium		–	7 (−11)
Silver	$AgCl_2^-$	3 (−9)	2.5 (−9)
Cadmium	$CdCl_2$	7.0 (−8)	7.9 (−8)
Indium	$In(OH)_2^+$	2 (−10)	1 (−10)
Tin	$SnO(OH)_3^-$	5 (−10)	6 (−10)
Antimony	$Sb(OH)_6^-$	2 (−7)	1.5 (−7)
Tellurium	Te(IV)	–	2 (−11)
	Te(VI)	–	5 (−11)
Iodine	I^-, IO_3^-	5.9 (−5)	5.8 (−5)
Xenon	Xe	6.5 (−7)	–
Cesium	Cs^+	3 (−10)	3.06 (−7)
Barium	Ba^{2+}	1.17 (−5)	1.5 (−5)
Lanthanum	$La(OH)_3$	4 (−9)	5.6 (−9)
Cerium	$Ce(OH)_3$	4 (−9)	1.7 (−9)
Praseodymium	$Pr(OH)_3$	6 (−10)	8.7 (−10)
Neodymium	$Nd(OH)_3$	4 (−9)	4.2 (−9)
Samarium	$Sm(OH)_3$	6 (−10)	8.4 (−10)
Europium	$Eu(OH)_3$	1 (−10)	2.1 (−10)
Gadolinium	$Gd(OH)_3$	8 (−10)	1.3 (−9)

(continued)

Table 2.16 (continued)

Element	Species	Abundance[b] (a)	(b)
Terbium	$Tb(OH)_3$	1 (−10)	2.1 (−10)
Dysprosium	$Dy(OH)_3$	1 (−9)	1.5 (−9)
Holmium	$Ho(OH)_3$	2 (−10)	4.5 (−10)
Erbium	$Er(OH)_3$	9 (−10)	1.3 (−9)
Thulium	$Tm(OH)_3$	2 (−10)	2.5 (−10)
Ytterbium	$Yb(OH)_3$	9 (−10)	1.5 (−9)
Lutetium	$Lu(OH)_3$	2 (−10)	3.2 (−10)
Hafnium		<8 (−9)	3.4 (−9)
Tantalum		<2.5 (−9)	<2.5 (−9)
Tungsten	WO_4^{2-}	<1 (−7)	1 (−7)
Rhenium	ReO_4^-	4 (−9)	8 (−9)
Iridium		−	2 (−12)
Platinum		−	2.7 (−10)
Gold	$AuCl_2^-$	1.1 (−8)	3 (−11)
Mercury	$HgCl_4^{2-}, HgCl_3^-$	6 (−9)	4.2 (−10)
Thallium	Tl^+	1.2 (−8)	1.4 (−8)
Lead	$PbCO_3$	1 (−9)	2.7 (−9)
Bismuth	$BiO^+, Bi(OH)_2^+$	1.0 (−8)	4.2 (−12)
Thorium	$Th(OH)_4$	< 7 (−10)	5 (−11)
Uranium	$UO_2(CO_3)_2^{4-}$	3.2 (−6)	3.2 (−6)

[a]No data exist for Osmium, Rhodium and most radioactive elements
[b]Powers of ten are shown in parentheses. (a) Quinby-Hunt and Turekian (1983) (b) Li (1991)

Table 2.17 Estimates of global fresh water resources (unit: 10^{12} m^3)[a]

Description	UNESCO (1978)	AGU (1979)	Baumgartner and Reichel (1975)	Berner and Berner (1987)	Residence time
Lakes[b]	176.4	280	225	125	17 a
Swamps	11.5	[d]	[d]	[d]	5 a
Rivers[c]	2.1	1.2	[e]	1.7	16 days
Ground water	10,530	60,000[f]	8,062	9,500	1,400a
Soil moisture	16.5	85	[g]	65	1 a
Permanent ice	24,064	24,000	27,820	29,000	>1,500 a
Atmosphere	12.9	14	13	13	8 d
Total fresh water	34,813	84,380	36,120	38,705	-

[a]Source: Gleick (1993), residence times from Hantel (2005)
[b]Volumes are about equal for fresh water lakes, and for saline lakes and inland seas
[c]Average instantaneous volume
[d]The values are either included in other categories or disregarded
[e]Included in the value for fresh water lakes
[f]Refers to water in the upper 5 km of Earth's crust, excluding chemically bound water
[g]Soil water included in ground water

Table 2.18 Present day land ice cover quantifying the relative size of glaciers, ice caps, and Greenland and Antarctic ice sheets[a]

	Glaciers	Ice caps	Glaciers and ice caps	Greenland ice sheet	Antarctic ice sheet
Area (10^{12} m^2)	0.43	0.24	0.68	1.71	13.9
Volume (10^{15} m^3)	0.08	0.10	0.18±0.04	2.85±0.14	26.37±1.32
Sea level rise[b] (m)	–	–	0.5	7.2	61.1

[a]Source: Hantel (2005)
[b]Global rise of sea level resulting from total melting of the ice mass

Table 2.19 Major rivers of the world: length, runoff rate, drainage area and suspended sediment discharge[a]

River	Length (km)	Runoff rate (10^9 m^3 a^{-1})	Drainage area (10^3 km^2)	Sediment discharge (10^9 kg a^{-1})
Africa				
Niger	4,160	192	1,210	40
Nile	6,670	202	2,870	[b]
Zaire-Congo	4,370	1,460	3,820	43
Zambesi	2,660	223	1,200	20
North America				
Columbia	1,950	267	669	8
Fraser		112	220	20
Mackenzie	4,240	350	1,800	100
Mississippi-Missouri	5,985	580	3,220	210
St. Lawrence	3,060	439	1,290	4
Yukon	3,000	207	852	60
South America				
Amazon	6,280	6,930	6,915	900
Magdalena	1,530	260	260	220
Orinoco	2,740	914	1,000	210
La Plata – Paraná	4,700	725	2,970	92
Eurasian Arctic				
Lena	4,400	532	2,490	12
Ob	3,650	395	2,990	16
Yenisei	3,490	610	2,580	13
Asia				
Amur	2,820	355	1,855	52
Ganges – Brahmaputra	3,000	1,400	1,730	1,670
Indus	3,180	220	960	100
Irrawaddy	2,300	486	410	265
Mekong	4,500	510	810	160
Salween	2,820	211	325	[c]
Xijiang	2,130	363	437	69
Yangzijiang	5,520	995	1,800	478
Europe				
Danube	2,860	214	817	67
Volga	3,350	254	1,360	[c]

[a]Data collected from various tables in Gleick (1993)
[b]Sediment discharge has discontinued due to construction of the Aswan High Dam
[c]No data given

Table 2.20 Estimates of long term mean annual run-off from the continents (excluding the Antarctic)[a]

	Area (10⁶ km²)	UNESCO (1978)	Baumgartner and Reichel (1975)	World Resources (1992)	Shiklomanov (1997)
Europe	10.46	2.960	2.950	3.264	2.900
Asia	43.48	14.088	12.000	14.088	13.508
Africa	30.10	4.605	3.431	3.793	4.040
North America	24.25	8.221	5.869	7.906	7.770
South America	17.86	11.805	11.020	10.502	12.030
Australasia	8.95	2.506	2.408	2.354	2.400
World	135.10	44.185	37.677	41.905	42.648

[a]Units: 10^{12} m³ a⁻¹; source: Hantel (2005)

Table 2.21 Estimates of continental water balances (mm a⁻¹)[a]

	Precipitation			Evaporation			Runoff		
	A	B	C	A	B	C	A	B	C
Africa	657	740	726	582	587	547	114	153	139
North America	645	756	670	403	418	383	242	339	287
South America	1,564	1,600	1,648	946	910	1,065	617	685	583
Asia	696	740	726	420	416	433	276	324	293
Australasia	803	791	736	534	511	510	269	280	229
Europe	657	790	734	375	507	415	282	283	319

[a]Key: *A* Baumgartner and Reichel (1975), *B* UNESCO (1978), *C* AGU (1979)

References

AGU, World Water Resources and Their Future. (English translation of a 1974 USSR publication, Nace, R., editor) (American Geophysical Union, Washington, DC, 1979)

Baumgartner, A., E. Reichel, *The World Water Balance: Mean Annual Global Continental and Maritime Precipitation, Evaporation and Runoff* (Elsevier, Amsterdam, 1975)

Berner, E.K.R., A. Berner, *The Global Water Cycle, Geochemistry and Environment* (Prentice Hall, Englewood Cliffs, NJ, 1987)

Gleick, P.H. (ed.), *Water in Crisis. A Guide to the World's Fresh Water Resources* (Oxford University Press, New York, 1993)

Hantel, M. (ed.), *Observed Global Climate*. Landolt-Börnstein, Numerical Data and Functional Relationships in Science and Technology, New Series, Group V: Geophysics, vol. 6. (Springer, Berlin, 2005)

Li, Y.-H., Geochim. Cosmochim Acta **55**, 3223–3240 (1991)

Livingstone, D.A., in: *Data of Geochemistry*, ed. by M. Fleischer, 6th edn. U.S. Geological Survey Profession Paper, pp. 440–446 (1963)

Meybeck, M., Rev. Géol. Dynam. Géogr. Phys. **21**, 215–246 (1979)

Murray, J.W., in: *Global Biogeochemical Cycles*, ed. by S.S. Butcher, R.J. Charlson, G.H. Orians, G.V. Wolfe (Academic, London, 1992), pp. 175–212

Quinby-Hunt, M.S., K.K. Turekian, EOS **64**, 130–131 (1983)

Shiklomanov, I.A., *Assessment of Water Resources and Water Availability in the World* (World Meteorological Organization and Stockholm Environment Institute, Geneva, 1997)

UNESCO, World Water Balance and Water Resources of the Earth. Studies and Report in Hydrology, No. 25, UNESCO, Paris (1978)
World Resources, *A Guide to the Global Environment* (Oxford University Press, New York, 1992)

2.7 The Biosphere

The biosphere, which encompasses all forms of life present on the globe, generally is taken to include all living organisms as well as dead (and decaying) organic material. The biosphere owes its existence largely to the energy captured from sunlight by organisms that make use of the photosynthetic process to combine carbon dioxide and water to form organic material. Other elements required by living organisms are nitrogen, sulfur, phosphorus and certain trace metals. Because of its activity, the biosphere as a whole takes part in the global cycle of all these elements, but carbon, hydrogen and oxygen are the major elements involved.

Table 2.22 Carbon content of the biosphere (Pg) compared to that in ocean and atmosphere[a]

Marine		Terrestrial	
Living biomass	~3	Assimilating parts	9.0×10^1
Dissolved organic carbon	~1×10^3	Structural parts	5.0×10^2
Particulate organic carbon	~3×10^1	Stem, branch, leaf litter	6.0×10^2
Total dissolved CO_2	3.7×10^4	Soil humus	~2×10^3
CO_2 in the upper mixed layer	6.7×10^2		
Atmosphere			
Current level	8.2×10^2	Preindustrial	6.0×10^2

[a]Data taken from Warneck (2000)

Comments: Living biomass in the marine environment is represented almost entirely by unicellular algae (phytoplankton) in near-surface waters. Rapid consumption by grazing species (zooplankton) and bacteria keeps the amount of phytoplankton relatively low in the face of a net primary production near 48 Pg a^{-1}. Dissolved organic carbon also arises from phytoplankton, both by direct exudation and by the decay of dead cells. The rapid turnover of organic carbon in the oceans causes the marine biosphere to be self-contained. In the terrestrial biosphere the main fraction of organic carbon occurs in the wooden parts of plants and in soil humus. The mass of carbon present in the terrestrial biosphere is comparable to that in the atmosphere, which causes annual cycles in the uptake and release of CO_2 by the biosphere to give rise to similar annual variations of the CO_2 mixing ratio in the atmosphere.

Table 2.23 Terrestrial biomass and net primary productivity[a]

Ecosystem type	Land Area (10^12 m^2)	Mean biomass Dry matter (kg m^-2)	Mean biomass Carbon (Pg)	Net primary production Biomass (kg m^-2 a^-1)	Net primary production Carbon (Pg a^-1)	Litter fall Biomass (kg m^-2 a^-1)	Litter fall Carbon (Pg a^-1)	Litter Biomass (kg m^-2)	Litter Carbon (Pg)	Organic soil Carbon (kg m^-2)	Organic soil (Pg)
Tropical rain forests	10	42	189	2.3	10.4	1.84	8.3	0.66	3.3	8	82
Tropical seasonal forests	4.5	25	51	1.6	3.24	1.3	2.6	0.84	1.9	9	41
Mangrove	0.3	30	4	1.0	0.14	0.6	0.1	10	1.5	8	3
Temperate forests	6	58	79	1.4	3.8	0.85	2.3	3	9	12	72
Boreal forests	9	22.8	92	0.8	3.2	0.6	2.4	3.5	15.8	15	135
Forest plantations	1.5	20	13.5	1.75	1.2	0.88	0.6	0.5	0.4	12	18
Woodland and shrubs	4.5	12	24	1.1	2.2	1.1	2.2	1.4	3.1	12	54
Semi-desert shrubs	21	16.5	7.4	0.15	1.4	0.12	1.2	0.1	1.1	8	168
Savanna	22.5	6.5	66	1.75	17.7	1.4	14.1	0.35	3.9	12	264
Temperate grassland	12.5	1.6	9	0.78	4.4	0.63	3.9	0.4	2.5	24	295
Tundra arctic/alpine	5.1	5.7	1.3	0.11	0.3	0.14	0.24	0.5	0.9	6.5	34
Scrub tundra	4.4	2.3	4.6	0.35	0.7	0.2	0.4	5.0	11	20	88
Cultivated land	16	4.2	3	0.95	6.8	0.45	3.1	0.05	0.4	8	128
Bogs, swamps, marshes	3.5	9.7	15	2.5	3.9	0.6	1.0	2.5	4.4	64	225
Human areas	2	4	1.4	0.5	0.2	0.3	0.2	0.3	0.2	5	10
Sums	122.8		560		59.9		42.6		59.5		1,635

[a]Source of data: Ajtay et al. (1979); 26.2×10^{12} [m^2] of the land area are ice covered or deserts with no appreciable input by the biosphere

Table 2.24   Global net primary production (NPP) of the biosphere (Pg a^{-1} of carbon)[a]

Terrestrial		Marine	
Seasonal Variation[b]			
January–March	11.2		11.3
April–June	15.7		10.9
July–September	18.0		13.0
October–December	11.5		12.3
Vegetation Classes[c]		*Biogeographic Categories*[d]	
Tropical rain forests	17.8	Oligotrophic	11.0
Broadleaf deciduous forests	1.5	Mesotrophic	27.4
Broadleaf and needle-leaf forests	3.1	Eutrophic	9.1
Needle-leaf evergreen forests	3.1	Macrophytes	1.0
Needle-leaf deciduous forests	1.4	Total NPP	48.5
Savannas	16.8		
Perennial grass lands	2.4		
Broadleaf shrubs with bare soil	1.0		
Tundra	0.8		
Desert	0.5		
Cultivated areas	8.0		
Total NPP	56.4		

[a]Net primary productivity is the amount of organic matter produced by photosynthesis in plants (difference between autotrophic photosynthesis and respiration). The fraction of carbon in green plant matter is approximately 45% (net dry weight). The data were derived by Field et al. (1998) from satellite indices of surface chlorophyll (oceans) and vegetation (terrestrial), combined with surface-based measurements of light utilization efficiencies by various methods
[b]NPP on land maximizes in the tropics in all seasons and shows a secondary maximum in the northern hemisphere in April–September, whereas in the oceans the seasonal dependence with latitude is slight
[c]Major vegetation categories as defined by DeFries and Townshend (1994)
[d]Categories determined by satellite-derived near-surface phytoplankton chlorophyll concentration (mg m^{-3}): oligotrophic $C_{sat} < 0.1$, mesotrophic $0.1 < C_{sat} < 1.0$, eutrophic $C_{sat} > 1$. The contribution of macrophytes (taken from Smith 1981) is not included in the seasonal totals

Comments: Previous estimates of terrestrial NPP were based on techniques such as harvests from small sample plots, dimensional analysis and census of forest stands, growth relations between different plant tissues (leaf or twig dry weight versus stem dry weight), and gas exchange measurements. Techniques for the determination of NPP of marine biota are mainly based on the uptake of radiocarbon. Two critical reviews gave a global NPP of 59.9 Pg a^{-1} for the terrestrial biosphere (Ajtay et al. 1979, see Table 2.23), and 43.5 Pg a^{-1} for marine biota (De Voos 1979). These values agree approximately with the more recent assessments summarized in Table 2.24. Because of extensive deforestation, especially in the tropics, the mass of carbon present in the terrestrial biosphere may have changed considerably since the middle of the past century, when the previous assessment was made. Net primary productivity has probably not changed much.

Table 2.25 Nitrogen in the biosphere, in terrestrial soils, and in the ocean[a]

Reservoir	Mass content (Tg)	Remarks
Continents		
Land plants	1.0 (4)	Based on C/N = 75
Litter and dead wood	1.5 (3)	Based on C/N = 60 and 90 Pg C
Animals	2.0 (2)	
Soil humus, organic fraction	2.0 (5)	Based on C/N = 10
Soil, inorganic fraction	1.6 (4)	Mainly fixed insoluble NH_4^+
Total	2.2 (5)	
Ocean		
Plankton	5.2 (2)	Based on C/N = 5.7
Animals	1.7 (2)	
Dead organic matter, particulates	5.3 (3)	Based on N/C = 5.7, 20 Pg C
Dissolved organic matter	3.7 (5)	Based on C/N = 2.7
Inorganic nitrogen	5.7 (5)	Mainly dissolved NO_3^-
Total	9.4 (5)	
Dissolved N_2	2.2 (7)	

[a]Orders of magnitude in parentheses; from Warneck (2000) reproduced with permission of Elsevier; the data were derived from carbon contents of the biosphere estimated in the 1970s and may have to be revised

Table 2.26 Global rates of nitrogen fixation (Tg a^{-1} as nitrogen)[a]

	1860	1993
Continents		
Natural ecosystems	128 (100–290)	107
Cultivated soils	15	31.5
Haber-Bosch synthesized	0	100
Subtotal	143	238.5
Atmosphere		
Lightning	~6	~6
Fossil fuel combustion	0.3	24.5
Stratosphere	0.6	0.6
Subtotal	~7	31
Oceans		
Pelagic regions	109 (85–141)	109
Continental shelves	12 (1.5–15)	12
Subtotal	121	121

[a]Source of data: Galloway et al. (2004)

Table 2.27 Global rates of denitrification (Tg a^{-1} as nitrogen)[a]

Land-based sources		
Soils	124	(65–175)
Ground water	44	(20–138)
Lakes and reservoirs	31	(19–43)
Rivers	35	(20–35)
Total	234	
Marine sources		
Estuaries	8	(3–10)
Continental shelves	250	(214–300)
Oceanic oxygen minimum zone	· 81	(50–113)
Total	339	

[a]Source of data: Seitzinger et al. (2006)

Table 2.28 Global atmospheric emissions of NO$_x$ and NH$_3$ (Tg a^{-1} as nitrogen)[a]

	1860		1993	
	NO$_x$	NH$_3$	NO$_x$	NH$_3$
Natural				
Lightning	5.4	–	5.4	–
Stratosphere	0.6	–	0.6	–
Soils and vegetation	2.9	6.0	2.9	0.6
Forest/bushfires	1.6	1.6	0.8	0.8
Ocean	–	5.6	–	5.6
Subtotal	10.5	13.3	9.7	11.0
Agriculture				
Agricultural soils and crops	–	0.2	2.6	4.0
Fertilizer loss	–	0	–	9.7
Animal waste	–	5.4	–	26
Biomass burning	2.0	1.0	6.4	4.6
Subtotal	2.0	6.6	9.0	44.3
Energy production				
Fossil fuel burning	0.3	0	20.4	0.1
Bio-fuel combustion	0.4	0.7	1.3	2.6
Industrial processes	0	0	5.1	0.2
Aircraft	0	–	0.5	–
Subtotal	0.6	0.7	27.2	2.9
Total emissions[b]	13.1	20.6	45.9	58.2

[a]Source of data Galloway et al. (2004)

[b]Wet and dry removal processes cause essentially all the emitted NO$_x$ and NH$_3$ to be deposited on the surface of Earth, only a very small fraction of ammonia is destroyed by chemical reactions in the atmosphere

Table 2.29 Summary of the global nitrogen cycle (Tg a^{-1} as nitrogen)[a]

	1860	1993
Continents		
N-fixation, natural and anthropogenic	143	239
Input of fixed nitrogen into rivers	70	118
River flux to inland systems	8	11
River flux to coastal areas	27	48
Aerial transport to coastal areas	9	33
Denitrification in soils	63[b]	67
Denitrification in rivers and lakes	37[b]	48
Denitrification to N$_2$O	8	11
Subtotal, denitrification	108	126
Storage in soils	–	60
Marine		
Input from the atmosphere	14	39
Input via rivers	27	48
Biological nitrogen fixation	121 (85–158)	121
Subtotal, input	162	208
Denitrification, estuaries and shelf	27	48
Denitrification, open ocean nitrate	145	145
Denitrification to N$_2$O	4	4
Subtotal, denitrification	176 (146–453)	197
Export to the sediments	9	16

[a]Source of data: Galloway et al. (2004)
[b]Partitioning estimated from the sum given for soils and rivers

Comments: The reduction of N$_2$ to NH$_3$/NH$_4^+$ by specialized bacteria in soils and aquatic systems is the dominant natural source of nitrogen in the biosphere (N-fixation). The products NH$_3$ and NH$_4^+$ are partly oxidized to NO$_2^-$ and NO$_3^-$. Most plants can utilize nitrate as well as ammonium to satisfy their nitrogen needs. The annual rate of nitrogen fixation is, in the long run, balanced by bacterial reduction of nitrate back to nitrogen (denitrification), which occurs in oxygen-poor soils and aquatic systems.

Table 2.30 Geological epochs and time scales[a]

Period or epoch	Time interval (Million years)	Key events of biological evolution	Climate related events	Atmospheric O$_2$ (%)
Cenozoic era				
Quaternary				
Holocene	10,000 a	Agriculture	Late medieval small ice age	~21
Pleistocene	0.01–1.75	Homo erectus breakout	Glaciation, northern hemisphere	
Tertiary				
Pliocene	1.8–5.3	Ape man fossils		~21–24
Miocene	5.3–24	Origin of grass	Continents are dispersing	
Oligocene	24–34	Mammals diversify		
Eocene	34–53	First hoofed mammals		~18–24
Paleocene	53–65	Earliest primates		
→ Mass extinction (e.g. dinosaurs, marine reptiles, ammonoids, diatoms, rudist bivalves) ←				
Mesozoic era				
Cretaceous	65–138	First flowering plants	Global warming, poles ice-free, meteoric impact at Chichuan Peninsula, Mexico	~12–18
Jurassic	138–208	Dinosaurs diversify, first birds	Break-up of Pangea, opening of N-Atlantic (~100 × 10^6 a)	~12–15
Triassic	203–250	First dinosaurs and mammals	Local evaporates in upper Triassic	~12–18
→ Mass extinction (90% of all marine and >75% of all terrestrial species) ←				
Paleozoic era				
Permian	250–292	Reptiles diversify, insect pollination	Extensive volcanism, evaporites	~15–30
Carboniferous	292–320	First conifers, first reptiles	Local glaciation	~17–30
	320–354	Amphibians, sudden fauna diversification	Gondwana and Euramerica form	
		Seed ferns	Supercontinent Pangea (350–180) × 10^6 a ←	
→ Mass extinction (~364 × 10^6 years, e.g. tabulate corals, brachiopods, trilobites jawless fishes) ←				

(continued)

Table 2.30 (continued)

Period or epoch	Time interval (Million years)	Key events of biological evolution	Climate related events	Atmospheric O_2 (%)
Devonian	354–410	Jawed fishes diversify, first amphibians	Gondwana and Euramerica on southern Hemisphere and equator, respectively	~13–25
Silurian	410–435	First vascular land plants ←	Laurentia and Baltica form Euramerica	~19–25
→ Mass extinction (e.g. bryozoans, brachiopods, trilobites) ←				
Ordovician	435–500	First animals ashore, first real vertebrates,(Agnatha), diversification of metazoan Families	Glaciation around South Pole (today NorthAfrica), evaporites (today Australia)	14–19
Cambrian	500–540	Increasing diversity of fauna and flora; Evolution of skeletons, first chordates, fishes	The proterozoic supercontinent breaks up forming Gondwana, Laurentia, Baltica mostly the in Southern hemisphere	~15–21
Precambrian era				
Proterozoic, late	540–1000	Ediacara fauna, soft-bodied metazoans	Local glaciation	~1–10
Middle	1000–1600		Supercontinent Rodinia (~1×10^9 years)	~1
Early	1600–2500		Earth ice-covered (~2.2–0.8×10^9 years) Abundant banded iron stones	~0.1–2
Archean	2500–3000	Photosynthetic bacteria	First known glacial event (2.8–2.5×10^9 years) Witwatersrand, South Africa	~0.01
	3000–3800	Most ancient fossils (3.4×10^9 years)	First permanent continental crust, oldest banded iron formation	~0.0001
Hadean	3800–4450	Earth formed ~4.56×10^9 [a] ago	Oldest minerals (4.2×10^9 years) and rocks, (3.8×10^9 years)	

[a]Assembled from data in Stanley (2009); amount of oxygen in the atmosphere from Berner (2006). Courtesy S. Brückner, Max-Planck-Institute for Chemistry, Mainz, Germany

References

Ajtay, G.L., P. Ketner, P. Duvigneaud, Chapter 5 in *The Global Carbon Cycle*, vol. **13**, ed. by B. Bolin, E.T. Degens, S. Kempe, P. Ketner (SCOPE, New York, 1979), pp. 129–181

Berner, R.A., Geochim. Cosmochim. Acta **70**, 5653–5664 (2006)

De Voos, C.G.N., Chapter 10, in *The Global Carbon Cycle*, vol. 13, ed. by B. Bolin, E.T. Degens, S. Kempe, P. Ketner (SCOPE, New York, 1979), pp. 259–292

DeFries, R.S., J.R.G. Townshend, Int. J. Remote Sens. **15**, 3567–3586 (1994)

Field, C.B., M.J. Behrenfeld, J.T. Randerson, P. Falkowski, Science **281**, 237–240 (1998)

Galloway, J.N., F.J. Dentener, D.G. Capone, E.W. Boyer, R.W. Howarth, S.P. Seitzinger, G.P. Asner, C.C. Cleveland, P.A. Green, E.A. Holland, D.M. Karl, A.F. Michaels, J.H. Porter, A.R. Townsend, C.J. Vörösmarty, Biogeochem. **70**, 153–226 (2004)

Seitzinger, S., J.A. Harrison, J.K. Böhlke, A.F. Bouwman, R. Lowrance, B. Peterson, C. Tobias, G. Van Drecht, Ecol. Appl. **16**, 2064–2090 (2006)

Smith, S.V., Science **211**, 838–840 (1981)

Stanley, S.M., *Earth System History*, 3rd edn. (Freeman, New York, 2009)

Warneck, P., *Chemistry of the Natural Atmosphere*, 2nd edn. (Academic (Copyright Elsevier), San Diego, 2000)

Chapter 3
Structure of the Atmosphere

Atmospheric pressure and density decrease quasi-exponentially with increasing altitude in accordance with the barometric law. Figure 3.1 presents an idealized vertical temperature profile of the atmosphere and the nomenclature adopted by international convention to identify different altitude regimes, which are distinguished by the prevailing sign of the temperature gradient.

In the troposphere the decrease of temperature with altitude arises from adiabatic cooling associated with convection. In the stratosphere and mesosphere, infrared radiation is the principal mode of heat transfer. The temperature increase in the middle atmosphere results from the input of solar ultraviolet energy due to absorption by ozone. In the thermosphere, photodissociation leads to low concentrations of infrared emitters such as CO_2 and H_2O, so that heat cannot be removed by radiation and must be conducted downward to regions where radiatively active species are more abundant and radiation to space is efficient.

The bulk chemical composition of dry air remains constant at altitudes up to ~90 km (homosphere). At higher altitudes, molecular diffusion replaces turbulent mixing as a mode of transport (as shown on the right side of Fig. 3.1). Individual components then are separated by diffusion in the gravitational field of Earth, so that the composition of air varies with altitude (heterosphere). See Chap. 9, The Upper Atmosphere.

The abundance of water vapor, which is an important component in the troposphere, is determined by the prevailing temperature, as water vapor condenses to form liquid drops when the vapor pressure exceeds the saturation point. Thus, the abundance of water vapor declines with altitude. The lowest temperatures occur at the tropical tropopause (~210 K), where the mole fraction of H_2O has declined to a few μmol mol^{-1}. This temperature determines in effect the abundance of water vapor in the stratosphere and higher atmospheric regions (some water is additionally produced in the stratosphere by the oxidation of methane).

P. Warneck and J. Williams, *The Atmospheric Chemist's Companion: Numerical Data for Use in the Atmospheric Sciences*, DOI 10.1007/978-94-007-2275-0_3, © Springer Science+Business Media B.V. 2012

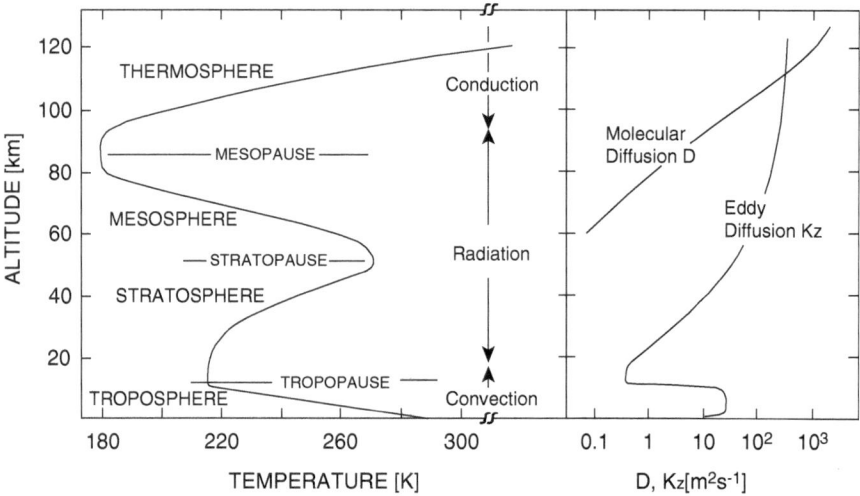

Fig. 3.1 Schematic representation of the vertical distribution of temperature (*left*) and coefficients for vertical transport by molecular and eddy diffusion in the atmosphere (*right*)

3.1 Basic Properties of Air

Table 3.1 Physico-chemical properties of air[a]

Molar mass (kg mol⁻¹)	$M_{air} = (\Sigma x_i M_i)/\Sigma x_i$	2.8964×10^{-2}
	$M_{moist\ air} = M_{dry\ air}/(1 - q + qM_{dry\ air}/M_w)$	See Table 3.2
Density (kg m⁻³)	$\rho = M_{air}\,p/R_g T$	1.2923
Specific heat (J kg⁻¹ K⁻¹)[b]	c_p (constant pressure)	1,005
	c_v (constant volume)	718
Speed of sound (m s⁻¹)	$v_s = [(c_p/c_v)R_g T/M_{air}]^{1/2}$	331.5
Thermal conductivity[c] (mW m⁻¹ K⁻¹)	$\kappa_T = aT^{3/2}/(T + b \times 10^{-12/T})$	24.10
Dynamic viscosity[c,d] (μPa s)	$\mu = \beta\,T^{3/2}/(T + S)$	17.16

[a] Formula expressions indicate the dependence on temperature and pressure. The numerical values refer to dry air at standard conditions ($T = 273.15$ K, $p = 1.01325 \times 10^5$ Pa)

[b] $c_p - c_v = R_g/M_{dry\ air}$

[c] The quantity is essentially independent of pressure. The formula and the associated coefficients approximate the empirical data in the temperature range $200 < T < 300$ K to better than 0.1% (COESA 1976)

[d] The kinematic viscosity is defined as the ratio of dynamic viscosity to the density: $\eta = \mu/\rho$

Parameters: x_i = mole fraction of the ith constituent; M_i = molar mass of the ith constituent; $M_w = 18.0153$ (g mol⁻¹) = molar mass of water; q = mass fraction of water; p = pressure (Pa); T = absolute temperature (K); $R_g = 8.3145$ (J K⁻¹ mol⁻¹) = gas constant; $c_p - c_v = R_g/M_{dry\ air}$; $a = 2.6464 \times 10^{-3}$ (J m⁻¹ s⁻¹ K^{1/2}), $b = 245.4$ (K); $\beta = 1.458 \times 10^{-6}$ (kg m⁻¹ s⁻¹ K^{1/2}); $S = 110.4$ (K) (Sutherland constant).

Table 3.2 Influence of water vapor on the density and molar mass of air

q (g kg^{-1})	0	5	10	15	20	25	30	35
$10^{-2}\,x_w$ (mol mol^{-1})[a]	0	0.808	1.624	2.448	3.281	4.123	4.973	5.831
$M_{\text{moist air}}$ (g mol^{-1})	28.964	28.877	28.789	28.703	28.617	28.531	28.446	28.361
$\rho_{\text{moist air}}$ (kg m^{-3})	1.2923	1.2883	1.2844	1.2806	1.2767	1.2729	1.2691	1.2653

[a] The relation between mole fraction and mass fraction is $x_w = (M_{\text{dry air}}/M_w)q/(1-q)$

Table 3.3 Bulk chemical composition of dry air (principal components and rare gases)[a]

Constituent	Molar mass (g mol^{-1})	Volume fraction (%) (a)	(b)
Nitrogen, N_2	28.0134	78.084 ± 0.004	78.093
Oxygen, O_2	31.9988	20.946 ± 0.002	20.933
Argon, Ar	39.948	0.934 ± 0.001	0.934
Neon, Ne	20.1797	$(1.818 \pm 0.004) \times 10^{-3}$	1.818×10^{-3}
Helium, He	4.002602	$(5.24 \pm 0.05) \times 10^{-4}$	5.24×10^{-4}
Krypton, Kr	83.798	$(1.14 \pm 0.01) \times 10^{-4}$	1.14×10^{-4}
Xenon, Xe	131.293	$(8.7 \pm 0.1) \times 10^{-6}$	8.7×10^{-6}
Carbon dioxide, CO_2	44.0095	0.0332[b]	0.0370[b]

[a] Data source: (a) Glueckauf (1951); the CO_2 content was extrapolated from the data reviewed by Callendar (1940) and did not correspond to the true value at that time. (b) recalculated for a CO_2 content of 370 ppm and oxygen consumption due to the combustion of fossil fuels, assuming 48% of CO_2 to have remained in the atmosphere (see comments)

[b] In addition to a global rise, CO_2 undergoes diurnal variations locally and annual variations globally due to assimilation and respiration by plants and the release from decaying biomass

Comments:

Volumetric gas analytical procedures involve the stepwise removal of CO_2 and O_2 to obtain the fraction of N_2 plus the rare gases ($79.0215 \pm 0.002\%$), and the further removal of nitrogen to obtain the fraction of argon (the predominant rare gas). Other rare gases were determined by mass spectrometry in the non-condensable fraction of liquefied air. All recent tables on rare gases, e.g. Ozima and Podosek (2001) rely on the previous measurements. In 1967–1970 Machta and Hughes (1970) re-determined atmospheric oxygen volume fractions and found 20.9458 ± 0.0017 over the continental shelf and open ocean, in agreement with earlier data. At this time the amount of atmospheric carbon dioxide derived from fossil fuel consumption had risen from ~300 μmol mol^{-1}, determined in the early 1900's (Benedict, 1912; Krogh 1919), to ~320 μmol mol^{-1}, but the associated loss of oxygen (~44 μmol mol^{-1}) would have escaped detection because of insufficient measurement precision. Keeling and Shertz (1992) and Keeling et al. (1996) have used an interferometer technique to demonstrate that, relative to nitrogen, oxygen does decrease at the rate predicted from fossil fuel consumption and the corresponding

CO_2 increase. Bender et al. (1996) have confirmed this by means of a mass spectrometric technique. The data under (b) in the above table are estimated by starting with CO_2 and O_2 in 1970 and the assumptions that the rate of oxygen consumption is 1.4 times that of CO_2 production, and that 48% of the excess CO_2 remains in the atmosphere (the remainder is either taken up by the ocean or assimilated by the biosphere).

References

Bender, M., T. Ellis, P. Tans, R. Francey, D. Lowe, Glob. Biogeochem. Cycles **10**, 9–21 (1996)
Benedict, F.G., Publication No. 166 (Carnegie Institute, Washington, DC, 1912)
Callendar, G.S., Quart. J. Roy. Meteor. Soc. **66**, 395–400 (1940)
Glueckauf, E., The composition of atmospheric air, in *Compendium of Meteorology*, ed. by T.F. Mahone (American Meteorological Society, Boston, 1951), pp. 3–12
Keeling, R.F., S.C. Piper, M. Heimann, Nature **381**, 218–221 (1996)
Keeling, R.F., S.R. Shertz, Nature **358**, 723–727 (1992)
Krogh, A., Kgl. Danske Videns. Sel. Math-Fys. Medd. **1**, 1–19 (1919)
Machta, L., E. Hughes, Science **168**, 1582–1584 (1970)
Ozima, M., F.A. Podosek, *Noble Gas Geochemistry*, 2nd edn. (Cambridge University Press, Cambridge, 2001)

3.2   Origin of the Atmosphere and Geochemical Cycles of the Principal Constituents

The atmosphere of Earth derived primarily by thermal outgassing of volatiles from virgin planetary material during and subsequent to the formation of the planet (by the aggregation of smaller bodies). The volatiles included water present in the ocean, nitrogen in the atmosphere, and carbon dioxide, now largely residing in the sediments as carbonate deposits.

Nitrogen: This element, while a primary outgassing product, is also vital to life. It enters the biosphere by bacterial nitrogen fixation, a process, which reduces N_2 to amino compounds and incorporates it directly into the living cell. The biosphere eventually releases nitrogen to the atmosphere by anaerobic bacterial denitrification (see also Tables 2.25–2.27).

Oxygen: This element is not a planetary outgassing product. Oxygen is a byproduct of the assimilation of CO_2 by phytoplankton and green plants and has evolved in Earth's history in conjunction with the development of the biosphere. The process is equivalent to the chemical reaction

$$CO_2 + H_2O \rightleftharpoons (CH_2O) + O_2 \quad \text{(photosynthesis/decay of organic matter)}$$

where the term (CH_2O) designates organic matter and the energy needed for the forward reaction is provided by solar radiation. The process is reversible, because the re-oxidation of organic matter consumes oxygen while converting organic carbon to CO_2. On a geological short time scale the production and consumption of O_2 is nearly balanced. Nevertheless, a small fraction of organic material escapes re-oxidation by incorporation in marine sediments. In the long run, organic matter accumulates in the sediments and a corresponding amount of oxygen remains to stay in the atmosphere. The present amount of atmospheric oxygen is understood to represent an excess over that used up in the oxidation of organic carbon and other reduced chemical compounds.

Carbon dioxide: This compound is readily absorbed by sea water. On a time scale of ~1,000 years atmospheric CO_2 is in equilibrium with that residing in the ocean. The bulk of CO_2 released by thermal outgassing has entered the sediments in the form of limestone deposits. The amount of CO_2 consumed annually in the weathering of continental silicate and carbonate rocks is subsequently restored and returns to the atmosphere when the eroded material has reached the ocean and is re-deposited in new sediments. The time scale for this process also is a few thousand years. The exchange of CO_2 with the biosphere (via photosynthesis and re-oxidation of organic material) occurs on a time scale of 10–100 years. The recent increase of atmospheric CO_2 is caused by the combustion of fossil fuels that are taken from deposits of organic carbon in the sediments.

Rare gases: The rare gases in the atmosphere have resulted from planetary outgassing. The dominant isotope of argon, ^{40}Ar, is a product of the radioactive decay of potassium-40. Helium-4, which originates from the radioactive decay of thorium and uranium, is a light element that can escape from the outer atmosphere to space.

Table 3.4 Cycles, fluxes through the atmosphere, and associated residence times of the main atmospheric constituents[a]

Constituent	Mass (kg)	Flux (Tg a^{-1})	Time scale (a)[b]	Remarks
N_2	3.9×10^{18}	2.1×10^2	1.8×10^7	N-fixation/denitrification
O_2	1.2×10^{18}	2.5×10^5	5×10^3	Exchange with biosphere
		2.7×10^2	4×10^6	Weathering of sediments
CO_2[c]	$\sim 7.7 \times 10^{14}$	4.7×10^4	15	Short term, land plants
		8×10^3	75	Long term, land plants
		3.8×10^3	~200	Fossil CO_2, transfer to ocean
		1×10^{-2}	7×10^4	Transfer to sediments
He	3.7×10^{12}	3.3×10^{-3}	1×10^6	Escape to space

[a] Source of data: Warneck (2000)
[b] Atmospheric residence or turnover time
[c] All entries as mass of carbon

3.3 Temperature, Pressure and Density as a Function of Altitude

Standard Atmosphere: The US Standard Atmosphere 1976 is an idealized (dry air) steady state representation of Earth's atmosphere from sea level up to high altitudes at 45° latitude. It is defined by sea level temperature and pressure, and a standard temperature-height profile COESA (Committee on Extension of US Standard Atmosphere) (1976).

The temperature profile is an interpolation of mean annual temperatures derived from measurements with balloons and rockets. To simplify the calculation of other parameters the temperature profile is divided into seven segments: 0–11, 11–20, 20–32, 32–47, 47–51, 51–71, 71–84.85 km altitude, with constant gradients:−6.5, 0.0, +1.9, +2.8, 0.0,−2.8,−2.0 K km^{-1}, respectively. The data for p, ρ and n are obtained from the hydrostatic equation $-dp/p = (g_0 M/R_g T)\, dz$ based on the ideal gas law: $g_0 = 9.80665$ (m s^{-2}) is the (standard) acceleration due to gravity, $M = 2.8964 \times 10^{-2}$ (kg mol^{-1}) is the molar mass of air, $R_g = 8.3145$ (J K^{-1} mol^{-1}) is the gas constant, and sea level pressure and temperature are set to the standard values $p_0 = 1013.25$ (hPa) and $T_0 = 288.15$ (K).

The scale height, defined by $H = (R_g/g_0 M)T = 29.272\, T$ (m), is directly proportional to T in the homosphere (0–85 km altitude), where g_0 and M remain constant. With increasing distance from Earth's surface one must set $g = g_0/(1 + z/r_0)^2$, $r_0 = 6371 \times 10^3$ m is Earth's mean radius, and one must make allowance for changes in M (see Chap. 9, The Upper Atmosphere).

Table 3.5 Temperature T, pressure p, density ρ, number concentration n, and scale height H as a function of geometric altitude in the homosphere[a]

z (km)	T (K)	p (hPa)	ρ (kg m^{-3})	n (m^{-3})	H (km)
0	288.15	1.01325 (3)	1.225	2.547 (25)	8.435
1	281.65	8.988 (2)	1.112	2.311 (25)	8.244
2	275.15	7.950 (2)	1.007	2.093 (25)	8.054
3	268.66	7.012 (2)	9.093 (−1)	1.891 (25)	7.864
4	262.17	6.166 (2)	8.194 (−1)	1.704 (25)	7.674
5	255.68	5.405 (2)	7.364 (−1)	1.531 (25)	7.484
6	249.19	4.722 (2)	6.601 (−1)	1.373 (25)	7.294
7	242.70	4.111 (2)	5.900 (−1)	1.227 (25)	7.104
8	236.22	3.565 (2)	5.258 (−1)	1.093 (25)	6.914
9	229.73	3.080 (2)	4.671 (−1)	9.711 (24)	6.725
10	223.25	2.650 (2)	4.135 (−1)	8.598 (24)	6.536
11	216.77	2.270 (2)	3.648 (−1)	7.585 (24)	6.345
12	216.65	1.940 (2)	3.119 (−1)	6.486 (24)	6.342

(continued)

Table 3.5 (continued)

z (km)	T (K)	p (hPa)	ρ (kg m^{-3})	n (m^{-3})	H (km)
13	216.65	1.658 (2)	2.666 (−19)	5.543 (24)	6.342
14	216.65	1.417 (2)	2.279 (−1)	4.736 (24)	6.342
15	216.65	1.211 (2)	1.948 (−1)	4.049 (24)	6.342
16	216.65	1.035 (2)	1.665 (−1)	3.461 (24)	6.342
17	216.65	8.850 (1)	1.423 (−1)	2.959 (24)	6.342
18	216.65	7.565 (1)	1.217 (−1)	2.529 (24)	6.342
19	216.65	6.467 (1)	1.040 (−1)	2.162 (24)	6.342
20	216.65	5.529 (1)	8.891 (−2)	1.849 (24)	6.342
21	217.58	4.729 (1)	7.572 (−2)	1.574 (24)	6.369
22	218.57	4.048 (1)	6.451 (−2)	1.341 (24)	6.398
23	219.57	3.467 (1)	5.501 (−2)	1.144 (24)	6.427
24	220.56	2.972 (1)	4.694 (−2)	9.759 (23)	6.456
25	221.55	2.549 (1)	4.008 (−2)	8.334 (23)	6.485
26	222.54	2.188 (1)	3.426 (−2)	7.123 (23)	6.514
27	223.54	1,880 (1)	2.930 (−2)	6.092 (23)	6.543
28	224.53	1.616 (1)	2.508 (−2)	5.214 (23)	6.572
29	225.52	1.390 (1)	2.148 (−2)	4.466 (23)	6.601
30	226.51	1.197 (1)	1.841 (−2)	3.828 (23)	6.630
31	227.50	1.031 (1)	1.579 (−2)	3.283 (23)	6.659
32	228.49	8.891	1.356 (−2)	2.818 (23)	6.688
33	230.97	7.673	1.157 (−2)	2.406 (23)	6.761
34	233.74	6.634	9.887 (−3)	2.056 (23)	6.842
35	236.51	5.746	8.463 (−3)	1.760 (23)	6.923
36	239.28	4.985	7.258 (−3)	1,509 (23)	7.004
37	242.05	4.332	6.236 (−3)	1.297 (23)	7.085
38	244.82	3.771	5.367 (−3)	1.116 (23)	7.166
39	247.58	3.288	4.627 (−3)	9.620 (22)	7.247
40	250.35	2.871	3.996 (−3)	8.308 (22)	7.328
41	253.11	2.511	3.456 (−3)	7.186 (22)	7.409
42	255.88	2.200	2.995 (−3)	6.227 (22)	7.450
43	258.64	1.930	2.599 (−3)	5.404 (22)	7.571
44	261.40	1.695	2.259 (−3)	4.697 (22)	7.652
45	264.16	1.491	1.966 (−3)	4.088 (22)	7.732
46	266.93	1.313	1.712 (−3)	3.564 (22)	7.813
47	269.68	1.159	1.497 (−3)	3.112 (22)	7.894
48	270.65	1.023	1.317 (−3)	2.738 (22)	7.922
49	270.65	9.034 (−1)	1.163 (−3)	2.418 (22)	7.922
50	270.65	7.978 (−1)	1.027 (−3)	2.135 (22)	7.922
51	270.65	7.046 (−1)	9.069 (−4)	1.886 (22)	7.922
52	269.03	6.221 (−1)	8.056 (−4)	1.675 (22)	7.875
54	263.52	4.834 (−1)	6.390 (−4)	1.329 (22)	7.714
56	258.02	3.736 (−1)	5.045 (−4)	1.049 (22)	7.553
58	252.52	2.872 (−1)	3.963 (−4)	8.239 (21)	7.391

(continued)

Table 3.5 (continued)

z (km)	T (K)	p (hPa)	ρ (kg m^{-3})	n (m^{-3})	H (km)
60	247.02	2.196 (−1)	3.097 (−4)	6.439 (21)	7.231
62	241.53	1.669 (−1)	2.407 (−4)	5.005 (21)	7.070
64	236.04	1.261 (−1)	1.861 (−4)	3.868 (21)	6.909
65	233.29	1.093 (−1)	1.632 (−4)	3.393 (21)	6.829
66	230.55	9.461 (−2)	1.440 (−4)	2.972 (21)	6.748
68	225.07	7.053 (−2)	1.091 (−4)	2.270 (21)	6.588
70	219.59	5.221 (−2)	8.283 (−5)	1.722 (21)	6.428
72	214.26	3.836 (−2)	6.237 (−5)	1.297 (21)	6.272
74	210.35	2.801 (−2)	4.639 (−5)	9.644 (20)	6.157
75	208.40	2.388 (−2)	3.992 (−5)	8.300 (20)	6.100
76	206.45	2.033 (−2)	3.431 (−5)	7.134 (20)	6.043
78	202.54	1.467 (−2)	2.524 (−5)	5.248 (20)	5.929
80	198.64	1.052 (−2)	1.846 (−5)	3.838 (20)	5.814
82	194.74	7.501 (−3)	1.342 (−5)	2.790 (20)	5.700
84	190.84	5.311 (−3)	9.694 (−6)	2.016 (20)	5.586
86	186.87	3.734 (−3)	6.958 (−6)	1.447 (20)	5.621

[a] Condensed from U.S. Standard Atmosphere 1976. Powers of 10 are shown in parentheses

Zonal mean temperatures:

The temperature values in the following tables for latitudes between 80° S and 80° N were derived by Barnett and Corney (1985) from a combination of satellite data above 30 hPa pressure and global atmospheric statistics from meteorological observations at altitudes below 50 hPa.

Table 3.6a Zonal mean temperature (K) as a function of latitude and height in January[a]

Height (km)	Southern latitudes									Northern latitudes							
	−80	−70	−60	−50	−40	−30	−20	−10	0	10	20	30	40	50	60	70	80
80	172.2	173.3	175.5	183.2	193.5	202.7	206.8	206.1	206.5	207.1	208.0	211.8	216.2	222.3	225.0	224.4	225.2
75	200.9	200.5	200.4	202.0	204.8	208.3	210.3	209.3	209.9	210.5	211.9	215.5	219.8	224.0	226.2	225.3	226.3
70	226.4	224.8	222.7	219.0	215.2	214.1	215.9	217.8	218.3	217.7	217.5	219.5	221.2	225.0	227.7	227.3	229.3
65	248.3	245.3	241.5	235.5	229.8	227.5	230.4	233.8	235.1	233.1	230.7	227.8	226.3	226.2	229.6	232.6	236.4
60	266.7	262.9	258.7	252.8	247.4	245.0	247.7	252.6	252.9	250.2	244.5	238.7	232.5	230.7	235.0	241.9	247.2
55	281.1	276.8	272.1	266.3	261.1	258.7	259.4	263.3	264.6	262.9	258.3	251.9	245.2	241.7	244.2	250.7	255.5
50	288.5	284.4	279.7	275.4	271.4	268.7	267.4	267.8	267.8	267.9	267.0	263.4	259.2	253.5	253.5	255.0	255.4
45	286.1	283.4	280.0	275.8	272.8	269.5	267.1	264.0	263.8	264.5	265.8	264.4	260.6	254.4	252.3	252.0	251.0
40	276.2	274.1	271.2	267.6	264.1	260.2	256.5	254.9	255.2	255.2	255.8	254.0	250.0	246.4	246.5	246.7	246.9
35	261.8	260.5	258.1	254.6	250.6	246.8	244.1	242.3	242.5	242.1	241.7	239.2	237.4	235.4	235.7	236.3	236.6
30	244.5	244.1	243.7	240.7	236.1	232.0	228.9	227.8	227.2	228.0	228.1	228.5	227.5	226.2	224.7	221.7	219.1
25	235.6	233.8	232.6	229.0	225.4	222.1	219.2	218.1	217.5	218.0	218.3	219.3	219.6	218.2	213.8	207.8	202.0
20	234.6	233.2	229.6	222.8	215.5	210.2	207.4	205.3	204.3	204.6	206.7	210.0	214.5	216.8	215.5	211.4	205.9
15	232.5	231.3	227.3	220.3	212.3	206.0	202.0	199.8	199.3	199.7	202.2	208.1	214.7	217.9	217.0	214.1	211.2
10	226.9	225.5	224.8	225.9	229.8	235.0	238.3	239.0	239.2	238.6	235.4	228.6	221.6	218.8	217.1	215.2	214.0
5	240.8	243.7	248.3	255.8	263.8	269.4	271.9	272.3	272.4	272.4	269.6	261.8	252.5	244.5	239.1	234.8	231.9
0	–	–	275.2	281.4	289.5	295.9	298.9	300.5	300.6	300.0	297.1	291.7	284.7	277.7	266.4	254.1	248.6

[a]Source: Barnett and Corney (1985)

Table 3.6b Zonal mean temperature (K) as a function of latitude and height in April[a]

Height (km)	Southern latitudes									Northern latitudes							
	−80	−70	−60	−50	−40	−30	−20	−10	0	10	20	30	40	50	60	70	80
80	228.2	224.6	218.3	214.7	211.4	208.2	207.2	210.7	211.3	209.1	208.5	207.5	204.2	201.6	202.0	200.4	199.5
75	229.7	227.2	223.7	218.4	214.8	212.2	209.7	209.6	208.7	208.1	210.9	214.1	214.7	216.3	217.2	215.6	214.5
70	232.9	230.8	228.5	222.5	219.7	218.5	215.9	213.3	212.6	213.9	219.3	222.2	223.9	226.0	227.1	226.9	226.4
65	241.4	237.9	234.0	228.5	228.3	230.4	228.6	226.1	225.9	226.9	229.6	231.9	233.7	234.9	235.1	234.0	232.8
60	249.9	245.5	241.0	238.5	239.2	240.3	242.0	242.6	243.1	242.1	240.1	241.2	243.1	244.4	244.7	244.9	244.4
55	254.7	250.9	247.3	248.7	252.2	254.5	257.0	259.2	259.7	258.2	255.2	255.4	257.5	258.4	257.8	258.8	259.2
50	258.9	257.2	255.5	257.7	263.4	265.9	267.1	269.2	269.7	269.2	268.2	268.2	269.6	269.5	269.3	269.1	268.7
45	252.8	252.3	252.2	256.1	260.9	263.8	267.0	269.5	270.2	269.7	269.0	269.2	270.4	270.9	268.7	265.5	263.2
40	237.0	236.2	237.9	243.8	248.6	252.3	257.0	260.4	261.6	261.1	259.5	259.0	260.9	260.0	256.2	252.0	249.2
35	220.0	219.5	223.6	229.6	235.3	239.4	243.0	246.5	247.8	247.5	246.0	245.1	245.2	243.5	239.9	235.5	233.3
30	210.3	211.9	215.0	220.6	225.1	228.4	230.6	231.5	231.0	230.4	231.6	230.7	228.4	226.6	225.8	225.2	225.1
25	203.6	209.0	216.3	219.7	221.2	221.6	221.5	221.1	220.3	220.1	220.6	220.4	219.4	219.8	221.5	224.3	227.0
20	212.1	214.6	216.9	217.4	214.9	211.3	208.0	206.4	206.0	206.4	208.5	211.6	215.1	218.6	221.1	222.9	224.1
15	217.1	219.1	219.8	217.6	213.1	207.4	202.5	199.5	199.1	199.8	203.0	208.6	214.6	218.8	221.4	222.9	223.5
10	218.3	219.7	220.6	222.0	226.0	231.6	237.0	239.3	239.7	238.9	235.5	229.4	223.5	221.2	221.1	221.5	221.4
5	233.7	237.8	244.1	252.5	260.7	266.9	271.2	272.4	272.5	272.4	270.2	264.9	257.6	250.8	245.2	240.9	237.9
0	–	–	274.5	280.8	288.9	295.0	298.8	301.0	301.1	300.5	297.8	292.6	286.0	280.3	273.1	262.2	253.2

[a] Source: Barnett and Corney (1985)

Table 3.6c Zonal mean temperature (K) as a function of latitude and height in July[a]

Height (km)	Southern latitudes									Northern latitudes							
	-80	-70	-60	-50	-40	-30	-20	-10	0	10	20	30	40	50	60	70	80
80	219.9	217.7	216.8	214.9	209.6	204.9	201.9	201.6	202.1	202.7	202.6	198.6	189.3	177.7	169.8	168.3	169.1
75	224.0	222.6	222.8	220.5	215.0	209.2	205.7	204.3	204.3	204.8	205.8	204.6	200.7	196.3	194.6	195.9	197.9
70	231.7	230.8	231.4	227.4	219.7	214.2	210.7	210.1	210.6	211.2	210.8	210.3	210.9	213.9	217.7	221.3	224.0
65	243.6	242.4	241.1	235.3	228.3	224.1	224.2	224.7	225.8	225.5	224.3	223.1	224.9	230.6	237.3	242.9	246.3
60	258.7	256.1	251.3	243.5	236.6	236.8	239.3	242.7	244.2	244.5	242.0	240.8	242.5	248.3	254.8	260.7	264.6
55	270.8	267.1	259.9	252.2	247.5	251.1	255.2	257.9	258.5	257.2	255.0	254.8	256.9	262.2	268.2	273.6	278.2
50	271.9	269.0	264.0	258.1	258.5	262.1	265.1	265.1	264.4	264.1	263.9	264.7	267.2	271.1	275.4	280.4	284.5
45	261.1	261.0	258.1	254.5	257.0	260.8	262.9	262.0	261.0	260.8	263.1	264.5	267.7	270.9	275.3	278.8	281.1
40	248.2	249.1	246.1	240.9	241.8	248.0	251.5	252.4	252.5	251.5	251.5	254.2	257.6	261.5	265.6	268.7	269.9
35	234.0	233.2	227.5	222.3	226.7	233.6	238.7	240.8	241.5	240.5	240.8	242.0	244.5	248.4	251.7	254.2	254.9
30	209.2	208.3	206.5	210.4	220.3	227.0	228.5	229.1	228.5	228.8	229.1	230.6	232.6	235.7	237.7	240.0	241.2
25	179.2	186.9	197.4	211.2	219.2	221.0	221.2	220.0	219.4	220.0	220.8	222.1	223.7	225.8	228.3	230.8	233.3
20	186.5	194.0	202.8	211.3	215.1	213.6	210.9	209.3	208.6	209.0	210.5	212.6	216.3	221.6	226.3	229.8	232.3
15	192.2	197.8	206.4	214.2	215.5	210.1	204.3	202.1	202.1	202.5	204.2	207.5	212.9	219.8	224.6	228.1	230.9
10	202.5	205.5	210.8	216.6	222.3	229.1	235.4	237.8	238.1	238.6	239.0	238.6	234.8	230.0	227.9	227.9	228.8
5	229.9	233.0	240.4	247.7	254.7	262.3	269.2	271.9	271.8	271.9	272.1	271.8	268.6	263.6	259.7	256.5	253.6
0	–	–	273.9	279.4	286.2	291.3	295.9	299.3	300.1	300.8	299.9	298.0	293.1	285.4	283.6	279.6	274.2

[a]Source: Barnett and Corney (1985)

Table 3.6d Zonal mean temperature (K) as a function of latitude and height in October[a]

Height (km)	Southern latitudes									Northern latitudes							
	−80	−70	−60	−50	−40	−30	−20	−10	0	10	20	30	40	50	60	70	80
80	200.2	201.9	203.8	202.7	204.7	205.9	206.3	208.3	211.3	211.0	208.4	208.8	210.7	213.4	215.1	220.8	223.8
75	215.6	217.3	219.1	217.5	215.5	212.7	208.9	207.2	208.4	209.6	210.6	212.5	214.3	216.8	220.6	223.6	225.9
70	228.2	228.8	229.1	227.6	224.7	221.4	217.9	212.9	211.8	212.9	216.9	218.2	219.0	221.0	225.6	228.0	230.5
65	236.5	236.4	236.6	235.9	233.9	231.7	228.8	226.0	224.7	225.3	228.1	229.7	227.7	227.4	231.8	236.1	240.2
60	251.3	248.3	245.2	243.3	242.0	241.1	240.0	241.6	241.9	241.5	241.2	239.7	239.0	238.3	240.0	244.0	249.2
55	267.2	261.7	256.5	254.7	254.6	255.1	255.4	257.5	258.6	258.1	256.3	254.4	252.4	249.7	247.8	249.7	253.3
50	279.3	273.1	267.7	264.9	266.3	267.5	268.1	268.1	268.6	268.0	266.5	266.0	263.7	259.0	256.5	255.8	256.3
45	284.0	277.1	269.8	266.2	266.1	267.5	268.3	268.5	268.8	268.1	266.4	263.9	261.3	257.1	252.5	249.2	247.8
40	283.6	274.1	261.6	255.0	254.6	255.7	258.3	260.0	260.3	259.3	256.4	252.3	249.1	245.4	239.2	232.6	228.9
35	269.0	260.0	247.5	240.0	239.2	241.7	245.0	247.1	247.2	245.9	242.9	239.3	235.8	231.8	227.0	219.2	212.9
30	242.8	238.5	232.0	226.7	225.8	228.9	231.1	230.8	231.4	231.5	230.7	228.0	225.1	222.1	218.5	215.6	211.4
25	216.6	220.2	225.8	224.6	222.9	221.5	220.3	219.6	218.9	219.2	219.9	220.2	219.5	218.2	216.4	213.7	210.6
20	209.0	212.0	216.4	219.6	218.4	213.9	210.1	208.3	208.1	208.1	209.4	211.7	214.5	217.2	218.4	217.1	215.1
15	203.1	207.6	214.0	218.5	216.8	210.2	203.9	200.5	200.1	200.4	202.1	206.3	212.0	216.9	220.0	220.6	220.2
10	205.6	209.1	214.4	220.2	224.8	230.2	235.8	238.4	238.7	238.7	237.5	233.4	227.3	223.2	221.4	220.6	220.5
5	233.1	236.9	242.9	249.8	257.4	264.4	269.4	271.9	272.1	272.4	271.8	268.2	262.0	254.8	248.4	243.3	239.0
0	–	–	276.4	280.5	286.4	292.1	296.5	299.3	300.1	300.9	300.1	297.0	291.1	283.5	277.0	268.3	261.6

[a]Source: Barnett and Corney (1985)

Table 3.7 Air masses (kg) in different regions of the dry atmosphere[a]

Total atmosphere[b]	5.13×10^{18}
Tropical troposphere	2.25×10^{18}
Extra-tropical troposphere	1.97×10^{18}
Total troposphere	4.22×10^{18}
Lower stratosphere (≤ 30 km)	8.48×10^{17}
Upper stratosphere (30–50 km)	5.80×10^{16}
Total stratosphere	9.06×10^{17}
Remaining upper atmosphere	4×10^{15}

[a]From Warneck (2000) reproduced with permission of Elsevier
[b]Trenberth and Guillemont (1994) derived 5.132×10^{18} kg for the dry air mass and 1.25×10^{16} kg for the mean mass of water vapor

Table 3.8 Location of the tropopause: pressure as a function of latitude and season[a]

Deg. latitude	0	10	20	30	40	50	60	70	80	85
Northern hemisphere										
January	91.1	93.1	100.0	158.0	239.2	262.7	278.0	288.4	280.8	274.8
April	93.8	97.1	103.5	172.4	219.9	247.9	278.3	303.5	293.1	280.3
July	102.1	103.1	105.4	115.4	144.5	214.3	238.7	258.3	279.1	287.3
October	102.8	103.1	103.7	124.4	180.8	227.2	259.5	276.0	284.4	289.1
Southern hemisphere										
January	91.1	93.5	101.4	110.5	161.4	240.7	278.5	288.9	305.2	310.2
April	93.8	96.7	102.3	138.8	201.0	237.2	268.8	282.4	291.7	286.4
July	102.1	103.3	110.8	168.7	235.8	239.1	236.5	220.4	212.3	213.2
October	102.8	103.5	107.8	156.2	220.5	246.1	240.1	212.6	195.0	187.1

[a]Derived from the ECHAM – Modular Earth Submodel System – Atmospheric Chemistry Model (Röckner et al. (2006), Jöckel et al. (2005); Courtesy Janeen Auld (Max-Planck-Institut für Chemie, Mainz)

Table 3.9 Mass mixing ratio q (g kg^{-1}) of water vapor in the troposphere[a]

z (km)	Tropics	Midlatitude	Polar	z (km)	Tropics	Midlatitude	Polar
Surface	12.13	3.37	0.775	8	0.913	0.113	0.030
1	10.16	2.66	0.764	9	0.560	0.0744	0.018
2	7.35	1.92	0.613	10	0.328	0.039	0.0113
3	5.35	1.32	0.436	11	0.175	0.026	0.007
4	3.98	0.90	0.273	12	0.105	0.020	–
5	2.93	0.568	0.166	13	0.060	–	–
6	2.10	0.335	0.094	14	0.032	–	–
7	1.41	0.194	0.050	15	0.017	–	–

[a]From radiosonde measurements; Source: Raschke and Stubenrauch (2005)

References

Barnett, J.J., M. Corney, *Handbook for Middle Atmosphere Program*, vol. 16, ed. by K. Labitzke, J.J. Barnett, B. Edwards (SCOSTEP, University of Illinois, Urbana, 1985)

Committee on Extension of US Standard Atmosphere (COESA) US Standard Atmosphere 1976 (U.S. Government Office, Washington, DC, 1996)

Jöckel, P., R. Sander, A. Kerkweg, H. Tost, J. Lelieveld, Atmos. Chem. Phys. **5**, 433–444 (2005)

Raschke, E., C. Stubenrauch, Water vapor in the atmosphere, in *Landolt-Börnstein, Group V: Geophysics*, vol. 6, ed. by M. Hantel (Springer, Berlin, 2005), pp. 5/1–5/18

Röckner, E., R. Brokopf, M. Esch, M. Giorgetta, S. Hagemann, L. Kornblueh, E. Manzini, U. Schlese, U. Schulzweida, J. Climate **19**, 3771–3791 (2006)

Trenberth, K.E., C.J. Guillemont, J. Geophys. Res. **99**, 23079–23088 (1994)

Warneck, P., *Chemistry of the Natural Atmosphere* (Academic Press, San Diego, 2000), Copyright Elsevier

3.4   Global Atmospheric Mean Circulation

The complex interaction of forces resulting from the rotation of the planet Earth and heat transport from the equator to the poles results in atmospheric circulation patterns such that the main direction of air flow in the troposphere (zonal circulation) is from east to west in subtropical latitudes (trade winds), and from west to east in middle latitudes (Westerlies), in both hemispheres. The Westerlies encircle the globe in a wavelike pattern with perturbations due to the build-up and decay of temporary, as well as stationary, cyclones and anticyclones, and the associated pressure fluctuations. The meridional circulation is dominated by the Hadley cells, which are characterized by uprising air in the Tropics and subsiding air in subtropical latitudes. The Hadley circulation, although much weaker than the zonal circulation, is responsible for the heat transport from equator to poles. The uprising air moves toward higher latitudes, where the descending branch of the Hadley cell gives rise to a high-pressure belt at about $30°$ latitude. The region of uprising air separating the northern and southern Hadley cells, called the inter-hemispheric tropical convergence zone (ITCZ), presents a barrier to the exchange of air between the northern and southern hemispheres.

The following tables present time averaged meridional cross sections of the zonal winds as compiled by Speth and Madden (1988) for selected pressure levels, latitudes and seasons for altitudes up to 30 km, and geostrophic winds calculated by Barnett and Corney (1985) from satellite-derived temperature fields for altitudes 15–85 km. Mean values of geopotential height are included to facilitate the conversion from pressure to altitude levels. The calculated geostrophic winds are expected to represent the zonal winds at altitudes below 60 km, whereas above that level increasing departures should occur due to forcing by tides and gravity waves.

An additional table presents approximate time constants for the exchange of air between the northern and southern hemispheres and between the stratosphere and

the troposphere Warneck (2000), as well as times constants associated with the
mean circulation patterns.

References

Barnett, J.J., M. Corney, *Handbook for Middle Atmosphere Program*, vol. 16, ed. by K. Labitzke,
    J.J. Barnett, B. Edwards (SCOSTEP, University of Illinois, Urbana, 1985)
Speth, P, R.A. Madden, in *Meteorology*, ed. by G. Fischer. Landolt-Börnstein, New Series V/4a
    (Springer, Berlin, 1988), pp. 140–438
Warneck, P., *Chemistry of the Natural Atmosphere*, 2nd edn. (Academic Press, San Diego, 2000)

Table 3.10 Time constants for the transport of trace species in the lower atmosphere

Air exchange between major reservoirs[a]		
Exchange between hemispheres	Troposphere	1.0 a
	Stratosphere	3.5 a
Stratosphere –troposphere	NH	1.3 a
Zonal transport		
500–300 hPa level (~6–8 km) circumpolar[b]		
40–60° N	Summer	22–53 d
	Winter	13–38 d
40–60° S	Summer	15–22 d
	Winter	13–21 d
50 hPa level (~20 km) circumpolar[b]		
40–60° N	Summer	3.5–7 m
	Winter	14–20 d
40–60° S	Summer	10–16 m
	Winter	6–12 d
Meridional transport		
(30–60° N, ~3300 km distance)[c]		
Troposphere, 800 hPa	Annual average	30 d
Stratosphere, 100 hPa	Summer	180 d
	Winter	40 d

[a]e-fold residence time (Adapted from Warneck 2000)
[b]Once around the globe
[c]Transport occurs predominantly by turbulent mixing

Table 3.11 Zonal mean wind (m s^{-1}) as a function of latitude and pressure level[a]

Pressure (hPa)	Southern latitudes										Northern latitudes									
	-80	-70	-60	-50	-45	-40	-35	-30	-25	-20	20	25	30	35	40	45	50	60	70	80
December – February																				
100	0.9	3.7	9.1	14.4	17.0	16.6	13.4	8.9	3.6	1.1	15.3	22.0	25.9	25.7	22.4	18.8	16.1	13.6	11.4	6.7
150											23.4	32.1	37.6	35.7	28.7	21.9	16.9	11.6	9.3	5.5
200	2.2	5.3	13.7	26.1	29.0	27.6	23.9	19.7	13.8	8.8	24.6	34.0	29.8	37.6	30.4	23.2	17.5	10.9	8.1	4.9
250	2.3	5.2	13.7	27.1	29.0	26.3	21.6	16.9	11.8	8.2	22.4	31.7	37.0	35.6	29.9	23.4	17.5	10.3	7.5	4.7
300	2.6	4.6	14.1	26.6	27.8	24.5	19.5	14.7	9.4	5.9	19.9	28.6	32.7	31.3	27.4	21.9	16.3	9.6	6.9	4.3
400											14.4	22.0	24.6	24.0	22.1	18.2	13.3	7.6	5.6	3.6
500	0.8	1.7	10.6	19.8	20.4	17.3	12.3	7.5	2.8	0.5	9.3	15.7	17.8	18.2	17.5	14.7	10.8	6.0	4.3	2.8
700	-0.9	-0.8	7.2	14.6	14.6	11.7	7.2	3.0	-0.1	-1.4	1.7	6.0	8.4	9.6	10.1	8.9	6.7	3.6	2.2	1.5
850	-2.4	-2.3	5.0	11.3	10.9	8.1	3.8	0.4	-2.0	-2.8	-1.8	1.1	3.8	5.2	5.7	5.2	4.1	2.0	1.2	0.7
1000	-3.7	-2.9	3.5	8.8	8.0	5.0	1.1	-2.2	-4.0	-4.3	-3.4	-0.8	1.5	2.8	2.8	2.4	2.0	0.2	0.3	0.1
March – May																				
100	7.1	12.4	19.7	19.9	19.6	18.7	17.7	16.2	13.0	8.7	11.3	15.7	18.2	18.1	16.0	13.3	10.4	6.9	5.5	3.2
150											20.1	15.8	28.5	26.8	22.9	18.0	13.3	8.0	5.9	3.3
200	5.0	10.1	20.3	26.9	26.9	26.1	26.6	27.3	24.2	18.1	21.1	27.4	30.2	29.0	25.5	20.8	15.7	9.4	7.0	3.7
250	4.8	9.6	19.7	27.2	26.8	25.1	24.2	24.2	21.4	16.7	18.9	24.7	27.1	27.0	25.0	21.1	16.4	10.2	7.3	3.8
300	3.7	8.3	19.1	26.2	25.3	23.0	21.6	20.9	18.0	13.9	16.0	21.6	23.8	23.9	22.8	19.7	15.5	10.1	7.7	3.8
400											10.8	15.8	17.6	18.3	18.1	15.8	12.3	8.0	6.1	3.1
500	0.9	4.3	14.3	19.5	18.6	16.1	13.3	11.0	7.6	4.6	6.4	10.8	12.7	13.8	14.1	12.7	9.8	6.3	4.7	2.5
700	-1.4	1.0	10.1	14.4	13.3	10.8	7.8	5.0	2.1	0.2	0.8	4.0	5.7	7.1	8.0	7.4	5.6	3.5	2.3	1.1
850	-3.3	-1.6	7.4	11.4	10.1	7.6	4.4	1.6	-0.7	-2.7	-1.7	0.4	2.0	3.5	4.4	4.1	3.1	1.8	0.8	0.3
1000	-5.1	-2.9	5.2	8.7	7.4	4.9	1.7	-1.1	-3.4	-5.2	-3.0	-1.7	-0.1	1.3	1.9	1.8	1.3	0.2	-0.5	-0.1

June – August

hPa																				
100	9.5	19.5	30.8	31.2	29.1	26.6	25.2	23.9	19.2	13.2	-8.9	-5.0	-0.4	4.7	8.5	9.4	7.5	2.8	1.5	1.4
150											-3.4	-0.5	4.9	12.4	17.7	17.7	13.6	5.6	3.6	2.5
200	5.5	12.5	22.1	26.2	27.5	30.2	35.3	39.0	35.6	27.4	-1.4	1.2	6.5	14.4	20.3	20.6	16.5	7.6	5.3	3.5
250	5.2	11.7	20.3	24.5	25.5	27.4	31.9	36.5	34.7	26.6	-1.2	1.3	6.1	13.4	18.8	19.4	16.1	8.3	6.3	4.4
300	4.0	9.4	18.3	22.9	23.7	25.1	28.5	31.8	29.8	23.1	-1.3	1.0	5.2	11.6	16.4	17.0	14.4	7.7	6.5	4.6
400											-1.7	0.3	3.7	8.5	12.3	13.2	11.3	5.8	5.3	3.9
500	1.2	5.0	13.3	16.3	16.7	16.9	17.2	17.3	14.7	10.3	-1.9	-0.3	2.4	6.1	9.3	10.4	9.1	4.4	4.1	3.1
700	-1.6	1.5	9.4	12.1	12.0	11.7	10.8	9.1	5.8	2.5	-2.5	-1.3	0.6	2.9	4.9	6.1	5.5	2.4	2.2	1.9
850	-3.2	-1.2	6.7	9.6	9.3	8.8	7.4	4.8	1.3	-1.8	-2.5	-2.0	-0.3	1.2	2.7	3.3	3.1	1.2	0.8	1.2
1000	-4.8	-2.5	4.6	7.2	6.8	6.2	4.5	1.6	-1.8	-5.8	-2.3	-2.3	-0.8	0.4	0.9	1.4	1.6	0.5	-0.3	0.8

September – November

hPa																				
100	7.4	19.3	30.6	27.7	23.9	20.1	18.0	17.5	14.1	8.9	1.7	6.1	10.8	14.7	16.5	16.4	14.6	10.6	7.7	4.2
150											7.0	12.2	18.2	22.6	23.9	22.3	18.8	12.0	7.9	4.1
200	4.3	12.3	23.5	28.0	27.8	27.6	29.4	31.7	29.2	23.2	7.8	13.2	19.3	24.0	25.7	24.4	21.0	13.2	8.3	4.2
250	4.3	11.0	21.3	26.7	25.9	25.2	26.6	29.2	27.5	21.6	6.6	12.1	17.7	22.1	24.2	23.8	21.2	13.5	8.4	4.4
300	3.1	8.9	20.0	25.2	24.8	24.0	24.7	25.9	23.4	18.1	5.1	10.3	15.4	19.3	21.7	22.0	19.6	12.7	8.0	4.1
400											2.4	7.1	11.3	14.5	17.0	17.7	15.9	10.1	6.3	3.3
500	0.6	4.9	15.1	18.3	17.8	16.6	15.4	14.4	11.2	7.3	0.5	4.2	7.6	10.1	13.2	14.2	12.9	8.1	4.8	2.4
700	-1.6	1.5	11.0	13.7	12.8	11.3	9.4	7.2	4.1	1.5	-2.0	0.3	2.5	4.9	7.2	8.5	8.1	5.0	2.4	1.1
850	-3.1	-1.1	8.2	10.9	9.8	8.2	5.9	2.9	0.3	-1.8	-3.1	-1.5	0.2	1.9	3.7	4.9	5.0	3.0	1.0	0.5
1000	-4.7	-2.2	6.1	8.5	7.3	5.6	2.9	-0.2	-2.8	-5.1	-3.4	-2.4	-0.7	0.6	1.5	2.3	2.8	1.3	0.0	0.0

[a]As given by Speth and Madden (1988)

Table 3.12a Mean zonal geostrophic wind (m s^{-1}) in the stratosphere and mesosphere

Pressure (hPa)	Southern latitudes						Northern latitudes					
	−70	−60	−50	−40	−30	−20	20	30	40	50	60	70
January												
0.0062	−9.8	−21.7	−20.3	−20.9	−12.3	3.4	21.2	31.4	31.3	20.5	9.3	19.4
0.0103	−13.6	−30.2	−35.5	−39.4	−27.1	−3.2	27.4	40.3	42.0	24.6	10.8	18.8
0.0169	−16.9	−37.3	−47.9	−54.1	−37.8	−6.4	34.2	48.5	51.9	28.2	11.7	18.0
0.0279	−19.4	−42.5	−56.3	−63.4	−44.5	−8.1	41.4	56.5	60.0	31.2	12.6	17.6
0.0460	−21.1	−45.3	−60.3	−67.1	−48.1	−11.2	47.4	63.0	66.1	34.5	14.0	18.1
0.0758	−21.9	−45.8	−59.7	−66.0	−50.2	−17.9	50.6	66.5	68.7	37.7	16.8	20.5
0.1250	−21.4	−44.1	−56.4	−62.9	−52.8	−27.8	43.5	65.3	67.1	40.4	21.7	25.5
0.2061	−19.6	−41.0	−52.3	−59.5	−56.0	−40.5	39.4	58.4	62.0	43.5	28.5	32.7
0.3398	−17.0	−37.5	−47.8	−55.9	−58.6	−53.4	25.8	48.0	55.7	46.5	35.3	40.1
0.5603	−13.7	−33.5	−42.7	−52.2	−59.4	−62.0	13.0	37.8	49.7	47.8	39.3	44.3
0.9237	−9.7	−29.2	−37.9	−48.4	−58.0	−64.4	5.9	30.1	43.2	45.7	40.0	45.2
1.52	−5.6	−25.1	−33.6	−44.1	−53.8	−59.5	4.6	24.4	35.3	40.6	38.9	44.6
2.51	−2.3	−21.0	−29.0	−38.8	−47.1	−50.6	5.3	18.4	27.3	36.0	28.5	43.8
4.14	0.5	−17.1	−24.2	−32.6	−38.7	−41.9	3.9	12.1	21.1	32.8	38.9	42.7
6.83	2.7	−13.4	−19.3	−26.0	−31.1	−34.0	0.5	7.1	17.3	30.3	38.6	40.4
11.25	3.9	−10.3	−14.2	−19.0	−23.8	−27.3	−0.9	4.9	15.1	27.8	36.3	35.6
18.55	0.3	−7.2	−9.6	−12.7	−16.8	−21.0	0.1	4.9	13.5	24.4	31.4	29.0
30.59	−2.4	−3.8	−5.4	−7.5	−10.7	−15.5	1.9	6.0	12.8	20.9	25.6	22.8
50.43	−1.1	0.6	0.7	−1.0	−4.7	−10.2	5.9	9.7	15.3	19.5	21.5	18.3
83.15	1.7	6.0	9.8	9.0	3.5	−2.9	15.0	18.8	22.1	10.6	19.4	15.1
137.09	4.8	11.4	19.3	20.6	14.4	6.7	28.7	32.3	30.0	21.8	17.3	12.1

Table 3.12b Mean zonal geostrophic wind (m s^{-1}) in the stratosphere and mesosphere

Pressure (hPa)	Southern latitudes						Northern latitudes					
	−70	−60	−50	−40	−30	−20	20	30	40	50	60	70
April												
0.0062	9.8	16.8	34.4	35.5	18.6	20.6	−22.2	−10.9	0.1	6.7	7.2	10.4
0.0103	14.5	22.2	39.0	40.5	22.0	14.6	−26.2	−18.5	−6.9	3.9	6.9	10.6
0.0169	18.3	27.5	44.3	45.5	27.1	13.4	−25.0	−21.0	−10.1	3.3	7.1	10.3
0.0279	20.4	32.4	50.1	50.2	32.8	17.1	−17.2	−18.8	−9.5	4.4	7.3	9.9
0.0460	23.7	37.5	56.0	54.4	38.2	24.1	−5.8	−14.8	−6.8	6.4	7.8	9.8
0.0758	27.8	43.4	61.3	56.9	42.5	33.3	5.9	−10.8	−4.0	8.2	8.7	9.8
0.1250	33.1	50.0	65.5	57.0	44.7	42.7	14.9	−6.8	−1.5	9.4	8.8	9.0
0.2061	38.6	55.4	68.3	57.2	44.7	47.0	17.9	−3.2	0.7	10.5	9.2	8.3
0.3398	43.5	58.8	69.0	57.6	43.5	45.1	16.2	0.1	3.3	11.7	10.2	8.5
0.5603	47.1	60.6	67.4	55.3	41.0	40.7	12.4	3.0	5.8	12.6	11.1	9.0
0.9237	48.2	60.6	63.7	50.3	38.5	36.9	9.9	5.1	7.4	12.9	11.3	8.7
1.52	46.8	56.9	57.4	43.8	33.9	31.0	8.9	6.7	8.8	12.2	9.3	6.4
2.51	44.6	50.6	49.4	36.8	26.6	21.5	7.1	8.0	10.3	10.1	5.4	2.6
4.14	42.1	43.0	40.7	29.4	18.1	10.0	3.3	8.8	10.7	6.6	0.8	−1.3
6.83	39.1	35.9	32.4	21.9	10.4	−0.2	−0.5	8.3	9.6	3.0	−3.4	−4.7
11.25	35.7	30.3	25.3	15.1	4.1	−7.1	−1.5	5.8	6.9	0.5	−5.4	−6.6
18.55	28.3	25.3	19.9	10.4	0.0	−9.9	−1.0	3.2	3.9	−0.5	−5.0	−6.2
30.59	20.2	21.0	17.1	8.8	−0.7	−9.4	−0.8	2.1	3.2	0.4	−2.8	3.9
50.43	16.0	18.2	16.6	10.7	2.2	−5.7	2.3	4.0	5.7	3.2	−0.1	−1.6
83.15	14.7	17.1	18.8	16.5	10.0	3.3	11.8	11.8	12.0	6.9	1.8	−0.3
137.09	13.5	17.5	22.7	24.5	21.3	16.7	26.6	23.8	19.6	10.9	3.8	0.6

Table 3.12c Mean zonal geostrophic wind (m s⁻¹) in the stratosphere and mesosphere

Pressure (hPa)	Southern latitudes						Northern latitudes					
	−70	−60	−50	−40	−30	−20	20	30	40	50	60	70
July												
0.0062	38.7	38.8	19.9	28.3	28.3	21.3	−7.7	−6.3	−13.6	−23.2	−30.3	−28.0
0.0103	40.6	40.2	24.2	36.3	36.3	25.7	−13.5	−20.3	−31.9	−38.1	−38.3	−31.9
0.0169	42.4	41.5	29.0	45.1	44.5	31.0	−15.3	−30.3	−46.7	−50.5	−45.0	−35.3
0.0279	44.2	43.1	34.8	54.8	53.7	37.6	−14.8	−36.2	−56.1	−58.8	−49.7	−37.7
0.0460	46.4	45.5	42.4	65.6	63.3	44.4	−14.7	−39.1	59.9	−62.7	−51.9	−39.3
0.0758	49.8	49.5	51.5	76.9	72.1	49.9	−16.9	−40.4	−59.1	−61.9	−51.6	−39.5
0.1250	55.1	55.7	61.5	87.2	78.3	51.8	−21.2	−42.2	−56.4	−58.3	−49.0	−38.1
0.2061	61.9	64.4	72.6	95.4	80.3	48.2	−27.8	−44.4	−53.2	−53.5	−45.1	−35.3
0.3398	68.4	73.7	83.4	101.2	78.9	40.7	−34.8	−46.0	−50.0	−48.3	−40.8	−31.8
0.5603	71.5	80.6	90.9	103.0	74.9	32.7	−40.0	−45.9	−46.7	−42.9	−36.2	−27.7
0.9237	71.1	84.0	93.5	100.1	69.9	28.0	−42.0	−44.4	−43.0	−38.2	−31.8	−23.3
1.52	69.3	85.3	92.1	92.7	63.1	25.9	−39.2	−40.8	−38.7	−33.8	−27.5	−19.1
2.51	67.5	86.9	90.3	82.8	53.6	23.1	−33.9	−35.2	−33.2	−28.9	−23.1	−15.9
4.14	65.7	89.0	88.1	71.3	41.3	16.5	−29.5	−28.8	−27.1	−23.4	−18.7	11.2
6.83	62.9	89.1	83.2	59.1	28.8	7.6	−26.3	−23.8	−21.6	−18.2	−14.7	−11.2
11.25	56.9	83.7	73.6	46.2	19.3	1.9	−24.0	−19.9	−16.6	−14.0	−11.7	−9.6
18.55	49.1	72.5	60.9	34.7	13.7	0.9	−20.9	−16.5	−12.9	−10.8	−9.3	−7.6
30.59	43.1	60.1	49.3	27.9	11.6	2.3	−17.6	−13.7	−10.1	−8.1	−6.8	−5.3
50.43	36.9	49.9	40.4	24.7	12.8	5.7	−13.9	−10.1	−5.9	−4.0	−3.4	−2.7
83.15	29.5	41.3	33.7	25.1	19.4	13.4	−8.4	−4.3	2.0	2.4	0.9	0.2
137.09	23.3	32.5	28.1	28.5	31.7	26.3	−1.0	3.5	12.8	9.8	5.1	3.1

Table 3.12d Mean zonal geostrophic wind (m s⁻¹) in the stratosphere and mesosphere

Pressure (hPa)	Southern latitudes						Northern latitudes					
	−70	−60	−50	−40	−30	−20	20	30	40	50	60	70
October												
0.0062	3.8	−13.3	−3.1	7.9	−6.6	−25.1	14.9	22.3	31.1	25.4	15.2	14.5
0.0103	2.2	−13.9	−5.3	2.6	−11.2	−30.7	8.6	23.7	34.7	28.3	18.8	18.9
0.0169	0.2	−14.2	−5.3	1.2	−10.7	−30.9	7.1	26.9	38.5	31.7	22.1	22.7
0.0279	−1.9	−14.4	−3.5	−3.8	−5.3	−24.5	10.5	30.8	41.8	36.0	25.4	25.5
0.0460	−3.3	−14.2	−0.9	8.3	1.7	−13.9	17.0	34.9	44.7	40.4	29.3	28.7
0.0758	−4.0	−13.6	1.4	12.9	8.1	−2.3	25.9	38.1	46.1	44.6	34.4	33.3
0.1250	−4.4	−13.2	3.0	16.6	13.7	7.3	35.0	39.6	45.4	48.1	40.3	39.0
0.2061	−3.2	−12.0	4.6	19.5	17.7	11.6	39.3	39.8	45.1	50.5	45.3	44.3
0.3398	0.8	−8.9	6.4	21.4	19.8	11.4	38.2	39.2	45.3	51.2	48.3	48.3
0.5603	6.4	−5.2	7.5	22.0	20.2	9.4	35.3	38.0	43.5	49.7	49.6	50.0
0.9237	12.6	−1.0	8.3	20.4	19.6	8.7	32.9	36.6	39.2	46.2	48.5	48.7
1.52	19.7	4.1	9.4	18.3	17.7	8.8	28.8	33.0	33.7	40.4	43.7	44.1
2.51	29.0	11.4	11.6	16.3	14.7	6.8	21.1	26.5	28.0	33.4	36.1	37.3
4.14	39.9	21.4	14.7	14.0	10.3	1.0	11.3	18.8	22.4	26.3	27.8	29.4
6.83	48.7	30.9	18.2	11.2	4.6	−5.8	2.1	11.6	16.7	20.1	20.8	21.9
11.25	52.4	37.7	21.7	8.2	−1.4	−10.3	−4.6	5.3	11.6	15.3	16.4	17.0
18.55	47.8	40.2	24.1	6.9	−4.8	−11.6	−7.9	0.9	8.1	11.7	13.3	13.5
30.59	37.6	39.0	25.6	8.5	−3.2	−9.5	−7.2	0.2	6.4	9.8	11.0	10.9
50.43	30.0	35.3	25.5	11.9	2.1	−4.4	−3.5	2.2	7.5	10.2	10.1	10.9
83.15	25.8	29.1	23.7	17.2	11.9	5.0	3.1	8.3	13.0	13.4	10.8	7.5
137.09	20.8	22.5	21.9	24.2	25.8	19.9	12.5	18.7	21.2	18.1	12.5	7.3

Chapter 4
Trace Gases

4.1 Overview

Trace gases include all those gaseous components in the atmosphere that – by virtue of their low concentrations – do not affect the bulk composition of air. This makes it convenient to quantify the local abundance of a trace gas by its molar mixing ratio (the chemical amount fraction). If reference is made to dry air, the influence of the variability of water vapor on the bulk composition of air is eliminated, and the molar mixing ratio becomes independent of changes in pressure and temperature. On the other hand, the presence of radicals, such as the hydroxyl radical (OH), are custom-arily reported in terms of number concentration, although radicals undoubtedly also represent trace gases.

Because the major components of air (N_2, O_2, H_2O and CO_2) are chemically rather inert, atmospheric chemistry is for the most part driven by trace gases. Photochemical smog, which arises from the oxidation of hydrocarbons in the presence of nitrogen oxides, causes the build-up of ozone and peroxyacetyl nitrate (PAN) to levels affecting the health of humans and leading to plant damage. Acid rain is caused by similar trace levels of SO_2 being oxidized to sulfuric acid. The biosphere emits trace gases prodigiously, primarily in reduced, chemically reactive forms. Some of the gases have been shown to be effective in improving the thermal tolerance of plants (e.g. isoprene), others permit communication between both plants and insects (e.g. monoterpenes). Most of the biogenic compounds are oxidized in the atmosphere by reactions following the interaction with OH radicals, whereby new trace gases are generated. The radiative properties of the atmosphere are also significantly influenced by trace gases such as CH_4 or N_2O that are thermally active in addition to CO_2 and H_2O, which are the major contributors to the radiation balance.

The concentration levels of trace gases in the atmosphere can be shown to adjust to a steady state between the sources and sinks and their global distributions. Substantial efforts have been made to identify individual sources and sinks and to

P. Warneck and J. Williams, *The Atmospheric Chemist's Companion: Numerical Data for Use in the Atmospheric Sciences*, DOI 10.1007/978-94-007-2275-0_4,
© Springer Science+Business Media B.V. 2012

quantify their contributions to the total budgets of many trace gases in the atmosphere. This aspect is a major theme in the tables presented below. For trace gases that are long-lived, with residence times of 2 years or longer, the general circulation of the atmosphere guarantees a fairly uniform mixing ratio. For trace gases that are short-lived, the mixing ratio is determined by the local sources and sinks, and it may undergo sizeable fluctuations. In this case, the global distribution of sources and sinks as well as advection with the winds becomes important.

The stratosphere is only weakly coupled to the troposphere. Much of the chemistry there is determined by the influx of long-lived trace gases from the troposphere. However, ozone is formed within the stratosphere by chemical reactions following the photodissociation of oxygen.

This chapter provides an overview of atmospheric trace gases including information on typical mixing ratios, major sources and sinks and the residence time; detailed atmospheric budgets of selected species; trace gas emissions as a function of source type (biogenic, anthropogenic, pyrogenic), latitude, country, and contrasting inventories; information concerning the removal rate of trace gases by OH, O_3 and NO_3; simulated and measured stratospheric trace gas distributions; as well as values of Ozone Depletion Potential (ODP), or and Global Warming Potential (GWP).

Table 4.1 Overview on important trace gases in the troposphere: approximate residence times, molar mixing ratios, global distribution, sources and sinks[a]

Trace gas	Distribution	
Residence time	Abundance[b]	Major sources and sinks (Tg a^{-1})[c]
Hydrogen, H_2 (2 a)	Uniform 0.5 ppm	CH_4 oxidation (+20), oxidation natural VOC (+18), biomass burning (+20), fossil fuel use (+15), reaction with OH (−16), uptake by soils (−75)
Methane, CH_4 (8.5 a)	Uniform 1.9 ppm	Rice paddy fields (+75), swamps/marshes (+120) domestic animals (+100), biomass burning (+50), fossil sources (+90), reaction with OH (−430), stratosphere (−40), uptake by soils (−30)
Carbon monoxide, CO (2 m)	150 ppb (NH) 50 ppb (SH)	Anthropogenic (+450), biomass burning (+700) CH_4 oxidation (+600), oxidation natural VOC (+800), reaction with OH (−2,000), stratosphere (−100), uptake by soils (−300)
Ozone, O_3 (2 m)	15–50 ppb Equator < poles	Influx from stratosphere (+600), photochemical production (+4,000), photochemical loss (−3,700), dry deposition (−800)
Nitrous oxide, N_2O (110 a)	Uniform 0.34 ppm	Emission from soils (+10), emissions from oceans (+6), anthropogenic (+9), stratosphere (−19)
Nitrogen oxides, NO_x (2 days)	30 ppt (M) 0.3–5 ppb (C)	Fossil fuel-derived (+21), biomass burning (+8), emission from soils (+7), lightning (+5), loss occurs by oxidation of NO_2 to HNO_3

(continued)

Table 4.1 (continued)

Trace gas	Distribution	
Residence time	Abundance[b]	Major sources and sinks (Tg a^{-1})[c]
Nitric acid, HNO$_3$ (6 days)	70 ppt (M) 0.1–2 ppb (C) 50–130 ppt (FT)	Oxidation of NO$_2$ (+43), dry deposition (−16), attachment to aerosol particles followed by wet deposition (−27)
Peroxyacetyl nitrate, CH$_3$C(O)OONO$_2$ (2–100 days)	120–180 ppt (FT) 7–10 ppt (M, SH) 10–90 ppt (M, NH)	(a major NO$_x$ reservoir in the upper troposphere) formed by association of CH$_3$C(O)O$_2$ radicals with NO$_2$, losses occur by thermal dissociation
Ammonia, NH$_3$ (~3 days)	50–90 ppt (M) 5 ppb (C)	Domestic animals (+22), emissions from oceans (+7), vegetation (+6), use of fertilizer (+6), dry deposition (−15), conversion to NH$_4^+$ aerosol followed by wet deposition (−30)
Hydrogen cyanide, HCN (5 m)	0.25 ppb (M)	Biomass burning (+1.6), uptake by oceans (−2)
Acetonitrile, CH$_3$CN (7 m)	50–150 ppt (M)	Biomass burning (+1.2), reaction with OH (−0.3) uptake by oceans (−1.2)
Carbonyl sulfide, OCS (7 a)	500 ppt Uniform	Soils, marshes (+0.3), emissions from the oceans (+0.3), oxidation of CS$_2$ and DMS (+0.6), uptake by vegetation (−0.5), reaction with OH (−0.1), loss to stratosphere (−0.1)
Hydrogen sulfide, H$_2$S (~3 day)	5–30 ppt (M) 50–100 ppt (C)	Anthropogenic sources (+3.3), emission from soils (+0.5), vegetation (+1.0), volcanoes (+1.0), the predominant sink is reaction with OH
Dimethyl sulfide, CH$_3$SCH$_3$ (2 days)	20–150 ppt	Emissions from the oceans (+50), soils and vegetation (+1), dominant sinks are the reactions with OH and NO$_3$
Carbonyl disulfide, CS$_2$ (7 days)	<30 ppt (M) 35–190 ppt (C)	Emission from the oceans (+0.4), anthropogenic sources (+0.6), reaction with OH is the major sink
Sulfur dioxide, SO$_2$ (4 days)	20–90 ppt (M) 0.1–2 ppb (C)	Fossil fuel-derived (+150), volcanoes (+16), oxidation of DMS (+40), dry deposition (−75), Conversion to SO$_4^{2-}$ and wet deposition (−120)
Methyl chloride, CH$_3$Cl (1.3 a)	550 ppt Uniform	Tropical vegetation (+2), emission from the oceans (+0.2), biomass burning (+0.7), reaction with OH provides the major sink
Methyl bromide CH$_3$Br (0.7 a)	6.8–9.2 ppt (M)	Emission from the oceans (+0.17), the major sink is reaction with OH
Methyl iodide, CH$_3$I (0.02 a)	1–2 ppt (M)	Emission from the oceans (+0.04), the major sink is photolysis
Ethane, C$_2$H$_6$ (60 days)	1.3 ppb (M, NH) 0.4 ppb (M, SH)	Biomass burning (+7), natural gas loss (+6), the major sink is reaction with OH
Propane, C$_3$H$_8$ (12 days)	1 ppb (M, NH) 0.2 ppb (M, SH)	Anthropogenic sources (+23), the major sink is reaction with OH

(continued)

Table 4.1 (continued)

Trace gas	Distribution	
Residence time	Abundance[b]	Major sources and sinks (Tg a^{-1})[c]
Benzene, C_6H_6 (10 days)	0.3 ppb (C) <0.1 (FT)	Industrial + fossil fuel (+1.5), Biofuel (+2) Biomass burning (+2.7), major sink is OH
Toluene, C_7H_8 (2 days)	0.6 ppb (C) <0.05 (FT)	Industrial + fossil fuel (+4.7), Biofuel (+1.1) Biomass burning (+1.8), major sink is OH
Isoprene, C_5H_8 (0.2 day)	0.2–5 ppb (C)	Emissions from deciduous trees (+570), the major sink is reaction with OH
Terpenes, $C_{10}H_{16}$ (0.4 day)	0.03–2 ppb (C)	Emissions from coniferous and deciduous trees (+140), sinks are reactions with OH and O_3
Methanol, CH_3OH (5 days)	0.5–10 ppb (NH) 0.2–2 ppb (SH)	Vegetation (+80), plant decay (+23)
Formaldehyde, HCHO (~10 h)	0.1–0.8 ppb (M) 0.4–5 ppb (C),	Oxidation of hydrocarbons is a major source Photolysis and reaction with OH provide sinks
Acetaldehyde, CH_3CHO (0.8 day)	0.1–0.2 (M) 0.2–0.8 (C) 0.08 (FT)	Oxidation of alkanes (+65), alkenes (+30), ethanol (+25), vegetation, (+35), ocean emissions (+23), the major sink is reaction with OH
Acetone, CH_3COCH_3 (15 days)	0.5–0.8 (M) 1–2 (C)	Oxidation of iso-alkanes (+30), vegetation (+35), Photolysis (−22), reaction with OH (−18), wet and dry deposition (−70)
Formic Acid, HCOOH (~3 days)	0.2–0.8 ppb (M) 0.2–2 ppb (C) 0.1 ppb (FT)	Oxidation of isoprene, etc. (+40), Vegetation (+4), Biomass burning (+2), the major sinks are wet and dry deposition
Acetic Acid,[b] CH_3COOH (~2.3 days)	0.3–0.8 ppb (M) 0.3–2 ppb (C) 0.09 ppb (FT)[c]	Oxidation of isoprene, etc. (+50), biomass burning (+12), vegetation (+4), the major sinks are wet and dry deposition

[a]From Warneck (2002) with additions and changes
[b]Abbreviations: *ppm* μmol mol^{-1}, *ppb* nmol mol^{-1}, *ppt* pmol mol^{-1}, *M* marine, *C* rural continental, *FT* free troposphere; the mixing ratios in urban regions generally are greater than elsewhere
[c]Sources are indicated by plus signs, sinks by minus signs

Table 4.2 Overview on halocarbons in the troposphere: approximate molar mixing ratios, global distributions, tropospheric mass contents, residence times[a]

Compound	Mixing ratio (pmol mol^{-1})	Ratio of NH–SH	G_T (Tg)	τ (year)	Remarks[b]
CH_3Cl	528	n. g.	4.4	1.0	See Table 4.12 for budget
CH_2Cl_2	25	~2	0.25	0.4	Anthropogenic
$CHCl_3$ (chloroform)	15	~2	0.13	0.5	Oceans, soils
CCl_4	96.4	1.01	2.9	26	Anthropogenic (−1%)
$CHClF_2$ (HCFC-22)	162	1.15	1.5	12	Anthropogenic (+3%)

(continued)

Table 4.2 (continued)

Compound	Mixing ratio (pmol mol^{-1})	Ratio of NH–SH	G_T (Tg)	τ (year)	Remarks[b]
CHCl$_2$F (HCFC-21)	~1	n. d.	0.015	1.7	
CCl$_3$F (CFC-11)	255	1.02	5.4	45	Anthropogenic (−0.7%)
CCl$_2$F$_2$ (CFC-12)	535	1.03	8.8	100	Anthropogenic
CClF$_3$ (CFC-13)	3	n. g.	0.05	640	Anthropogenic
CF$_4$ (FC-14)	75	n. g.	0.96	5 × 10^4	Anthropogenic (+2%)
CH$_2$F$_2$ (HFC-32)	–	–	–	4.9	Replacement for HCFC22
CHF$_3$ (HFC-23)	~4	n. d.	0.05	270	Replacement for HCFC22
CH$_3$Br	10	1.3	0.14	0.7	Anthropogenic, ocean
CH$_2$Br$_2$	2–3	n. d.	0.02	0.33	Ocean
CHBr$_3$ (bromoform)	2–3	n. g.	0.02	0.07	Ocean
CH$_2$BrCl	1–2	n. g.	0.003	0.41	Ocean
CHBrCl$_2$	~1	n. g.	0.003	0.19	Ocean, anthropogenic
CHBr$_2$Cl	~1	n. g.	0.03	0.21	Ocean
CBrF$_3$ (Halon 1301)	2.6	n. g.	0.07	65	(Fire extinguisher) (+2%)
CBrClF$_2$ (Halon 1211)	4.15	1.1	0.17	16	(Fire extinguisher) (+1%)
CBr$_2$F$_2$ (Halon 1202)	0.038	n. d.	0.001	2.9	Anthropogenic (−5%)
CH$_3$I	1–3	n. g.	0.01	0.005	Ocean
CH$_2$I$_2$	0.1–0.5	–	–	Minutes	Ocean
CH$_2$ClI	0–0.1	–	–	Hours	Ocean
CF$_3$CH$_2$F (HFC-134a)	~2	~3	0.03	14	Increasing in importance
CHF$_2$CF$_3$ (HFC-125)	–	–	–	29	Important in future
CF$_3$CF$_3$ (FC-116)	4	n. g.	0.08	~1 × 10^4	Very long-lived
CCl$_2$FCClF$_2$ (CFC-113)	84	1.01	2.2	79.5	Anthropogenic
CClF$_2$CClF$_2$ (CFC-114)	14.7	n. g.	0.5	300	Anthropogenic
CF$_3$CClF$_2$ (CFC-115)	8.3	n. g.	0.1	1,700	Anthropogenic
CH$_3$CH$_2$Cl	12	~1.6	0.02	0.08	Anthropogenic
CH$_2$ClCH$_2$Cl	4	~10	0.06	0.19	Industrial
CH$_3$CCl$_3$ (methyl chloroform)	22	1.02	2.6	5.0	Anthropogenic (−18%)
CH$_2$ClCH$_2$Cl	25	2.6	0.36	0.4	Anthropogenic
CHClCCl$_2$	8	~10	0.005	0.01	Anthropogenic
CCl$_2$CCl$_2$	12	~3.5	0.1	0.27	Anthropogenic
CH$_3$CFCl$_2$ (HCFC-141b)	17.2	~2	0.06	9.3	CFC replacement, rising
CH$_3$CF$_2$Cl (HCFC-142b)	15.4	~1.3	0.1	17.9	CFC replacement, rising
CHCl$_2$CF$_3$ (HCFC-123)	0.064	n. d.	0.002	1.3	Rising (+6%), low ODP

(continued)

Table 4.2 (continued)

Compound	Mixing ratio (pmol mol^{-1})	Ratio of NH–SH	G_T (Tg)	τ (year)	Remarks[b]
CHClFCF$_3$ (HCFC-124)	1.64	n. d.	0.03	5.8	Increase slowing
CBrF$_2$CBrF$_2$ (Halon-2402)	0.48	n. d.	0.02˙	20	Anthropogenic
SF$_5$CF$_3$	0.12	n. d.	0.003	650–950	Very high GWP
SF$_6$	5.5	n. d.	0.14	3,200	Transformer insulator
NF$_3$	0.5	n. d.	0.005	740	Microelectronics
COCl$_2$	~20	n. d.	0.29	0.2	Oxidation of atmospheric Chlorine compounds

[a]Compiled from Singh (1995), Graedel and Keene (1995), Fabian and Singh (1999), Clerbaux et al. (2007), Law et al. (2007); *n. g.* no gradient, *n. d.* not determined

[b]Major sources are indicated; numbers in parentheses indicate observed rates of rise or decrease of mixing ratios in percent per year

Comments: Regarding naming conventions for CFCs, HCFCs and HFCs: originally all organic molecules that contained chlorine and fluorine were termed CFCs, but today these species are subdivided into CFCs, HCFCs and HFCs.

The system used for identification is as follows: "CFC-01234n" where 0 = number of double bonds (omitted if zero); 1 = number of carbon atoms minus one; 2 = number of hydrogen atoms plus one; 3 = number of fluorine atoms; 4 = number of chlorine atoms replaced by bromine; n = a letter added to identify isomers (the normal isomer in any number has the smallest mass difference on each carbon, n = a, b, or c are added as the masses diverge from normal. If the compound is cyclic, then the number is prefixed with "C". Chemical names are frequently used in place of the numbers for common materials - such as trichloroethylene or chloroform. The originally specified prefixes were FC (FluoroCarbon), or R (Refrigerant), but today most are prefixed by the specific classifications CFC, HCFC, and HFC.

The term *Halon* is used when bromine is a component. The numbering of these compounds follows a separate system (the prefix "Halon" is always used): "Halon-0123" where 0 = the number of carbon atoms, 1 = the number of fluorine atoms, 2 = the number of chlorine atoms, 3 = the number of bromine atoms; examples are Halon-1211 = bromochlorodifluoromethane (CBrClF$_2$) and Halon-2402 = 1,2-dibromo-1,1,2,2-tetrafluoroethane.

Table 4.3 Life times, ozone depletion potential, ODP, radiative efficiency (W m^{-2} ppb^{-1}), and global warming potential, GWP, for three time horizons[a]

Compound	Life time (a)	ODP	Radiative efficiency	GWP (time horizon) 20 a	100 a	500 a
Carbon dioxide (CO_2)	~200[b]		1.4 (−5)[c]	1	1	1
Methane (CH_4)	12		1.4 (−5)[c]	72	25	7.6
Nitrous oxide (N_2O)	114		3.03 (−3)[c]	289	298	153
Substances controlled by the Montreal Protocol						
CFC-11 (CCl_3F)	45	1.0	0.25	6,730	4,750	1,620
CFC-12 (CCl_2F_2)	100	1.0	0.32	11,000	10,900	5,200
CFC13 ($CClF_3$)	640	0.8	0.25	10,800	14,400	16,400
CFC-113 (CCl_2FCClF_2)	85	1.0	0.3	6,540	10,000	8,730
CFC-114 ($CClF_2CClF_2$)	300	0.6	0.31	8,040	10,000	8,730
CFC-115 ($CClF_2CF_3$)	1,700	0.4	0.18	5,310	7,370	9,990
Halon-1301 ($CBrF_3$)	65	10.0	0.32	8,480	7,140	2,760
Halon-1211 ($CBrClF_2$)	16	3.0	0.3	4,750	1,890	575
Halon-2402 ($CBrF_2CBrF_2$)	20	6.0	0.33	3,680	1,640	503
Carbon tetrachloride (CCl_4)	26	1.1	0.33	2,700	1,400	435
Methyl bromide (CH_3Br)	0.7	0.6	0.01	17	5	1
Methylchloroform (CH_3CCl_3)	5	0.1	0.06	506	146	45
HCFC-22 ($CHClF_2$)	12	0.055	0.2	5,160	1,810	549
HCFC-123 ($CHCl_2CF_3$)	1.3	0.02	0.14	273	77	24
HCFC-124 ($CHClFCF_3$)	5.8	0.022	0.22	2,070	609	185
HCFC-141b (CH_3CCl_2F)	9.3	0.11	0.14	2,250	725	220
HCFC-142b (CH_3CClF_2)	17.9	0.065	0.2	5,490	2,310	705
HCFC-225ca ($CHCl_2CF_2CF_3$)	1.9	0.025	0.2	429	122	37
HCFC-225cb ($CHClFCF_2CClF_2$)	5.8	0.033	0.32	2,030	595	181
Hydrofluorocarbons						
HFC-23 (CHF_3)	270		0.19	12,000	14,800	12,200
HFC-32 (CH_2F_2)	4.9		0.11	2,330	675	205
HFC-125 (CHF_2CF_3)	29		0.23	6,350	3,500	1,100
HFC-134a (CH_2FCF_3)	14		0.16	3,830	1,430	435
HFC-143a (CH_3CF_3)	52		0.13	5,890	4,470	1,590
HFC-152a (CH_3CHF_2)	1.4		0.09	437	124	38
HFC-227ea (CF_3CHFCF_3)	34.2		0.26	5,310	3,220	1,040
HFC-236fa ($CF_3CH_2CF_3$)	240		0.28	8,100	9,810	7,660
HFC-245fa ($CHF_2CH_2CF_3$)	7.6		0.28	3,380	1,030	314
HFC-365mfc ($CH_3CF_2CH_2CF_3$)	8.6		0.21	2,520	794	241
Perfluorinated compounds						
Sulfur hexafluoride (SF_6)	3,200		0.52	16,300	22,800	32,600
Nitrogen trifluoride (NF_3)	740		0.21	12,300	17,200	20,700
PFC-14 (CF_4)	50,000		0.10	5,210	7,390	11,200
PFC-116 (C_2F_6)	10,000		0.26	8,630	12,200	18,200
PFC-218 (C_3F_8)	2,600		0.26	6,310	8,830	12,500
PFC-318 (C_4F_8)	3,200		0.32	7,310	10,300	14,700
PFC-3-1-10 (C_4F_{10})	2,600		0.33	6,330	8,860	12,500
PFC-4-1-12 (C_5F_{12})	4,100		0.41	6,510	9,160	13,300
PFC-5-1-14 (C_6F_{14})	3,200		0.49	6,600	9,300	13,300

(continued)

Table 4.3 (continued)

Compound	Life time (a)	ODP	Radiative efficiency	GWP (time horizon)		
				20 a	100 a	500 a
PFC-9-1-18 ($C_{10}F_{18}$)	>1,000		0.56	>5,500	>7,500	>9,500
Trifluoromethyl sulfur pentafluoride (SF_5CF_3)	800		0.57	13,200	17,700	21,200
Fluorinated ethers						
HFE-125 (CHF_2OCF_3)	136		0.44	13,800	14,900	8,490
HFE-134 (CHF_2OCHF_2)	26		0.45	12,200	6,320	1,960
HFE-143a (CH_3OCF_3)	4.3		0.27	2,630	756	230
PFPMIE ($CF_3OCF(CF_3)CF_2OCF_2OCF_3$)	800		0.65	7,620	10,300	12,400
Other compounds						
Dimethylether (CH_3OCH_3)	0.015		0.02	1	1	«1
Methylene chloride (CH_2Cl_2)	0.38		0.3	31	8.7	2.7
Methyl chloride (CH_3Cl)	1	0.02	0.01	45	13	4

Source of data: Daniel et al. (2007)
[a]Units: 1 ppb = 1 nmol mol^{-1}
[b]The life time for CO_2 relates to the transport of excess CO_2 into the deep ocean
[c]Numbers in parentheses indicate powers of ten

Definitions: The ozone depletion potential (ODP) of a chemical compound is the relative amount of degradation to the ozone layer it can cause, with trichlorofluoromethane (CFC-11) being fixed at an ODP of 1.0.

The global-warming potential (GWP) compares the amount of heat trapped by a certain mass of the gas in question to the amount of heat trapped by a similar mass of carbon dioxide over a specific time interval, commonly 20, 100 or 500 years. It is expressed as a factor with the GWP of CO_2 being taken as unity. The GWP depends on the absorption of infrared radiation by a given species, the spectral location of its absorbing wavelengths, and the atmospheric lifetime of the species.

References

Clerbaux, C., D.M. Cunnold (Lead Authors), 16 coauthors, Long-lived compounds, Chapter 1, in *Scientific Assessment of Ozone Depletion: 2006*, Global Ozone Research and Monitoring Project, Report No. **50**, 1.1–1.63 (2007)

Daniel, J.S., G.J.M. Velders (Lead Authors), 7 coauthors, Halocarbon scenarios, ozone depletion potentials, and global warming potentials, Chapter 8, in *Scientific Assessment of Ozone Depletion: 2006*, Global Ozone Research and Monitoring Project, Report No. **50**, 8.1–8.63 (2007)

Fabian, P., O.N. Singh (eds.), *Reactive Halogen Compounds in the Atmosphere*, O. Hutzinger (editor in chief) The Handbook of Environmental Chemistry, vol. 4/E (Springer, Berlin, 1999)

Graedel, T.E., W.C. Keene, Glob. Biogeochem. Cycles **9**, 47–77 (1995)

Law, K.S., W.T. Sturges (Lead Authors), 18 coauthors, Halogenated very short-lived substances, Chapter 2, in *Scientific Assessment of Ozone Depletion: 2006*, Global Ozone Research and Monitoring Project, Report No. **50**, 2.1–2.57 (2007)

Singh, H.B., Halogens in the atmospheric environment, in *Composition, Chemistry, and Climate of the Atmosphere*, ed. by H.B. Singh (Van Nostrand Reinhold, New York, 1995), pp. 216–250

Warneck, P., in *Encyclopedia of Physical Science and Technology*, ed. by R.A. Meyers, vol. 17, 3rd edn. (Academic Press, San Diego, 2002), pp. 153–174

4.2   Detailed Global Budgets of Trace Gases

Table 4.4 Estimates for the global budget of hydrogen, H_2, in the troposphere (Tg a^{-1})[a]

Type of source or sink	Conrad and Seiler (1980)	Novelli et al. (1999)	Ehhalt (1999)	Warneck (2000)	Rhee et al. (2006) Global	NH	SH	Ehhalt and Rohrer (2009)
Sources								
Anthropogenic emissions	20±10	15±10	20±10	17	15±6	13	1.5	11±4
Biomass burning	10±10	16±5	10±5	19	16±3	8.4	7.5	15±6
Oceans	4±2	3±2	3±2	4	6±5	2.4	3.6	6±3
Methane oxidation	15±5	26±9	15±5	20	b	b	b	23±8
Oxidation of NMHC	25±10	14±7	20±10	18	64±12	42	23	18±7
Biological N_2 fixation	3±2	3±1	30±2	3	6±5	3.6	2.4	3±2
Volcanoes	–	–	–	0.2	–	–	–	–
Total sources	87±38	77±16	71±20	81	107±15	69	38	76±14
Sinks								
Oxidation by OH radicals	8±3	19±5	25±5	16	19±3	9.4	9.7	19±5
Uptake by soils	90±20	56±41	40±30	70	88±11	62	26	60–90
Total sinks	98±23	75±41	65±30	86	107±11	72	35	59–109

[a]Source references see at the end of the section
[b]Included with non-methane hydrocarbons (NMHC), next line

Table 4.5 Estimates for the global budget of methane, CH_4, in the troposphere (Tg a^{-1})

Type of source or sink	Bolle et al. (1986)	Khalil et al. (1993)	Ehhalt (1999)	Wuebbles and Hayhoe (2002)	Mikaloff-Fletcher et al. (2004)
Emissions					
Ruminants etc.	70–100	55–90	85 (65–100)	81 (65–100)	91
Termites	2–5	15–35	40 (10–55)[b]	20 (2–22)	29
Rice paddy fields	70–170	55–90	30 (10–50)	60 (25–90)	54
Natural wetlands	25–70	110	115 (55–150)	100 (92–232)	231
Tundra	2–15	[c]	–	–	–
Ocean	1–7	4	–	4 (0.2–2)	–
Domestic sewage	–	27–80	25 (15–80)	[d]	[d]
Landfills	10	11–32	40 (20–70)	61	35
Animal waste	–	20–30	25 (20–30)	[d]	[d]
Coal mining	35	25–50	30 (15–45)	46 (15–64)	30
Natural gas leakage	30–40	30	40 (25–50)	30 (25–50)	52
Biomass burning	55–100	50	40 (20–80)	50 (27–80)	88
Total emission rate	552	402–601	530 (405–645)	503 (312–640)	610
Removal processes					
Reaction with OH	260	440	490 (405–575)	445 (360–530)	507
Loss to the stratosphere	60	15	40 (32–48)	40 (32–48)	40
Removal by soils	20	30	30 (15–45)	30 (15–45)	30
Total removal rate	340	485	560 (452–668)	515 (407–623)	577

[a]The IPCC Fourth Assessment, Denman et al. (2007), reports total emission and removal rates of 582 and 581 (Tg a^{-1}), respectively
[b]Includes other insects
[c]Included in the estimate for natural wetlands
[d]Included in Ruminants etc.

Table 4.6 Estimates for the global budget of carbon monoxide, CO, in the troposphere (Tg a^{-1})[a]

Type of source or sink	Logan et al. (1981) Global	NH	SH	Seiler and Conrad (1987)	Khalil and Rasmussen (1990)	Pacyna and Graedel (1995)	Holloway et al. (2000)	Duncan et al. (2007)
Sources								
Fossil fuel combustion[b]	450	425	25	640±200	500	440±150	300	464–487
Biomass burning	655	415	240	1,000±600	680	700±200	586	451–573
Biofuels	–	–	–	–	–	–	162	189
Oxidation of human-made HC[a]	90	85	5	–	90	–	–	–
Oxidation of natural HC[a]	560	380	180	900±500	600	800±400	672	354–379
Ocean emissions	40	13	27	100±90	40	50±40	–	–
Emissions from vegetation	130	90	40	75±25	100	75±25	–	–
Oxidation of methane	810	405	405	600±300	600	600±20	760	778–861
Total source strength	2,735	1,813	922	3,315±1,700	2,600	2,700±1,000	2,491	2,236–2,489
Sinks								
Reaction with OH radicals	3,170	1,890	1,280	2,000±600	2,200	2,000±600	2,491	–
Consumption by soils	250	210	40	390±140	250	250±100	–	–
Flux into the stratosphere	–	–	–	110±30	100	110±30	–	–
Total sink strength	3,420	2,100	1,320	2,500±770	2,550	2,400±750	2,491	2,231–2,366

[a]Source references see at the end of the section; *HC* hydrocarbons
[b]Includes related activities such as emissions from petroleum industries

Table 4.7 Modeled global budgets of tropospheric ozone (Tg a^{-1})[a]

Influx from stratosphere	Chemical production[b]	Chemical loss[b]	Dry deposition	Global burden (Tg)	Residence time (days)	Model[c]
570	3,310	3,170	710	350	33	TM3 (1)
470	4,900	4,300	1,070	320	22	GEOS-Chem (2)
593	4,895	4,498	990	322	21	CHASER (3)
340	5,260	4,750	860	360	23	MOZART-2 (4)
540	4,560	4,750	860	290	19	MATCH-MPIC(5)
523	4,486	3,918	1,090	296	22	LMDz-INCA (6)
395	4,980	4,420	950	273	19	STOCHEM (7)
520	4,090	3,850	760	283	22	FRSGC/UCI (8)
715	4,436	3,890	1,261	303	21	LMDz-INCA (9)
770±400	3,420±770	3,470±520	770±180	300±30	24±2	11 models (10)
520±200	5,060±570	4,560±720	1,010±220	340±40	22±2	25 models (11)

[a]Source references: (1) Lelieveld and Dentener (2000), (2) Bey et al. (2001), (3) Sudo et al. (2003), (4) Horowitz et al. (2003), (5) von Kuhlmann et al. (2003), (6) Hauglustaine et al. (2004), (7) Stevenson et al. (2004), (8) Wild et al. (2003), (9) Folberth et al. (2006), (10) Houghton et al. (2001), (11) Stevenson et al. (2006)

[b]The major chemical reactions acting as sources are $HO_2 + NO \rightarrow NO_2 + OH$ and $CH_3O_2 + NO \rightarrow CH_3O + NO_2$, both followed by $NO_2 + h\nu + O_2 \rightarrow NO + O_3$; the major chemical loss reactions are $O_3 + h\nu \rightarrow O_2 + O(^1D)$ and $HO_2 + O_3 \rightarrow OH + 2\ O_2$ and $OH + O_3 \rightarrow HO_2 + O_2$

[c]All models are three-dimensional chemical transport models based on observed wind fields, where meteorology and chemistry are decoupled. The number of species included in the chemistry varies with the model from 30 to 85

Table 4.8 Estimates for the global budget of NO_x in the troposphere (Tg a^{-1} as nitrogen)[a]

Type of source or sink	Ehhalt and Drummond (1982)	Logan (1983)	Lee et al. (1997)	Ehhalt (1999)	Müller and Stravrakou (2005)	Denman et al. (2007)
Production						
Fossil-fuel combustion	13.5±5.1	19.9 (14–28)	22 (13–31)	21±5	22.8	25.6
Biomass burning	11.5±5.4	12.0 (4–24)	7.9 (3–15)	7.5±3	4.4	5.9
Lightning discharges	5.0±3.0	8.0 (2–20)	5.0 (2–20)	7±3	2.8	1.1–6.4
Release from soils	5.5±4.0	8.0 (4–16)	7.0 (4–12)	5.5±2.2	12.1	7.3
NH$_3$ oxidation	3.1±1.9	(0.10)	0.9 (0.3–1.2)	3±1	–	–
Stratosphere	0.6±0.3	0.5	0.64 (0.4–1.0)	0.15±0.05	–	–
High-flying aircraft	0.3±0.1	–	0.85	0.45±0.1	–	–
Total production rate	39±20	48.4 (25–99)	44 (23–81)	45±7	42.1	41.8–47.1
Removal						
Wet deposition, land	17±7	19 (8–30)	–	–	–	–
Wet deposition, oceans	8±6	8 (2–12)	–	–	–	–
Dry deposition	–	16 (12–22)	–	20	–	–
Total removal rate	25±13	43 (23–64)	–	–	–	–

[a]Emissions occur primarily as NO, the conversion of NO to NO_2 and HNO_3 occurs subsequently in the atmosphere. The removal of NO_x proceeds by oxidation to HNO_3 and wet deposition in the form of nitrates. Dry deposition includes gaseous NO_2 and HNO_3. The inflow from the stratosphere consists predominantly of HNO_3

Table 4.9 Estimates for the global budget of nitrous oxide, N_2O (Tg a^{-1})[a]

	Khalil and Rasmussen (1992)		Bouwman et al. (1995)		Denman et al. (2007)
Natural sources					
Tropical forests	11.6	(5.3–17.9)	5.2		–
Temperate forests	1.2	(1.6–0.8)	0.8		–
Grasslands	0.16		2.3		–
Arable land	–		1.5		–
Total soils	12.3	(5.7–18.9)	10.7	(10.4–11.0)	6.6
Oceans	3.1	(1.6–4.7)	5.7	(4.4–8.9)	3.8
Total natural	15.4	(7.3–23.6)	16.4	(14.8–19.9)	11.0
Anthropogenic sources					
Fossil fuel combustion	0.8	(0.1–2.2)	0.47	(0.16–0.94)	0.7
Adipic acid Production	0.7		0.47	(0.30–0.61)	–
Nitric acid production	–		0.3	(0.12–0.47)	–
Nitrogen fertilizer	1.0	(0.4–3.0)	1.6	(0.64–2.6)	2.8
Domestic animal excreta	0.5	(0.3–1.0)	1.6		–
Biomass burning	1.6	(0.2–3.0)	0.36		0.7
Land use change	0.7		0.6		–
NH_3 conversion[b]	–		0.63	(0.47–1.9)	0.6
Total anthropogenic[c]	7.9	(5–10)	9.4	(4.3–8.9)	6.7
Total identified sources	23.4	(12.2–33.6)	25.8	(19.1–28.8)	17.7
Major sink and rate of increase					
Stratospheric sink	16.5	(11.8–21.2)	19.3	(14.1–25.1)	–
Atmospheric increase	5.5	(4.7–6.3)	6.1	(4.9–7.4)	–

[a]Source references see at the end of the section
[b]Oxidation of ammonia by OH radicals in the atmosphere
[c]Includes additional small sources

Table 4.10 Estimates of sources and sinks of ammonia, NH_3 (Tg a^{-1} as nitrogen)

Process	Böttger et al. (1978)	Stedman and Shetter (1983)	Warneck (1988)	Schlesinger and Hartley (1992)	Dentener and Crutzen (1994)	Denman et al. (2007)
Sources						
Coal combustion	0.03	<2	≤2	2	–	2.5
Automobiles	0.2–0.3	–	0.2	0.2	–	–
Biomass burning	–	–	2–8	5	2	5.4
Domestic animals	20–30	23	22	32	22	35
Wild animals	–	3	4	–	2.5	–
Human excrements	–	1.5	3	4	–	2.6
Soil/plant emissions	1	(51)[a]	15	10	5.1	2.4
Fertilizer losses	1.2–2.4	3.5	3	9	6.4	–
Oceans	–	–	–	13	7	8.2
Sum of sources	22–34	83	54	75	45	56.1

(continued)

Table 4.10 (continued)

Process	Böttger et al. (1978)	Stedman and Shetter (1983)	Warneck (1988)	Schlesinger and Hartley (1992)	Dentener and Crutzen (1994)	Denman et al. (2007)
Sinks						
Precipitation (continents)	15±7	50	30	30	13.6[c]	–
Precipitation (oceans)	6±6	10	8	16	16[c]	–
Dry deposition (land)	[b]	14	10	10	13.6[c]	–
Reaction with OH	3	9	1	1	1.8[c]	–
Sum of sinks	24±13	83	49	57	45	–

[a]Adopted to balance the budget
[b]Included in the figure for wet precipitation
[c]Derived from three-dimensional model calculations

Table 4.11 Global atmospheric sulfur budget (Gmol a^{-1})[a]

Type of source or sink	Langner and Rodhe (1991)	Pham et al. (1996)	Feichter et al. (1996)	Chin et al. (1996)
Dimethyl sulfide (CH$_3$SCH$_3$)				
Emissions from ocean	500	625	528	681
Oxidation to SO$_2$	500	584	528	647
Oxidation to methane sulfonate	–	38	–	34
Dry deposition of MSA	–	16	–	3
Wet deposition of MSA	–	22	–	31
Sulfur dioxide (SO$_2$)				
Anthropogenic emissions	2,078	2,875	2,425	2,034
Volcanoes	266	288	109	209
Biomass burning	78	91	78	72
Oxidation of DMS[b]	500	584	528	673
Total sources	2,922	3,838	3,140	2,962
Gas-phase oxidation of SO$_2$	244	203	525	234
Oxidation in clouds of SO$_2$	1,312	1,734	1,078	1,300
Dry deposition of SO$_2$	953	1,719	1,256	831
Wet deposition of SO$_2$	444	182	281	622
Aerosol sulfate (SO$_4^{2-}$)				
Anthropogenic emissions	109	–	–	–
Gas-phase oxidation of SO$_2$	244	203	525	234
In-cloud oxidation of SO$_2$	1,312	1,734	1,078	1,300
Total sources	1,665	1,937	1,603	1,534
Dry deposition of sulfate	268	531	209	175
Wet deposition of sulfate	1,397	1,406	1,394	1,359
Residence times (day)				
Residence time of DMS	3.0	0.9	2.2	1.0
Residence time of MSA	–	6.1	–	6.2
Residence time of SO$_2$	1.2	0.6	1.5	1.3
Residence time of sulfate	5.3	4.7	4.3	3.9

[a]Recalculated from Chin et al. (1996)
[b]Chin et al. (1996) include some sulfur dioxide from the oxidation of OCS and H$_2$S

Table 4.12 Sources and sinks for methyl chloride (Gg a^{-1})[a]

Type of source or sink	Lee-Taylor et al. (2001)	Yoshida et al. (2004)	Keppler et al. (2005)	Clerbaux and Cunnold (2007)
Sources				
Tropical/subtropical plants	2,380	2,900	910	820–8,200
Tropical senescing leaves	–	–	–	30–2,500
Biomass burning	733	611	911	325–1,125
Oceans	477	508	600	380–500
Salt marshes	170	170	170	65–440
Fungi	128	–	160	43–470
Wetlands	48	48	40	48
Rice paddies	–	–	5	2.4–4.9
Fossil fuel burning	–	–	105	5–205
Waste incineration	162	162	45	15–75
Industrial processes	–	–	10	10
All sources	4,098	4,399	2,956	1,743–13,578
Sinks				
Reaction with OH radicals	3,850	3,994	3,180	3,800–4,100
Loss to stratosphere	–	–	200	100–300
CI reaction	–	–	370	180–550
Uptake by soils	256	256	890	100–1,600
Polar downdraft to deep ocean	–	149	75	93–145
All sinks	4,106	4,399	4,715	4,273–6,695

[a]References are given at the end of the section

Table 4.13 Global budget for methanol, CH$_3$OH (Tg a^{-1})[a]

Type of source or sink	Heikes et al. (2002)	Galbally and Kirstine (2002)	von Kuhl-mann et al. (2003)	Jacob et al. (2005)	Millet et al. (2008)
Sources					
Terrestrial plant growth	280 (50 to >280)	100 (37–212)	77	128 (100–160)	80
Plant decay	20 (10–40)	13 (5–31)	–	23 (5–40)	23
Biomass burning	12 (2–32)	13 (5–31)	15	13 (10–20)	12
Atmospheric production	30 (18–30)	19 (6–19)	28	38 (50–100)	37
Urban sources	8 (5–11)	4 (3–5)	2	4 (1–10)	5
Oceans, biogenic	–	–	–	–	85
Total sources	350 (90–490)	149 (83–260)	123	206 (170–330)	242
Sinks					
Gas-phase oxidation by OH	100 (25–150)	109 (60–203)	77	129	88
In-cloud oxidation by OH	10 (5–20)	5 (2–15)	–	<1	<1
Dry deposition (land)	70 (25–150)	24 (11–43)	37	55	40
Wet deposition	10 (4–36)	11 (5–20)	9	12	13
Uptake by oceans	50 (20–150)	0.3 (0.2–0.6)	–	10	101
Total sinks	270 (160–570)	149 (82–273)	123	206	242

(continued)

Table 4.13 (continued)

Type of source or sink	Heikes et al. (2002)	Galbally and Kirstine (2002)	von Kuhl-mann et al. (2003)	Jacob et al. (2005)	Millet et al. (2008)
Corollary					
Atmospheric inventory (Tg)	3.9	3.4	–	4	3.1
Atmospheric residence time (d)	9	8	–	7	4.9

[a]Source references are given at the end of the section

Table 4.14 Global sources and sinks of formic acid and acetic acid (Tg a^{-1})[a]

	HCOOH			CH_3COOH		
Type of source or sink	Warneck (2000)	Paulot et al. (2011)	Other authors	Warneck (2000)	Paulot et al. (2011)	Other authors
Sources						
Atmospheric reactions			17[b,c]			75[b], 42[c]
Biogenic precursors	37.5	42.2		44.1	57.3	
Anthropogenic precursors	–	6.4		–	1.26	
Emissions						
Anthropogenic	0.002	0.16	–	0.006	0.42	–
Biomass burning	2.9	1.5	8.0[b], 8.4[c]	12.6	11.2	14.6[b], 16.7[c]
Biofuel burning	–	0.3	1.2	–	6.9	16.4[c]
Cattle	–	1.8	–	–	2.4	–
Vegetation	4.1	2.6	5.6	4.2	2.6	–
Soils	1.7	1.8	–	1.4	3.4	–
Total sources	46.2	56.7	30.6[b], 27[c]	62.3	85.6	93[b], 73[c]
Sinks						
Reaction with OH radicals	–	10.5	–	–	24.8	–
Dry deposition	157	24.7	–	79.8	31.3	–
Dust	–	1.4	–	–	2.4	–
Wet deposition	69.5	20.1	–	42.0	27.1	–
Total sinks	213	56.7	–	182	85.6	–

[a]The data were originally given in Gmol a^{-1}
[b]von Kuhlmann et al. (2003)
[c]Ito et al. (2007)

Comments: The major sources of both acids are chemical reactions within the troposphere. Atmospheric reactions include reactions of ozone with unsaturated hydrocarbons, primarily isoprene, the oxidation of glycol aldehyde and hydroxy-propanone initiated by OH radicals, and the reaction of acetyl peroxy radicals with HO_2 radicals.

Table 4.15 Global atmospheric budget of acetaldehyde and acetone (Tg a^{-1})[a]

Type of source or sink	CH$_3$CHO			CH$_3$COCH$_3$	
	Singh et al. (2004)	Millet et al. (2010)	Jacob et al. (2002)	Singh et al. (2004)	Marandino et al. (2005, 2006)
Sources					
Anthropogenic emissions	0–1	2	1.1 ± 0.5	2	–
Hydrocarbon oxidation[b]	15–45	128	28 ± 5	20–36	–
Biomass burning	5–15	3	4.5 ± 1.6	7–11	–
Vegetation, growth and decay	20–50	23	35 ± 9	25–75	–
Oceans[c]	75–175	57	27 ± 6	0–15	–
Total sources	200 (115–286)	213	95 ± 15	95 (57–148)	–
Sinks					
Photolysis	–	22	46	–	22 ± 8
Reaction with OH radicals	–	188	27	–	18 ± 7
Dry and wet deposition	–	3	23	–	71 ± 25
Total sinks	–	213	95	–	111 ± 27

[a]Source references are given at the end of the section
[b]The hydrocarbons that are dominant as sources of acetaldehyde are alkanes, alkenes and ethanol contributing approximately 65, 30, and 25 Tg a^{-1}, respectively. In the case of acetone *iso*-alkanes such as propane are major sources followed by monoterpenes, 21 and 7 Tg a^{-1}, respectively, according to Jacob et al. (2002)
[c]Jacob et al. (2002) suggested that the oceans provided a global source of acetone, whereas Marandino et al. (2005, 2006) concluded that the oceans represent a net sink

References

Bey, I., D.J. Jacob, R.M. Yantosca, J.A. Logan, B.D. Field, A.M. Fiore, Q.B. Li, H.G. Liu, L.J. Mickley, M.G. Schultz, J. Geophys. Res. **106**, 23073–23095 (2001)

Bolle, H.-J., W. Seiler, B. Bolin, *SCOPE* **29**, 157–203 (1986)

Böttger, A., D.H. Ehhalt, G. Gravenhorst, *Atmosphärische Kreisläufe von Stickoxiden und Ammoniak*. Report Number 1558, Kernforschungsanlage Jülich, Germany (1978)

Bouwman, A.F., K.W. Van der Hoek, J.G.J. Olivier, J. Geophys. Res. **100**, 2785–2800 (1995)

Chin, M., D.J. Jacob, G.M. Gardner, M.S. Foreman-Fowler, P.A. Spiro, D.L. Savoie, J. Geophys. Res. **101**, 18667–18690 (1996)

Clerbaux, C., D.M. Cunnold (Lead Authors), 16 co-authors, Long-lived compounds, Chapter 1, in *Scientific Assessment of Ozone Depletion: 2006*, Global Ozone Research and Monitoring Project, Report No. **50**, 1–63 (2007)

Conrad, R., W. Seiler, J. Geophys. Res. **85**, 5493–5498 (1980)

Denman, K.L., G. Brasseur, A. Chidthaisong, P. Ciais, P.M. Cox, R.E. Dickinson, D. Hauglustaine, C. Heinze, E. Holland, D. Jacob, U. Lohmann, S. Ramachmandran, P.L. Da Silva Dias, S.C. Wofsy, X. Zhang, in *Climate Change 2007: The Physical Science Basis*, ed. by S.D. Solomon, D. Qin, M. Manning, Z. Chen, M. Marquis, K.B. Averyt, M. Tignor, H.L. Miller. Contribution of Working Group I to the Fourth Assessment Report of the Intergovernmental Panel on Climate Change, Cambridge University Press, New York, 2007), pp. 499–588

Dentener, F.J., P.J. Crutzen, J. Atmos. Chem. **19**, 331–369 (1994)

Duncan, B.N., J.A. Logan, I. Bey, I.A. Megretskaia, R.M. Yantosca, P.C. Novelli, N.B. Jones, C.P. Rinsland, J. Geophys. Res. **112**, D22301 (2007). doi:10.1029/ 2007JD008459

Ehhalt, D.H., in *Global Aspects of Atmospheric Chemistry*, ed. by R. Zellner (Steinkopff Verlag, Darmstadt, 1999), pp. 21–109

Ehhalt, D.H., J.W. Drummond, in *Chemistry of the Unpolluted and Polluted Troposphere*, ed. by H.W. Georgii, W. Jaeschke. NATO ASI Series **C 96**, 219–251 (1982)

Ehhalt, D.H., F. Rohrer, Tellus **B 61**, 500–535 (2009)

Feichter, J., E. Kjellstrom, H. Rodhe, F. Dentener, J. Lelieveld, G.-J. Roelofs, Atmos. Environ. **30**, 1693–1708 (1996)

Folberth, G.A., D.A. Hauglustaine, J. Lathière, F. Brocheton, Atmos. Chem. Phys. **6**, 2273–2319 (2006)

Galbally, I.E., W. Kirstine, J. Atmos. Chem. **43**, 195–229 (2002)

Hauglustaine, D.A., F. Hourdin, S. Walters, L. Jourdain, M.-A. Filiberti, J.-F. Lamarque, E.A. Holland, J. Geophys. Res. **109**, D04314 (2004). doi:10.1029/2003JD003957

Heikes, B.G., W. Chang, M.E.Q. Pilson, E. Swift, H.B. Singh, A. Guenther, D.J. Jacob, B.D. Field, R. Fall, D. Riemer, L. Brand, Glob. Biogeochem. Cycles **16**, 1133 (2002). doi: 10.1029/2002GB001895

Holloway, T., H. Levy II, P. Kasibhatla, J. Geophys. Res. **105**, 12123–12147 (2000)

Horowitz, L.W., S. Walters, D.L. Mauzerall, L.K. Emmons, P.J. Rasch, C. Granier, X.X. Tie, J.F. Lamarque, M.G. Schultz, G.S. Tyndall, J.J. Orlando, G.P. Brasseur, J. Geophys. Res. **108**, 4784 (2003). doi :10.1029/2002JD002853

Houghton, J.T., Y. Ding, D.J. Griggs, M. Noguer, P.J. van der Linden, X. Dai, K. Maskell, C.A. Johnson (eds.) *Climate Change 2001: The Scientific Basis*, Third Assessment Report of the Intergovernmental Panel on Climate Change (IPCC) (Cambridge University Press, New York, 2001)

Ito, A., S. Sillman, J.E. Penner, J. Geophys. Res. **112**, D06309 (2007). doi: 10.1029/2005JD006556

Jacob, D.J., B.D. Field, E.M. Jin, I. Bey, Q. Li, J.A. Logan, R.M. Yantosca, J. Geophys. Res. **107**, D4100 (2002). doi:10.1029/2001JD000694

Jacob, D.J., B.D. Field, Q. Li, D.R. Blake, J. de Gouw, C. Warneke, A. Hansel, A. Wisthaler, H.B. Singh, A. Guenther, J. Geophys. Res. **110**, D08303 (2005). doi:10.1029/2004JD005172

Keppler, F., D.B. Harper, T. Röckmann, R.M. Moore, J.T.G. Hamilton, Atmos. Chem. Phys. **5**, 2403–2411 (2005)

Khalil, M.A.K., R.A. Rasmussen, Chemosphere **20**, 227–242 (1990)

Khalil, M.A.K., R.A. Rasmussen, J. Geophys. Res. **97**, 14651–14660 (1992)

Khalil, M.A.K., M.J. Shearer, R.A. Rasmussen, in *Atmospheric Methane: Sources, Sinks, and Role in Global Change*, ed. by M.A.K. Khalil. NATO ASI Series, vol. I 13 (Springer, Berlin, 1993), pp. 168–179, 180–198

Langner, J., H. Rodhe, J. Atmos. Chem. **13**, 225–263 (1991)

Lee, D.S., I. Köhler, E. Grobler, F. Rohrer, R. Sausen, L. Gallardo-Klenner, J.G.J. Oliver, F.J. Dentener, A.F. Bouwman, Atmos. Environ. **31**, 1735–1749 (1997)

Lee-Taylor, J., G. Brasseur, Y. Yokouchi, J. Geophys. Res. **106**, 34221–34233 (2001)

Lelieveld, J., F.J. Dentener, J. Geophys. Res. **105**, 3531 (2000). doi:10.1029/1999JD901011

Logan, J.A., J. Geophys. Res. **88**, 10785–10807 (1983)

Marandino, C.A., W.J. De Bruyn, S.D. Miller, M.J. Prather, E.S. Saltzman, Geophys. Res. Lett. **32**, L15806 (2005). doi:10.1029/2005GL023285

Mikaloff-Fletcher, S.E., P.P. Tans, L.M. Bruhwiler, J.B. Miller, M. Heimann, Glob. Biogeochem. Cycles **18**, GB4005 (2004)

Millet, D.B., D.J. Jacob, T.G. Custer, J.A. de Gouw, A.H. Goldstein, T. Karl, H.B. Singh, B.C. Sive, R.W. Talbot, C. Warneke, J. Williams, Atmos. Chem. Phys. **8**, 6887–6905 (2008)

Millet, D.B., A. Guenther, D.A. Siegel, N.B. Nelson, H.B. Singh, J.A. de Gouw, C. Warneke, J. Williams, G. Eerdekens, V. Sinha, T. Karl, F. Flocke, E. Apel, D.D. Riemer, P.I. Palmer, M. Barkley, Atmos. Chem. Phys. **10**, 3405–3425 (2010)

Müller J.-F, T. Stavrakou, Atmos. Chem. Phys. **5**, 1157–1186 (2005)

Novelli, P.C., P.M. Lang, K.A. Masari, D.F. Hurst, R. Myers, J.W. Elkins, J. Geophys. Res. **104**, 30427–30444 (1999)

Pacyna, J.M., T.E. Graedel, Annu. Rev. Energy Environ. **20**, 265–300 (1995)

Paulot, F., D. Wunch, J.D. Crounse, G.C. Toon, D.B. Millet, P.F. DeCarlo, C. Vigouroux, N.M. Deutscher, G. González Abad, J. Notholt, T. Warneke, J.W. Hannigan, C. Warneke, J.A. de Gouw, E.J. Dunlea, M. De Mazière, D.W.T. Griffith, P. Bernath, J.L. Jimenez, P.O. Wennberg, Atmos. Chem. Phys. **11**, 1989–2013 (2011)
Pham, M., J.-F. Müller, G.P. Brasseur, C. Granier, G. Mégie, Atmos. Environ. **30**, 1815–1822 (1996)
Rhee, T.S., C.A.M. Brenninkmeijer, T. Röckmann, Atmos. Chem. Phys. **6**, 1611–1625 (2006)
Schlesinger, W.H., A.E. Hartley, Biogeochemistry **15**, 191–211 (1992)
Seiler, W., R. Conrad, in *Geophysiology of Amazonia*, ed. by R. Dickinson (Wiley, New York, 1987), pp. 133–162
Singh, H.B., L.J. Salas, R.B. Chatfield, E. Czech, A. Fried, J. Walega, M.J. Evans, B.D. Field, D.J. Jacob, D. Blake, B. Heikes, R. Talbot, G. Sachse, J.H. Crawford, M.A. Avery, S. Sandholm, H. Fuelberg, J. Geophys. Res. **109**, D15S07 (2004). doi:10.1029/2003JD003883
Stedman, D.H., R.E. Shetter, in *Trace Atmospheric Constituents, Properties, Transformations and Fates*, ed. by S.E. Schwartz (Wiley, New York, 1983), pp. 411–454
Stevenson, D.S. (39 coauthors), J. Geophys. Res. **111**, D08301 (2006). doi:10.1029/2005JD006338
Stevenson, D.S., R.M. Doherty, M.G. Sanderson, W.J. Collins, C.E. Johnson, R.G. Derwent, Faraday Discuss. **130**, 41–57 (2004)
Sudo, K., M. Takahashi, H. Akimoto, Geophys. Res. Lett. **30**, 2256 (2003). doi:1029/2003GL018526
von Kuhlmann, R., M.G. Lawrence, P.J. Crutzen, P.J. Rasch, J. Geophys. Res. **108**, 4294 (2003). doi:10.1029/2002JD002893; 4729, doi:10.1029/2002JD003348
Warneck, P., *Chemistry of the Natural Atmosphere* (Academic Press, San Diego, 1988)
Warneck, P., *Chemistry of the Natural Atmosphere*, 2nd edn. (Academic Press, San Diego, 2000), Copyright Elsevier.
Wild, O., J.K. Sunder, M.J. Prather, I.S.A. Isaksen, H. Akimoto, E.V. Browell, S.J. Oltmans, J. Geophys. Res. **108**, 8826 (2003). doi:10.1029/2002JD003285
Wuebbles D.J., K. Hayhoe, Earth-Sci. Rev. **57**, 3–4, 177–210 (2002)
Yoshida, Y., Y. Wang, T. Zeng, R. Yantosca, J. Geophys. Res. **109**, D24309 (2004). doi:10.1029/2004JD004951

4.3 Emissions of Trace Gases

Table 4.16 Global emissions of hydrocarbons and other organic volatiles[a]

Type of source	Emissions (Tg a^{-1})	Remarks
Anthropogenic sources		
Petroleum related sources and chemical industry	36–62	Mainly automobiles, alkanes, alkenes, aromatic compounds
Natural gas	2–14	Mainly light alkanes
Organic solvent use	8–20	Higher alkanes and aromatic compounds
Biomass burning	25–80	Mainly light alkanes and alkenes
Total anthropogenic sources	71–175	
Biogenic sources		
Emissions from foliage	175–503	Isoprene
	127–480	Monoterpenes
	510	Higher alkanes, alkenes, alcohols, aldehydes, ketones, esters
Grasslands	<26	
Soils	<3	
Ocean waters	2.5–6	Light alkanes, alkenes, isoprene
	<26	C_9–C_{28} alkanes
Total biogenic sources	815–1,530	–

[a]From Warneck (2000) reproduced with kind permission by Elsevier

Table 4.17 Detailed speciation of biogenic global emissions (Tg a^{-1})[a]

Isoprene	530	Acetone	29	Acetaldehyde	11
Methanol	160	Ethene	17	Propene	6.2
Monoterpenes	97	Sesquiterpenes	12	Formaldehyde	4.2
Other compounds	10				

[a]Source: MEGAN database, courtesy A. Guenther, based on Guenther et al. (2006)

Table 4.18 Latitudinal distribution of emissions of volatile organic compound (VOC) in three categories: biogenic, anthropogenic, pyrogenic (Tg a^{-1} as carbon)[a]

Northern hemisphere				Southern hemisphere			
Latitude	Biogenic	Anthrop.	Pyrogenic	Latitude	Biogenic	Anthrop.	Pyrogenic
0–5°	46.9811	7.7841	9.9521	0–5°	49.5898	7.2686	2.1484
5–10°	40.4644	13.5240	9.2551	5–10°	74.3794	9.1709	4.8609
10–15°	35.7347	7.2484	1.9758	10–15°	100.0151	8.0117	8.9837
15–20°	36.7917	8.1005	6.7501	15–20°	94.9112	5.5507	5.4287
20–25°	51.5559	13.7136	1.3138	20–25°	91.0440	3.7539	1.9399
25–30°	37.1990	15.9859	0.2644	25–30°	63.6977	2.6618	1.3779
30–35°	37.3487	15.9305	0.4137	30–35°	40.8052	1.6015	0.7970
35–40°	35.4609	15.9305	0.6077	35–40°	28.2480	1.1766	0.2116
40–45°	33.2779	12.9066	1.1009	40–45°	13.2296	0.2494	0.0512
45–50°	35.5788	11.6435	1.6045	45–50°	4.9126	0.1302	0.0009
50–55°	22.3897	9.6988	6.0589	50–55°	2.2822	0.0949	0.0000
55–60°	5.3128	5.2902	0.8804	55–60°	0.6066	0.0020	
60–65°	0.2183	4.3334	1.9366	60–65°	0.0743	0.0000	
65–70°	0.0029	0.9773	0.0295	65–70°	0.0390	0.0000	
70–75°	0.0000	0.0703	0.0000	70–75°	0.0000		
75–80°	0.0000	0.0002	0.0000	75–80°	0.0000		
80–85°	0.0000	0.0000	0.0000	80–85°	0.0000		
				Total emissions	1,000.19	182.802	67.944

[a]Sources of data: Biogenic: Guenther et al. (1995, 2006); Anthropogenic: Oliver and Berdowski (2001); Pyrogenic: Andreae and Merlet (2001) with modification by Yokelson et al. (2008)

Table 4.19 Biomass burning emission factors for selected gases produced by different materials (g kg^{-1} referring to dry fuel burned)[a]

Species	Savanna and grass land	Tropical forest	Extra-tropical forest	Biofuel burning	Charcoal burning	Agricultural residues
CO_2	1,613±95	1,580±90	1,569±131	1,550±95	2,611±241	1,515±177
CO	65±20	104±20	107±37	78±31	200±38	92±84
CH_4	2.3±0.9	6.8±2.0	4.7±1.9	6.1±2.2	6.2±3.3	2.7
NMHC	3.4±1.0	8.1±3.0	5.7±4.6	7.3±4.7	2.7±1.9	(7.0)
C_2H_2	0.29±0.27	0.21−0.59	0.27±0.09	0.51−0.9	0.05−0.13	(0.36)
C_2H_4	0.79±0.56	1.0−2.9	1.12±0.55	1.8±0.6	0.46±0.33	(1.4)
C_2H_6	0.32±0.16	0.5−1.9	0.6±0.15	1.2±0.6	0.53±0.48	(0.97)
C_3H_4	0.022±0.014	0.013	0.04−0.06	(0.024)	(0.06)	(0.032)
C_3H_6	0.26±0.14	0.55	0.59±0.16	0.5−1.9	0.13−0.56	(1.0)
C_3H_8	0.09±0.03	0.15	0.25±0.11	0.2−0.8	0.07−0.30	(0.52)
1-Butene	0.09±0.06	0.13	0.09−0.16	0.1−0.5	0.02−0.20	(0.13)
iso-Butene	0.03±0.012	0.11	0.05−0.11	0.1−0.5	0.01−0.16	(0.08)
2-Butene	0.045±0.014	0.09	0.018−0.18	0.1−0.48	0.02−0.09	(0.09)
n-Butane	0.019±0.09	0.041	0.069±0.038	0.03−0.13	0.02−0.10	(0.06)
Isoprene	0.020±0.012	0.016	0.10	0.15−0.42	0.017	(0.05)
Benzene	0.23±0.11	0.39−0.41	0.49±0.08	1.9±1.0	0.3−1.7	0.14
Toluene	0.13±0.06	0.21−0.29	0.40±0.10	1.1±0.7	0.08−0.61	0.026
Methanol	(1.3)	(2.0)	2.0±1.4	(1.5)	(3.8)	(2.0)
HCHO	0.26−0.44	(1.4)	2.2±0.5	0.13±0.05	(2.6)	(1.4)
CH_3CHO	0.5−0.39	(0.65)	0.48−0.52	0.14±0.05	(1.2)	(0.65)
Acetone	0.25−0.62	(0.62)	0.52−0.59	0.01−0.04	(1.2)	(0.63)
C_6H_5CHO	0.029	0.027	(0.036)	0.02−0.03	(0.07)	0.009
Furan	0.095	(0.48)	0.4−0.45	(0.65)	(0.9)	(0.5)
2-Mefuran	0.044−0.048	0.17	0.47	(0.18)	(0.46)	0.012
3-Mefuran	0.006−0.011	0.029	0.05	(0.023)	(0.06)	0.003
Furfural	(0.23)	(0.37)	0.29−0.63	0.22	(0.72)	(0.37)
HCOOH	(0.7)	(1.1)	2.9±2.4	0.13	(2.0)	0.22
CH_3COOH	(1.3)	(2.1)	3.8±1.8	0.4−1.4	(4.1)	0.8

H$_2$	0.97±0.38	3.6–4.0	1.8±0.5	(1.8)	(4.6)	(2.4)
NO$_x$ [b]	3.9±2.4	1.6±0.7	3.0±1.4	1.1±0.6	3.9	2.5±1.0
N$_2$O	0.21±0.10	(0.2)	0.26±0.07	0.06	(0.2)	0.07
NH$_3$	0.6–1.5	–	1.4±0.8	–	–	–
HCN	0.025–0.031	–	–	–	–	–
CH$_3$CN	0.11	–	0.19	–	–	–
SO$_2$	0.35±0.16	0.57±0.23	1.0	0.27±0.30	–	–
OCS	0.015±0.009	–	0.030–0.036	–	–	0.065±0.077
CH$_3$Cl	0.075±0.029	0.02–0.18	0.05±0.032	0.04–0.07	0.012	0.24±0.14
CH$_3$Br	0.0021±0.001	0.0078±0.0035	0.0032±0.0012	–	–	–
CH$_3$I	0.0005±0.0005	0.0068	0.0006	–	–	–

[a]Source of data: Andreae and Merlet (2001), reproduced (modified) with permission of the American Geophysical Union; values in parentheses are based on emission ratios to CO; values for charcoal production are CO$_2$ 440, CO 70, CH$_4$ 10.7, NMHC 2.0, CH$_3$OH 0.16, acetone 0.02, furfural 0.12, HCOOH 0.2, CH$_3$COOH 0.98, NO$_x$ 0.04, N$_2$O 0.03, NH$_3$ 0.09 (g kg^{-1}); *NMHC* non-methane hydrocarbons
[b]NO$_x$ is reported as NO

Table 4.20 Global emissions of gaseous species resulting from biomass burning (Tg a^{-1})[a]

Species	Savanna grassland	Tropical forests	Extra-tropical forests	Biofuel burning	Charcoal making	Charcoal burning	Agricultural residues[b]	Total
Dry matter burned	3,160	1,330	640	2,701	158	38	540	8,600
CO_2	5,096	2,101	1,004	4,187	70	99	818	13,400
CO	206	139	68	209	11	7.6	50	690
H_2	3.1	5.1	1.2	4.8	–	0.17	1.1	15.3
CH_4	7.4	9.0	3.0	16.5	1.7	0.24	1.5	39
C_2H_2	0.92	0.53	0.17	1.90	0.01	0.004	0.20	3.7
NMHC[c]	10.7	10.8	3.6	19.6	0.3	0.1	3.5	49
CH_3OH	3.8	2.6	1.3	3.9	0.02	0.14	0.9	12.7
HCHO	1.1	1.8	1.4	0.4	–	0.10	0.7	5.5
CH_3CHO	1.6	0.86	0.32	0.37	–	0.05	0.31	3.5
CH_3COCH_3	1.4	0.83	0.35	0.06	–	0.05	0.30	3.0
HCOOH	2.1	1.4	1.8	0.36	0.03	0.08	0.1	5.9
CH_3COOH	4.2	2.8	2.5	2.4	0.15	0.15	0.4	12.6
N_2	9.8	4.1	2.0	8.4	–	0.12	1.7	26.0
NO_x	12.2	2.2	1.9	2.9	0.01	0.15	1.3	20.7
N_2O	0.67	0.27	0.17	0.16	–	0.008	0.04	1.31
NH_3	3.4	1.7	0.88	3.5	0.01	0.05	0.7	10.3
HCN	0.09	0.20	0.10	0.41	0.02	0.006	0.08	0.9
CH_3CN	0.33	0.24	0.12	0.49	–	0.007	0.10	1.3
SO_2	1.1	0.76	0.64	0.74	–	0.015	0.22	3.5
COS	0.05	0.05	0.02	0.11	0.01	0.002	0.03	0.27
CH_3Cl	0.24	0.10	0.03	0.14	–	5 (–4)	0.13	0.65
CH_3Br	0.006	0.010	0.002	0.008	–	1.1 (–4)	0.002	0.029
CH_3I	0.0016	0.0090	4 (–4)	0.0027	–	4 (–5)	5 (–4)	0.014
Hg	3 (–4)	1 (–4)	6 (–5)	3 (–4)	–	–	1 (–4)	8 (–4)

[a]Source of data: Andreae and Merlet (2001), reproduced with permission of the American Geophysical Union; values in parentheses indicate powers of ten
[b]Values exclude agricultural waste used as bio-fuel
[c]NMHC non-methane hydrocarbons

Table 4.21 Summary of natural sources of sulfur in the atmosphere (Gmol a^{-1} of sulfur)[a]

Source	H_2S	CH_3SCH_3	CS_2	OCS	SO_2
Oceans	<9	500–1,300	2.4–9.5	2.8–7.8	–
Coastal wetlands	0.2–30	0.2–18	0.2–1.2	2.3–8.7	–
Soils and plants	2–56	3–24	0.4	–	–
Volcanoes	16–47	–	0.2–2.4	0.1–1.5	230–300
Biomass burning	–	–	–	0.7–4.3	81
Atmospheric reactions[b]	–	–	–	4.5–14.8	–
Sums	18–133	503–1,342	3.3–14.1	10.4–37.1	311–381

[a]Source: Warneck (2000), Reproduced with kind permission of Elsevier
[b]The reactions include $OH + CS_2$ and $OH + CH_3SCH_3$

Comment: Natural emissions of SO_2 are dwarfed by anthropogenic emissions; the EDGAR (2000) emissions dataset (see below) gives 150 Tg a^{-1}, or 2,344 Gmol a^{-1} globally, which is comparable to the data in Table 4.11 for the 1980s.

Table 4.22 Regional contributions to anthropogenic emissions of CO, NO_x (as nitrogen) and non-methane hydrocarbons (NMHC) for the year 2000 (Tg a^{-1})[a]

	CO		NO_x		NMHC	
Asia	256.93	(48.4)	11.18	(36.9)	57.00	(41.8)
Africa	80.39	(15.2)	1.58	(5.23)	14.03	(10.3)
North/Central America[b]	93.55	(17.6)	6.31	(20.8)	24.86	(18.2)
South America	22.32	(4.21)	1.26	(4.17)	8.14	(5.97)
Europe	69.20	(13.1)	6.53	(21.5)	30.51	(22.4)
Australia/Oceania	4.99	(0.94)	0.54	(1.77)	1.41	(1.03)
Sea/Oceans	3.00	(0.57)	2.90	(9.57)	0.42	(0.31)
Total emissions	530.38		30.30		136.35	

[a]Courtesy T. Butler, based on the EDGAR 2,000 emissions dataset Olivier and Berdowski (2001), Olivier et al. (2005); Numbers in parentheses indicate percent contributions
[b]Includes the Caribic Sea

Table 4.23 Contributions of transport, domestic and industrial sectors to the total anthropogenic emissions of CO, NO_x (as nitrogen) and NMHC (Tg a^{-1}) according to different inventories[a]

	CO		NO_x		NMHC	
Sector	Global	Megacities	Global	Megacities	Global	Megacities
EDGAR						
Transport	199.1 (37.5)	27.2 (51.3)	13.4 (44.2)	1.3 (40.3)	38.2 (28.1)	4.5 (37.9)
Domestic	265.0 (49.9)	17.5 (33.0)	3.1 (10.2)	0.3 (7.4)	28.4 (20.9)	1.9 (15.5)
Industrial	66.9 (12.6)	8.3 (15.7)	13.8 (45.6)	1.7 (52.3)	69.4 (51.0)	5.6 (46.5)
IPCC						
Transport	195.0 (41.4)	26.1 (60.7)	16.1 (57.8)	1.7 (53.8)	47.8 (41.2)	5.7 (44.2)
Domestic	238.7 (50.7)	13.5 (31.3)	1.5 (5.3)	0.1 (4.6)	28.9 (24.9)	1.8 (13.7)
Industrial	37.7 (8.0)	3.4 (7.9)	10.2 (36.9)	1.3 (41.6)	39.4 (34.0)	5.5 (42.1)
RETRO						
Transport	193.2 (40.5)	22.8 (48.6)	13.3 (48.2)	1.4 (46.5)	24.2 (15.9)	3.5 (19.2)
Domestic	279.5 (58.6)	23.6 (50.3)	3.3 (12.0)	0.4 (11.8)	26.4 (17.4)	2.2 (12.3)
Industrial	4.3 (0.9)	0.5 (1.0)	10.9 (39.8)	1.2 (41.8)	101.4 (66.7)	12.3 (68.5)

[a]Source of data: Butler et al. (2008); numbers in parentheses indicate percent contributions

Comments: The EDGAR (Emissions Database for Global Atmospheric Research) emissions inventory was developed by Olivier and Berdowski (2001), Olivier et al. (2005), and is available from http://www.mnp.nl/edgar/model/v32ft2000edgar; The IPCC inventory is described by Dentener et al. (2005) and has been used in the IPCC fourth Assessment report; it can be downloaded from ftp://ftp-ccu.jrc.it/pub/dentener/IPCC-AR4/2000; The RETRO inventory (RE analysis of the TROpospheric chemical composition over the past 40 years), which focuses on ozone precursors, was developed by Pulles et al. (2007) and can be downloaded from http://retro.enes.org.

Table 4.24 Contributions of transport, domestic and industrial sectors to anthropogenic emissions of CO, NO_x (as nitrogen) and non-methane hydrocarbons (NMHC) for the year 2000 (10^7 kg a^{-1}) in different countries[a]

CO

Country	Trans.	Dom.	Ind.	(%)[b]	Country	Trans.	Dom.	Ind.	(%)[b]
China	1,636	6,081	947	(16.3)	Canada	620	47.4	70.9	(1.39)
United States	6,059	653	478	(13.6)	Iran	523	117	57.2	(1.31)
India	480	5,110	294	(11.1)	Tanzania	5.28	575	50.4	(1.19)
Indonesia	480	1,391	95	(3.71)	Malaysia	225	349	38.0	(1.15)
Nigeria	185	1,455	268	(3.60)	South Korea	350	60.0	180	(1.11)
Russian Federation	193	804	381	(2.60)	Germany	245	147	185	(1.09)
Ukraine	845	157	267	(2.39)	Vietnam	61.2	478	16.0	(1.05)
Japan	546	18.0	568	(2.13)	Saudi Arabia	340	18.4	102	(0.87)
Sudan	9.91	894	191	(2.06)	Italy	294	39.2	103	(0.82)
Brazil	555	135	271	(1.81)	United Kingdom	303	42.5	87.8	(0.82)
Mexico	707	165	49.9	(1.74)	France	255	72.3	102	(0.81)
Pakistan	69.7	751	34.0	(1.61)	Australia	209	16.9	180	(0.77)
Bangladesh	14.0	736	43.1	(1.50)					

NO_x

Country	Trans.	Dom.	Ind.	(%)[b]	Country	Trans.	Dom.	Ind.	(%)[b]
United States	219	22.5	245	(16.1)	Germany	28.2	3.97	22.1	(1.79)
China	62.9	54.2	296	(13.6)	Ukraine	23.5	3.23	27.4	(1.79)
Sea/Oceans	287	2.42	0.64	(9.56)	South Korea	23.0	2.25	28.7	(1.78)
India	44.1	50.7	96.7	(6.32)	Mexico	29.3	2.27	19.1	(1.67)
Russian Federation	17.9	14.9	113	(4.80)	Australia	18.1	0.50	27.0	(1.51)
Japan	43.0	2.82	37.8	(2.76)	Indonesia	16.3	13.2	13.6	(1.42)
Spain	56.1	1.06	17.9	(2.48)	South Africa	13.0	2.66	26.8	(1.40)
Canada	41.0	2.82	22.5	(2.19)	France	22.8	3.23	12.3	(1.27)
United Kingdom	29.2	3.23	28.5	(2.01)	Italy	21.2	2.55	13.4	(1.23)
Brazil	38.0	2.09	17.0	(1.88)	Iran	19.0	3.63	13.5	(1.19)

Poland	8.06	1.17	2.27	(1.07)
Taiwan	8.69	1.05	18.6	(0.94)
Turkey	11.6	0.81	14.8	(0.90)
NMHC				
United States	672	75.2	964	(12.5)
China	191	627	332	(8.43)
India	167	600	129	(6.57)
Russian Federation	42	53.8	695	(5.80)
Japan	186	0.98	317	(3.69)
Ukraine	404	11.4	76.7	(3.61)
Saudi Arabia	61.0	2.06	417	(3.52)
Indonesia	167	160	114	(3.23)
Nigeria	33.0	157	165	(2.60)
Canada	133	5.46	200	(2.48)
Mexico	124	19.8	190	(2.45)
Brazil	118	15.7	160	(2.15)
United Kingdom	67.7	2.50	214	(2.09)

Thailand	14.0	1.22	10.7	(0.86)
Saudi Arabia	14.2	1.88	7.93	(0.79)
Iran	71.2	13.56	194	(2.05)
Germany	35.3	9.11	142	(1.37)
France	68.6	7.22	102	(1.31)
Italy	77.3	4.41	96.1	(1.30)
Venezuela	51.2	0.07	179	(1.69)
Malaysia	79.4	41.0	79.8	(1.47)
Norway	7.22	0.62	168	(1.29)
Iraq	21.6	5.83	121	(1.09)
South Korea	87.9	5.76	54.1	(1.08)
Taiwan	45.7	9.97	84.1	(1.03)
Pakistan	20.5	88.3	18.4	(0.93)
Sea/Oceans	3.55	31.5	6.51	(0.31)

[a] Courtesy N. Butler, based on the EDGAR 2000 emissions dataset, Olivier and Berdowski (2001), Olivier et al. (2005); Countries are listed in the order of decreasing total emissions; *Trans.* Transport, *Dom.* Domestic, *Ind.* Industrial

[b] Percent of global emissions

References

Andreae, M.O., P. Merlet, Glob. Biogeochem. Cycles **15**, 955–966 (2001)

Butler, T.M., M.G. Lawrence, B.R. Gurjar, J. van Aardenne, M. Schultz, J. Lelieveld, Atmos. Environ. **42**, 703–719 (2008)

Dentener, F., D. Stevenson, J. Cofala, R. Mechler, M. Amann, P. Bergamaschi, F. Raes, R. Derwent, Atmos. Chem. Phys. **5**, 1731–1755 (2005)

Guenther, A., C.N. Hewitt, D. Erickson, R. Fall, C. Geron, T. Graedel, P. Hartley, L. Klinger, M. Lerdau, A. McKay, T. Pierce, B. Scholes, R. Steinbrecher, R. Tallamraju, J. Taylor, P. Zimmerman, J. Geophys. Res. **100**, 8873–8892 (1995)

Guenther, A., T. Karl, P. Harley, C. Wiedinmyer, P.I. Palmer, C. Geron, Atmos. Chem. Phys. **6**, 3181–3210 (2006)

Olivier, J.G.J., J.J.M. Berdowski, Global emissions, sources and sinks, in *The Climate System*, ed. by J. Berdowski, R. Guicherit, B.J. Heij (A. A. Balkema Publishers/Swets & Zeitlinger Publishers, Lisse, 2001), pp. 33–78

Olivier, J.G.J., J.A. van Aardenne, F. Dentener, L. Ganzeveld, J.A.H.W. Peters, Environ. Sci. **2**, 81–99 (2005)

Pulles, T., M. ven het Bolscher, R. Brand, A. Visschedijk, *Assessment of Global Emissions from Fuel Combustion in the Final Decades of the 20th Century; Application of the Emission Inventory Model TEAM*, Technical Report A-R0132B. Netherlands Organisation for Applied Research (TNO), Apeldoorn, The Netherlands (2007)

Warneck, P., *Chemistry of the Natural Atmosphere*, 2nd edn. (Academic Press, San Diego, 2000), Copyright Elsevier

Yokelson, R.J., T.J. Christian, T.G. Karl, A. Guenther, Atmos. Chem. Phys. **8**, 3509–3527 (2008)

4.4   Concentrations of Oxidizing Species

Table 4.25 The vertical distribution of ozone: mid-latitude model distribution of atmospheric ozone[a]

Altitude z (km)	Number density n_s (cm^{-3})	Variability[b] σ (%)	Partial pressure p_s (hPa)	Mixing ratio[c] n_s/n	Mixing ratio[c] ρ_s/ρ	Overhead column abundance α (mol m^{-2})	DU[d]
2	6.8 (11)	56	2.6 (−5)	3.26 (−8)	5.4 (−8)	1.539 (−1)	345
4	5.8 (11)	50	2.1 (−5)	3.38 (−8)	5.6 (−8)	1.516 (−1)	340
6	5.7 (11)	53	1.9 (−5)	4.10 (−8)	6.8 (−8)	1.497 (−1)	335
8	6.5 (11)	90	2.1 (−5)	5.97 (−8)	9.9 (−8)	1.478 (−1)	331
10	1.13 (12)	109	3.5 (−5)	1.32 (−7)	2.18 (−7)	1.457 (−1)	327
12	2.02 (12)	78	6.0 (−5)	3.11 (−7)	5.16 (−7)	1.419 (−1)	318
14	2.35 (12)	63	7.0 (−5)	4.95 (−7)	8.21 (−7)	1.352 (−1)	303
16	2.95 (12)	48	8.8 (−5)	8.51 (−7)	1.41 (−6)	1.274 (−1)	285
18	4.04 (12)	30	1.21 (−4)	1.60 (−6)	2.65 (−6)	1.176 (−1)	264
20	4.77 (12)	21	1.43 (−4)	2.58 (−6)	4.27 (−6)	1.042 (−1)	234
22	4.86 (12)	17	1.47 (−4)	3.62 (−6)	6.00 (−6)	8.836 (−2)	198
24	4.54 (12)	14	1.38 (−4)	4.69 (−6)	7.77 (−6)	7.223 (−2)	162
26	4.03 (12)	14	1.24 (−4)	5.67 (−6)	9.39 (−6)	5.715 (−2)	128
28	3.24 (12)	14	1.00 (−4)	6.16 (−6)	1.02 (−5)	4.377 (−2)	98.1
30	2.52 (12)	13	7.88 (−5)	6.58 (−6)	1.09 (−5)	3.306 (−2)	74.1
32	2.03 (12)	17	6.40 (−5)	7.18 (−6)	1.19 (−5)	2.469 (−2)	55.3
34	1.58 (12)	17	5.10 (−5)	7.66 (−6)	1.27 (−5)	1.794 (−2)	40.2
36	1.22 (12)	14	4.03 (−5)	8.09 (−6)	1.34 (−5)	1.269 (−2)	28.4
38	8.73 (11)	13	2.95 (−5)	7.84 (−6)	1.30 (−5)	8.635 (−3)	19.4
40	6.07 (11)	13	2.10 (−5)	7.30 (−6)	1.21 (−5)	5.735 (−3)	12.9
42	3.98 (11)	11	1.40 (−5)	6.40 (−6)	1.06 (−5)	3.719 (−3)	8.33
44	2.74 (11)	18	9.89 (−6)	5.84 (−6)	9.67 (−6)	2.398 (−3)	5.37
46	1.69 (11)	21	6.23 (−6)	4.74 (−6)	7.86 (−6)	1.489 (−3)	3.34
48	1.03 (11)	17	3.85 (−6)	3.76 (−6)	6.23 (−6)	9.269 (−4)	2.08
50	6.64 (10)	17	2.48 (−6)	3.11 (−6)	5.15 (−6)	5.852 (−4)	1.31
52	3.84 (10)	18	1.43 (−6)	2.29 (−6)	3.80 (−6)	3.648 (−4)	0.818
54	2.55 (10)	27	9.28 (−7)	1.92 (−6)	3.18 (−6)	2.373 (−4)	0.532
56	1.61 (10)	32	5.74 (−7)	1.56 (−6)	2.58 (−6)	1.527 (−4)	0.342
58	1.12 (10)	26	3.90 (−7)	1.36 (−6)	2.25 (−6)	9.937 (−5)	0.223
60	7.33 (9)	34	2.50 (−7)	1.13 (−6)	1.88 (−6)	6.216 (−5)	0.139
62	4.81 (9)	38	1.60 (−7)	9.60 (−7)	1.59 (−6)	3.778 (−5)	0.085
64	3.17 (9)	38	1.03 (−7)	8.21 (−7)	1.36 (−6)	2.179 (−5)	0.049
66	1.72 (9)	38	5.5 (−8)	5.80 (−7)	9.6 (−7)	1.129 (−5)	0.025
68	7.5 (8)	68	2.4 (−8)	3.32 (−7)	5.5 (−7)	5.583'(−6)	0.013
70	5.4 (8)	57	1.6 (−8)	3.08 (−7)	5.1 (−7)	3.083 (−6)	0.007
72	2.2 (8)	77	6.5 (−9)	1.69 (−7)	2.8 (−7)	1.292 (−6)	0.003
74	1.7 (8)	63	4.9 (−9)	1.75 (−7)	2.9 (−7)	5.422 (−7)	0.001

[a] Source: US Standard Atmosphere (COESA 1976); Powers of ten are shown in parentheses
[b] The natural variability is defined by $\sigma = 100 \Delta n_s/n_s$, where n_s is the average number density of ozone
[c] Chemical amount fraction n_s/n and mass fraction ρ_s/ρ; where n and ρ are the total number density and the density of air, respectively
[d] The Dobson Unit is defined as a 10^{-5} m thick layer of ozone at $T = 273.15$ K and $p = 101.325$ hPa; 1 DU = 4.4615 µmol m^{-2}

Table 4.26 Distribution of ozone concentrations in the troposphere (nmol mol^{-1})[a]

P (hPa)	1,000	700	500	300	200	150	1,000	700	500	300	200	150
z (km)	0	3	5.5	9.5	12.5	14	0	3	5.5	9.5	12.5	14
	January						April					
85–75°N	27.8	41.7	48.2	110.3	391.0	752.7	22.6	48.2	55.3	162.7	681.5	1,039
75–65°N	24.4	39.1	46.2	95.3	364.5	670.0	26.2	53.1	60.6	184.0	673.9	932.2
65–55°N	26.3	43.6	49.7	92.3	402.5	676.1	35.7	57.1	64.9	130.4	515.2	731.9
55–45°N	28.1	44.8	52.9	90.0	293.3	542.7	40.2	58.0	65.7	109.2	414.6	617.4
45–35°N	24.9	42.0	46.5	79.9	203.4	338.2	41.4	55.1	59.2	88.3	298.3	447.2
35–25°N	22.7	41.1	44.8	57.7	105.0	131.3	39.4	55.7	60.5	76.8	124.4	190.0
25–15°N	19.8	40.0	45.7	47.8	59.3	80.5	28.1	44.1	52.8	56.9	65.7	87.1
15°–5°N	19.7	38.3	43.3	46.3	56.7	78.1	25.4	39.8	48.0	52.7	61.6	76.7
5°N–5°S	16.2	33.3	42.9	40.5	50.6	63.5	13.9	30.7	43.0	46.3	48.3	63.8
5°–15°S	13.9	32.3	44.5	41.1	55.0	74.6	14.2	31.3	42.4	43.4	45.1	58.8
15–25°S	14.0	33.9	44	47.7	67.9	87.3	16.7	37.3	45.9	51.8	57.3	64.5
25–35°S	20.3	33.8	45	54.2	79.3	103.7	18.5	31.6	38.4	49.8	85	116.7
35–45°S	18.0	26.6	38	54.4	123.3	174.0	16.3	27.8	35.0	46.9	99.2	173.6
45–55°S	12.5	25.5	35.3	55.1	143.9	199.5	13.3	31.1	39.9	51.4	102.2	215.3
55–65°S	13.1	19.8	29.5	72.3	251.1	356.5	21.3	28.9	36.5	61.1	184	327.4
65–75°S	22.4	23.2	32.3	82.1	284.2	420.6	27.9	27.5	35.3	73.1	238.7	387.5
75–85°S	27.3	27.3	31.5	92.3	290.1	427.5	26.8	26.8	33.7	83.4	256	420.7
	July						October					
85–75°N	21.5	40.4	57.9	162.7	462.1	539.1	27.4	39.4	50.2	120.0	290.1	503.5
75–65°N	21.2	45.6	59.3	153.6	427.3	533.8	26.9	43.4	50.5	97.5	273.0	458.1
65–55°N	24.6	48.7	67.3	110.7	355.4	470.1	26.1	46.6	55.7	92.2	221.3	362.3
55–45°N	38.4	53.1	65.8	94.7	260.5	377.0	28.7	50.3	56.7	74.3	144.5	257.3
45–35°N	41.7	54.5	62.6	75.8	127.3	203.0	29.1	45.1	50.2	57.2	107.9	179.2
35–25°N	25.9	52.3	55.5	50.8	64.3	90.9	24.9	40.6	47.2	47.3	64.4	83.8
25–15°N	16.1	43.7	42.8	40.3	50.3	71.8	18.1	36.6	44.6	41.4	51.6	66.2
15°–5°N	16.1	43.3	41.0	40.5	48.8	68.4	18.8	36.6	44.7	44.1	53.2	67.0
5°N–5°S	21.5	39.0	44.4	51.7	65.5	80.1	20.8	43.0	52.8	55.0	60.7	69.0
5°–15°S	22.6	32.9	43.8	46.0	60.8	75.9	21.7	45.5	56.6	54.5	59.3	69.0
15–25°S	22.9	40.4	46.1	50.2	66.7	84.4	21.0	48.7	63.5	76.4	76.4	90.9
25–35°S	16.0	36.2	41.4	50.7	114.5	172.7	20.4	43.5	57.1	79.0	113.5	158.0
35–45°S	14.9	32.9	37.3	56.5	205.2	308.0	20.0	36.0	44.4	76.8	228.3	360.9
45–55°S	14.5	35.7	39.2	67.6	255.7	367.6	18.3	36.1	43.4	76.8	238.4	381.9
55–65°S	21.5	34.9	38.1	55.6	246.4	435.0	26.6	38.9	46.5	78.7	268.3	433.9
65–75°S	30.7	32.1	37.5	49.0	172.1	405.3	36.9	39.3	48.8	76.5	251.4	416.9
75–85°S	33.2	33.2	38.0	49.6	166.4	395.4	34.2	34.2	41.2	63.7	206.1	344.6

[a]Source of data: Fortuin and Kelder (1998)

Table 4.27 Zonally and monthly averaged OH concentrations (10^5 molecule cm^{-3}) as a function of altitude, latitude and season in the troposphere[a]

P (hPa)	January							April						
	1,000	900	800	700	500	300	200	1,000	900	800	700	500	300	200
90°N	–	–	–	–	–	–	–	0.4	0.6	0.8	1.0	0.8	0.9	1.3
84°N	–	–	–	–	–	–	–	0.3	0.5	0.6	0.8	1.0	1.1	1.5
76°N	–	–	–	–	–	–	–	0.5	0.7	0.9	1.4	1.4	1.6	2.1
68°N	0.0	0.0	0.0	0.0	0.0	0.0	0.0	3.9	3.5	3.4	3.2	2.2	2.3	2.7
60°N	0.2	0.2	0.2	0.2	0.2	0.2	0.2	3.8	4.1	5.0	4.4	3.5	3.2	3.3
52°N	0.4	0.4	0.8	0.6	0.6	0.7	0.8	4.9	5.3	7.1	6.0	5.1	4.2	4.0
44°N	0.7	0.9	1.8	1.4	1.6	1.6	1.7	7.8	8.7	11.0	8.8	7.7	5.6	5.0
36°N	1.7	2.1	3.6	3.3	3.7	3.5	3.5	9.0	11.9	13.8	12.7	11.3	7.5	6.4
28°N	4.3	5.1	6.2	7.1	6.4	4.5	4.6	12.4	14.8	17.1	18.1	14.8	7.9	6.9
20°N	7.3	9.2	11.4	12.2	10.0	6.4	5.9	15.1	18.9	21.4	23.7	17.8	9.6	8.0
12°N	10.0	13.7	16.3	16.0	14.3	8.5	7.2	14.3	19.4	22.1	23.5	20.4	10.4	8.4
4°N	7.0	11.2	15.4	18.5	20.7	11.2	9.2	7.7	14.1	18.8	22.4	23.5	12.3	10.0
4°S	7.1	10.9	15.5	20.2	23.0	12.8	9.6	7.1	13.4	16.0	20.8	22.3	12.7	10.6
12°S	8.9	13.1	18.4	22.8	26.1	14.2	11.1	7.3	13.6	15.9	19.1	19.8	12.0	10.7
20°S	10.2	16.0	24.5	26.6	25.6	14.6	11.2	8.7	12.8	16.8	18.4	16.7	11.0	9.6
28°S	10.5	15.8	24.4	26.1	24.3	14.8	11.4	7.9	10.4	12.9	14.1	13.6	9.7	8.6
36°S	9.6	13.8	18.4	21.1	21.1	14.7	11.8	5.8	6.8	8.0	9.6	10.5	8.0	6.9
44°S	6.2	8.1	11.0	16.2	18.9	14.2	11.4	2.7	3.1	3.9	5.5	7.5	6.1	5.6
52°S	4.1	5.1	7.4	12.5	15.5	12.7	10.2	1.1	1.4	1.7	2.8	4.3	4.0	3.4
60°S	3.0	3.7	5.2	9.9	12.0	11.0	9.3	0.4	0.5	0.7	1.3	2.1	2.1	1.8
68°S	4.5	7.2	7.3	8.6	8.9	9.6	9.0	0.2	0.3	0.3	0.5	0.7	0.8	0.7
76°S	4.0	5.0	5.0	6.8	7.3	8.5	8.9	0.1	0.1	0.1	0.1	0.2	0.1	0.2
84°S	4.7	5.0	5.0	6.4	6.6	7.9	8.9	0.0	0.0	0.0	0.1	0.1	0.0	0.0
90°S	–	–	–	6.5	6.6	7.7	8.6	–	–	–	–	–	–	–

(continued)

Table 4.27 (continued)

P (hPa)	July							October						
	1,000	900	800	700	500	300	200	1,000	900	800	700	500	300	200
90°N	5.5	7.4	8.3	10.0	9.7	6.9	8.8	–	–	–	–	–	–	–
84°N	3.8	5.2	6.4	8.6	10.9	7.7	9.5	0.0	0.0	0.0	0.0	0.0	0.0	0.0
76°N	3.7	5.1	6.9	9.7	11.5	7.6	8.4	0.1	0.1	0.1	0.1	0.1	0.1	0.1
68°N	7.4	11.6	13.9	15.4	13.2	8.2	7.7	0.5	0.5	0.5	0.6	0.5	0.4	0.6
60°N	6.2	10.6	16.2	17.0	15.3	9.2	8.0	0.8	0.9	1.2	1.2	1.5	1.3	1.5
52°N	6.2	11.8	19.5	19.4	17.5	10.6	8.6	1.7	1.9	2.9	2.6	3.1	2.8	2.9
44°N	12.3	23.1	26.9	25.1	22.1	12.6	10.0	4.4	4.7	6.5	5.3	5.7	4.9	5.1
36°N	14.9	24.7	27.9	28.0	25.0	14.4	11.3	6.3	8.3	10.1	9.2	9.0	7.3	7.5
28°N	12.6	17.9	22.7	24.7	23.3	12.3	8.8	9.4	12.0	13.5	13.2	11.3	7.0	6.5
20°N	13.0	17.6	21.9	26.5	24.8	13.7	10.0	10.1	13.9	15.3	16.0	14.6	8.8	8.0
12°N	10.9	16.4	20.1	24.4	24.8	13.6	10.0	9.9	15.3	17.4	19.1	20.1	10.6	8.8
4°N	10.3	14.6	19.2	22.3	23.2	12.9	10.2	8.4	13.5	19.1	23.7	25.1	12.4	9.9
4°S	10.4	15.4	20.4	21.3	20.1	12.4	10.9	11.4	18.5	24.5	28.7	28.0	14.3	11.2
12°S	9.9	13.8	17.8	17.8	15.9	10.5	9.9	13.7	19.5	25.5	29.6	26.7	13.9	11.1
20°S	7.9	10.9	12.7	13.1	11.3	8.6	7.8	13.1	19.5	23.8	27.1	22.2	13.6	10.4
28°S	4.5	5.6	6.2	7.1	7.1	5.8	5.8	9.2	13.0	16.0	19.3	17.2	10.6	9.1
36°S	2.6	2.8	3.2	4.0	4.6	3.7	3.3	8.0	9.4	11.0	13.5	13.4	8.9	7.4
44°S	0.9	1.0	1.2	1.9	2.6	2.2	1.9	4.9	5.6	7.0	10.2	10.6	7.2	6.1
52°S	0.3	0.3	0.4	0.7	1.1	1.0	0.9	3.0	3.5	4.5	6.9	7.5	6.0	5.3
60°S	0.1	0.1	0.1	0.2	0.3	0.2	0.2	2.7	2.9	3.2	4.5	5.4	5.6	5.1
68°S	0.0	0.0	0.0	0.0	0.0	0.0	0.0	2.7	3.4	3.3	3.7	3.6	5.3	5.2
76°S	–	–	–	–	–	–	–	1.2	1.6	1.0	1.7	1.9	3.4	3.4
84°S	–	–	–	–	–	–	–	0.0	0.9	0.9	1.0	1.3	2.3	2.2
90°S	–	–	–	–	–	–	–	–	–	–	0.8	1.1	1.7	1.6

[a] Source of data: Spivakovsky et al. (2000)

Table 4.28   Average concentration of OH radicals (10^5 molecule cm^{-3})[a]

	32–90°N	0–32°N	NH	0–32°S	32–90°S	SH	Global
January	1.5	10.6	7.0	17.8	12.1	15.3	11.3
July	16.5	17.9	17.3	12.3	1.8	8.0	12.7
Annual	7.3	14.7	11.6	15.4	6.4	11.6	11.6

Source of data: Spivakovsky et al. (2000) reproduced with permission from AGU
[a] Integrated with respect to the mass of air from the surface to 100 hPa within 0°–32° latitude and to 200 hPa outside that region

Table 4.29   Summary of reported NO_3 and N_2O_5 measurements (pmol mol^{-1})[a, b]

Method	NO_3	N_2O_5	Location/Date	Ref[c]
Urban polluted				
LP-DOAS	0–355		Los Angeles basin, California (Aug-Sep 79), GS	(1)
LP-DOAS	0–78		Deuselbach, Julich, Germany (Apr-Aug 80) GS	(2)
LP-DOAS	0–80		Riverside, California (Sep 83) GS	(3)
SS-DOAS	0–90		Cambridge, England (Jul-Oct 95) GS	(4)
CRDS	0–15	0–200	Boulder, Colorado (Apr 01) GS	(5)
CRDS	0–70	0–3,000	Boulder, Colorado (Oct–Nov 01) GS	(6)
LP-DOAS	5–31		Houston, Texas (Aug–Sep 00) GS	(7)
LP-DOAS, MP	0–100		Houston, Texas (Aug–Sep 00) GS	(8)
CIMS		0–600	Boulder, Colorado (Oct 02) GS	(9)
SS-DOAS	40–360		Heidelberg, Germany (Apr–Aug 99) GS	(10)
LIF		0–200	San Francisco Bay Area, Calif. (Jan 04) GS	(11)
LIF		0–160	Tokyo, Japan (Dec 03) GS	(12)
LP-DOAS	0–175		Heidelberg, Germany (Jun 04) GS	(13)
CRDS	0–380	0–3,100	Northeast U.S.A. (Jul–Aug 04) Aircraft	(14)
LP-DOAS/MP	0–220		Phoenix, Arizona (Jun 01) GS	(15)
CRDS	0–10	0–200	Tampa, Florida (Mar 04) Aircraft	(16)
CIMS		0–40	Mexico City (Mar 06) GS	(17)
CRDS	0–400	0–3,800	Houston, Texas (Sep–Oct 06) Aircraft	(18)
CRDS, CIMS	0–60	0–2,000	Boulder, Colorado (Feb 09) GS	(19)
Lunar/LP-DOAS	0–250		Table Mountain, Calif. (May 03–Sep04) GS	(20)
LP-DOAS	0–800		Jerusalem, Israel (Jul 05–Sep 07) GS	(21)
CRDS	0–80	0–1,800	Erie, Colorado (Oct 04) tall tower	(22)
BBCEAS		3–166 av 55–798 max	London, England (Oct–Nov 07) tall tower	(23)
LP-DOAS/MP	0–150		Houston, Texas (Sep 06) GS	(24)

(continued)

Table 4.29 (continued)

Method	NO$_3$	N$_2$O$_5$	Location/Date	Ref^c
Rural semi-polluted				
Lunar/LP-DOAS	0–70		Fritz Peak, Colorado (Aug–Oct 79) GS	(25)
LP-DOAS	0–50		Fritz Peak, Colorado (Oct 84) GS	(26)
LP-DOAS	0–80		San Joaquin Valley, Calif. (Jul–Aug 90) GS	(27)
LP-DOAS	0–20		Key Biscane, Florida (Jul 89) GS	(28)
SS-DOAS	0–20		Fritz Peak, Colorado (Aug–Oct 93) GS	(29)
LP-DOAS	0–70 (10 av)		Pabstthum, Germany (Jul–Aug 98) GS	(30)
LP-DOAS	5 av		Lindenberg, Germany (Feb–Sep 98) GS	(30)
LP-DOAS	0–34		Fraser Valley, British Columbia (Aug 01) GS	(31)
CRDS	0.5	0–15	Northeast U.S.A. (Jul–Aug 04) Aircraft	(32)
LIF	3–23	3–24	Tokyo, Japan (Jun 04) GS	(33)
CRDS	0–140	0–1,600	U.S. East Coast (Jul–Aug 02) Ship	(34)
CRDS		0–4	U.S. East Coast (Jul–Aug 04) Ship	(35)
LP-DOAS	0–70		Appledore Island, N. H. (Jul 04) GS	(36)
LIF		0–20	Toyokawa, Japan (Feb 06) GS	(37)
CRDS	0–40	0–800	U.S. Gulf Coast (Aug–Sep 06) Ship	(38)
CRDS	0–70	0–600	U.S. East Coast (Jul–Aug 04) Ship	(39)
CIMS		0–100	U.S. East Coast (Mar 06) Ship	(40)
CRDS	0–50	0–500	Kleiner Feldberg, Germany (May 08) GS	(41)
LP-DOAS	0–70		Saturna Island, B.C. (Jul–Aug 05) GS	(42)
CRDS		0–80	Fairbanks, Alaska (Nov 09) GS	(43)
Less Polluted				
Lunar/LP-DOAS	0.3		Mauna Loa, Hawaii (Nov 81) GS	(44)
MI-ESR	0–10		Schauinsland, Germany (Aug 90) GS	(45)
LP-DOAS	8 av, 80 max		Rugen, Germany (Apr 93– Jun 94) GS	(46)
LP-DOAS	8 av, 20 max		Tenerife, Canary Islands (May 94) GS	(47)
LP-DOAS	10 av, 25 max		North Norfolk, U. K. (Spring, Autumn 94) GS	(47)
LP-DOAS	4–25		North Norfolk, U. K. (Winter 94, Sum 95) GS	(48)
LP-DOAS	1–40		Helgoland, Germany (Oct 96) GS	(49)
LP-DOAS	1–40		Mace Head, Ireland (Summer 96 and 97) GS	(50)
LP-DOAS	1–20		Tenerife, Canary Islands (Summer 97) GS	(50)
SS-DOAS	8–40		England, Ireland, Tenerife, Norway, Australia, various (96–00) GS	(51)
CRDS		0–20	Fairbanks, Alaska (Dec 03) GS	(52)
LP-DOAS	0–36 (2.2 av)		Crete, Greece (Jul–Aug 01) GS	(53)

(continued)

Table 4.29   (continued)

Method	NO_3	N_2O_5	Location/Date	Ref[c]
LP-DOAS	1–6 av		Crete, Greece (Jun 01–Sep 03) GS	(54)
LP-DOAS[d]	0–25	0–25	Mace Head, Ireland (Jul–Aug 02) GS	(55)
CRDS		0–250	Fairbanks, Alaska (Dec 02, Feb 04) GS	(56)
CRDS		0–80	Fairbanks, Alaska (Nov 07) GS	(57)
LP-DOAS	0–72		Roscoff, France (Aug–Sept 06) GS	(58)

[a] The data were kindly provided by Steven S. Brown, NOAA
[b] Abbreviations: *DOAS* Differential Optical Absorption Spectroscopy, *LP-DOAS* Long Path DOAS (artificial light source), *SS-DOAS* Long Path DOAS (scattered sunlight as light source), *Lunar/ LP-DOAS* Long Path DOAS (Moon light as light source), *MI-ESR* Matrix Isolation Electron Spin Resonance, *CRDS* Cavity Ring Down Spectroscopy, *BBCRDS* Broad Band CRDS, *BBCEAS* Broad Band Cavity Enhanced Absorption Spectroscopy, *LIF* Laser Induced Fluorescence, *CIMS* Chemical Ionization Mass Spectrometry, *MP* multiple path, *GS* ground site, *av* average
[c] References: (1) Platt et al. (1980), (2) Platt et al. (1981), (3) Pitts et al. (1984), (4) Aliwell and Jones (1998), (5) Brown et al. (2001), (6) Brown et al. (2003), (7) Geyer et al. (2003), (8) Stutz et al. (2004), (9) Slusher et al. (2004), (10) von Friedeburg, et al. (2002), (11) Wood et al. (2005), (12) Matsumoto et al. (2005), (13) Kern et al. (2006), (14) Brown et al. (2006), (15) Wang et al. (2006), (16) Dubé et al. (2006), (17) Zheng et al. (2008), (18) Brown et al. (2009) (19) Thornton et al. (2010), (20) Chen et al. (2011), (21) Asaf et al. (2010), (22) Brown et al. (2007), (23) Benton et al. (2010), (24) Stutz et al. (2010), (25) Noxon et al. (1980), (26) Perner et al. (1985) (27) Smith et al. (1995), (28) Yvon et al. (1996), (29) Weaver et al. (1996), (30) Geyer et al. (2001a, b), (31) McLaren et al. (2004), (32) Brown et al. (2005), (33) Matsumoto et al. (2006), (34) Aldener et al. (2006), (35) Osthoff et al. (2006), (36) Ambrose et al. (2007), (37) Nakayama et al. (2008), (38) Osthoff et al. (2008), (39) Sommariva et al. (2009), (40) Kercher et al. (2009), (41) Crowley et al. (2010), (42) McLaren et al. (2010), (43) Huff et al. (2010) (44) Noxon (1983), (45) Mihelcic et al. (1993), (46) Heintz et al. (1996), (47) Carslaw et al. (1997a, b), (48) Allan et al. (1999) (49) Martinez et al. (2000), (50) Allan et al. (2000). (51) Allan et al. (2002), (52) Simpson (2003), (53) Vrekoussis et al. (2004), (54) Vrekoussis et al. (2006), (55) Saiz-Lopez et al. (2006), (56) Ayers and Simpson (2006), (57) Apodaca et al. (2008), (58) Mahajan et al. (2009)
[d] The BBCEAS technique was also applied

Comments: NO_3 is formed primarily via the slow reaction of NO_2 with O_3, and it can react further with NO_2 to form N_2O_5. The three species NO_2, NO_3 and N_2O_5 are nearly always in thermal equilibrium. By day, NO_3 rapidly undergoes photolysis. In urban conditions, direct losses of NO_3 occur through reaction with NO and anthropogenic unsaturated volatile organic compounds, with further indirect losses to sulfate aerosol via the uptake of N_2O_5. In rural conditions direct NO_3 loss occurs via reactions with unsaturated biogenic organic compounds (e.g. limonene). Under less polluted conditions NO_3 may have a longer lifetime, although in marine environments a rapid reaction with dimethyl sulfide may occur and indirect loss via N_2O_5 uptake to sea salt aerosol can also be significant.

References

Aldener, M., S.S. Brown, H. Stark, E.J. Williams, B.M. Lerner, W.C. Kuster, P.D. Goldan, P.K. Quinn, T.S. Bates, F.C. Fehsenfeld, A.R. Ravishankara, J. Geophys. Res. 111, D23, D23S73 (2006). doi:10.1029/2006JD007252
Aliwell, S.R., R.L. Jones, J. Geophys. Res. 103, 5719–5727 (1998)
Allan, B.J., N. Carslaw, H. Coe, R.A. Burgess, J.M.C. Plane, J. Atmos. Chem. 33, 129–154 (1999)
Allan, B., G. McFiggans, J.M.C. Plane, H. Coe, G. McFadyen, J. Geophys. Res. 105, 24191–24204 (2000)
Allan, B.J., J.M.C. Plane, H. Coe, J. Shillito, J. Geophys. Res. 107, 4588 (2002). doi:10.1029/2002JD002112
Ambrose, J.L., H. Mao, H.R. Mayne, J. Stutz, R. Talbot, B.C. Sive, J. Geophys. Res. 112, D21, D21302 (2007). doi:10.1029/2007JD008756
Apodaca, R.L., D.M. Huff, W.R. Simpson, Atmos. Chem. Phys. 8, 7451–7463 (2008)
Asaf, D., E. Tas, D. Pederesen, M. Peleg, M. Luria, Environ. Sci. Technol. 44(15), 5901–5907 (2010)
Ayers, J.D., W.R. Simpson, J. Geophys. Res. 111, D14, D14309 (2006). doi:10.1029/2006JD007070
Benton, A.K., J.M. Landridge, S.M. Ball, W.J. Bloss, M. Dall'Osto, E. Nemitz, R.M. Harrison, R.L. Jones, Atmos. Chem. Phys. 10, 9781–9795 (2010)
Brown, S.S., H. Stark, S.J. Ciciora, A.R. Ravishankara, Geophys. Res. Lett. 28, 3227–3230 (2001)
Brown, S.S., H. Stark, T.B. Ryerson, E.J. Williams, D.K. Nicks, M. Trainer, F.C. Fehsenfeld, A.R. Ravishankara, J. Geophys. Res. 108, 4299 (2003). doi:10.1029/2002JD002917
Brown, S.S., H.D. Osthoff, H. Stark, W.P. Dubé, T.B. Ryerson, C. Warneke, J.A. de Gouw, A.G. Wollny, D.D. Parrish, F.C. Fehsenfeld, A.R. Ravishankara, J. Photochem. Photobiol. A 176(1–3), 270–278 (2005)
Brown, S.S., T.B. Ryerson, A.G. Wollny, C.A. Brock, R. Peltier, A.P. Sullivan, R.J. Weber, W.P. Dubé, M. Trainer, J.F. Meagher, F.C. Fehsenfeld, A.R. Ravishankara, Science 311, 67–70 (2006)
Brown, S.S., W.P. Dubé, H.D. Osthoff, D.E. Wolfe, W.M. Angevine, A.R. Ravishankara, Atmos. Chem. Phys. 7, 139–149 (2007)
Brown, S.S., W.P. Dubé, H. Fuchs, T.B. Ryerson, A.G. Wollny, C.A. Brock, R. Bahreini, A.M. Middlebrook, A.J. Neuman, E. Atlas, J.M. Roberts, H.D. Osthoff, M. Trainer, F.C. Fehsenfeld, A.R. Ravishankara, J. Geophys. Res. 114, D00F10 (2009). doi:10.1029/2008JD011679
Carslaw, N., J.M.C. Plane, H. Coe, E. Cuevas, J. Geophys. Res. 102, 10613–10622 (1997)
Carslaw, N., L.J. Carpenter, J.M.C. Plane, B.J. Allan, R.A. Burgess, K.C. Clemitshaw, H. Coe, S.A. Penkett, J. Geophys. Res. 102, 18917–18933 (1997)
Chen, C.M., R.P. Cageao, L. Lawrence, J. Stutz, R.J. Salawitch, L. Jourdain, Q. Li, S.P. Sander, Atmos. Chem. Phys. 11, 963–978 (2011)
Committee on Extension of US Standard Atmosphere (COESA) US Standard Atmosphere 1976, (Washington, DC, 1976)
Crowley, J.N., G. Schuster, N. Pouvesle, U. Parchatka, H. Fischer, B. Bonn, H. Bingemer, J. Lelieveld, Atmos. Chem. Phys. 10, 2795–2812 (2010)
Dubé, W.P., S.S. Brown, H.D. Osthoff, M.R. Nunley, S.J. Ciciora, M.W. Paris, R.J. McLaughlin, A.R. Ravishankara, Rev. Sci. Instrum. 77, 034101 (2006)
Fortuin, P.J.F., H. Kelder, J. Geophys. Res. 103, 31709–31734 (1998)
Geyer, A., R. Ackermann, R. Dubois, B. Lohrmann, R. Müller, U. Platt, Atmos. Environ. 35, 3619–3631 (2001)
Geyer, A., B. Alicke, S. Konrad, T. Schmitz, J. Stutz, U. Platt, J. Geophys. Res. 106, 8013–8025 (2001)
Geyer, A., B. Alicke, R. Ackermann, M. Martinez, H. Harder, W. Brune, P. di Carlo, E. Williams, T. Jobson, S. Hall, R. Shetter, J. Stutz, J. Geophys. Res. 108, 4368 (2003). doi:10.1029/2002JD002967

Heintz, F., U. Platt, H. Flentje, R. Dubois, J. Geophys. Res. **101**, 22891–22910 (1996)

Huff, D.M., P.L. Joyce, G.J. Fochesatto, W.R. Simpson, Atmos. Chem. Phys. Discuss. **10**, 25329–25354 (2010)

Kercher, J.P., T.P. Riedel, J.A. Thornton, Atmos. Meas. Tech. **2**, 193–204 (2009)

Kern, C., S. Trick, B. Rippel, U. Platt, Appl. Opt. **45**, 2077–2088 (2006)

Mahajan A.S., H. Oetjen, A. Saiz-Lopez, J.D. Lee, G.B. McFiggans, J.M.C. Plane, Geophys. Res. Lett. **36**, L16803 (2009). doi:10.1029/2009GL038018

Martinez, M., D. Perner, E.M. Hackenthal, S. Kulzer, L. Schutz, J. Geophys. Res. **105**, 22685–22695 (2000)

Matsumoto, J., H. Imai, N. Kosugi, Y. Kaji, Atmos. Environ. **39**, 6802–6811 (2005)

Matsumoto, J., K. Imagawa, H. Imai, N. Kosugi, M. Ideguchi, S. Kato, Y. Kajii, Atmos. Environ. **40**, 6294–6302 (2006)

McLaren, R., R.A. Salmon, J. Liggio, K.L. Hayden, K.G. Anlauf, W.R. Leaitch, Atmos. Environ. **38**, 5837–5848 (2004)

McLaren, R., P. Wojtal, D. Majonis, J. McCourt, J.D. Halla, J. Brook, Atmos. Chem. Phys. **10**, 4187–4206 (2010)

Mihelcic, D., D. Klemp, P. Musgen, H.W. Patz, A. Volzthomas, J. Atmos. Chem. **16**, 313–335 (1993)

Nakayama, T., T. Ide, F. Taketani, M. Kawai, K. Takahashi, Y. Matsumi, Atmos. Environ. **42**, 1995–2006 (2008)

Noxon, J.F., J. Geophys. Res. **88**, 11017–11021 (1983)

Noxon, J.F., R.B. Norton, E. Marovich, Geophys. Res. Lett. **7**, 125–128 (1980)

Osthoff, H.D., R. Sommariva, T. Baynard, A. Pettersson, E.J. Williams, B.M. Lerner, J.M. Roberts, H. Stark, P.D. Goldan, W.C. Kuster, T.S. Bates, D. Coffman, A.R. Ravishankara, S.S. Brown, J. Geophys. Res. **111**, D23S14 (2006). doi:10.1029/2006JD007593

Osthoff, H.D., J.M. Roberts, A.R. Ravishankara, E.J. Williams, B.M. Lerner, R. Sommariva, T.S. Bates, D. Coffman, P.K. Quinn, J.E. Dibb, H. Stark, J.B. Burkholder, R.K. Talukdar, J. Meagher, F.C. Fehsenfeld, S.S. Brown, Nat. Geosci. **1**, 324–328 (2008)

Perner, D., A. Schmeltekopf, R.H. Winkler, H.S. Johnston, J.G. Calvert, C.A. Cantrell, W.R. Stockwell, J. Geophys. Res. **90**, 3807–3812 (1985)

Platt, U., D. Perner, A.M. Winer, G.W. Harris, J.N. Pitts Jr., Geophys. Res. Lett. **7**, 89–92 (1980)

Platt, U., D. Perner, J. Schröder, C. Kessler, A. Toennissen, J. Geophys. Res. **86**, 11965–11970 (1981)

Pitts Jr., J.N., H.W. Biermann, R. Atkinson, A.M. Winer, Geophys. Res. Lett. **11**, 557–560 (1984)

Saiz-Lopez, A., J.A. Shillito, H. Coe, J.M.C. Plane, Atmos. Chem. Phys. **6**, 1513–1528 (2006)

Simpson, W.R., Rev. Sci. Instrum. **74**, 3442–3452 (2003)

Slusher, D.L., L.G. Huey, D.J. Tanner, F.M. Flocke, J.M. Roberts, J. Geophys. Res. **109**, D19, D19315 (2004). doi:10.1029/2004JD004670

Smith, N., J.M.C. Plane, C.F. Nien, P.A. Solomon, Atmos. Environ. **29**, 2887–2897 (1995)

Sommariva, R., H.D. Osthoff, S.S. Brown, T.S. Bates, T. Baynard, D. Coffman, J.A. de Gouw, P.D. Goldan, W.C. Kuster, B.M. Lerner, H. Stark, C. Warneke, E.J. Williams, F.C. Fehsenfeld, A.R. Ravishankara, M. Trainer, Atmos. Chem. Phys. **9**, 3075–3093 (2009)

Spivakovsky, C.M., J.A. Logan, S.A. Montzka, Y.J. Balkanski, M. Foreman-Fowler, D.B.A. Jones, L.W. Horowitz, A.C. Fusco, C.A.M. Brenninkmeijer, M.J. Prather, S.C. Wofsy, M.B. McElroy, J. Geophys. Res. **105**, 8931–8980 (2000)

Stutz, J., B. Alicke, R. Ackermann, A. Geyer, A. White, E. Williams, J. Geophys. Res. **109**, D12306 (2004). doi:10.1029/2003JD004209

Stutz, J., K.W. Wong, L. Lawrence, L. Ziemba, J.H. Flynn, B. Rappenglück, B. Lefer, Atmos. Environ. **44**(33), 4099–4106 (2010)

Thornton, J.A., J.P. Kercher, T.P. Riedel, N.L. Wagner, J. Cozic, J.S. Holloway, W.P. Dubé, G.M. Wolfe, P.K. Quinn, A.M. Middlebrook, B. Alexander, S.S. Brown, Nature **464**, 271–274 (2010)

von Friedeburg, C., T. Wagner, A. Geyer, N. Kaiser, B. Vogel, H. Vogel, U. Platt, J. Geophys. Res. **107**, D13, 4168 (2002). doi:10.1029/2001JD000481

Vrekoussis, M., M. Kanakidou, N. Mihalopoulos, P.J. Crutzen, J. Lelieveld, D. Perner, H. Berresheim, E. Baboukas, Atmos. Chem. Phys. **4**, 169–182 (2004)

Vrekoussis, M., E. Liakakou, N. Mihalopoulos, M. Kanakidou, P.J. Crutzen, J. Lelieveld, Geophys.
    Res. Lett. **33**, L05811 (2006). doi:10.1029/2005GL025069
Wang, S., R. Ackermann, J. Stutz, Atmos. Chem. Phys. **6**, 2671–2693 (2006)
Weaver, A., S. Solomon, R.W. Sanders, K. Arpag, H.L. Miller, J. Geophys. Res. **101**, 18605–18612
    (1996)
Wood, E.C., T.H. Bertram, P.J. Wooldridge, R.C. Cohen, Atmos. Chem. Phys. **5**, 483–491 (2005)
Yvon, S.A., J.M.C. Plane, C.F. Nien, D.J. Cooper, E.S. Saltzman, J. Geophys. Res. **101**, 1379–
    1386 (1996)
Zheng, J., R. Zhang, E.C. Fortner, R.M. Volkamer, L. Molina, A.C. Aiken, J.L. Jimenez, K. Gaeggeler,
    J. Dommen, S. Dusanter, P.S. Stevens, X. Tie, Atmos. Chem. Phys. **8**, 6823–6838 (2008)

4.5   Trace Gases in the Stratosphere

Satellite Data: Measured stratospheric number densities of ozone, O_3, nitrogen dioxide, NO_2, and bromine oxide, BrO, from the SCIAMACHY limb observations are given for the altitudes 20 km, 25 km and 30 km, as a function of latitude and season (January, April, July and October) for 2005. Pressure and temperature are provided for the calculation of mixing ratios. The SCIAMACHY observations are performed on a sun synchronous orbit, with a local solar time of approx. 10.00 am for the equator crossing. For low and mid-latitudes the local time is slightly different: observations are performed at approx. 10.30 am for 30°N, 9.30 am for 30°S, and approx. 11.00 am for 60°N and 9.00 am for 60°S. For higher latitudes it differs significantly. For comparison to other observations and simulations the actual local time needs to be considered since NO_2 and BrO show a strong diurnal cycle. The retrieval uncertainty for the altitudes of 20, 25 and 30 km is estimated to be about 5%, 5%, 5–10% for O_3, 20%, 10–20%, 10% for NO_2 and 25%, 30% and 40% for BrO. For information on the SCIAMACHY instrument and limb measurements the reader is referred to Bovensmann et al. (1999). For details on the retrieval algorithms see: Rozanov et al. (2005), Kühl et al. (2008), Pukite et al. (2010). The data were kindly provided: for O_3 by A. Rozanov, C. v Savigny, IUP Bremen; for NO_2 by A. Rozanov, IUP Bremen; for BrO by S. Kühl, J. Pukite, T. Wagner, MPI Chemie, Mainz.

Simulated data: Simulated stratospheric mixing ratios of selected trace gases are given in order to provide a more comprehensive global suite of data. The data for the subsequent tables are from the ECHAM/MESSy Atmospheric Chemistry (EMAC) model. ECHAM: Röckner et al. (2006); MESSy (Modular Earth Submodel System): Jöckel et al. (2005, 2010). This model describes atmospheric chemistry and meteorological processes (here using a 90 layer model up to 0.01 hPa) and has performed well in recent model comparisons (Gettelman et al. 2010). The species O_3, N_2O, HNO_3, H_2O, CH_4, CO_2, SF_6, $CFCl_3$, ClO, HCl, $ClONO_2$ and temperature are given at four altitudes: 15, 18, 25, 40 km, and as a function of latitude and season. The tables are based only on 1 year, which was chosen arbitrarily (2006). The altitude was estimated from the barometric equation using surface density and pressure. The data were interpolated onto a 0.5 km vertical grid, provided courtesy of Peter Hoor (University Mainz) and Patrick Jöckel (DLR), and assembled by Janeen Auld (Max-Planck-Institut für Chemie).

Table 4.30 Zonal mean ozone column densities (DU)[a]

Φ	Jan	Feb	Mar	April	May	June	July	Aug	Sep	Oct	Nov	Dec
80° N	–	–	470	465	414	371	332	308	291	–	–	–
75° N	–	433	460	462	416	370	332	308	302	299	–	–
70° N	–	436	459	455	415	368	334	313	308	309	314	–
65° N	395	432	451	444	410	367	338	320	312	315	332	–
60° N	392	428	441	431	406	372	346	327	317	317	332	358
55° N	390	426	433	421	402	375	350	330	318	317	327	353
50° N	387	418	420	410	394	372	346	326	313	312	322	349
45° N	376	402	401	395	382	360	335	319	307	302	311	338
40° N	354	374	377	373	363	341	321	310	300	291	297	320
35° N	322	338	347	348	342	323	310	303	295	283	284	299
30° N	292	303	316	325	324	311	302	298	290	280	276	281
25° N	269	278	291	304	307	301	296	291	284	275	270	267
20° N	254	261	271	287	291	290	289	286	279	270	263	256
15° N	248	251	260	275	279	282	284	283	279	268	261	252
10° N	246	246	254	267	271	275	280	281	279	267	260	251
5° N	247	248	254	261	264	268	274	277	278	263	258	251
0°	251	250	255	259	260	263	268	273	276	263	259	253
5° S	255	254	257	258	258	259	262	268	272	265	264	257
10° S	260	258	259	259	257	256	259	264	270	269	270	264
15° S	266	262	261	260	258	258	261	266	273	277	278	272
20° S	271	265	264	263	264	264	268	274	282	287	286	279
25° S	277	270	269	271	271	273	279	288	295	301	298	287
30° S	286	278	277	278	281	289	295	306	313	317	311	297
35° S	295	286	284	284	291	306	315	327	333	336	323	307
40° S	306	294	289	289	303	319	331	343	348	354	335	318
45° S	319	303	296	297	312	327	340	353	360	371	350	332
50° S	334	313	305	306	318	328	342	355	367	387	366	347
55° S	344	322	312	314	322	328	338	351	368	402	381	358
60° S	344	325	315	318	323	337	344	339	353	402	390	365
65° S	338	324	317	319	322	–	340	325	324	374	388	366
70° S	331	317	312	313	–	–	–	307	291	333	376	364
75° S	324	306	305	302	–	–	–	294	267	297	357	358
80° S	320	299	299	–	–	–	–	–	253	274	346	356

[a] Four year average of data (1978–1982) from the TOMS/instrument on Nimbus 7, Keating et al. (1989); Dobson unit: 1 DU = 446.149 μmol m^{-2}

Table 4.31a Number densities of O_3, NO_2 and BrO at 20 km from satellite measurements

z 20 km	O_3 (10^{12} molecule cm^{-3})				NO_2 (10^8 molecule cm^{-3})				BrO (10^6 molecule cm^{-3})			
	Jan	Apr	Jul	Oct	Jan	Apr	Jul	Oct	Jan	Apr	Jul	Oct
80° N	–	5.6	3.7	2.9	–	14.4	24.6	4.2	–	13.5	14.4	21.1
75° N	–	5.5	3.9	3.8	–	13.1	23.2	3.2	–	18.9	15.6	29.3
70° N	–	5.5	3.9	4.1	1.2	11.7	21.5	2.9	19.2	21.1	16.2	31.1
65° N	5.6	5.3	3.9	4.0	1.3	10.5	19	3.3	25.4	22.6	15.9	29.6
60° N	5.6	5.3	3.8	3.9	1.3	9.8	15.3	4.0	28.5	23.3	16.2	27.5
55° N	5.2	5.1	3.7	3.8	1.5	9.5	12.7	4.5	28.9	24.4	17.2	25.7
50° N	4.9	4.8	3.5	3.8	1.9	9.3	10.8	4.8	27.5	23.4	16.4	23.5
45° N	4.7	4.6	3.2	3.5	2.5	8.9	9.2	5.1	28.0	22.0	15.3	21.7
40° N	4.4	4.3	2.9	3.0	3.1	8.6	8.4	5.2	26.8	20.8	15.2	19.5
35° N	4.1	3.6	2.6	2.6	3.6	7.9	7.8	5.1	23.5	18.7	13.9	17.0
30° N	3.6	3.1	2.4	2.3	3.9	7.6	7.5	5.0	20.6	16.0	11.8	14.4
25° N	2.9	2.6	2.2	2.0	4.3	7.2	6.9	4.9	16.1	13.4	10.5	12.0
20° N	2.3	2.1	2.2	1.9	4.4	5.7	6.0	4.0	13.2	11.4	10.8	11.2
15° N	1.8	1.7	2.0	1.9	4.2	4.7	5.3	3.2	11.2	10.4	10.3	10.6
10° N	1.6	1.6	1.9	1.8	3.1	3.6	3.9	2.5	9.4	8.6	8.8	9.4
5° N	1.6	1.6	1.9	1.8	2.3	2.2	2.7	1.9	8.1	7.1	8.4	8.5
0	1.6	1.6	1.8	1.7	1.7	1.5	2.0	1.6	8.2	7.0	8.6	8.8
5° S	1.7	1.6	1.7	1.6	1.6	1.4	1.7	1.6	8.3	7.1	8.3	8.6
10° S	1.7	1.6	1.6	1.6	1.7	1.4	1.8	1.8	8.8	7.5	8.8	9.4
15° S	1.7	1.6	1.7	1.8	2.1	1.6	1.9	1.9	9.3	8.0	9.9	10.8
20° S	1.8	1.6	2.2	2.4	2.8	1.9	2.0	2.3	10.6	9.1	12.6	12.5
25° S	1.8	1.8	2.7	3.0	4.0	2.6	2.7	4.8	11.8	11.7	14.8	15.1
30° S	2.1	2.2	3.2	3.6	5.0	3.4	3.3	6.0	12.3	13.5	17.7	17.3
35° S	2.3	2.5	3.8	3.9	6.5	4.1	3.3	6.5	13.9	15.4	20.5	19.7
40° S	2.6	2.9	4.3	4.4	7.8	4.5	2.9	6.8	16.1	18.6	23.6	21.9
45° S	3.0	3.3	4.5	4.6	9.1	4.9	2.6	7.0	17.7	21.7	26.5	23.1
50° S	3.4	3.6	4.7	4.7	10.5	4.9	2.2	7.8	19.4	24.0	28.5	24.2
55° S	3.9	3.9	5.0	4.7	11.9	4.6	1.9	9.1	21.4	25.9	29.2	25.2
60° S	4.3	4.2	5.4	4.4	13.6	4.1	2.0	10.1	21.3	27.5	29.4	24.5
65° S	4.5	4.4	–	3.8	17.3	3.7	2.5	11.2	21.2	28.3	27.6	24.3
70° S	4.6	4.6	–	3.0	23.2	3.4	–	10.6	20.7	28.1	24.0	23.4
75° S	4.5	–	–	2.1	25.0	3.6	–	9.4	20.8	27.0	–	20.9
80° S	4.4	–	–	1.6	26.6	3.0	–	9.2	20.2	26.5	–	15.9

Table 4.31b Pressure and temperature at 20 km from satellite measurements

z 20 km	Pressure (Pa)				Temperature (K)			
	Jan	Apr	Jul	Oct	Jan	Apr	Jul	Oct
80° N	4,063	5,246	5,709	4,878	192	226	229	213
75° N	4,177	5,238	5,704	4,947	195	225	228	214
70° N	4,342	5,228	5,692	5,026	199	223	227	215
65° N	4,544	5,224	5,680	5,105	204	222	225	216
60° N	4,751	5,227	5,668	5,177	209	220	224	216

(continued)

Table 4.31b (continued)

z 20 km	Pressure (Pa)				Temperature (K)			
	Jan	Apr	Jul	Oct	Jan	Apr	Jul	Oct
55° N	4,940	5,239	5,660	5,244	213	218	222	216
50° N	5,087	5,257	5,653	5,302	215	217	219	215
45° N	5,187	5,279	5,638	5,349	216	216	216	213
40° N	5,250	5,306	5,620	5,389	215	214	213	212
35° N	5,286	5,327	5,589	5,412	213	211	210	210
30° N	5,297	5,332	5,543	5,410	210	209	209	209
25° N	5,296	5,324	5,480	5,390	207	206	207	208
20° N	5,292	5,315	5,431	5,363	205	204	207	206
15° N	5,287	5,307	5,394	5,334	203	202	206	205
10° N	5,285	5,302	5,367	5,309	203	202	205	204
5° N	5,286	5,302	5,353	5,293	203	202	205	203
0	5,286	5,303	5,348	5,287	203	202	204	203
5° S	5,285	5,303	5,355	5,295	203	202	205	203
10° S	5,285	5,304	5,368	5,313	203	202	207	205
15° S	5,296	5,316	5,381	5,335	203	203	208	206
20° S	5,316	5,333	5,385	5,352	203	204	209	208
25° S	5,340	5,349	5,380	5,367	204	206	210	209
30° S	5,360	5,357	5,367	5,382	206	209	212	212
35° S	5,372	5,349	5,338	5,389	209	211	214	214
40° S	5,376	5,319	5,271	5,378	213	213	216	217
45° S	5,373	5,266	5,153	5,331	217	215	215	219
50° S	5,366	5,189	4,960	5,226	221	216	213	220
55° S	5,358	5,090	4,688	5,060	225	216	207	220
60° S	5,356	4,974	4,372	4,842	228	215	200	220
65° S	5,359	4,846	4,070	4,598	230	214	193	218
70° S	5,379	4,727	3,833	4,365	232	212	188	216
75° S	5,400	4,614	3,661	4,166	234	209	184	211
80° S	5,414	4,515	3,552	4,021	235	207	181	208

Table 4.32a   Number densities of O_3, NO_2 and BrO at 25 km from satellite measurements

z 25 km	O_3 (10^{12} molecule cm^{-3})				NO_2 (10^8 molecule cm^{-3})				BrO (10^6 molecule cm^{-3})			
	Jan	Apr	Jul	Oct	Jan	Apr	Jul	Oct	Jan	Apr	Jul	Oct
80° N	–	3.7	2.4	1.6	–	17.6	22.6	11.0	–	11.6	10.3	8.1
75° N	–	3.7	2.7	2.3	–	16.4	23.1	9.1	–	11.1	10.2	10.5
70° N	–	3.7	2.9	3.0	3.2	15.4	23.4	8.2	10.2	10.9	9.7	11.0
65° N	3.1	3.7	3.2	3.2	3.6	14.8	23.6	8.9	11.9	10.7	9.7	11.4
60° N	3.5	3.8	3.6	3.4	4.0	14.9	22.0	9.7	12.2	10.3	9.9	11.3
55° N	3.9	3.9	3.9	3.6	4.6	15.3	20.6	10.2	12.4	10.2	10.2	11.3
50° N	4.1	4.0	4.1	3.8	4.9	15.4	19.8	10.9	12.4	10.0	10.4	11.0
45° N	4.2	4.2	4.2	3.9	5.4	15.4	18.8	11.7	12.0	10.0	10.3	10.8
40° N	4.3	4.3	4.4	4.0	6.0	15.6	18.1	12.1	11.6	9.9	10.0	10.6
35° N	4.3	4.4	4.5	4.2	6.5	15.9	17.1	11.6	10.8	9.9	9.8	10.2

(continued)

Table 4.32a (continued)

z 25 km	O$_3$ (10^{12} molecule cm^{-3})				NO$_2$ (10^8 molecule cm^{-3})				BrO (10^6 molecule cm^{-3})			
	Jan	Apr	Jul	Oct	Jan	Apr	Jul	Oct	Jan	Apr	Jul	Oct
30° N	4.5	4.6	4.6	4.3	7.4	15.4	16.5	10.8	10.9	9.9	9.9	10.2
25° N	4.6	4.6	4.6	4.3	8.3	13.9	14.9	9.6	10.6	9.9	10.0	10.1
20° N	4.6	4.6	4.4	4.3	8.2	12.1	13.2	8.5	10.2	9.6	9.9	10.2
15° N	4.5	4.4	4.4	4.3	7.1	10.3	11.2	7.2	10.1	9.6	9.7	10.0
10° N	4.2	4.3	4.4	4.4	6.1	8.6	8.6	6.1	9.8	9.5	9.9	10.0
5° N	3.9	4.1	4.3	4.3	5.6	6.7	6.4	5.2	9.7	9.6	9.8	9.8
0	3.9	4.0	4.2	4.3	5.3	5.5	5.4	4.8	9.8	9.6	9.9	9.8
5° S	4.0	4.1	4.2	4.3	5.6	5.2	4.9	4.9	9.7	9.7	9.8	9.8
10° S	4.2	4.2	4.2	4.2	6.7	5.1	4.9	5.7	9.9	9.7	9.9	9.7
15° S	4.4	4.4	4.2	4.2	8.2	5.6	5.5	7.8	10.1	10.0	10.2	9.8
20° S	4.5	4.5	4.4	4.4	10.6	6.5	7.4	12.3	9.9	10.2	10.4	10.1
25° S	4.5	4.6	4.5	4.5	13.1	8.0	8.9	16.7	10.0	10.6	11.0	10.2
30° S	4.5	4.5	4.5	4.5	15.6	9.5	8.5	17.0	10.0	10.8	11.4	10.4
35° S	4.2	4.3	4.4	4.2	17.7	10.7	7.5	16.2	10.5	11.0	11.6	10.7
40° S	3.9	4.0	4.2	4.1	19.0	11.3	6.6	15.2	10.5	11.1	11.9	10.8
45° S	3.8	3.8	3.9	4.1	20.1	11.4	6.1	14.7	10.7	11.5	12.3	11.0
50° S	3.7	3.6	3.7	3.9	20.9	11.4	6.4	14.8	10.6	11.6	12.4	10.9
55° S	3.7	3.4	3.3	3.8	22.1	10.8	6.5	14.8	11.0	11.7	11.2	11.1
60° S	3.6	3.3	3.3	3.8	23.7	9.9	6.4	15.1	10.7	11.5	10.8	11.1
65° S	3.3	3.1	–	3.5	27.0	9.4	6.5	15.7	10.4	11.1	12.1	10.9
70° S	3.0	3.0	–	3.2	28.6	9.4	–	16.4	10.2	10.1	11.7	10.6
75° S	2.7	–	–	2.9	27.8	10.7	–	17.3	10.2	9.5	–	9.9
80° S	2.5	–	–	2.6	27.2	12.6	–	18.1	9.8	10.0	–	9.3

Table 4.32b Pressure and temperature at 25 km from satellite measurements

z 25 km	Pressure (Pa)				Temperature (K)			
	Jan	Apr	Jul	Oct	Jan	Apr	Jul	Oct
80° N	1,693	2,474	2,730	2,173	196	226	231	210
75° N	1,761	2,459	2,722	2,214	198	225	230	212
70° N	1,858	2,441	2,709	2,259	201	223	229	213
65° N	1,976	2,424	2,693	2,307	204	222	228	215
60° N	2,100	2,411	2,677	2,350	208	220	227	216
55° N	2,216	2,405	2,661	2,387	212	219	225	217
50° N	2,309	2,408	2,644	2,416	215	218	224	217
45° N	2,372	2,418	2,626	2,439	218	218	223	217
40° N	2,408	2,433	2,607	2,459	219	219	222	217
35° N	2,426	2,446	2,586	2,473	219	219	221	217
30° N	2,429	2,451	2,561	2,475	219	220	221	218
25° N	2,424	2,448	2,528	2,465	219	220	220	218
20° N	2,419	2,439	2,495	2,448	219	220	219	218
15° N	2,411	2,425	2,465	2,425	218	219	218	217
10° N	2,399	2,408	2,441	2,402	217	217	217	216
5° N	2,387	2,395	2,428	2,390	216	216	216	216
0	2,381	2,390	2,424	2,387	215	216	216	216
5° S	2,385	2,396	2,432	2,393	215	216	216	216

(continued)

Table 4.32b (continued)

z 25 km	Pressure (Pa)				Temperature (K)			
	Jan	Apr	Jul	Oct	Jan	Apr	Jul	Oct
10° S	2,397	2,410	2,445	2,405	217	217	217	216
15° S	2,413	2,428	2,460	2,420	218	219	218	216
20° S	2,430	2,446	2,469	2,432	219	220	218	216
25° S	2,447	2,459	2,471	2,443	219	220	219	216
30° S	2,463	2,464	2,466	2,457	220	220	219	217
35° S	2,478	2,459	2,450	2,470	221	219	218	217
40° S	2,495	2,440	2,410	2,476	222	219	217	219
45° S	2,513	2,409	2,337	2,467	224	218	215	220
50° S	2,532	2,366	2,221	2,433	226	218	211	222
55° S	2,549	2,312	2,059	2,370	229	216	206	225
60° S	2,566	2,249	1,872	2,283	231	215	199	227
65° S	2,584	2,180	1,696	2,180	233	213	193	230
70° S	2,605	2,115	1,563	2,076	234	211	188	232
75° S	2,622	2,053	1,474	1,978	235	209	186	233
80° S	2,632	1,999	1,421	1,898	235	208	184	232

Table 4.33a Number densities of O_3, NO_2 and BrO at 30 km from satellite measurements

z 30 km	O_3 (10^{12} molecule cm^{-3})				NO_2 (10^8 molecule cm^{-3})				BrO (10^6 molecule cm^{-3})			
	Jan	Apr	Jul	Oct	Jan	Apr	Jul	Oct	Jan	Apr	Jul	Oct
80° N	–	1.7	1.5	0.6	–	13.3	18.1	9.1	–	5.5	5.3	4.6
75° N	–	1.7	1.7	1.0	–	13.5	18.8	9.1	–	5.3	5.0	4.4
70° N	–	1.9	1.9	1.4	3.8	13.7	19.9	9.2	3.8	5.1	4.9	4.7
65° N	1.3	2	2.1	1.6	4.7	14.0	20.9	10.3	4.3	4.6	5.0	4.9
60° N	1.5	2.1	2.3	1.8	5.8	14.5	21.0	11.4	4.7	4.5	5.2	4.7
55° N	1.7	2.2	2.5	2.0	7.5	15.0	20.5	12.5	5.0	4.4	5.2	4.8
50° N	1.8	2.3	2.7	2.2	8.6	15.5	20.3	13.5	5.1	4.5	5.3	4.8
45° N	2.1	2.5	2.9	2.3	9.9	16.1	19.9	14.2	5.1	4.5	5.2	4.9
40° N	2.2	2.8	3.1	2.5	11.0	16.8	19.8	14.7	4.9	4.7	4.9	4.8
35° N	2.3	3.1	3.3	2.9	11.6	16.7	19.3	14.9	4.8	4.8	4.9	5.0
30° N	2.4	3.2	3.5	3.1	12.4	15.8	17.8	14.2	4.6	5.0	5.1	5.1
25° N	2.6	3.4	3.6	3.4	12.9	14.7	16.4	13.3	4.8	5.0	5.1	5.1
20° N	2.8	3.6	3.7	3.7	12.5	13.4	15.3	12.4	4.5	5.0	5.1	5.1
15° N	3.2	3.8	3.8	3.7	11.6	12.2	13.6	11.8	4.8	5.2	5.2	5.2
10° N	3.6	4.1	3.9	3.9	10.6	11.3	12.0	11.7	5.0	5.2	5.2	5.3
5° N	3.8	4.2	3.9	3.8	9.9	10.0	10.7	12.0	5.1	5.3	5.2	5.2
0	3.9	4.1	3.9	3.9	10.0	9.1	10.4	12.6	5.3	5.3	5.1	5.2
5° S	3.9	4.1	3.9	4.0	10.7	9.2	10.2	12.2	5.0	5.3	5.1	5.1
10° S	3.7	4.0	3.9	4.0	11.7	9.9	10.2	11.3	5.2	5.4	5.3	5.0
15° S	3.6	3.8	3.8	4.1	13.6	11.2	10.1	11.3	5.1	5.4	5.3	5.1
20° S	3.4	3.6	3.5	3.7	16.8	12.8	10.8	14.0	5.1	5.2	5.3	5.1
25° S	3.2	3.3	3.1	3.3	18.5	14.1	11.8	16.0	5.1	5.3	5.1	5.0
30° S	3.1	2.9	2.8	3.1	19.2	14.9	11.8	16.0	4.9	5.3	5.2	5.1
35° S	2.9	2.7	2.6	2.8	19.4	15.0	11.4	15.9	5.2	5.2	5.1	5.1
40° S	2.7	2.4	2.4	2.7	19.9	15.0	10.8	15.4	5.3	5.1	5.2	5.3
45° S	2.6	2.2	2.1	2.5	20.0	14.3	10.4	14.7	5.5	5.2	5	5.1

(continued)

Table 4.33a (continued)

z 30 km	O$_3$ (10^{12} molecule cm^{-3})				NO$_2$ (10^8 molecule cm^{-3})				BrO (10^6 molecule cm^{-3})			
	Jan	Apr	Jul	Oct	Jan	Apr	Jul	Oct	Jan	Apr	Jul	Oct
50° S	2.4	2.0	1.8	2.4	20.2	13.9	9.6	14.1	5.5	4.8	4.6	5.2
55° S	2.3	1.8	1.5	2.3	20.1	12.9	8.4	13.5	5.4	4.8	4.2	5.2
60° S	2.2	1.6	1.6	2.3	20.8	11.8	6.8	12.8	5.0	4.5	3.4	5.2
65° S	2.0	1.5	–	2.1	21.6	10.9	4.9	12.6	5.1	4.1	3.3	5.0
70° S	1.8	1.4	–	1.9	20.8	10.5	–	12.2	5.1	3.8	3.2	4.7
75° S	1.6	–	–	1.7	19.8	10.9	–	12.2	4.9	3.5	–	4.1
80° S	1.5	–	–	1.5	19.5	10.5	–	12.5	4.9	3.4	–	3.7

Table 4.33b Pressure and temperature at 30 km from satellite measurements

z 30 km	Pressure (Pa)				Temperature (K)			
	Jan	Apr	Jul	Oct	Jan	Apr	Jul	Oct
80° N	743	1,162	1,329	968	210	224	237	211
75° N	777	1,154	1,323	992	211	224	237	212
70° N	824	1,143	1,313	1,020	212	224	236	214
65° N	882	1,135	1,302	1,050	213	224	235	216
60° N	945	1,128	1,290	1,078	215	224	235	218
55° N	1,008	1,126	1,277	1,103	217	224	234	220
50° N	1,061	1,129	1,264	1,122	219	225	233	222
45° N	1,102	1,137	1,250	1,136	222	226	232	223
40° N	1,129	1,149	1,236	1,147	225	228	231	223
35° N	1,142	1,160	1,223	1,154	226	229	230	224
30° N	1,145	1,167	1,211	1,156	226	230	229	224
25° N	1,143	1,169	1,196	1,155	226	231	229	224
20° N	1,140	1,166	1,182	1,152	226	231	229	226
15° N	1,136	1,158	1,170	1,150	226	231	229	229
10° N	1,131	1,150	1,160	1,149	226	231	230	231
5° N	1,127	1,143	1,156	1,149	226	230	230	233
0	1,125	1,140	1,154	1,149	227	230	231	233
5° S	1,127	1,142	1,154	1,149	227	230	229	232
10° S	1,133	1,149	1,156	1,147	227	230	228	230
15° S	1,142	1,158	1,160	1,147	227	230	226	228
20° S	1,151	1,167	1,164	1,146	228	231	226	227
25° S	1,161	1,171	1,164	1,148	229	230	226	226
30° S	1,172	1,170	1,160	1,152	230	229	226	225
35° S	1,182	1,163	1,148	1,157	231	228	224	225
40° S	1,195	1,150	1,119	1,161	233	226	221	225
45° S	1,207	1,130	1,068	1,159	235	225	217	226
50° S	1,220	1,101	995	1,150	236	222	212	228
55° S	1,234	1,064	904	1,132	238	219	209	231
60° S	1,250	1,022	807	1,103	239	216	206	235
65° S	1,264	979	721	1,067	240	213	205	240
70° S	1,279	939	658	1,028	241	210	204	244
75° S	1,290	904	619	990	242	209	204	248
80° S	1,295	876	597	956	242	208	205	251

Table 4.34a Simulated temperature and concentrations of O_3, N_2O and HNO_3 at 15 km

z 15 km	Temperature (K)				O_3 (nmol mol^{-1})				N_2O (nmol mol^{-1})				HNO_3 (nmol mol^{-1})			
	Jan	Apr	Jul	Oct	Jan	Apr	Jul	Oct	Jan	Apr	Jul	Oct	Jan	Apr	Jul	Oct
85° N	214.0	215.8	229.6	220.7	1292.6	1231.7	763.28	752.79	275.04	277.48	298.63	300.17	3.649	3.033	1.529	2.047
80° N	214.5	217.5	228.5	220.3	1281.7	1252.5	752.51	734.10	275.48	276.99	298.85	300.78	3.613	3.049	1.516	1.987
70° N	215.8	220.3	225.8	219.7	1215.5	1246.5	705.02	683.16	278.12	277.96	300.25	302.51	3.422	2.966	1.439	1.816
60° N	216.2	220.3	223.1	218.5	1048.0	1055.6	630.87	595.14	285.11	285.37	302.76	305.40	2.909	2.464	1.309	1.550
50° N	216.1	217.8	218.4	214.8	860.27	848.99	488.05	450.51	292.59	292.95	307.47	310.02	2.328	1.954	0.999	1.132
40° N	214.0	214.0	209.7	209.5	523.28	631.5	272.37	277.63	305.50	300.84	314.46	315.26	1.321	1.414	0.533	0.624
30° N	207.3	208.1	204.4	204.1	207.74	332.28	175.58	144.14	316.06	311.73	317.44	318.74	0.392	0.650	0.335	0.248
20° N	201.7	201.4	201.0	200.3	98.233	130.65	119.84	84.928	318.91	318.3	318.87	319.88	0.128	0.205	0.212	0.109
10° N	198.8	198.5	199.2	199.0	72.032	76.825	68.627	62.198	319.42	319.61	319.65	320.03	0.091	0.108	0.072	0.057
0	198.3	198.5	199.5	199.3	63.923	65.987	63.486	66.053	319.44	319.66	319.64	319.94	0.082	0.093	0.060	0.072
10° S	198.7	198.5	199.9	199.4	82.138	60.665	69.913	73.30	318.74	319.55	319.55	319.79	0.119	0.076	0.071	0.078
20° S	200.4	200.0	202.0	201.5	109.84	81.149	91.278	114.50	318.08	319.19	319.16	318.87	0.182	0.118	0.112	0.135
30° S	203.6	205.2	209.3	208.1	158.39	157.15	213.1	251.32	316.39	316.94	315.03	313.46	0.255	0.287	0.434	0.421
40° S	208.8	211.5	215.9	214.6	284.56	291.15	529.36	488.73	310.20	311.52	301.41	301.96	0.498	0.627	1.337	1.019
50° S	218.2	216.6	214.8	216.7	552.67	460.01	753.59	652.19	295.16	303.71	290.36	289.51	1.081	1.056	2.058	1.372
60° S	224.4	219.2	207.4	210.4	735.73	647.79	842.29	597.00	284.47	294.35	284.39	280.77	1.482	1.575	2.791	1.117
70° S	226.5	218.2	198.9	200.4	795.00	753.46	740.78	374.59	281.43	288.59	288.4	285.73	1.582	1.927	2.497	0.569
80° S	228.0	216.2	194.8	194.7	861.65	865.49	640.21	252.14	277.22	282.36	293.19	293.65	1.718	2.293	1.210	0.311
85° S	228.5	215.3	194.2	194.1	868.10	866.53	620.35	233.52	277.43	282.02	294.14	294.24	1.718	2.31	0.416	0.31

Table 4.34b Simulated temperature and concentrations of O_3, N_2O and HNO_3 at 18 km

z 18 km	Temperature (K)				O_3 (nmol mol^{-1})				N_2O (nmol mol^{-1})				HNO_3 (nmol mol^{-1})			
	Jan	Apr	Jul	Oct	Jan	Apr	Jul	Oct	Jan	Apr	Jul	Oct	Jan	Apr	Jul	Oct
80° N	216.3	215.1	228.4	219.7	2423.0	2133.5	1509.1	1569.8	230.3	250.4	274.1	272.2	6.382	4.458	2.746	3.952
70° N	216.7	218.7	225.9	219.4	2257.9	2015.9	1419.5	1506.3	238.0	256.1	277.4	274.5	5.896	4.128	2.589	3.710
60° N	216.4	219.6	222.7	218.6	1948.2	1748.6	1246.8	1353.0	251.4	265.8	284.1	280.3	4.993	3.506	2.275	3.205
50° N	214.9	217.0	217.3	215.3	1609.7	1429.3	1025.2	1080.3	265.5	276.4	291.6	290.3	4.012	2.796	1.835	2.421
40° N	210.8	212.0	208.1	209.4	1031.1	1032.7	689.6	727.6	288.5	289.7	302.5	302.6	2.397	1.927	1.164	1.480
30° N	201.6	203.8	200.6	201.0	477.4	586.4	474.5	404.9	308.2	304.9	309.2	312.6	0.914	0.938	0.756	0.689
20° N	192.9	194.9	195.8	194.4	216.6	303.5	369.2	250.7	316.3	314.1	312.2	316.9	0.265	0.379	0.548	0.334
10° N	188.5	190.1	193.6	193.0	147.8	195.3	313.6	217.7	318.1	317.3	313.7	318.0	0.131	0.191	0.400	0.219
0	187.0	188.6	193.1	193.5	134.6	147.0	203.3	203.1	318.2	318.6	316.9	318.3	0.102	0.121	0.200	0.157
10° S	188.4	189.6	193.8	193.6	170.6	152.5	186.6	227.3	317.0	318.2	317.1	316.8	0.143	0.138	0.216	0.199
20° S	192.2	193.7	197.1	197.0	239.3	208.9	255.2	341.6	314.8	316.2	314.5	311.7	0.238	0.247	0.412	0.454
30° S	197.9	202.2	206.3	205.8	367.2	392.8	539.9	637.8	309.3	308.8	303.0	299.1	0.464	0.622	1.108	1.125
40° S	207.0	210.5	214.8	214.4	651.6	693.9	1043.2	1043.6	295.1	295.1	281.8	281.4	1.034	1.312	2.325	2.041
50° S	218.5	216.2	215.4	217.9	1117.9	1044.9	1450.8	1321.5	270.0	277.5	262.6	264.1	2.014	2.220	3.464	2.599
60° S	225.8	218.1	206.7	212.8	1465.2	1371.2	1689.2	1224.8	251.8	259.7	243.5	242.7	2.772	3.219	4.623	2.341
70° S	229.0	216.6	195.8	201.6	1608.7	1607.5	1622.9	760.6	245.6	245.6	240.1	239.8	2.984	4.090	5.248	1.490
80° S	231.0	214.6	190.6	193.3	1695.9	1796.8	1506.9	366.8	241.2	233.6	246.0	248.2	3.134	4.834	3.776	0.748
85° S	231.7	214.0	189.7	192.0	1701.9	1826.6	1467.2	303.6	242.2	231.5	248.4	247.6	3.141	4.959	2.603	0.661

Table 4.34c Simulated temperature and concentrations of O_3, N_2O and HNO_3 at 25 km

z 25 km	Temperature (K)				O_3 (nmol mol^{-1})				N_2O (nmol mol^{-1})				HNO_3 (nmol mol^{-1})			
	Jan	Apr	Jul	Oct	Jan	Apr	Jul	Oct	Jan	Apr	Jul	Oct	Jan	Apr	Jul	Oct
85° N	215.0	212.1	229.6	213.6	5178.6	4639.2	3329.5	3794.4	141.9	154.1	176.7	159.3	9.805	7.442	6.212	9.018
80° N	215.2	213.1	229.2	213.6	5186.4	4647.2	3407.2	3731.4	142.1	158.4	177.4	156.9	9.790	7.313	6.103	9.023
70° N	214.6	216.2	227.3	214.4	5182.1	4761.2	3613.2	3892.8	151.8	170.6	179.1	161.2	9.419	6.825	5.835	8.639
60° N	213.9	218.3	224.5	215.1	5147.6	4802.8	4022.0	4034.6	159.2	181.8	186.1	173.6	9.103	6.306	5.661	8.002
50° N	212.4	217.4	220.9	215.1	5001.5	4679.8	4186.9	4118.9	174.7	195.1	196.3	190.7	8.378	5.706	5.337	7.112
40° N	209.9	215.3	217.1	214.8	4668.4	4617.1	4216.2	4121.4	197.8	199.1	211.0	214.7	7.191	5.429	4.802	5.811
30° N	208.0	213.4	214.0	213.4	4145.4	4447.9	4097.6	3791.8	222.8	210.9	230.8	250.3	5.557	4.793	4.040	3.948
20° N	208.2	211.2	212.2	211.8	3305.3	3725.9	3829.3	3632.3	263.3	259.3	259.8	270.4	3.109	2.818	2.948	2.892
10° N	208.2	211.1	212.3	210.6	2949.6	3548.7	3967.8	3658.5	291.7	281.5	274.7	279.5	1.620	1.938	2.516	2.398
0	208.0	213.4	212.4	210.1	2884.7	3869.8	3803.2	3450.0	295.2	283.9	282.2	286.0	1.359	1.904	2.276	1.949
10° S	208.5	211.1	212.3	210.6	3252.0	3562.7	3530.1	3540.9	279.2	279.0	270.5	256.2	1.833	1.973	2.696	2.870
20° S	209.9	211.0	211.6	211.5	3621.8	3440.3	3763.8	4051.1	238.8	255.6	224.0	204.0	3.110	2.728	4.699	4.905
30° S	211.9	213.1	212.2	213.9	3808.9	3738.6	4265.8	4454.4	197.8	215.4	194.6	172.9	4.290	4.463	6.467	6.320
40° S	216.4	214.5	214.0	217.6	3988.9	3959.8	4626.5	4702.3	165.3	170.2	181.0	160.9	5.419	6.623	7.483	7.069
50° S	222.5	215.1	212.2	220.0	4116.5	4013.4	4815.3	4774.7	145.4	151.2	164.2	145.9	6.026	7.698	8.540	7.205
60° S	227.7	214.1	201.9	218.3	4017.6	4023.2	4643.8	4242.8	133.4	133.4	125.9	115.8	6.340	8.824	10.000	6.017
70° S	231.4	211.1	188.9	211.3	3637.1	3764.2	3843.7	3000.0	125.1	117.7	100.3	90.9	6.558	9.981	6.650	3.888
80° S	233.3	208.6	183.4	202.2	3189.9	3432.6	3649.8	1925.1	120.5	105.9	93.5	78.5	6.876	10.855	1.260	2.358
85° S	233.6	207.8	182.3	199.4	2985.0	3360.2	3627.8	1728.1	118.0	103.2	93.3	77.3	7.030	11.060	0.933	2.051

Table 4.34d Simulated temperature and concentrations of O_3, N_2O and HNO_3 at 40 km

z 40 km	Temperature (K)				O_3 (nmol mol⁻¹)				N_2O (nmol mol⁻¹)				HNO_3 (nmol mol⁻¹)			
	Jan	Apr	Jul	Oct	Jan	Apr	Jul	Oct	Jan	Apr	Jul	Oct	Jan	Apr	Jul	Oct
85° N	235.4	241.8	254.9	226.1	5698.4	5941.6	5879.5	5330.6	8.6	3.7	7.1	3.9	0.413	0.311	0.351	0.565
80° N	235.3	241.3	254.4	226.7	5504.5	6264.6	5888.0	5403.5	8.6	3.8	7.0	4.2	0.418	0.309	0.355	0.558
70° N	235.0	240.6	253.5	228.1	5307.2	6956.5	6187.1	5817.7	8.2	6.0	5.6	5.1	0.423	0.328	0.342	0.544
60° N	234.9	240.8	252.6	229.8	5521.6	7412.6	6996.3	6556.8	7.7	10.3	3.9	7.0	0.421	0.380	0.299	0.515
50° N	236.0	243.5	250.2	232.2	6254.6	7639.1	7321.0	7349.8	8.8	14.4	6.9	10.2	0.403	0.411	0.363	0.508
40° N	238.3	244.4	247.2	234.7	7044.0	7767.7	7631.0	7875.0	11.1	19.6	12.6	25.2	0.380	0.460	0.442	0.563
30° N	240.1	244.2	243.7	237.1	7565.5	7891.6	7996.8	8135.5	16.7	26.7	22.9	56.9	0.426	0.503	0.521	0.635
20° N	238.6	244.6	240.3	239.7	7915.6	7833.7	8393.8	8157.1	35.1	41.3	43.5	78.7	0.581	0.547	0.595	0.631
10° N	236.4	244.2	237.2	243.3	8374.9	7865.2	8846.8	7960.9	59.5	59.6	73.5	84.9	0.676	0.587	0.647	0.599
0	235.4	242.6	235.6	245.3	8969.1	8094.1	8894.6	7804.2	88.2	81.7	83.3	80.8	0.678	0.633	0.694	0.584
10° S	236.7	243.7	236.8	243.8	9150.6	7907.8	8420.3	7998.0	95.1	83.8	66.2	80.0	0.641	0.624	0.681	0.595
20° S	240.0	241.8	238.6	242.0	8799.8	7912.9	7815.9	8236.8	75.3	74.9	38.9	54.4	0.583	0.628	0.592	0.562
30° S	244.8	238.4	237.9	240.3	8143.2	8030.7	7343.2	8308.1	41.6	46.8	22.5	30.5	0.502	0.609	0.511	0.506
40° S	249.6	235.8	234.6	239.3	7482.0	7695.2	6904.0	7926.0	23.1	24.9	20.8	28.8	0.433	0.568	0.544	0.494
50° S	253.4	232.0	228.6	241.4	6909.2	7235.8	6657.3	7417.6	13.0	17.3	17.3	25.9	0.374	0.579	0.578	0.462
60° S	256.1	226.7	221.2	247.6	6368.1	6560.6	5901.8	6873.0	9.7	8.4	7.5	19.8	0.346	0.615	0.474	0.368
70° S	258.0	222.1	214.7	255.2	5689.9	5171.5	4426.4	6290.7	10.8	3.0	1.4	12.3	0.355	0.709	0.442	0.253
80° S	259.6	220.8	210.9	261.0	5468.4	4390.7	4113.2	5624.3	11.6	2.1	1.2	6.8	0.338	0.806	0.498	0.173
85° S	260.1	220.7	209.9	262.3	5473.8	4332.0	4130.8	5490.1	10.4	1.9	1.3	4.9	0.325	0.821	0.554	0.149

Table 4.35a Simulated concentrations of CH_4, H_2O, CO_2 and SF_6 at 15 km

z 15 km	CH_4 (μmol mol^{-1})				H_2O (μmol mol^{-1})				CO_2 (μmol mol^{-1})				SF_6 (pmol mol^{-1})			
	Jan	Apr	Jul	Oct	Jan	Apr	Jul	Oct	Jan	Apr	Jul	Oct	Jan	Apr	Jul	Oct
85° N	1.567	1.578	1.653	1.660	3.416	3.123	2.770	3.597	375.7	376.3	378.6	379.3	5.228	5.328	5.546	5.600
80° N	1.569	1.576	1.654	1.663	3.433	2.995	2.801	3.656	375.7	376.3	378.7	379.4	5.233	5.327	5.549	5.607
70° N	1.579	1.580	1.660	1.670	3.414	2.815	2.934	3.836	375.8	376.4	378.9	379.5	5.257	5.341	5.565	5.631
60° N	1.606	1.608	1.670	1.683	3.395	2.665	3.216	4.187	376.2	377.0	379.3	379.8	5.316	5.404	5.594	5.671
50° N	1.636	1.637	1.691	1.703	3.256	2.665	4.459	4.851	376.7	377.5	380.1	380.0	5.385	5.468	5.656	5.739
40° N	1.687	1.668	1.726	1.727	3.060	3.072	9.491	6.051	377.5	378.3	381.5	380.2	5.515	5.538	5.773	5.825
30° N	1.732	1.709	1.743	1.744	3.097	4.026	12.181	7.444	378.6	379.7	382.1	379.8	5.657	5.652	5.833	5.895
20° N	1.749	1.732	1.751	1.751	4.328	6.452	10.686	9.169	379.3	381.0	382.3	379.5	5.713	5.740	5.861	5.922
10° N	1.751	1.737	1.751	1.749	6.307	7.852	9.428	9.603	379.7	381.5	382.3	379.6	5.726	5.766	5.865	5.918
0	1.744	1.731	1.747	1.743	6.959	7.887	8.524	8.709	379.7	381.2	382.1	379.9	5.718	5.758	5.849	5.901
10° S	1.730	1.724	1.742	1.739	7.439	8.204	7.306	8.470	379.3	380.7	381.9	380.1	5.690	5.745	5.834	5.889
20° S	1.724	1.723	1.739	1.736	7.354	7.218	5.866	7.571	379.1	380.5	381.9	380.2	5.673	5.734	5.820	5.874
30° S	1.719	1.718	1.717	1.714	5.592	4.782	3.499	5.347	378.8	380.1	381.0	380.3	5.650	5.698	5.741	5.795
40° S	1.696	1.700	1.659	1.666	4.303	3.265	2.701	4.074	378.2	379.3	379.1	379.7	5.575	5.630	5.574	5.653
50° S	1.639	1.670	1.617	1.615	3.429	2.753	2.767	3.489	377.1	378.4	377.9	378.7	5.419	5.546	5.467	5.527
60° S	1.597	1.634	1.594	1.580	3.077	2.654	3.333	3.355	376.5	377.5	377.4	377.9	5.322	5.454	5.412	5.444
70° S	1.585	1.611	1.609	1.598	3.031	2.716	4.202	3.359	376.3	377.0	377.5	378.1	5.294	5.398	5.435	5.478
80° S	1.569	1.586	1.627	1.628	3.031	2.846	4.433	3.622	376.1	376.6	377.9	378.6	5.259	5.339	5.472	5.539
85° S	1.569	1.585	1.631	1.630	3.035	2.882	4.481	3.535	376.1	376.5	377.9	378.7	5.260	5.334	5.480	5.543

Table 4.35b Simulated concentrations of CH_4, H_2O, CO_2 and SF_6 at 18 km

z 18 km	CH_4 (μmol mol^{-1})				H_2O (μmol mol^{-1})				CO_2 (μmol mol^{-1})				SF_6 (pmol mol^{-1})			
	Jan	Apr	Jul	Oct	Jan	Apr	Jul	Oct	Jan	Apr	Jul	Oct	Jan	Apr	Jul	Oct
85° N	1.400	1.473	1.560	1.551	3.370	3.141	2.756	2.904	373.5	374.8	376.5	376.7	4.968	5.143	5.335	5.345
80° N	1.400	1.477	1.561	1.550	3.370	3.128	2.755	2.909	373.5	374.9	376.5	376.7	4.971	5.151	5.336	5.342
70° N	1.429	1.498	1.573	1.559	3.346	3.054	2.710	2.945	373.9	375.2	376.7	376.9	5.014	5.191	5.362	5.362
60° N	1.479	1.534	1.599	1.581	3.300	2.910	2.624	3.006	374.5	375.7	377.2	377.4	5.096	5.259	5.418	5.412
50° N	1.532	1.575	1.627	1.621	3.234	2.725	2.621	3.178	375.3	376.4	378.0	378.3	5.186	5.338	5.483	5.502
40° N	1.619	1.625	1.671	1.671	3.036	2.469	2.901	3.611	376.5	377.3	379.4	379.6	5.353	5.443	5.600	5.634
30° N	1.697	1.683	1.713	1.715	2.478	2.201	3.402	3.876	377.7	378.6	380.5	380.3	5.548	5.580	5.687	5.780
20° N	1.732	1.717	1.718	1.735	1.942	2.190	3.601	3.825	378.5	379.9	381.1	380.3	5.658	5.676	5.727	5.853
10° N	1.741	1.728	1.718	1.738	1.715	2.194	3.303	3.728	378.9	380.5	381.3	380.3	5.690	5.716	5.741	5.869
0	1.738	1.729	1.728	1.737	1.560	1.933	2.694	3.640	379.0	380.7	381.6	380.6	5.691	5.731	5.774	5.866
10° S	1.726	1.725	1.725	1.730	1.709	2.095	2.780	3.787	378.8	380.4	381.3	380.7	5.662	5.717	5.763	5.839
20° S	1.715	1.718	1.712	1.707	2.262	2.149	2.668	3.703	378.4	380.0	380.8	380.5	5.628	5.687	5.720	5.766
30° S	1.695	1.691	1.666	1.655	2.488	2.214	2.476	3.146	378.0	378.9	379.3	379.6	5.564	5.598	5.593	5.619
40° S	1.640	1.637	1.585	1.584	2.770	2.476	2.582	2.819	377.1	377.5	377.3	378.1	5.422	5.461	5.407	5.459
50° S	1.542	1.568	1.513	1.518	2.992	2.745	2.781	2.754	375.7	376.3	376.0	376.8	5.214	5.313	5.265	5.333
60° S	1.472	1.499	1.441	1.437	3.057	2.932	2.990	2.620	374.7	375.3	374.9	375.5	5.083	5.182	5.140	5.193
70° S	1.448	1.446	1.427	1.425	3.071	3.042	3.135	2.452	374.4	374.5	374.7	375.3	5.042	5.086	5.106	5.166
80° S	1.431	1.400	1.449	1.454	3.076	3.114	3.150	2.352	374.1	373.9	374.9	375.6	5.014	5.008	5.137	5.205
85° S	1.435	1.392	1.458	1.452	3.076	3.127	3.130	2.310	374.1	373.8	375.0	375.6	5.017	4.993	5.150	5.200

Table 4.35c Simulated concentrations of CH_4, H_2O, CO_2 and SF_6 at 25 km

z 25 km	CH_4 (μmol mol^{-1})				H_2O (μmol mol^{-1})				CO_2 (μmol mol^{-1})				SF_6 (pmol mol^{-1})			
	Jan	Apr	Jul	Oct	Jan	Apr	Jul	Oct	Jan	Apr	Jul	Oct	Jan	Apr	Jul	Oct
85° N	1.089	1.134	1.213	1.149	3.873	3.787	3.636	3.765	371.0	371.7	372.7	372.5	4.687	4.772	4.880	4.860
80° N	1.090	1.150	1.216	1.140	3.871	3.752	3.633	3.783	371.0	371.8	372.7	372.5	4.688	4.786	4.880	4.854
70° N	1.126	1.198	1.222	1.156	3.796	3.658	3.623	3.752	371.3	372.2	372.7	372.6	4.719	4.833	4.884	4.867
60° N	1.154	1.240	1.247	1.201	3.738	3.579	3.581	3.669	371.4	372.6	372.9	372.9	4.742	4.874	4.903	4.902
50° N	1.211	1.287	1.283	1.263	3.615	3.497	3.523	3.551	371.8	373.0	373.1	373.3	4.789	4.916	4.933	4.957
40° N	1.295	1.300	1.335	1.349	3.437	3.478	3.445	3.377	372.4	373.1	373.6	374.1	4.861	4.928	4.984	5.050
30° N	1.383	1.341	1.407	1.478	3.254	3.422	3.320	3.053	373.2	373.5	374.4	375.5	4.954	4.971	5.077	5.222
20° N	1.526	1.514	1.516	1.553	2.983	3.270	3.113	2.834	374.9	375.6	375.7	376.5	5.145	5.193	5.244	5.341
10° N	1.630	1.597	1.573	1.589	2.829	3.215	3.099	2.726	376.6	376.8	376.5	377.0	5.309	5.309	5.324	5.400
0	1.643	1.606	1.600	1.613	2.868	3.160	3.086	2.621	377.0	376.9	376.9	377.3	5.333	5.317	5.358	5.442
10° S	1.584	1.586	1.552	1.497	2.924	3.149	3.121	2.892	376.1	376.6	376.3	375.9	5.237	5.286	5.283	5.267
20° S	1.431	1.493	1.377	1.307	3.084	3.172	3.312	3.336	373.9	375.3	374.3	373.9	5.005	5.145	5.047	5.015
30° S	1.277	1.340	1.272	1.200	3.309	3.300	3.466	3.548	372.2	373.4	373.2	373.0	4.815	4.939	4.922	4.903
40° S	1.158	1.173	1.226	1.162	3.535	3.544	3.538	3.637	371.2	371.7	372.8	372.7	4.696	4.755	4.874	4.872
50° S	1.085	1.104	1.167	1.109	3.697	3.677	3.635	3.663	370.7	371.2	372.3	372.4	4.641	4.693	4.820	4.834
60° S	1.043	1.040	1.021	0.986	3.801	3.816	3.834	3.512	370.5	370.8	371.3	371.7	4.618	4.649	4.706	4.744
70° S	1.016	0.983	0.914	0.868	3.873	3.948	3.560	3.167	370.4	370.6	370.6	371.0	4.603	4.617	4.620	4.659
80° S	0.998	0.938	0.886	0.801	3.914	4.052	2.841	2.845	370.4	370.4	370.5	370.5	4.595	4.597	4.603	4.604
85° S	0.987	0.928	0.885	0.793	3.938	4.076	2.509	2.778	370.3	370.4	370.5	370.5	4.592	4.593	4.602	4.599

Table 4.35d Simulated concentrations of CH_4, H_2O, CO_2 and SF_6 at 40 km

z 40 km	CH_4 (μmol mol^{-1})				H_2O (μmol mol^{-1})				CO_2 (μmol mol^{-1})				SF_6 (pmol mol^{-1})			
	Jan	Apr	Jul	Oct	Jan	Apr	Jul	Oct	Jan	Apr	Jul	Oct	Jan	Apr	Jul	Oct
85° N	0.422	0.312	0.429	0.283	5.158	5.289	5.227	5.570	369.8	369.7	370.5	370.4	4.462	4.394	4.577	4.529
80° N	0.420	0.317	0.424	0.289	5.150	5.288	5.235	5.555	369.8	369.7	370.5	370.4	4.455	4.400	4.572	4.532
70° N	0.409	0.370	0.384	0.318	5.151	5.244	5.299	5.491	369.7	369.9	370.3	370.5	4.439	4.460	4.536	4.543
60° N	0.404	0.456	0.339	0.368	5.171	5.127	5.368	5.377	369.7	370.3	370.1	370.5	4.441	4.539	4.496	4.563
50° N	0.430	0.527	0.425	0.441	5.152	5.018	5.225	5.218	369.8	370.6	370.4	370.7	4.472	4.597	4.560	4.593
40° N	0.462	0.608	0.529	0.617	5.094	4.864	5.019	4.845	369.9	371.0	370.7	371.1	4.492	4.649	4.610	4.671
30° N	0.540	0.685	0.639	0.891	4.952	4.706	4.790	4.269	370.2	371.2	371.0	372.1	4.541	4.687	4.651	4.812
20° N	0.752	0.808	0.795	1.034	4.531	4.459	4.477	3.962	370.9	371.7	371.5	372.9	4.654	4.741	4.721	4.910
10° N	0.948	0.936	0.975	1.073	4.117	4.202	4.115	3.879	371.5	372.0	372.2	373.2	4.738	4.791	4.819	4.946
0	1.090	1.063	1.044	1.049	3.833	3.956	3.977	3.932	372.1	372.4	372.5	373.2	4.797	4.840	4.862	4.941
10° S	1.099	1.083	0.960	1.040	3.824	3.921	4.158	3.952	372.1	372.5	372.3	373.1	4.794	4.848	4.835	4.935
20° S	0.946	1.028	0.774	0.861	4.132	4.017	4.554	4.332	371.3	372.2	371.9	372.5	4.699	4.806	4.769	4.850
30° S	0.721	0.803	0.623	0.683	4.583	4.443	4.865	4.706	370.4	371.0	371.5	372.0	4.580	4.665	4.712	4.769
40° S	0.591	0.615	0.604	0.664	4.862	4.838	4.902	4.739	370.0	370.4	371.5	372.0	4.533	4.585	4.702	4.759
50° S	0.498	0.528	0.541	0.625	5.086	5.040	4.996	4.801	369.8	370.2	371.2	371.9	4.506	4.557	4.654	4.737
60° S	0.454	0.388	0.351	0.539	5.191	5.376	5.222	4.939	369.7	369.9	370.4	371.7	4.495	4.513	4.485	4.686
70° S	0.463	0.278	0.209	0.429	5.168	5.652	5.299	5.114	369.8	369.6	369.8	371.4	4.502	4.480	4.302	4.619
80° S	0.472	0.242	0.201	0.342	5.144	5.748	5.256	5.253	369.8	369.6	369.8	371.2	4.507	4.472	4.272	4.564
85° S	0.458	0.235	0.203	0.308	5.179	5.767	5.277	5.304	369.8	369.6	369.8	371.1	4.502	4.470	4.281	4.541

Table 4.36a Simulated concentrations of CFCl₃, HCl, ClO and ClONO₂ at 15 km

z 15 km	CFCl₃ (nmol mol⁻¹)				HCl (nmol mol⁻¹)				ClO (nmol mol⁻¹)				ClONO₂ (nmol mol⁻¹)			
	Jan	Apr	Jul	Oct	Jan	Apr	Jul	Oct	Jan	Apr	Jul	Oct	Jan	Apr	Jul	Oct
85° N	0.176	0.182	0.212	0.213	0.630	0.597	0.363	0.340	0.000	0.002	0.001	0.000	0.116	0.103	0.021	0.031
80° N	0.177	0.181	0.213	0.214	0.625	0.604	0.358	0.331	0.000	0.002	0.001	0.000	0.115	0.103	0.020	0.030
70° N	0.181	0.183	0.215	0.217	0.593	0.591	0.335	0.305	0.001	0.003	0.001	0.001	0.107	0.100	0.019	0.025
60° N	0.192	0.194	0.219	0.222	0.509	0.503	0.295	0.261	0.003	0.002	0.001	0.001	0.082	0.071	0.017	0.018
50° N	0.204	0.206	0.227	0.230	0.414	0.409	0.219	0.187	0.002	0.001	0.000	0.000	0.058	0.046	0.011	0.010
40° N	0.225	0.218	0.239	0.239	0.240	0.304	0.102	0.098	0.001	0.001	0.000	0.000	0.020	0.025	0.004	0.003
30° N	0.243	0.235	0.243	0.245	0.073	0.143	0.050	0.033	0.000	0.000	0.000	0.000	0.002	0.007	0.002	0.001
20° N	0.247	0.246	0.246	0.247	0.021	0.035	0.024	0.009	0.000	0.000	0.000	0.000	0.000	0.001	0.000	0.000
10° N	0.248	0.248	0.247	0.247	0.010	0.009	0.008	0.004	0.000	0.000	0.000	0.000	0.000	0.000	0.000	0.000
0	0.248	0.248	0.247	0.247	0.008	0.007	0.008	0.005	0.000	0.000	0.000	0.000	0.000	0.000	0.000	0.000
10° S	0.247	0.248	0.247	0.246	0.018	0.006	0.010	0.007	0.000	0.000	0.000	0.000	0.000	0.000	0.000	0.000
20° S	0.246	0.247	0.246	0.245	0.030	0.014	0.018	0.026	0.000	0.000	0.000	0.000	0.000	0.000	0.000	0.000
30° S	0.244	0.244	0.240	0.237	0.060	0.056	0.089	0.118	0.000	0.000	0.000	0.000	0.001	0.001	0.003	0.004
40° S	0.234	0.235	0.220	0.220	0.158	0.145	0.290	0.289	0.000	0.000	0.001	0.001	0.005	0.004	0.025	0.019
50° S	0.213	0.224	0.203	0.202	0.377	0.262	0.409	0.427	0.001	0.001	0.005	0.004	0.022	0.014	0.067	0.053
60° S	0.197	0.210	0.195	0.190	0.524	0.390	0.384	0.443	0.001	0.001	0.020	0.022	0.038	0.034	0.097	0.094
70° S	0.193	0.201	0.201	0.198	0.568	0.463	0.262	0.287	0.001	0.002	0.026	0.057	0.041	0.049	0.037	0.075
80° S	0.187	0.192	0.208	0.209	0.627	0.539	0.207	0.122	0.001	0.001	0.000	0.107	0.045	0.067	0.004	0.037
85° S	0.187	0.192	0.210	0.210	0.626	0.543	0.200	0.086	0.001	0.001	0.000	0.132	0.044	0.068	0.001	0.030

Table 4.36b Simulated concentrations of CFCl$_3$, HCl, ClO and ClONO$_2$, at 18 km

z 18 km	CFCl$_3$ (nmol mol^{-1})				HCl (nmol mol^{-1})				ClO (nmol mol^{-1})				ClONO$_2$ (nmol mol^{-1})			
	Jan	Apr	Jul	Oct	Jan	Apr	Jul	Oct	Jan	Apr	Jul	Oct	Jan	Apr	Jul	Oct
85° N	0.111	0.138	0.170	0.164	1.112	0.904	0.724	0.732	0.001	0.005	0.003	0.001	0.295	0.235	0.080	0.117
80° N	0.112	0.139	0.170	0.163	1.109	0.897	0.720	0.735	0.001	0.005	0.002	0.003	0.294	0.229	0.079	0.119
70° N	0.122	0.147	0.176	0.167	1.035	0.843	0.672	0.705	0.004	0.006	0.002	0.005	0.261	0.199	0.072	0.108
60° N	0.141	0.161	0.187	0.177	0.900	0.747	0.578	0.630	0.008	0.005	0.002	0.004	0.204	0.146	0.056	0.087
50° N	0.161	0.178	0.199	0.195	0.750	0.631	0.471	0.496	0.007	0.003	0.001	0.002	0.147	0.097	0.037	0.052
40° N	0.196	0.199	0.218	0.216	0.487	0.468	0.304	0.315	0.003	0.002	0.001	0.001	0.060	0.049	0.016	0.021
30° N	0.229	0.223	0.229	0.234	0.207	0.257	0.196	0.149	0.000	0.001	0.000	0.000	0.011	0.013	0.008	0.005
20° N	0.243	0.238	0.234	0.241	0.071	0.112	0.145	0.072	0.000	0.000	0.000	0.000	0.001	0.003	0.004	0.001
10° N	0.246	0.243	0.237	0.243	0.038	0.057	0.121	0.054	0.000	0.000	0.000	0.000	0.000	0.001	0.002	0.001
0	0.246	0.246	0.242	0.243	0.034	0.033	0.065	0.048	0.000	0.000	0.000	0.000	0.000	0.000	0.001	0.001
10° S	0.244	0.245	0.243	0.241	0.055	0.039	0.059	0.071	0.000	0.000	0.000	0.000	0.001	0.000	0.001	0.001
20° S	0.240	0.242	0.239	0.234	0.094	0.074	0.103	0.154	0.000	0.000	0.000	0.000	0.001	0.001	0.002	0.004
30° S	0.232	0.230	0.221	0.215	0.182	0.195	0.278	0.340	0.000	0.000	0.001	0.001	0.005	0.005	0.017	0.020
40° S	0.211	0.210	0.189	0.188	0.389	0.397	0.561	0.570	0.001	0.001	0.003	0.003	0.021	0.024	0.071	0.062
50° S	0.176	0.184	0.161	0.164	0.717	0.628	0.753	0.745	0.003	0.003	0.009	0.006	0.072	0.065	0.151	0.130
60° S	0.150	0.159	0.137	0.138	0.936	0.830	0.737	0.882	0.005	0.005	0.048	0.023	0.123	0.125	0.221	0.226
70° S	0.141	0.139	0.134	0.136	1.016	0.972	0.510	0.771	0.004	0.007	0.059	0.081	0.139	0.185	0.107	0.201
80° S	0.135	0.123	0.143	0.147	1.071	1.087	0.403	0.622	0.005	0.006	0.000	0.132	0.149	0.236	0.013	0.136
85° S	0.136	0.121	0.146	0.147	1.063	1.107	0.395	0.578	0.005	0.002	0.000	0.178	0.146	0.246	0.005	0.115

Table 4.36c Simulated concentrations of CFCl$_3$, HCl, ClO and ClONO$_2$ at 25 km

z 25 km	CFCl$_3$ (nmol mol^{-1})				HCl (nmol mol^{-1})				ClO (nmol mol^{-1})				ClONO$_2$ (nmol mol^{-1})			
	Jan	Apr	Jul	Oct	Jan	Apr	Jul	Oct	Jan	Apr	Jul	Oct	Jan	Apr	Jul	Oct
85° N	0.019	0.026	0.028	0.018	1.721	1.668	1.800	1.852	0.003	0.024	0.026	0.004	0.724	0.702	0.415	0.539
80° N	0.019	0.027	0.029	0.018	1.719	1.638	1.783	1.880	0.002	0.024	0.026	0.011	0.722	0.696	0.425	0.529
70° N	0.023	0.033	0.031	0.020	1.658	1.539	1.716	1.810	0.021	0.029	0.023	0.018	0.696	0.677	0.475	0.551
60° N	0.027	0.039	0.036	0.027	1.608	1.452	1.562	1.680	0.032	0.032	0.026	0.023	0.679	0.649	0.549	0.557
50° N	0.036	0.049	0.045	0.038	1.505	1.362	1.440	1.519	0.035	0.031	0.027	0.024	0.636	0.599	0.550	0.541
40° N	0.051	0.053	0.060	0.058	1.357	1.315	1.294	1.313	0.032	0.029	0.026	0.024	0.561	0.598	0.519	0.479
30° N	0.075	0.066	0.082	0.098	1.191	1.221	1.123	1.016	0.025	0.026	0.023	0.021	0.453	0.551	0.436	0.327
20° N	0.122	0.115	0.112	0.119	0.887	0.881	0.898	0.841	0.016	0.019	0.021	0.019	0.244	0.305	0.302	0.248
10° N	0.155	0.135	0.118	0.127	0.623	0.712	0.795	0.763	0.014	0.019	0.022	0.019	0.136	0.218	0.266	0.223
0	0.160	0.130	0.125	0.135	0.584	0.705	0.736	0.704	0.014	0.022	0.023	0.018	0.122	0.232	0.236	0.195
10° S	0.141	0.132	0.120	0.113	0.733	0.742	0.824	0.920	0.017	0.020	0.022	0.020	0.185	0.221	0.256	0.300
20° S	0.103	0.116	0.082	0.065	1.061	0.931	1.140	1.253	0.021	0.020	0.026	0.026	0.338	0.276	0.449	0.568
30° S	0.067	0.079	0.054	0.038	1.359	1.244	1.326	1.425	0.026	0.025	0.033	0.033	0.504	0.434	0.598	0.733
40° S	0.039	0.040	0.040	0.028	1.586	1.571	1.411	1.491	0.031	0.029	0.038	0.038	0.631	0.608	0.665	0.793
50° S	0.024	0.027	0.029	0.022	1.740	1.706	1.497	1.569	0.036	0.030	0.039	0.041	0.682	0.664	0.740	0.836
60° S	0.016	0.017	0.015	0.013	1.869	1.835	1.412	1.763	0.035	0.028	0.112	0.045	0.672	0.706	0.813	0.882
70° S	0.011	0.010	0.009	0.009	2.043	2.004	1.039	1.910	0.030	0.022	0.225	0.086	0.592	0.687	0.190	0.869
80° S	0.009	0.005	0.007	0.007	2.173	2.170	1.209	1.979	0.031	0.011	0.000	0.137	0.498	0.634	0.006	0.834
85° S	0.008	0.005	0.007	0.007	2.231	2.209	1.258	1.969	0.030	0.004	0.000	0.175	0.461	0.623	0.001	0.823

Table 4.36d Simulated concentrations of CFCl₃, HCl, ClO and CIONO₂ at 40 km

z 40 km	CFCl₃ (nmol mol⁻¹)				HCl (nmol mol⁻¹)				ClO (nmol mol⁻¹)				ClONO₂ (nmol mol⁻¹)			
	Jan	Apr	Jul	Oct	Jan	Apr	Jul	Oct	Jan	Apr	Jul	Oct	Jan	Apr	Jul	Oct
85° N	0.000	0.000	0.000	0.000	2.546	2.519	2.622	2.500	0.008	0.611	0.528	0.091	0.422	0.104	0.050	0.570
80° N	0.000	0.000	0.000	0.000	2.558	2.503	2.620	2.482	0.008	0.593	0.528	0.197	0.417	0.130	0.054	0.552
70° N	0.000	0.000	0.000	0.000	2.566	2.489	2.571	2.460	0.160	0.504	0.556	0.277	0.416	0.191	0.077	0.481
60° N	0.000	0.000	0.000	0.000	2.543	2.510	2.474	2.462	0.240	0.428	0.568	0.308	0.397	0.219	0.141	0.409
50° N	0.000	0.000	0.000	0.000	2.514	2.528	2.494	2.470	0.283	0.385	0.476	0.312	0.356	0.226	0.184	0.363
40° N	0.000	0.000	0.000	0.000	2.492	2.545	2.515	2.503	0.317	0.344	0.400	0.279	0.313	0.231	0.215	0.314
30° N	0.000	0.000	0.000	0.000	2.500	2.554	2.524	2.526	0.316	0.314	0.342	0.225	0.286	0.231	0.237	0.254
20° N	0.000	0.000	0.000	0.000	2.551	2.569	2.512	2.509	0.258	0.273	0.287	0.206	0.272	0.220	0.242	0.216
10° N	0.000	0.000	0.000	0.000	2.537	2.567	2.456	2.514	0.223	0.239	0.241	0.203	0.251	0.209	0.234	0.194
0	0.000	0.000	0.000	0.000	2.457	2.519	2.436	2.526	0.215	0.213	0.218	0.211	0.225	0.200	0.238	0.189
10° S	0.000	0.000	0.000	0.000	2.420	2.534	2.492	2.516	0.227	0.203	0.219	0.218	0.213	0.192	0.255	0.194
20° S	0.000	0.000	0.000	0.000	2.452	2.553	2.542	2.510	0.266	0.207	0.242	0.266	0.215	0.210	0.280	0.222
30° S	0.000	0.000	0.000	0.000	2.511	2.534	2.541	2.505	0.330	0.245	0.262	0.317	0.214	0.267	0.311	0.248
40° S	0.000	0.000	0.000	0.000	2.551	2.527	2.549	2.530	0.380	0.272	0.236	0.318	0.192	0.314	0.344	0.241
50° S	0.000	0.000	0.000	0.000	2.570	2.515	2.506	2.548	0.427	0.271	0.219	0.339	0.164	0.355	0.411	0.215
60° S	0.000	0.000	0.000	0.000	2.601	2.460	2.431	2.545	0.461	0.276	0.242	0.401	0.128	0.438	0.468	0.167
70° S	0.000	0.000	0.000	0.000	2.683	2.450	2.499	2.528	0.474	0.242	0.194	0.489	0.063	0.541	0.415	0.113
80° S	0.000	0.000	0.000	0.000	2.700	2.473	2.579	2.519	0.476	0.128	0.004	0.586	0.040	0.677	0.349	0.066
85° S	0.000	0.000	0.000	0.000	2.692	2.478	2.567	2.505	0.488	0.049	0.004	0.628	0.038	0.716	0.368	0.049

References

Bovensmann, H., J.P. Burrows, M. Buchwitz, J. Frerick, S. Nöel, V.V. Rozanov, K.V. Chance, A.P.H. Goede, J. Atmos. Sci. **56**, 127–150 (1999)

Gettelman, A. (37 coauthors), J. Geophys. Res. **115**, D00M08 (2010). doi:10.1029/2009JD013638

Jöckel, P., R. Sander, A. Kerkweg, H. Tost, J. Lelieveld, Atmos. Chem. Phys. **5**, 433–444 (2005)

Jöckel, P., A. Kerkweg, A. Pozzer, R. Sander, H. Tost, H. Riede, A. Baumgärtner, S. Gromov, B. Kem, Technical note: Development cycle 2 of the Modular Earth Submodel System (MESSy2). Geosci. Model Dev. **3**, 717–752 (2010)

Keating, G.M., M.C. Pitts, D.F. Young, in: *Middle Atmosphere Program, Handbook for MAP*, ed. by G.M. Keating, Reference Models of Trace Species, vol. 31 (International Council of Scientific Unions, Paris, 1989), pp. 1–36

Kühl, S., J. Pukite, T. Deutschmann, U. Platt, T. Wagner, Adv. Space Res. **42**, 1747–1764 (2008)

Pukite, J., S. Kühl, T. Deutschmann, S. Dörner, P. Jöckel, U. Platt, T. Wagner, Atmos. Meas. Tech. **3**, 1155–1174 (2010)

Röckner, E., R. Brokopf, M. Esch, M. Giorgetta, S. Hagemann, L. Kornblueh, E. Manzini, U. Schlese, U. Schulzweida, J. Clim. **19**, 3771–3791 (2006)

Rozanov, A., H. Bovensmann, A. Bracher, S. Hrechanyy, V. Rozanov, M. Sinnhuber, F. Stroh, J.P. Burrows, Adv. Space Res. **36**, 846–854 (2005)

Chapter 5
The Atmospheric Aerosol

The term aerosol refers to a suspension of fine particles in air. The principal sources of the aerosol in the troposphere are: (a) surface agitation by wind force, which generates soil dust, sea salt particles, and biogenic material; (b) high temperature processes including volcanic eruptions, forest fires and anthropogenic combustion processes; (c) chemical reactions leading to compounds with low vapor pressures such as the oxidation of SO_2 to sulfuric acid, NO_2 to nitric acid, or glyoxal to oxalic acid, and the neutralization of the acids by ammonia or alkaline minerals to form water-soluble salts. Particles in the size range above ~0.1 μm act as cloud condensation nuclei (CCN). While most clouds evaporate again so that the aerosol particles are recycled, precipitation transfers the material to the earth surface. The particles are incorporated in rain water during the formation of precipitating clouds (nucleation scavenging) and by attachment to rain drops (below-cloud scavenging). Dry deposition also contributes to the removal of particles from the atmospheric boundary layer (primarily in the size range >5 μm). Directly emitted particles usually retain their original chemical character in the vicinity of sources, but during transport in the troposphere all particles undergo chemical modification both by the coagulation of small particles and by the accumulation of material resulting from chemical reactions occurring in the gas phase and in clouds. In the stratosphere the principal source of aerosol particles is the chemical conversion of sulfur compounds to sulfuric acid.

P. Warneck and J. Williams, *The Atmospheric Chemist's Companion: Numerical Data for Use in the Atmospheric Sciences*, DOI 10.1007/978-94-007-2275-0_5, © Springer Science+Business Media B.V. 2012

5.1   Characterization of the Tropospheric Aerosol

Table 5.1 Typical mass and particle concentrations for tropospheric aerosols near Earth's surface and the corresponding mean particle diameter[a]

	Urban	Continental (rural)	Marine (remote)	Arctic (summer)
Total mass concentration (μg m^{-3})	~100	30–50	~10[b]	~1
Total number concentration (cm^{-3})	10^5–10^6	$(1$–$2)\times10^4$	300–600	25
Mean particle diameter (μm)	0.03	0.07	0.16	0.17

[a]Adapted from Warneck (2000) assuming an average density of 1.8 kg m^{-3}
[b]Includes 8 μg m^{-3} sea salt

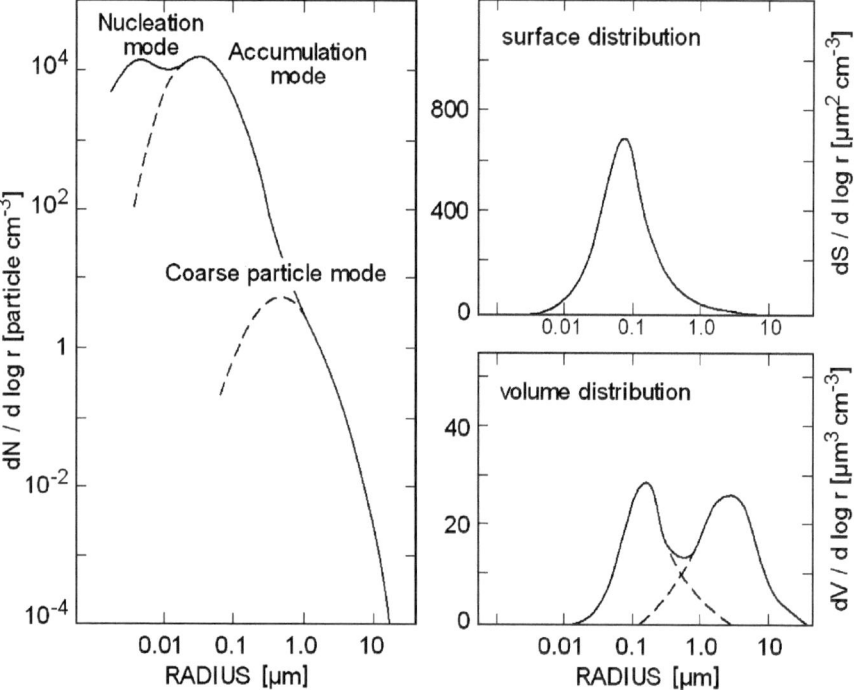

Fig. 5.1 Generalized size distributions for the concentrations of number, surface and volume of the tropospheric aerosol

The principal parameters to characterize the aerosol are number concentration, mass concentration, size distribution and chemical composition. These parameters vary appreciably depending on location and other factors.

Concentration: Typical mass and number concentrations in different regions are summarized in Table 5.1.

Size range: Approximately 2 nm–20 μm; larger particles in the atmosphere are rare because they undergo gravitational sedimentation at rates that increase rapidly with particle size.

Size distributions: It is necessary to distinguish distributions for concentrations of mass, number and surface. The general appearance of these distributions, which is shown in Fig. 5.1, indicates the existence of three modes termed nucleation mode, accumulation mode and coarse particle mode, respectively.

Size distribution of mass concentration: Bimodal; *accumulation mode* (0.05–1.0 μm) and *coarse particle mode* (>1 μm). Particles < 0.05 μm in size do not contribute significantly to the concentration of aerosol mass.

Size distribution of number concentration: Bimodal; *nuclei mode* (1–50 nm), *accumulation mode* (0.05–1.0 μm). The number concentration of particles in the coarse mode is $\leq 1\,cm^{-3}$, which is negligible compared to the total, but these particles, because of their volume, contribute significantly to the total mass concentration. Desert aerosols contain a larger fraction of particles in the coarse mode.

Bulk chemical composition: Table 5.2 provides typical concentrations and mass fractions of chemical constituents in rural continental and marine aerosols.

Size distribution of chemical composition: Particles resulting from primary emissions of mineral dust, sea salt and biogenic material are found throughout the entire size range >0.1 μm, but they contribute to total mass mainly in the coarse particle mode. Newly produced particles originating either from direct emissions of combustion products or from gas-phase condensation reactions in the air occur in the nucleation mode. The accumulation mode arises from several sources: the coagulation of particles present in the nucleation mode, the condensation of products from gas-phase chemical reactions onto pre-existing particles, and chemical reactions in the aqueous phase of clouds. The last process is effective because particles in the 0.05–1.0 μm size range are favored cloud condensation nuclei; they retain the newly acquired material when clouds dissipate and the particles are released to the environment. Electrolytes and organic compounds contribute appreciably to the mass of particles in the accumulation range.

Dependence on altitude: Above the boundary layer the aerosol becomes fairly uniformly mixed in the air, so that number and mass concentrations generally follow the decrease of air density with increasing altitude. In the main cloud region, the concentration of large particles may be significantly reduced due to uptake by cloud drops. The size distribution usually changes little with altitude, except in the coarse particle range, where losses occur. Because the upper troposphere receives air by high rising currents originating from the continents, it is generally assumed that particles in tropospheric regions outside the boundary layer (especially above the cloud level) represent an aged aerosol of continental origin.

Stratospheric aerosol: The stratospheric aerosol occurs as a layer at about 25 km altitude. It consists mainly of sulfuric acid particles with an admixture of nitrosyl sulfates and solid granules containing silicates. Sulfuric acid is produced by the oxidation of carbonyl sulfide, which has its origin in the troposphere, and from sulfur dioxide injected into the stratosphere together with ash particles by high-reaching volcanic eruptions.

Table 5.2 Typical chemical composition of tropospheric aerosols (at ground level)[a]

Constituent	Marine Concentration (μg m^{-3})	Marine Mass fraction (%)	Rural continental Concentration (μg m^{-3})	Rural continental Mass fraction (%)
Total electrolytes	12.2	93.3	10.0	40
Cl$^-$	5.80 (−0.7)[b]	47.5	0.14	2.0
SO$_4^{2-}$	1.92 (+1.0)[b]	34.9	5.54	76.9
NO$_3^-$	0.05	0.7	1.51	21.0
NH$_4^+$	0.16 (+0.16)[b]	4.3	2.05	73.2
Na$^+$	3.60	77.6	0.08	2.8
K$^+$	0.13	2.9	0.12	4.3
Mg^{2+}	0.43	10.7	0.25	8.9
Ca^{2+}	0.16 (+0.02)[b]	4.5	0.30	10.7z
Insoluble minerals	<0.1[c]	<0.8[c]	4.3	17.0
SiO$_2$	–	–	2.45	57.0
Al$_2$O$_3$	–	–	0.77	17.9
CaO	–	–	0.35	8.1
Fe$_2$O$_3$	–	–	0.73	17.0
Organic compounds[d]	0.80	6.7	7.5	30
Biogenic material[e]	0.40	3.4	1.7	7
Elemental carbon	0.15	–	1.5	6
Total	13.6		25.0	

[a] From Georgii and Warneck (1999) with changes and additions
[b] Excess compared with sea water is indicated in parentheses
[c] Minerals can make a significant contribution in terrestrial dust plumes over the oceans
[d] The number of organic compounds involved is too large to be given in detail here; for more information see Sect. 5.4
[e] Biogenic material includes plant debris, pollen, fungal spores in the continental atmosphere, and cellular material from the surface waters over the oceans

Table 5.3 Characteristics of marine aerosol particles, global annual average meridional distribution[a]

Latitude belt	Number density N_{nucl} (cm^{-3})	Number density N_{acc} (cm^{-3})	Mean diameter D_{nucl} (nm)	Mean diameter D_{acc} (nm)	Mass concentration $C_{seasalt}$ (μg m^{-3})	Mass concentration $C_{nsssulf}$ (μg m^{-3})	C_{MSA} a	C_{MSA} b
75–90° N	160	60	45 ± 1.5	170 ± 1.6	0.38 ± 0.17	0.572	11.5	22.3
60–75° N	–	–	–	–	7.38 ± 1.40	0.293	2.9	42.0
45–60° N	230	110	31 ± 1.5	200 ± 1.6	14.20 ± 3.00	0.315	3.1	22.3
30–45° N	210	250	45 ± 1.5	180 ± 1.4	6.49 ± 0.90	0.322	18.0	43.2
15–30° N	250	170	44 ± 1.4	170 ± 1.4	8.07 ± 0.73	0.757	26.8	37.0
0–15° N	280	240	46 ± 1.6	160 ± 1.5	7.28 ± 0.60	1.66ß	28.7	20.3
0–15° S	150	160	47 ± 1.5	170 ± 1.5	8.35 ± 0.73	1.900	26.8	17.1
15–30° N	390	220	40 ± 1.4	160 ± 1.4	12.20 ± 0.95	1.980	36.9	14.9
30–45° S	600	200	36 ± 1.4	150 ± 1.4	5.77 ± 0.52	2.750	34.7	–
45–60° S	300	110	31 ± 1.4	140 ± 1.5	14.80 ± 3.82	1.240	37.2	–
60–75° S	310	70	35 ± 1.4	150 ± 1.6	14.40 ± 5.05	0.886	85.6	13.6
75–90° S	–	–	–	–	0.32 ± 0.14	0.672	44.6	–

[a] Number concentration of particles in the nucleation mode N_{nucl}, in the accumulation mode N_{acc}, geometric mean diameter for the nucleation mode D_{nucl} and the accumulation mode D_{acc}, mass concentration of sea salt $C_{seasalt}$, non-seasalt sulfate $C_{nsssulf}$; methane sulfonate C_{MSA} (a Jan–March, b July–September); Source of data: Heintzenberg et al. (2000)

Representation of size distributions by log-normal functions: Observed aerosol size distributions are approximated mathematically by a series of three log-normal size distributions, representing size ranges for the nucleation mode (i = 1), accumulation mode (i = 2) and coarse particle mode (i = 3):

$$\frac{dN(r)}{d\log r} = \sum_{i=1}^{i=3} \frac{1}{\sqrt{2\pi}} \frac{n_i}{\log s_i} \exp\left[-\frac{\left(\log r / R_i\right)^2}{2\left(\log s_i\right)^2}\right]$$

Here, r (μm) is the particle radius, $N(r)$ (particle cm^{-3}) is the cumulative number density distribution for particles $> r$, R_i (μm) is the mean particle radius, n_i is the integral of the ith normal function, and $\log s_i$ is a measure of particle polydispersity.

Table 5.4 Model parameters for aerosol size distributions of number concentration based on a sum of three log-normal functions representing size ranges of (1) nucleation, (2) accumulation and (3) coarse modes[a]

Region	mode, i	n_i (cm^{-3})	R_i (μm)	$\log s_i$
Urban	1	9.93×10^4	6.51×10^{-3}	2.45×10^{-1}
	2	1.11×10^3	7.14×10^{-3}	6.66×10^{-1}
	3	3.64×10^4	2.48×10^{-2}	3.37×10^{-1}
Rural	1	6.65×10^3	7.39×10^{-3}	2.25×10^{-1}
	2	1.74×10^2	2.69×10^{-2}	5.57×10^{-1}
	3	1.99×10^3	4.19×10^{-2}	2.66×10^{-1}
Remote continental	1	3.20×10^3	1.00×10^{-2}	1.61×10^{-1}
	2	2.90×10^3	5.80×10^{-2}	2.17×10^{-1}
	3	3.00×10^{-1}	9.00×10^{-1}	3.80×10^{-1}
Desert dust storm	1	7.26×10^2	1.00×10^{-3}	2.47×10^{-1}
	2	1.14×10^3	1.88×10^{-2}	7.70×10^{-1}
	3	1.87×10^{-1}	1.08×10^1	4.38×10^{-1}
Background	1	1.29×10^2	3.60×10^{-3}	6.45×10^{-1}
	2	5.97×10^1	1.27×10^{-1}	2.53×10^{-1}
	3	6.35×10^{-1}	2.59×10^{-1}	4.25×10^{-1}
Marine	1	1.33×10^2	3.60×10^{-3}	6.57×10^{-1}
	2	6.66×10^1	1.33×10^{-1}	2.10×10^{-1}
	3	3.06×10^0	2.90×10^{-1}	3.96×10^{-1}
Polar	1	2.17×10^1	6.89×10^{-2}	2.45×10^{-1}
	2	1.86×10^{-1}	3.57×10^{-1}	3.00×10^{-1}
	3	3.04×10^{-4}	4.29×10^0	2.91×10^{-1}
Stratosphere[b]	2	4.49×10^0	2.17×10^{-1}	2.48×10^{-1}

[a] Source: Jaenicke (1998). For a regional characterization see the comments below
[b] Only one mode is applied for the stratospheric aerosol

Regional characterization:

The *remote continental aerosol* is assumed to occur in regions remaining essentially unaffected by humans.

The *rural aerosol* is a continental aerosol modified by a moderate impact due to anthropogenic sources.

Desert aerosols are greatly influenced by dust storms, which raise number and mass concentrations of particles >1 μm. Because of their potential for long-range transport, desert aerosols deserve a separate treatment.

The *urban aerosol* is associated with the high population density that creates a unique urban environment. A large fraction of the urban aerosol is due to emissions from automobiles traffic (exhaust emissions, road dust, etc.), home heating and industrial activities. Especially noteworthy is the high number concentration of Aitken particles from automobile exhaust.

The *tropospheric background aerosol* occurs in the free troposphere, that is, in the region above the main cloud layer. It is assumed to represent an aged continental aerosol that has lost much of the coarse particle fraction.

The *maritime aerosol*, characteristic of the marine boundary layer, is composed of sea salt particles modified in the accumulation range by nitrate resulting from the interaction with nitric acid, ammonium sulfate from the oxidation of sulfur compounds, and a possible admixture of the tropospheric background aerosol. A mineral component occurs only in regions affected by long-range transport of desert dust.

Polar aerosols deserve a separate treatment. The Antarctic and the Arctic are regions of low particle concentrations but with a significant marine component. At least the Arctic, however, receives an input of aged aerosol particles from the adjacent continents, which in the winter can accumulate to form an Arctic haze.

References

Georgii, H.W., P. Warneck, in *Global Aspects of Atmospheric Chemistry*, ed. by R. Zellner. Topics in Physical Chemistry, vol. 6 (Steinkopff, Darmstadt, 1999), pp. 111–179

Heintzenberg, J., D.C. Covert, R. Van Dingenen, Tellus **52B**, 1104–1122 (2000)

Jaenicke, R., in *Atmospheric Particles, IUPAC Series on Analytical and Physical Chemistry of Environmental Systems*, ed. by R.M. Harrison, R. Van Grieken, vol. 5 (Wiley, Chichester, 2000), pp. 1–28

Warneck, P., *Chemistry of the Natural Atmosphere*, 2nd edn. (Academic Press, San Diego, 2000), Copyright Elsevier

5.2   Global Production/Emission Rates

Table 5.5 Overview on global mass production rates of particulate matter from various sources

Type of source[a]	1995 Summary[b]	More recent estimates[c]
Natural sources		
Primary emissions:		
Mineral dust (Pg a^{-1})	1.5 (1.0–3.0)	(1.5±0.7) (1), (1.8±0.2) (2), 1.8 (3), 1.8 (4)
Sea salt (Pg a^{-1})	1.3 (1–10)	5.2 (5), 5.9 (6), 10.1 (7), 2.78 (8), 3.53 (9), 6.5 (10)
Volcanoes (Tg a^{-1})	33 (4–100)	20 (11)
Biological particles (Tg a^{-1})	50 (26–80)	
Secondary sources (Tg a^{-1}):		
Sulfate from biogenic gases	90 (60–110)	33 (12), 16.7–39.8 (13)
Sulfate from volcanic SO$_2$	12 (4–40)	13 (11), 4.7–13.4 (13)
Products from biogenic NMHC	55 (40–200)	18.5 (14), 7.8 (15)
Nitrates from NO$_x$	22 (10–40)	
Anthropogenic sources		
Primary emissions (Tg a^{-1}):		
Industrial dust, etc.	100 (40–130)	
Organic carbon, fossil fuels	10 (5–25)	28.5 (15), 8.9 (17), 3.1 (18)
Black Carbon (a) industrial	10 (5–20)	6.6 (15), 4.7 (17), 2.8 (18)
(b) biomass burning	~10	5.6 (15), 4.8 (16), 3.3 (17), 5.8 (18), 5.2 (19)
Secondary sources (Tg a^{-1}):		
Sulfate from SO$_2$	140 (120–180)	111 (12), 100.4–156.7 (13)
Biomass burning (w/o BC)[d]	80 (50–140)	45 (15), 77 (16), 32 (18), 53 (18), 46 (19)
NMHC oxidation products	10 (5–25)	
Nitrates from NO$_x$	36 (20–50)	

[a] Only particles <10 μm in size that remain airborne and do not precipitate in the vicinity of the source area. See further comments below. NMHC = non-methane hydrocarbons
[b] Andreae (1995), best estimate and range
[c] Key: (1) Tegen et al. (1996); (2) Dentener et al. (1996); (3) Ginoux et al. (2001); (4) Tegen et al. (2002); (5) Warneck (2000); (6) Tegen et al. (1997); (7) Gong et al. (2002); (8) Erickson et al. (1999); (9) Takemura et al. (2000); (10) Grini et al. (2002); (11) Mather et al. (2003); (12) Chin and Jacob (1996); (13) Koch et al. (1999); (14) Griffin et al. (1999); (15) Liousse et al. (1996); (16) Andreae and Merlet (2001); (17) Bond et al. (2004); (18) Ito and Penner (2005); (19) Generoso et al. (2003)
[d] This includes primary and secondary aerosol except black carbon

Comments:

Mineral dust: Recent estimates are obtained by comparing global dust distributions observed from satellites with results of global transport models based on prescriptions for dust emissions, global distribution of arid regions, and local wind force. As the largest particles are re-deposited in the source region, the emission rate depends critically on the assumed upper cutoff limit of the particle size.

Sea salt: The estimate of Warneck (2000) is based on precipitation data; the other estimates were obtained by combining a suitable source prescription with global wind fields. Again, the assumed upper size limit of the particles is critical in determining the mass emission rate.

Volcanoes: Small eruptions emit fine ash particles remaining in the troposphere. Violent eruptions can inject material also into the stratosphere, but here the effect of sulfur dioxide is more important. A fraction of ~50% of the average annual rate of SO_2 emission from volcanoes, quiescent and eruptive, is converted to sulfate in the troposphere, whereas the remainder is lost by wet and dry deposition. The return rate of sulfate from the stratosphere is small in comparison.

Biological particles: This category includes seeds, pollen, spores and bacteria as well as leaf waxes, resins, and other material shed by leaves. Continental aerosols contain roughly 10% by volume.

Sulfate from biogenic gases: The major contributor is the oxidation of dimethylsulfide (DMS), which leads mainly to sulfur dioxide as end product with smaller contributions from intermediate species such as methane sulfonate. As in the case of volcanic SO_2, 51% are assumed to be oxidized to form sulfate.

Products from biogenic non-methane hydrocarbons: Typical hydrocarbons in this category are isoprene and the terpenes. Their oxidation leads partly to the formation of non-volatile products that condense to become associated with aerosol particles.

Nitrates: The oxidation of NO_x in the atmosphere produces nitric acid, which can react with ammonia and alkaline materials present in soil dust and sea salt in addition to undergoing dry and wet deposition. The fraction entering into chemical reactions to form nitrate is highly uncertain. The 25% conversion factor assumed by Andreae (1995) is probably too large.

Industrial dust: This category comprises dust production by human activities such as transportation, power plants, cement manufacturing and metallurgy, waste incineration, etc. As these sources affect environmental quality, they have become widely regulated and emissions have been reduced significantly in developed countries. Mainly coarse particles are produced that are deposited close to the sources.

Combustion: Many individual sources contribute to the emission of primary organic particles from combustion processes. Major contributors are the open burning of biomass (forest, savanna, crop residuals), and the contained combustion of fossil fuels, wood and wastes. A fraction of primary particular material consists of nearly pure elemental carbon, with some oxygen and hydrogen attached, in a layered, hexagonal structure, called 'Black Carbon'. The material is formed by charring of organic material during combustion or by condensation from the gas phase in reducing flames as soot. It is the principal organic substance in aerosols strongly absorbing light, so that it can be considered separately. The oxidation in the atmosphere of anthropogenic non-methane hydrocarbons emitted as a by-product from combustion engines and other sources may also lead to the production of aerosols, but the emission rate is much lower than that of natural hydrocarbons, and the conversion efficiency also is lower. The open combustion of biomass produces copious amounts of secondary aerosols in the form of tar condensates and photochemically generated organic sulfate and nitrate compounds, in addition to primary aerosol composed of ash and charcoal particles. In this case, primary and secondary aerosols cannot be differentiated except for black carbon.

Table 5.6 Global emissions (Gg a^{-1}) of atmospheric trace metals from natural sources, median and range[a]

Trace metal	Wind-borne soil particles	Sea salt	Volcanoes	Forest fires	Biogenic processes		Total
					Continental[b]	Marine	
As	2.6	1.7	3.8	0.19	1.6	2.3	12
	(0.3–5.0)	(0.19–3.1)	(0.15–7.5)	(0.0–0.38)	(0.23–3.0)	(0.16–4.5)	(0.86–23)
Cd	0.21	0.06	0.82	0.11	0.19	0.05	1.3
	(0.01–0.4)	(0.0–0.11)	(0.14–1.5)	(0.0–0.22)	(0.0–0.83)	(0.0–0.1)	(0.15–2.6)
Co	4.1	0.07	0.96	0.31	0.58	0.08	6.1
	(0.6–7.5)	(0.0.14)	(0.02–1.9)	(0.02–0.6)	(0.05–0.12)	(0.0–0.15)	(0.69–11)
Cr	27	0.07	15	0.09	1.05	0.06	44
	(3.6–50)	(0.03–1.4)	(0.81–29)	(0.0–0.18)	(0.1–2.0)	(0.0–0.12)	(4.5–8.3)
Cu	8.0	3.6	9.4	3.8	2.9	0.39	28
	(0.9–15)	(0.23–6.9)	(0.9–18)	(0.1–7.5)	(0.1–5.6)	(0.02–0.75)	(2.3–54)
Hg	0.05	0.02	1.0	0.02	0.63	0.77	2.5
	(0.0–0.1)	(0.0–0.04)	(0.03–2.0)	(0.0–0.05)	(0.02–1.24)	(0.04–1.5)	(0.1–4.9)
Mn	221	0.86	42	23	28.3	1.5	317
	(42–400)	(0.02–1.7)	(4.2–80)	(1.2–45)	(4–50)	(0.08–3.0)	(52–582)
Mo	1.3	0.22	0.4	0.57	0.46	0.08	3.0
	(0.12–2.5)	(0.01–0.43)	(0.04–0.75)	(0.04–1.1)	(0.04–0.75)	(0.0–0.15)	(0.14–5.8)
Ni	11	1.3	14	2.3	0.61	0.12	30
	(1.8–20)	(0.01–2.6)	(0.93–28)	(0.1–4.5)	(0.1–1.0)	(0.01–0.45)	(3.0–57)

(continued)

Table 5.6 (continued)

Trace metal	Wind-borne soil particles	Sea salt	Volcanoes	Forest fires	Biogenic processes[b] Continental	Marine	Total
Pb	3.9	1.4	3.3	1.9	1.5	0.24	12
	(0.3–7.5)	(0.02–2.8)	(0.54–6.0)	(0.06–3.8)	(0.03–2.9)	(0.02–0.45)	(0.97–23)
Sb	0.78	0.56	0.71	0.22	0.24	0.05	2.4
	(0.06–1.5)	(0.0–1.19)	(0.01–1.4)	(0.0–0.45)	(0–1.2)	(0.0–0.1)	(0.07–4.7)
Se	0.18	0.55	0.95	0.26	3.7	4.7	9.3
	(0.01–0.35)	(0–1.1)	(0.1–1.8)	(0.0–0.52)	(0.15–5.3)	(0.4–9.0)	(0.66–18)
V	16	3.1	5.6	1.8	1.05	0.16	28
	(1.2–30)	(0.14–7.2)	(0.21–11)	(0.02–3.6)	(0.04–2.05)	(0.02–0.3)	(1.6–54)
Zn	19	0.44	9.6	7.6	5.1	3.0	45
	(3.0–35)	(0.02–0.86)	(0.31–19)	(0.3–15)	(0.3–10)	(0.04–6.0)	(4.0–86)

[a]Source of data: Niagru (1989)
[b]Primarily particles shed by plants

Table 5.7 Global emissions (Gg a^{-1}) of atmospheric trace metals from anthropogenic sources[a]

Trace metal	Coal combustion		Oil combustion		Non-ferrous metals	Steel manufacture	Refuse incineration	Phosphate fertilizers	Wood combust.	Cement production	Misc.	Median value
	Electric power	Industry and domestic	Electric power	Industry and domestic								
As	0.23–1.55	0.2–1.98	0.006–0.03	0.007–0.072	9.55–15.08	0.36–2.48	0.17–0.45	–	0.06–0.3	0.18–0.89	1.25–2.80	18.82
Cd	0.08–0.39	0.1–0.5	0.023–0.17	0.02–0.072	2.66–8.20	0.03–0.28	0.06–1.44	0.07–0.27	0.06–0.18	0.009–0.534	–	7.57
Cr	1.24–7.75	1.68–11.9	0.09–0.58	0.36–1.79	–	2.84–28.40	0.25–1.43	–	0.89–1.78	–	–	30.48
Cu	0.93–3.10	1.39–4.95	0.35–2.32	0.18–1.07	15.13–32.57	0.14–2.84	1.01–2.14	0.14–0.68	0.6–1.2	0.02–14.24	–	35.37
Hg	0.16–0.54	0.49–2.97	–	–	0.05–0.22	–	0.16–2.16	–	0.06–0.30	–	–	3.56
In	–	–	–	–	0.011–0.039	–	–	–	–	–	–	0.025
Mn	–	1.49–11.88	0.06–0.58	0.36–1.79	2.3–33.5	–	5.25–10.0	–	–	–	–	38.27
Mo	0.23–2.32	0.40–2.48	0.06–0.41	0.11–0.54	–	–	–	–	–	–	–	3.27
Ni	1.4–9.3	1.98–14.85	3.84–14.5	7.16–28.64	8.78	0.04–7.10	0.13–0.60	0.14–0.69	0.6–1.8	0.09–0.89	–	55.65
Pb	0.78–4.65	0.99–9.90	0.23–1.74	0.72–2.15	30.06–69.64	1.07–14.2	1.64–3.10	0.06–0.27	1.2–3.0	0.02–14.24	252[b]	332.35
Sb	0.16–0.78	0.2–1.48	–	–	0.69–2.38	0.004–0.007	0.44–0.90	0.0004–0.001	–	–	–	3.51
Se	0.11–0.78	0.79–1.98	0.04–0.29	0.11–0.54	0.74–2.10	0.001–0.002	0.03–0.10	–	–	–	–	3.79
Sn	0.16–0.76	0.10–0.99	0.35–2.32	0.29–3.58	0.43–1.70	–	0.16–1.46	–	–	–	–	6.14
Tl	0.16–0.62	0.50–0.99	–	–	–	–	–	–	–	2.67–5.34	–	5.14
V	0.31–4.65	1.0–9.9	6.96–52.20	21.48–71.60	0.04–0.09	0.07–1.42	0.3–2.0	–	–	–	–	86.00
Zn	1.09–7.75	1.49–11.88	0.17–1.28	0.36–2.51	51.03–93.83	7.10–31.95	2.95–8.85	1.37–6.85	–	1.2–6.0	1.72–4.78	131.88

[a]Partly rounded values of estimates reported by Niagru and Pacina (1988)
[b]Predominantly emissions from gasoline-powered mobile sources

Table 5.8 Global emissions of particulate matter (PM) resulting from biomass burning (Tg a^{-1})[a]

Species[b]	Savannah and grassland	Tropical forest	Extra-tropical forest	Bio-fuel burning	Charcoal production	Charcoal burning	Agricultural residues[c]	Total
Dry matter burned	3,160	1,330	640	2,701	158	38	540	8,600
Total PM	26.2	11.3	11.3	25.5	0.63	0.46	7.0	82.4
PM$_{2.5}$	16.1	12.0	8.3	19.4	–	0.34	2.1	58.3
Organic carbon	10.6	7.0	5.8	10.7	–	0.18	1.8	36.1
Black carbon	1.5	0.88	0.36	1.6	–	0.06	0.37	4.8
Total carbon	11.7	8.7	5.3	14.0	–	0.24	2.2	42.2
Potassium, K	1.09	0.39	0.16	0.14	–	0.02	0.15	1.9
CN	1.1 (28)	4.5 (27)	2.2 (27)	9.2 (27)	–	1.3 (26)	1.8 (27)	2.9 (28)
CCN (1% S)	6.3 (27)	2.7 (27)	1.7 (27)	5.4 (27)	–	7.6 (25)	1.1 (27)	1.7 (28)
NP(>0.12 μm)	3.7 (27)	1.3 (27)	6.4 (26)	2.7 (27)	–	3.8 (25)	5.4 (26)	9.0 (27)

[a]Source of data: Andreae and Merlet (2001); values in parentheses indicate powers of ten

[b]Abbreviations: $PM_{2.5}$ particles in the size range ≤ 2.5 μm, CN condensation nuclei, CCN cloud condensation nuclei activated at 1% supersaturation, $NP(>0.12$ μm) number of particles produced in the size range >0.12 μm

[c]Value excludes agricultural waste used as bio-fuel

References

Andreae, M.O,. in *World Survey of Climatology*, ed. by A. Henderson-Sellers. Future Climates of the World, vol. 16 (Elsevier, Amsterdam, 1995), pp. 341–392

Andreae, M.O., P. Merlet, Glob. Biogeochem. Cycles **15**, 955–966 (2001)

Bond, T.C., D.G. Streets, K.F. Yarber, S.M. Nelson, J.-H. Woo, Z. Klimont, J. Geophys. Res. **109**, D14203, 1–43 (2004)

Chin, M., D. Jacob, J. Geophys. Res. **101**, 18691–18699 (1996)

Dentener, F.J., G.R. Carmichael, Y. Zhang, J. Lelieveld, P. Crutzen, J. Geophys. Res. **101**, 22869–22889 (1996)

Erickson III, D.J., C. Seuzaret, W.C. Keene, S.L. Gong, J. Geophys. Res. **104**, 8347–8372 (1999)

Generoso, S., F.M. Bréon, Y. Balkanski, O. Boucher, M. Schultz, Atmos. Chem. Phys. **3**, 1211–1222 (2003)

Ginoux, P., M. Chin, I. Tegen, J.M. Prospero, B. Holben, O. Dubovik, S.-J. Lin, J. Geophys. Res. **106**, 20255–20273 (2001)

Gong, S.L., L.A. Barrie, M. Lazare, J. Geophys. Res. **107**(D24), 4779, 1–14 (2002)

Griffin, R.J., D.R. Cocker III, J.H. Seinfeld, D. Dabdub, Geophys. Res. Lett. **26**, 2721–2724 (1999)

Grini, A., G. Myhre, J.K. Sundet, I.S.A. Isaksen, J. Clim. **15**, 1717–1730 (2002)

Ito, A., J.E. Penner, Glob. Biogeochem. Cycles **19**, GB2028, 1–14 (2005)

Koch, D., D. Jacob, I. Tegen, D. Rind, M. Chin, J. Geophys. Res. **104**, 23799–23822 (1999)

Liousse, C., J.E. Penner, C. Chuang, J.J. Walton, H. Eddleman, H. Cachier, J. Geophys. Res. **101**, 19411–19432 (1996)

Mather, T.A., D.M. Pyle, C. Oppenheimer, in *Volcanism and the Earth's Atmosphere*, ed. by A. Robock, C. Oppenheimer. Geophysical Monograph, vol. 137 (American Geophysical Union, Washington, DC, 2003), pp. 189–212

Niagru, J.O., Nature **338**, 47–49 (1989)

Niagru, J.O., J.M. Pacina, Nature **333**, 134–139 (1988)

Takemura, T., H. Okamoto, Y. Marujama, A. Numaguti, A. Higurashi, T. Nakajima, J. Geophys. Res. **105**, 17853–17873 (2000)

Tegen, I., A.A. Lacis, I. Fung, Nature **380**, 419–422 (1996)

Tegen, I., P. Hollrig. M. Chin, I. Fung, D. Jacob, J. Penner, J. Geophys. Res. **102**, 23895–23915 (1997)

Tegen, I., S.P. Harrison, K. Kohfeld, I.C. Prentice, J. Geophys. Res. **107**, D21, 4576 (2002). doi:10.1029/2001JD000963

Warneck, P., *Chemistry of the Natural Atmosphere*, 2nd edn. (Academic Press, San Diego, 2000)

5.3 Elemental Composition of Atmospheric Particles

Soil dust, volcanoes, fly ash from power plants, and other combustion products are significant sources of particulate matter in the atmosphere. Their contribution can be assessed by studying the elemental composition of aerosol and source materials. The data in the following tables have been selected from an extensive literature to provide a comprehensive listing of elemental concentrations. Subsequent tables list concentrations observed in various regions of the world.

Table 5.9 Elemental abundances (mg kg^{-1}) in Saharan desert soils: geometric average and range, and ratio X/Al, for bulk material and for particles smaller than 16 μm radius[a]

	0.5 < r < 400 μm			r < 16 μm		
	Abundance		X/Al[b]	Abundance		X/Al[b]
Na	3,144	(1,199–8,246)	0.138	6,244	(4,745–8,216)	0.101
Mg	5,327	(1,914–14,825)	0.234	18,116	(11,725–27,991)	0.292
Al	22,729	(9,418–54,853)	1	62,063	(47,394–81,273)	1
Si	284,747	(134,013–602,900)	1.253	122,113	(51,143–291,562)	1.968
Cl	52.8	(28.3–98.5)	2.32 (−3)	80.4	(23.9–270.3)	1.30 (−3)
K	7,550	(3,970–14,360)	0.332	13,510	(10,069–18,126)	0.218
Ca	7,967	(1,852–34,276)	0.351	22,837	(11,135–46,836)	0.368
Sc	4.14	(1.72–9.96)	1.82 (−4)	14.5	(9.52–22.1)	2.34 (−4)
Ti	2,488	(1,140–5,431)	0.109	10,339	(6,500–16,449)	0.167
V	34.3	(12.4–94.8)	1.51 (−3)	130.7	(101.8–168)	2.11 (−3)
Cr	25.4	(10.5–61.3)	1.11 (−3)	89.3	(61.6–129.6)	1.44 (−3)
Mn	190.6	(67.5–538.2)	8.39 (−3)	685.7	(540.1–870.6)	0.011
Fe	11,872	(4,647–30,330)	0.522	38,026	(28,188–51,296)	0.613
Co	4.21	(1.43–12.4)	1.85 (−4)	14	(9.09–21.5)	2.25 (−4)
Cu	23.3	(7.27–74.4)	1.03 (−3)	123	(58.7–253.4)	1.98 (−3)
Zn	43.9	(17.6–109.7)	1.93 (−3)	152.6	(89.4–260.5)	2.46 (−3)
Ga	4.94	(2.08–11.8)	2.17 (−4)	14.2	(9.23–21.9)	2.29 (−4)
As	1.52	(0.78–2.94)	6.69 (−5)	4.28	(1.79–10.2)	6.90 (−5)
Rb	32.6	(15.4–68.9)	1.43 (−3)	65.8	(46.6–92.8)	1.06 (−3)
Sr	94.3	(56.1–158.5)	4.15 (−3)	106.5	(52.1–217.5)	1.72 (−3)
Zr	365	(187–713)	0.016	1,097	(369–3,262)	0.0177
Ag	0.36	(0.15–0.84)	1.59 (−5)	2.37	(0.76–7.39)	3.82 (−5)
Sb	0.32	(0.15–0.71)	1.41 (−5)	0.75	(0.19–2.9)	1.21 (−5)
I	1.34	(0.63–2.85)	5.90 (−5)	3.28	(1.69–6.37)	5.28 (−5)
Cs	0.8	(0.41–1.54)	3.52 (−5)	2.14	(1.32–3.47)	3.45 (−5)
Ba	231.3	(131.4–407.1)	0.0102	329.1	(323.8–474.9)	5.30 (−3)
La	20.2	(12.4–32.7)	8,89 (−4)	89.4	(41.8–191.1)	1.44 (−3)
Ce	41.3	(18.8–90.7)	1.82 (−3)	143.2	(57.6–356.1)	2.31 (−3)
Nd	15.8	(8.93–27.9)	6.95 (−4)	73.8	(27.3–199.2)	1.19 (−3)
Sm	3.06	(1.79–5.22)	1.35 (−4)	12.9	(6.36–26.2)	2.08 (−4)
Eu	0.65	(0.34–1.24)	2.86 (−5)	2.19	(1.44–3.32)	3.53 (−5)
Tb	0.37	(0.22–0.63)	1.63 (−5)	1.6	(0.9–2.84)	2.58 (−5)
Dy	2.76	(1.54–4.94)	1.21 (−4)	13.6	(7.3–26.6)	2.19 (−4)
Yb	2.05	(1.1–3.81)	9.02 (−5)	8.38	(3.56–19.7)	1.35 (−4)
Lu	0.35	(0.18–0.67)	1.54 (−5)	1.32	(0.62–2.81)	2.13 (−5)
Hf	16.4	(6.6–40.6)	7.21 (−4)	51.7	(10.1–263.6)	8.33 (−4)
Ta	0.81	(0.31–2.1)	3.56 (−5)	2.69	(1.62–4.48)	4.33 (−5)
W	0.58	(0.24–1.38)	2.55 (−5)	1.72	(0.8–3.69)	2.77 (−5)
Au	0.004	(0.0021–0.0076)	1.76 (−7)	0.061	(0.023–0.17)	9.83 (−7)
Th	5.29	(3.13–8.59)	2.33 (−4)	26.1	(8.73–78.3)	8.52 (−7)

[a]Composite of 4 Sahara soil samples from different locations. Source: Jaenicke and Schütz (1988), numerical data communicated by Schütz
[b]Values in parentheses indicate powers of ten

Comments: The percentage contributions of major elements are approximately:

	Na	Mg	Al	Si	K	Ca	Ti	Mn	Fe	Zr	Ba
Bulk material	0.93	1.57	6.71	84.04	2.23	2.35	0.73	0.06	3.50	0.11	0.07
Fine fraction	2.11	6.13	20.98	41.29	4.57	7.72	3.50	0.23	12.86	0.37	0.11

The comparison shows that alkali and earth alkaline elements as well as aluminum and iron are enriched in the < 20 μm fraction. The major minerals involved are quartz and alumino-silicates; the former is dominant in the coarse fraction and the latter are dominant in the fine fraction.

Table 5.10 Elemental abundances (mg kg^{-1}) in re-suspended Chinese Loess samples: average and range, ratio X/Al, for total suspended matter and for particles smaller than 10 μm radius[a]

	TSP[b]			r < 10 μm		
	Abundance		X/Al[c]	Abundance		X/Al[c]
Na	1,637	(705–3,130)	0.0293	4,234	(356–16,263)	0.0463
Mg	1,668	(1,026–2,311)	0.0299	3,520	(1,790–4,926)	0.0385
Al	55,792	(43,928–63,636)	1	91,356	(69,416–110,262)	1
Si	175,425	(151,426–201,247)	3.144	274,143	(233,983–342,000)	3.000
P	328	(197–559)	5.88 (−3)	509	(245–609)	5.57 (−3)
S	295	(174–514)	5.29 (−3)	509	(375–722)	5.57 (−3)
Cl	587	(550–868)	0.0105	757	(566–924)	8.29 (−3)
K	17,200	(14,341–20,351)	0.308	25,184	(20,043–27,724)	0.276
Ca	67,789	(44,302–102,251)	1.215	98,332	(63,566–150,331)	1.076
Ti	2,980	(2,582–3,371)	0.0534	4,099	(3,324–4,861)	0.0449
V	107	(77–133)	1.92 (−3)	161	(119–207)	1.76 (−3)
Cr	71	(58–83)	1.27 (−3)	95	(69–123)	1.04 (−3)
Mn	723	(596–879)	0.0130	1,060	(841–1,208)	0.0116
Fe	31,216	(25,407–37,396)	0.560	43,955	(33,336–47,585)	0.481
Co	60	(38–88)	1.08 (−3)	86	(66–103)	9.41 (−4)
Ni	32	(21–36)	5.74 (−4)	48	(41–60)	5.25 (−4)
Cu	31	(25–40)	5.55 (−4)	50	(43–62)	5.47 (−4)
Zn	97	(67–124)	1.74 (−3)	163	(113–291)	1.78 (−3)
Ga	–	–	–	1	(0–3)	1.09 (−5)
As	15	(10–17)	2.69 (−4)	28	(19–41)	3.06 (−4)
Se	2	(0–5)	3.58 (−5)	2	(1–5)	2.19 (−5)
Br	3	(1–6)	5.38 (−5)	2	(1–5)	2.19 (−5)
Rb	79	(66–100)	1.42 (−3)	108	(95–135)	1.18 (−3)
Sr	166	(124–214)	2.98 (−3)	232	(181–317)	2.54 (−3)
Y	24	(18–27)	4.30 (−4)	39	(21–60)	4.27 (−4)
Zr	86	(80–94)	1.54 (−3)	112	(83–127)	1.23 (−3)
Mo	10	(0–20)	1.79 (−4)	12	(0–29)	1.31 (−4)
Pd	5	(2–12)	8.96 (−5)	11	(2–31)	1.20 (−4)
Ag	8	(3–16)	1.43 (−4)	11	(0–29)	1.20 (−4)
Cd	16	(8–23)	2.87 (−4)	20	(3–48)	2.19 (−4)

(continued)

Table 5.10 (continued)

	TSP[b]			$r < 10$ μm		
	Abundance		X/Al[c]	Abundance		X/Al[c]
In	12	(4–18)	2.15 (–4)	23	(4–36)	2.52 (–4)
Sn	20	(7–25)	3.58 (–4)	12	(0–24)	1.31 (–4)
Sb	4	(0–5)	7.17 (–5)	5	(0–24)	5.47 (–5)
Ba	239	(99–371)	4.28 (–3)	298	(158–464)	3.26 (–3)
La	64	(0–141)	1.15 (–3)	164	(47–376)	1.80 (–3)
Au	3	(0–11)	5.38 (–5)	1	(0–4)	1.10 (–5)
Hg	6	(5–9)	1.08 (–4)	10	(4–14)	1.10 (–4)
Tl	5	(1–11)	8.96 (–5)	8	(0–19)	8.76 (–5)
Pb	45	(32–58)	8.07 (–4)	69	(34–99)	7.55 (–4)
U	4	(1–11)	7.17 (–5)	4	(0–12)	4.38 (–5)

[a] Composite of samples from five sites on the Chinese Loess Plateau (Cao et al. 2008); range is for averages at different sites
[b] Total suspended particles derived from samples passed through 400 mesh sieve, nominal geometric diameter <38.5 μm
[c] Numbers in parentheses indicate powers of ten

Table 5.11 Elemental composition of fly-ash and plume aerosol from coal-fired power plants[a]

	NBS Standard	Fly ash[b]		Plume aerosol[b]		
		X/Al[c]	Wangen (1981)	X/Al[c]	Wangen (1981)	X/Al[c]
Na%	0.32±0.04	0.025	1.27±0.03	0.091	2186.6	0.175
Mg%	1.8±0.4	0.142	0.99±0.25	0.071	2740.5	0.219
Al%	12.7±0.5	1	13.9±0.64	1	12493.7	1
Si%	21±2	1.654	–	–	–	–
Cl	42±10	3.3×10^{-4}	–	–	139.3	0.011
K%	1.61±0.15	0.127	0.82±0.04	0.059	3491.6	0.279
Ca%	4.7±0.6	0.370	1.74±0.04	0.125	5271.9	0.422
Sc	27±1	2.1×10^{-4}	13.2±1.2	9.5×10^{-5}	1.8	1.5×10^{-4}
Ti%	0.74±0.03	0.058	0.6±0.08	0.043	683.8	0.055
V	235±13	0.0019	94.3±6.4	6.8×10^{-4}	14.6	0.0012
Cr	127±6	0.001	28.5±2.7	2.1×10^{-4}	114.9	0.0092
Mn	496±19	0.0039	149±4	0.0011	45.8	0.0027
Fe%	6.2±0.3	0.488	2.56±0.23	0.184	4811.0	0.385
Co	41.5±1.2	3.3×10^{-4}	9.6±0.7	6.9×10^{-5}	13.1	0.0011
Ni	98±9	7.7×10^{-4}	–	–	–	–
Cu	120	9.4×10^{-4}	65±10	4.7×10^{-4}	12.1	9.7×10^{-4}
Zn	216±25	0.0017	79.6	5.7×10^{-4}	70.5	0.0038
Ga	49	3.9×10^{-4}	33.0±5.1	2.4×10^{-4}	6.2	4.9×10^{-4}
As	58±4	4.6×10^{-4}	15.4±1.5	1.1×10^{-4}	4.3	3.5×10^{-4}
Se	10.2±1.4	8.0×10^{-5}	6.0±1.6	4.3×10^{-5}	7.1	5.7×10^{-4}
Br	12±4	9.4×10^{-5}	2.1	1.5×10^{-5}	3.4	2.7×10^{-4}
Rb	125±10	9.8×10^{-4}	–	–	–	–
Sr	1,700±300	0.013	488±47	0.0035	84.2	0.0067

(continued)

Table 5.11   (continued)

		Fly ash[b]		Plume aerosol[b]		
	NBS Standard	X/Al[c]	Wangen (1981)	X/Al[c]	Wangen (1981)	X/Al[c]
Zr	301 ± 20	0.0024	–	–	–	–
Cd	1.45 ± 0.06	1.1×10^{-5}	–	–	–	–
In	–	–	–	–	0.06	4.7×10^{-6}
Sb	6.9 ± 0.6	5.4×10^{-5}	3.4 ± 0.6	2.4×10^{-5}	0.86	6.9×10^{-5}
I	94.6	7.5×10^{-4}	–	–	1.6	1.3×10^{-4}
Cs	8.6 ± 1.1	6.8×10^{-5}	4.0 ± 0.5	2.9×10^{-5}	1.1	1.3×10^{-4}
Ba	$2,700 \pm 200$	0.021	$1,130 \pm 40$	0.0081	224.0	0.018
La	82 ± 2	6.5×10^{-4}	59.6 ± 1.3	4.3×10^{-4}	10.0	8.0×10^{-4}
Ce	146 ± 15	0.0011	110 ± 7.4	7.9×10^{-4}	18.3	0.0015
Sm	12.4 ± 0.9	9.7×10^{-5}	–	–	–	–
Hf	10.8	8.5×10^{-5}	9.4 ± 0.9	6.8×10^{-4}	3.0	2.4×10^{-4}
Ta	1.8 ± 0.3	1.4×10^{-5}	2.4 ± 0.3	1.7×10^{-5}	1.4	1.1×10^{-4}
W	–	–	2.7	1.9×10^{-5}	0.6	4.9×10^{-5}
Hg	0.14	1.1×10^{-6}	–	–	–	–
Pb	75 ± 5	5.9×10^{-4}	–	–	–	–
Th	24.8 ± 2.2	2.0×10^{-4}	27.0 ± 2.7	1.9×10^{-4}	3.6	2.9×10^{-4}
U	12.0 ± 0.5	9.4×10^{-5}	9.8	7.0×10^{-5}	0.7	5.7×10^{-5}

[a] National Bureau of Standard (NBS) fly-ash standard as reported by Klein et al. (1975), except Iodine from Gladney et al. (1975); fly-ash in electrostatic precipitator and aerosol concentrations in power plant plume (San Juan Basin, New Mexico), Wangen (1981)
[b] Fly ash concentrations in (mg kg^{-1}) except (%) where indicated; plume aerosol concentrations in (ng m^{-3})
[c] Relative composition with Aluminum as reference element

Table 5.12 Elemental composition of ash (or lava) and particulate matter emitted from volcanoes[a]

	St. Augustine				Mt. Etna				Kilauea	
	Ash (mg kg⁻¹)	X/Al[b]	Plume (ng m⁻³)	X/Al[b]	Lava (mg kg⁻¹)	X/Al[b]	Plume (µg m⁻³)	X/Al[b]	Plume (ng m⁻³)	X/Al[b]
Na	3.00±0.05%	0.333	240±30	0.33	2.71%	2.82	70±36	0.21	9,500	2.02
Mg	1.14±0.22%	0.093	150±100	0.13	3.07%	0.32	21±4	0.062	4,800	1.02
Al	9.01±0.24%	1	720±40	1	9.6%	1	341±59	1	4,700	1
S	1,740±190	0.019	7,700±800	10.7	2,800	0.029	358±214	1.05	9.4	2.0 (−3)
Cl	1,390±45	0.015	9,090±240	12.6	2,300	0.024	186±131	0.55	21	4.5 (−3)
K	0.84±0.04%	0.093	93±9	0.13	1.54%	0.16	97±55	0.28	2,700	0.57
Ca	4.05±0.19%	0.45	343±65	0.48	7.42%	0.77	61.8±19.7	0.18	–	–
Sc	15.4±0.1	1.7 (−4)	0.144±0.007	2.0 (−4)	20	2.1 (−4)	0.037±0.011	1.1 (−4)	3.8	8.1 (−4)
Ti	0.34±0.02%	0.038	46±22	0.064	0.63%	0.066	11.6±10.2	0.034	1,800	0.38
V	147±6	1.6 (−3)	1.52±0.2	0.0021	305	3.2 (−3)	0.45±0.40	1.3 (−3)	22	4.7 (−3)
Cr	24.3±0.3	2.4 (−4)	<13	–	32	3.3 (−4)	0.15±0.11	4.4 (−4)	60	0.013
Mn	765±15	8.5 (−3)	7.0±0.9	0.0097	1,000	0.01	2.9±1.3	8.4 (−3)	140	0.03
Fe	3.36±0.03%	0.373	389±9	0.54	7.72%	0.80	30±18	0.088	11,000	2.34
Co	11.4±0.2	1.3 (−4)	0.12±0.03	1.7 (−4)	29.5	3.1 (−4)	0.23±0.05	6.8 (−4)	4.7	1.0 (−3)
Ni	–	–	–	–	54	5.6 (−4)	0.07±0.02	2.1 (−4)	–	–
Cu	46±2	5.1 (−4)	15.0±1.1	0.021	90	9.4 (−4)	2.6±1.3	7.7 (−3)	12	2.5 (−3)
Zn	51.2±0.8	5.7 (−4)	2.20±0.15	0.0031	110	1.1 (−3)	0.84±0.65	2.5 (−3)	270	5.7 (−3)
As	1.11±0.01	1.2 (−5)	5.2±0.4	0.0073	1.0	1.0 (−5)	0.31±0.07	9.1 (−4)	250	0.057
Se	0.34±0.09	3.8 (−6)	2.35±0.04	0.0033	0.05	5.2 (−7)	0.84±0.65	2.5 (−3)	970	0.21
Br	5.9±0.8	6.5 (−5)	57±2	0.079	4.5	4.7 (−5)	5.9±4.5	0.017	310	0.066
Rb	22.4±0.8	2.5 (−4)	<26	–	40.3	4.2 (−4)	0.47±0.29	1.4 (−3)	–	–
Sr	410±70	0.0046	6.5±5.9	0.009	1,270	0.013	0.62±0.23	1.8 (−3)	–	–
Ag	–	–	–	–	0.9	9.4 (−6)	0.0048±0.003	1.4 (−5)	–	–
Cd	0.064±0.021	7.1 (−7)	1.07±1.07	0.0015	0.85	8.9 (−6)	0.16±0.08	4.9 (−4)	82	0.017
Sb	0.061±5	6.8 (−7)	0.69±0.04	9.5 (−4)	0.12	1.3 (−6)	5.0±5.5	0.015	2.6	5.5 (−4)

Cs	0.36±0.03	4.0(−6)	<0.0037	—	0.8	8.3(−6)	0.023±0.015	6.8(−5)	—	—
Ba	578±6	0.0064	7.6±2.6	0.012	680	7.1(−3)	0.33±0.13	9.8(−4)	—	—
La	12.8±0.4	1.4(−4)	0.156±0.007	2.1(−4)	72.5	7.5(−4)	0.053±0.021	1.5(−4)	2.8	6.0(−4)
Ce	25.9±0.1	2.9(−4)	0.333±0.013	4.6(−4)	105	1.1(−3)	0.10±0.04	2.9(−4)	—	—
Sm	3.64±0.01	4.0(−5)	0.035±0.0017	4.9(−5)	9.2	9.6(−5)	0.009±0.004	2.6(−5)	0.7	1.5(−4)
Eu	0.91±0.02	1.0(−5)	0.0091±0.0011	1.3(−5)	2.4	2.5(−5)	0.0023±0.001	6.7(−6)	—	—
Yb	2.81±0.03	3.1(−5)	0.024±0.006	3.3(−5)	1.3	1.4(−5)	0.0016±0.0007	4.7(−6)	—	—
Lu	0.41±0.004	4.6(−6)	0.0035±0.0013	4.9(−6)	0.31	3.2(−6)	0.0002±0.00009	6.5(−7)	—	—
Hf	3.57±0.03	4.0(−5)	0.031±0.004	4.3(−5)	4.3	4.5(−5)	0.002±0.0006	6.0(−6)	—	—
Ta	0.12±0.02	1.3(−6)	0.0037±0.0011	4.2(−6)	—	—	—	—	—	—
W	1.1±0.3	1.2(−5)	0.044±0.022	6.1(−5)	1.0	1.0(−5)	0.0021±0.0008	6.2(−6)	2.8	6.0(−4)
Au	0.0047±0.0007	5.2(−7)	0.00087±0.00017	1.2(−6)	0.1	1.0(−6)	0.0017±0.0002	4.9(−6)	2.0	4.3(−4)
Hg	not detected	—	>240	—	0.04	4.2(−7)	0.0053±0.0037	1.5(−5)	—	—
Pb	9.1±0.5	1.0(−4)	1.82±0.20	2.5(−3)	8	8.3(−5)	0.59±0.21	1.7(−3)	—	—
Th	2.54±0.02	2.8(−5)	0.024±0.002	3.3(−5)	8.6	9.0(−5)	0.007±0.003	2.1(−5)	—	—

[a] St. Augustine: Ash and aircraft samples taken during the eruption in February 1976 (Lepel et al. 1978); Mt. Etna: Average of lava and three samples taken in the plume of the Bocca Nuova summit crater at its rim in May and June 2001 before the eruption in July (Aiuppa et al. 2003); Kilauea: Aircraft sample taken in the eruption plume in March 1984 when Kilauea and Mauna Loa were in action simultaneously (Crowe et al. 1987)

[b] Relative composition with Aluminum as reference element; powers of ten in parentheses when fractions <0.01

Comment: Trace elements in particular matter emitted from volcanoes often are enriched compared to crustal abundances. Thermodynamic modeling suggests that in the high temperature environment of volcanoes these elements can be volatilized in the form of fluorides, chlorides or sulfides. Upon cooling in the atmosphere they are deposited as sublimates and become mainly attached to particles in the volcanic plumes.

Table 5.13 Elemental abundances (percent PM_{10} mass) of California soil dusts: re-suspended samples of paved road dust (PRD), unpaved road dust (UNPRD), agricultural fields (AGRI), and construction sites (CONSTR)[a]

	PRD[b]	UNPRD[c]	AGRI[d]	CONSTR[e]
TC	7.89±4.68	2.96±1.10	3.168±1.754	3.26±1.17
EC	0.99±0.95	0.098±0.267	0.207±0.545	0.359±0.374
C_{carb}	1.31±0.59	1.199±0.217	0.900±0.695	0.531±0.574
Na	0.156±0.083	0.387±0.413	0.239±0.181	0.615±0.089
Mg	0.783±0.052	0.665±0.426	0.763±0.169	0.727±0.170
Al	10.00±3.01	9.071±5.093	10.97±3.28	12.33±3.75
Si	28.17±8.96	30.57±13.58	33.09±10.49	38.21±12.29
P	0.388±0.354	0.093±0.064	0.099±0.069	0.080±0.046
S	0.352±0.21	0.424±0.226	0.228±0.252	0.132±0.053
Cl	0.101±0.142	0.169±0.292	0.006±0.0513	0±0.074
K	2.821±0.549	2.811±0.952	2.702±0.593	3.304±0.673
Ca	3.485±1.177	5.541±2.980	2.449±0.799	5.165±1.162
Ti	0.455±0.135	0.403±0.092	0.437±0.056	0.456±0.087
V	0.0047±0.0267	0.0053±0.0676	0.007±0.0415	0.0183±0.0504
Cr	0.0008±0.0083	0.0006±0.020	0.0026±0.0125	0.0039±0.0136
Mn	0.0759±0.0054	0.097±0.030	0.0893±0.0174	0.111±0.014
Fe	5.225±1.043	5.274±0.601	5.156±0.498	4.595±0.569
Co	0.0022±0.0823	0.001±0.083	0.0002±0.081	0.0052±0.0721
Ni	0.0013±0.0021	0.002±0.0055	0.0029±0.0044	0.0041±0.0044
Cu	0.0168±0.0119	0.0153±0.0072	0.0177±0.011	0.013±0.006
Zn	0.0965±0.0467	0.039±0.021	0.056±0.083	0.0283±0.0088
Ga	0.0014±0.0018	0±0.0047	0.0004±0.0028	0±0.0043
As	0.0016±0.0027	0.0021±0.0049	0.0015±0.0029	0.0008±0.0048
Se	0.0002±0.001	0.0004±0.0026	0.0002±0.0015	0.001±0.0024
Br	0.0016±0.0012	0.0015±0.0026	0.0008±0.0014	0.0006±0.0023
Rb	0.0139±0.0046	0.0156±0.0031	0.0137±0.0032	0.0162±0.0019
Sr	0.0305±0.0016	0.0523±0.0109	0.0294±0.0082	0.0537±0.0041
Y	0.0025±0.0013	0.0029±0.0029	0.0024±0.0017	0.0027±0.0021
Zr	0.0146±0.0057	0.012±0.003	0.0109±0.0028	0.0122±0.0027
Mo	0.0004±0.026	0±0.066	0.001±0.0039	0.0022±0.0059
Pd	0±0.016	0±0.04	0.0004±0.0237	0.0017±0.0372
Ag	0±0.047	0±0.047	0.0017±0.0279	0±0.00435
Cd	0±0.02	0.0014±0.0494	0.0027±0.0294	0±0.0458

(continued)

Table 5.13 (continued)

	PRD[b]	UNPRD[c]	AGRI[d]	CONSTR[e]
In	0.0002±0.022	0±0.056	0.0001±0.0332	0±0.0515
Sn	0.008±0.028	0.0074±0.0703	0.005±0.0413	0.0024±0.0647
Sb	0.0116±0.0321	0.0076±0.082	0.0103±0.0486	0.0132±0.0754
Ba	0.1215±0.0779	0.090±0.276	0.0827±0.157	0.0862±0.2517
La	0.0111±0.1524	0.0948±0.392	0.052±0.232	0.0024±0.359
Au	0.0003±0.0048	0.019±0.0086	0.0004±0.006	0.0005±0.0078
Hg	0.0008±0.0022	0.0014±0.0057	0.0008±0.0034	0.0017±0.0052
Tl	0.0003±0.0022	0.0003±0.0055	0.0005±0.0033	0.0005±0.005
Pb	0.0109±0.0074	0.0058±0.0073	0.0042±0.005	0.0084±0.0055
U	0.0011±0.0026	0.0017±0.0055	0.0012±0.0037	0.0018±0.0054

[a] Source: Chow et al. (2003); values represent averages and uncertainties (the higher of either root mean squared error or standard deviation of the average) of samples taken from the San Joaquin Valley, California; PM_{10} particles with aerodynamic diameter <10 μm, *TC* total carbon, *EC* elemental carbon, C_{carb} carbonate carbon
[b] Average of two urban and two rural roads
[c] Unpaved agricultural roads, composite of samples from three sites
[d] Agricultural soils: composite of 29 samples from almond orchard, cotton field, grape vineyard, safflower field, tomato field, after harvesting and land preparation
[e] Building and roadway construction/earth moving soil, two samples

Table 5.14 Chemical composition (mass percent) of aerosol particles from biomass burning, water-soluble species and elements[a]

	Tropical forest		Cerrado	
	Flaming	Smoldering	Flaming	Smoldering
Na	0.014±0.008	0.014±0.009	0.022±0.016	0.024±0.021
Mg	0.025±0.022	0.025±0.028	0.025±0.025	0.032±0.019
Al	0.2±0.2	0.6±0.5	0.6±0.5	0.7±1.2
Si	–	–	4.2±1.6	2.3±1.3
P	0.045±0.007	0.036±0.014	0.052±0.020	0.042±0.012
Cl	0.3±0.2	0.2±0.2	2.2±1.6	1.0±1.3
K	0.8±0.6	0.4±0.3	3.0±2.1	1.3±1.7
Ca	0.08±0.03	0.06±0.04	0.10±0.09	0.09±0.09
Ti	–	0.008±0.003	0.043±0.060	0.020±0.024
Cr	–	0.014±0.005	0.013±0.002	0.023±0.021
Mn	0.005±0.002	0.003±0.001	0.009±0.008	0.006±0.003
Fe	0.031±0.009	0.048±0.029	0.077±0.043	0.045±0.026
Ni	–	–	0.006±0.002	0.002±0.001
Cu	0.004±0.001	0.003±0.002	0.005±0.003	0.004±0.002
Zn	0.007±0.004	0.004±0.002	0.020±0.015	0.010±0.014
Se	0.0028±0.0024	0.0020±0.0009	0.0034±0.0007	0.0012±0.0003
Br	0.051±0.020	0.029±0.007	0.057±0.028	0.042±0.020
Rb	0.012±0.003	0.008±0.002	0.015±0.005	0.013±0.005
Sr	–	0.003±0.002	0.006±0.004	0.004±0.002
Zr	–	–	0.018±0.006	–

(continued)

Table 5.14 (continued)

	Tropical forest		Cerrado	
	Flaming	Smoldering	Flaming	Smoldering
NH_4^+	0.09±0.06	0.06±0.06	0.10±0.08	0.05±0.04
NO_3^-	0.12±0.09	0.12±0.07	0.62±0.33	0.33±0.25
SO_4^-	0.90±0.68	0.39±0.33	0.72±0.30	0.350.37
$HCOO^-$	0.019±0.009	0.025±0.014	0.024±0.012	0.023±0.013
CH_3COO^-	0.25±0.09	0.26±0.09	0.28±0.12	0.25±0.10
$C_2O_4^{2-}$	0.04±0.04	0.04±0.02	0.07±0.05	0.06±0.05
BC	7.3±5.5	3.9±1.9	12.6±6.8	6.5±7.6

[a] Aerosol particles collected in the plumes of tropical forest and cerrado biomass burning fires in the Amazon basin, Yamasoe et al. (2000)

Table 5.15 Elemental composition of particles from the combustion of oil and its derivatives

	Gasoline powered vehicles[a] exhaust concentrations (ng m^{-3})		Diesel engine exhaust (mass%)			Diesel fuel (mg dm^{-3})	Oil-fired boiler (mass%)	
	<2.6 μm	2.5–10 μm	HMC[b]	NIST[c]	Wang[d]	Wang[d]	HMC[e]	DSL[f]
Na	700±1,600	630±1,000	0.17	–	–	–	0.25	1.46
Mg	2,500±6,200	1,400±1,600	–	–	0.497	7.12	0.013	0.12
Al	4,100±5,400	2,400±3,200	0.052	–	2.14	32.8	0.42	1.4
Si	–	–	0.59	0.016	1.66	46.0	0.89	3.7
S	–	–	0.07	1.45	–	–	10.3	–
Cl	16,000±48,000	1,600±1,700	0.28	0.012	–	–	0.056	–
K	770±1,900	860±1,100	0.056	–	–	–	0.002	0.06
Ca	–	–	0.16	0.23	2.00	41.2	0.006	1.26
Ti	–	–	0.03	–	0.151	4.07	0.001	–
V	–	–	0.007	–	0.088	1.03	0.001	4.66
Cr	110±320	480±480	0.018	–	0.585	4.40	0.002	0.03
Mn	49±77	35±46	0.009	0.0015	0.123	1.04	0	0.04
Fe	2,400±4,400	5,300±5,900	0.13	0.069	3.13	27.8	0.13	3.55
Co	–	–	–	–	0.122	2.04	–	0.023
Ni	–	–	0.027	0.005	0.38	2.61	0	1.28
Cu	–	–	0	0.005	0.332	2.78	0.001	0.021
Zn	1,200±3,600	420±440	0	0.087	0.335	5.63	0.001	0.12
Br	1,600±6,500	160±170	0.011	0.00035	–	–	0.005	–
Sr	–	–	0	–	0.173	0.71	0	0.008
Cd	–	–	–	–	0.043	0.53	–	0.003
Sb	4.6±8.8	3.0±4.1	–	–	0.115	0.97	–	0.15
Ba	–	–	0	–	0.081	1.12	0	0.03
Mo	–	–	–	–	0.741	4.27	–	–
Pb	–	–	0	0.0023	0.072	2.04	0.032	0.05

[a] Average of 49 vehicles of US, Japanese and European origin; Huang et al. (1994); also reported La 1.1±1.3, 2.9±3.4, Sm 0.29±0.28, 0.07±0.15
[b] Heavy-duty diesel trucks (>33.7% carbonaceous material, <2.5 μm), Hildemann et al. (1991)
[c] NIST SRM 1,650 (~98% carbonaceous material, total PM) Huggins et al. (2000)
[d] Diesel exhaust (total PM) explored and fuel used by Wang et al. (2003)
[e] Industrial boiler operated with No. 2 fuel (<2.5 μm), Hildemann et al. (1991)
[f] Elemental composition of fly-ash from Porcheville (France) oil-fired power station; Desboeufs et al. (2005) and private communication

Table 5.16 Elemental composition of desert aerosols relative to Aluminum[a]

	Central Atlantic[b]	Syria[c]	Sudan[d]	Enewetak[e]	Zhenbeitai[f]
Na	0.052	0.45	0.158[†]	77±14*	0.15
Mg	0.216	–	0.205±0.039	9.6±15*	0.35
Si	–	1.79	2.59±0.14	–	1.9
P	–	–	0.033±0.008	–	–
S	–	7.38	0.33±0.23	–	–
Cl	–	–	0.208±0.176	120±15*	–
K	0.23	0.67	0.25±0.08	3.4±0.014*	0.32
Ca	0.33	1.29	0.73±0.15	3.92±0.014*	1.0
Sc	2.1 (−4)	3.8 (−4)	2.08 (−4)[†]	2.0±0.19 (−4)	5.5 (−4)
Ti	0.081	0.058	0.12±0.01	–	0.051
V	1.57 (−3)	9.2 (−3)	3.3±0.83 (−3)	2.10±0.02 (−3)	0.011*
Cr	1.35 (−3)	–	1.61±0.58 (−3)	1.77±0.02 (−3)	0.019*
Mn	0.013	0.02	0.019±0.002	0.010±0.019	0.015
Fe	0.66	0.58	1.01±0.10	0.658±0.019	0.59
Co	2.9 (−4)	9.2 (−3)	3.5 (−4)[†]	2.66±0.19 (−4)	8.3 (−4)
Ni	2.33 (−3)	–	1.9±0.8 (−3)	–	3.0 (−3)*
Cu	<0.015	–	7.64±0.35 (−3)	9.1±0.23 (−4)	6.2 (−3)*
Zn	2.17 (−3)	0.036	0.023±0.035	2.78±0.02 (−3)	0.035*
Ga	3.9 (−4)	–	–	–	–
As	1.9 (−4)	4.0 (−4)	<3.3 (−4)[†]	–	–
Se	6.5 (−5)	–	9.9 (−5)[†]	1.9±0.02 (−3)	–
Br	–	7.0 (−3)	4.9±2.8 (−3)	0.25±0.016*	–
Rb	1.23 (−3)	–	8.5±1.7 (−4)	2.78±0.02 (−3)	–
Sr	–	4.8 (−3)	4.9±1.1 (−3)	–	5.0 (−3)
Zr	4.8 (−3)	–	–	–	–
Mo	1.1 (−4)	–	–	–	4.7 (−4)*
Ag	2.1 (−5)	–	3.5 (−4)[†]	5.0±2.4 (−5)	7.0 (−5)*
Cd	<5.3 (−4)	–	–	5.8±3.5 (−5)	3.7 (−4)
In	<1.3 (−5)	2.8 (−5)	–	–	–
Sb	1.5 (−5)	2.5 (−4)	1.8 (−4)[†]	6.6±2.4 (−5)	6.5 (−4)*
I	–	2.7 (−3)	–	0.056±0.019	–
Cs	5.8 (−5)	1.08 (−3)	<1.6 (−5)[†]	1.64±0.2 (−4)	–
Ba	0.013	0.031	–	6.7±20 (−3)	9.0 (−3)
La	7.1 (−4)	9.9 (−4)	7.35 (−4)[†]	–	6.1 (−4)
Ce	1.5 (−3)	–	8.6 (−4)[†]	1.13±0.02 (−3)	1.3 (−3)
Sm	1.3 (−4)	9.9 (−5)	–	–	1.1 (−4)
Eu	2.8 (−5)	4.6 (−4)	2.9 (−5)[†]	1.3±2.0 (−5)	2.5 (−5)
Tb	2.4 (−5)	–	–	–	–
Dy	6.2 (−5)	–	–	–	8.4 (−5)
Yb	4.1 (−5)	–	–	–	–
Lu	8.9 (−6)	–	–	–	–
Hf	7.9 (−5)	–	–	4.9±1.9 (−5)	–
Ta	3.8 (−5)	–	–	1.5±1.9 (−5)	–
W	4.9 (−5)	1.9 (−4)	–	–	–
Au	5.2 (−7)	1.1 (−5)	–	–	–
Pb	–	0.044	0.010±0.004	1.64±0.02 (−3)	2.5 (−2)*
Th	1.8 (−4)	–	1.2 (−4)[†]	2.1±0.02 (−4)	2.5 (−4)

[a] Powers of ten in parentheses
[b] Rahn et al. (1979) Shipboard measurements of African dust plume about 20°N latitude
[c] Cornille et al. (1990) Arid region 40 km north-west of Damascus
[d] Eltayeb et al. (1993), Khartoum; dagger indicates data from Penkett et al. (1979), Gezira region, dry north winds
[e] Duce et al. (1983) at Enewetak Atoll, Pacific Ocean, April-May during the occurrence of Asian dust storms; asterisks indicate elements dominated by marine sources
[f] Arimoto et al. (2004) at Zhenbeitai, China. The city lies downwind of the Mu Us Desert, a regional source of desert dust. An asterisk indicates that the element is enriched compared to certified loess reference material

Table 5.17 Elemental composition of aerosols (ng m^{-3}) in remote regions of the troposphere[a]

	South Pole[b] Winter	South Pole[b] Summer	High Arctic[c] Open sea	High Arctic[c] Pack ice	Norwegian Arctic[d] Winter	Norwegian Arctic[d] Summer	Alaska[e]	Chacaltaya[f]	Amazon Basin[g] <2 μm	Amazon Basin[g] 2–10 μm
Na	19.0±7.2	7.2±3.2	490	20	230	66	17.8±3.8	59.7	19.4±7.4	33.4±17.5
Mg	2.9±1.1	1.6±0.4	49	8.5	48	18.2	2.4±2.5	29.0	14.1±5.2	34.7±24.6
Al	0.24±0.04	0.51±0.07	7.8	13.5	40.0	8.3	6.3±5.1	253	–	–
Si	–	–	28	22	134	31	35.0±4.2	550	–	66.2±39.8
P	–	–	1.7	1.26	<14	<10	0.7±1.3	–	6.62±2.67	8.89±1.33
S	6.0±3.0	28±16	165	32	830	101	101±2.2	172	208±139	42.3±19.0
Cl	27±4	7.7±1.3	670	17.6	85	40	1.6±5.9	67.0	8.16±0.88	80.8±28.9
K	1.0±0.7	0.60±0.18	22	2.1	31	3.7	9.6±2.2	79.6	111±114	33.1±13.3
Ca	1.2±0.85	0.78±0.23	23	3.0	34	7.3	9.5±2.1	83.3	8.96±4.82	16.9±14.1
Sc	3.1±0.8 (–5)	5.4±0.7 (–5)	<4 (–4)	<3 (–3)	4.3 (–3)	1.18 (–4)	–	0.043	–	–
Ti	0.16±0.08	0.29±0.09	0.27	0.44	1.0	<0.6	1.31±2.4	15.5	146±0.49	3.52±3.75
V	4.2±0.6 (–4)	1.1±0.4 (–3)	0.043	0.043	0.54	0.022	0.35±1.7	0.36	1.04±0.39	–
Cr	0.011±0.005	0.015±0.002	<1.1	<1.0	<0.4	0.56	0.3±1.8	0.60	2.59±0.95	3.41±0.63
Mn	3.1±0.7 (–3)	5.5±0.8 (–3)	0.085	0.26	0.77	0.07	0.35±2.1	2.53	0.79±0.42	0.51±0.34
Fe	0.25±0.06	0.44±0.12	5.9	10.6	17.8	5.6	11.5±2.4	181	11.1±5.2	21.5±14.6
Co	3.4±1.5 (–4)	4.7±2.4 (–4)	0.014	0.019	9.6 (–3)	<4 (–3)	–	0.91	–	–
Ni	–	–	<0.2	<0.18	0.29	<0.2	0.09±1.8	0.54	0.75±0.18	0.44±0.07
Cu	0.13±0.08	0.19±0.13	<0.15	0.16	<0.9	<0.3	0.26±	1.27	1.15±0.40	–
Zn	0.11±0.06	0.12±0.01	0.19	0.18	3.9	<0.15	0.89±3.4	4.34	2.61±1.02	1.81±1.16
Ge	–	–	0.024	0.023	–	–	–	–	–	–
As	6.4±1.9 (–3)	5.3±1.1 (–3)	<0.06	0.022	0.520.01	0.09±1.8	–	–	–	–
Se	4.1±0.6 (–3)	6.8±2.2 (–3)	0.088	<0.2	0.156	0.035	0.07±1.7	1.61	–	–
Br	0.085±0.02	0.61±0.37	1.26	0.25	8.8	0.73	.40±3.0	0.18	0.96±0.39	2.14±1.15
Rb	1.3±0.4 (–3)	2.4±07 (–3)	0.052	0.042	0.083	<0.08	0.16±2.1	1.05	0.65±0.42	–
Sr	<0.15*	0.031±0.012*	0.22	0.10	<1.1	<0.5	0.18±2.0	–	–	–
Zr	–	0.066	0.056	–	–	025±1.8	–	–	0.84±±0.28	–
Mo	–	<0.3	<0.16	<0.14	<0.08	0.58±1.6	–	–	–	–

Ag	5.1±3.8(−4)	8.6±5.9(−4)	<0.14	<0.13	<0.018	<0.018	—	—	—
Cd	0.05±0.04	0.11±0.06	<0.5	<0.3	0.08	<0.11	—	1.09	—
In	1.7±0.7(−4)	3.3±1.7(−4)	<5.5(−3)	<2(−3)	1.6(−3)	<1.2(−3)	—	0.013	—
Sb	3.1±1.1(−3)	3.7±1.8(−3)	<5(−3)	0.019	0.092	2.4(−3)	—	0.87	—
I	0.13±0.04	0.26±0.06	0.72	0.55	0.9	0.28	—	0.25	—
Cs	4.0±2.0(−5)	6.0±3.0(−5)	<0.04	<0.04	8.9(−3)	3.0(−3)	—	0.072	—
Ba	0.05±0.03	0.04±0.02	<5	<4	<1.5	<0.7	—	—	—
La	4.3±1.1(−4)	5.6±2.1(−4)	0.03	0.013	0.0137	5.5(−3)	—	0.16	—
Ce	6.5±3.8(−4)	8.8±2.9(−4)	<0.3	<0.3	<0.05	<0.02	—	—	—
Sm	5±3(−5)	5±3(−5)	<3(−3)	4.9(−4)	2.7(−3)	7.2(−4)	—	0.027	—
Eu	7±3(−6)	2±2(−6)	0.016	<0.01	1.16(−3)	<1(−4)	—	—	—
Lu	<1(−3)*	7±2(−6)*	<7(−3)	<6(−3)	<2.8(−3)	<4(−4)	—	—	—
Mo	4.3±1.7(−3)	4.3±1.8(−3)	—	—	<0.2	<0.05	—	—	—
Yb	4±4(−5)	5±3(−5)	—	—	—	—	—	—	—
Hf	2.2±0.7(−5)	4.0±1.0(−5)	—	—	—	—	—	—	—
Ta	2.0±1.0(−5)	3.0±1.0(−5)	—	—	—	—	—	—	—
W	8.7±2.9(−4)	9.5±1.6(−4)	<0.09	<0.03	<0.04	<4(−3)	—	0.72	—
Au	3.0±1.0(−5)	4.0±1.0(−5)	8.2(−4)	4.6(−4)	<9(−4)	<1.5(−4)	—	—	—
Pb	—	0.027±0.01	0.3	0.2	3.0	<0.7	0.53±2.9	5.25	0.97±0.12
Th	6.0±2.0(−5)	1.1±0.6(−4)	<0.035	<0.03	3.7(−3)	1.73(−3)	—	0.054	—

[a] Powers of ten in parentheses

[b] South Pole, base line data for austral winter and summer (1979–1983): Tuncel et al. (1989), asterisk indicates data from Cunningham and Zoller (1981), the value for Pb is from Maenhaut et al. (1979)

[c] High Arctic: 70–90°N, onboard the Swedish Icebreaker Oden, August–October 1991, over the sea and over pack-ice, Maenhaut et al. (1996a)

[d] Ny Ålesund, Spitsbergen, median values for winters 1983/1984, 1986, summer 1984, Maenhaut et al. (1989)

[e] Alaska: Denali National Park, geometric average of data 1988–1993, Polissar et al. (1996)

[f] Chacaltaya Mountain, Bolivia: 5,220 m altitude, Adams et al. (1977)

[g] Bacia Modelo Tower 70 km north of Manaus, Brazil, above tropical forest canopy, Artaxo et al. (1988)

Table 5.18 Elemental composition of aerosols (ng m⁻³) in rural regions

	Montana[a]	California[b]	Vermont[c]	Germany[d]	England[e]	Spain[f]	Norway[g]
Na	90±55	–	42.1±3.2	171	809	800	350
Mg	–	53±34	3.46±2.58	67	331	200	–
Al	930±860	357±168	8.33±4.42	301	239	212	73
Si	2,700±2,100	1,377±520	44.1±2.6	570	–	514	250
P	–	120±38	0.94±1.45	<1.9	–	15	–
S	550±640	1,060±672	702±2.2	1,065	–	2,400	880
Cl	68±50	281±176	0.97±1.64	14	2,145	200	380
K	280±250	177±162	31.7±1.9	113	723	200	76
Ca	390±310	681±272	15.1±2.3	154	907	500	67
Sc	0.13±0.12	–	–	–	0.070	0.1	0.013
Ti	34±23	106±9	2.02±2.88	23.3	<22	12	71
V	1.4±1.1	32±1	0.82±2.51	1.5	11.0	8	1.9
Cr	1.4±1.1	15±4	0.35±1.77	<3.5	2.9	0.6	0.68
Mn	9.3±8.0	20±6	3.46±2.58	4.2	14.7	5	4.6
Fe	410±360	739±249	17.1±2.4	188	282	200	83
Co	0.17±0.11	18±4	–	–	0.34	0.4	0.1
Ni	0.57±0.85	11±2	0.39±3.36	1.0	6.5	5	1.9
Cu	2.0±1.1	71±7	0.99±2.03	3.0	18.4	77	2.4
Zn	6.5±3.9	39±12	6.12±1.96	18.2	98.1	54	18
As	1.8±2.0	5±1	0.23±2.46	1.7	3.6	1	0.63
Se	0.27±0.23	7±1	0.28±2.72	0.98	1.59	–	0.37
Br	2.8±1.7	4±1	1.49±2.15	3.0	44.1	–	5.3
Rb	1.2±1.1	7±1	0.13±2.26	–	<2.5	1	–
Sr	–	12±3	0.18±1.85	1.03	–	2	1.0
Mo	–	70±11	0.58±1.41	–	<0.86	0.6	–
Cd	–	46±8	–	0.6	<2.5	0.1	0.14
In	–	13±1	–	0.004	0.072	–	–
Sb	0.14±0.10	76±2	–	–	2.2	1	0.51
Cs	0.096±0.093	–	–	–	0.18	0.1	–
Ba	8.0±6.4	–	–	3.9	–	1	–
Pb	14±12	37±5	2.69±1.95	10.2	127	45	13
Th	0.15±0.13	–	–	–	0.055	0.7	–

[a] Near Colstrip, Montana, average for the period May–September 1975. Crecelius et al. (1980). Also reported La 0.5±0.46, Sm 0.073±0.064, Eu 0.016±0.012, Tb 0.02±0.04, Hf 0.044±0.037, Ta 0.015±0.014
[b] At Mira Loma, ~90 km east of downtown Los Angeles, $PM_{2.5}$, Na et al. (2004). Also observed Ga 48±13, Y 6±1, Pd 19±1, Ag 15±3, Sn 67±2, Tl 5±1
[c] At Underhill, Vermont, geometric average of $PM_{2.5}$ samples for the period 1988–1995, Polissar et al. (2001). Also observed Zr 0.23±2.05
[d] Erzgebirge, eastern Germany, for the period 1994–1997. Matschullat et al. (2000). Also reported Ga 0.39, Ge <0.12, I 0.78
[e] At Chilton (Harwell) average for the period 1972–1981, Cawse (1987). Also observed Ce 0.43, Ag 0.29, Au 0.011, La 0.59, Sm 0.026, Eu 0.01
[f] At Mt. Bartolo, eastern Spain, Querol et al. (2001); also observed: Ga 0.1, Y 0.1, Zr 3, Sn 1, La 0.2, Ce 0.2, Tb 0.03, Yb 0.02, Hf 0.2
[g] At Birkenes, southern Norway, 1985–1986 Amundsen et al. (1992), also observed: Ag 0.05, La 0.06; 1987–1989 Kemp (1993)

Table 5.19 Elemental composition of urban aerosols (ng m^{-3}) in North America[a]

	Philadelphia[b]		Houston[c]		New York[d]	Los Angeles[e]	Calexico[f]
	<2.5 μm	2.5–10 μm	<2.5 μm	2.5–15 μm	<2.5 μm	<10 μm	<10 μm
TM[a]	28.7	11.4	38.6	24.8	16.1	67.4	61.9±43.1
Na	146±15	–	317±379	278±238	61.1±77.1	1,632	279±201
Mg	–	–	–	–	6.7±20.1	335	399±276
Al	53±6	550±170	123±60	1,093±366	14.4±39.1	758	3,829±2,664
Si	103±21	1,610±390	210±32	2,990±965	107.1±88.5	2,040	11,268±7,698
P	–	–	28±12	61±34	3.7±6.9	187	26.6±31.5
S	4,200±290	239±140	4,834±279	397±430	1,400±1,000	3,353	863±390
Cl	3±5	69±13	32±6	366±111	49.3±183.3	1,119	1,298±1,525
K	101±8	151±23	119±6	170±20	52.1±56.4	237	1,548±1,122
Ca	40±5	360±40	155±8	2,775±144	53.5±25.3	585	3,330±2,375
Ti	15±9	65±12	30±18	69±18	5.4±4.1	76.7	153±118
V	13±1.4	7±4	19±12	16±11	6.1±4.0	5.2	9.6±8.4
Cr	1.5±0.2	3±3	4±3	7±3	0.7±0.6	2.7±2.7	13±2
Mn	6±0.7	11±3	14±2	21±3	2.1±1.7	32.7	29.8±22.4
Fe	91±7	490±40	162±9	604±34	117.6±59.9	836	1,593±1,239
Co	0.48±0.06	–	1±2	0±2	0.7±1.0	0	–
Ni	11±1.5	4±1	4±1	4±1	18.4±12.8	4.6	1.8±1.4
Cu	–	–	28±3	18±3	4.0±2.8	22.4	10.8±11.0
Zn	82±6	30±4	84±5	58±4	32.1±21.3	113.8	39.6±30.8
Ga	7.6±8.4 (−2)	–	0±1	0±1	–	–	0.4±0.5
Ge	–	–	1±1	1±1	–	–	–
As	0.63±0.22	–	2±3	0±2	0.9±0.9	6.9	0.8±1.0
Se	1.8±0.2	0.0±0.5	1±1	0±0	1.4±1.6	8.1	1.8±1.3
Br	29±2	15±1.4	55±3	36±3	4.5±3.4	16.3	12.6±10.8
Rb	–	–	0±1	1±1	–	–	4.9±3.9

(continued)

Table 5.19 (continued)

	Philadelphia[b]		Houston[c]		New York[d]	Los Angeles[e]	Calexico[f]
	<2.5 μm	2.5–10 μm	<2.5 μm	2.5–15 μm	<2.5 μm	<10 μm	<10 μm
Sr	0.4±0.6	2±0.6	1±1	8±1	0.8±1.0	17.9	21.4±17.2
Zr	–	–	–	–	–	–	5.9±4.5
Mo	0.88±0.36	–	–	–	0.8±1.0	–	0.7±0.7
Ag	0.17±0.12	–	–	–	–	–	1.3±1.6
Cd	2.2±0.7	0±2	0±2	0±2	1.0±1.5	–	1.7±3.1
In	7±4 (−3)	–	–	–	–	–	1.4±2.8
Sn	5±4	1±3	2±3	1±3	–	–	3.2±5.3
Sb	79±8	181±13	6±3	2±3	–	–	6.7±10.8
Ba	–	–	48±13	91±16	18.6±12.0	70.1	51.3±57.3
La	1.12±0.13	–	–	–	–	–	12.3±14.6
W	0.23±0.06	–	0±2	2±2	–	–	–
Au	1.7±0.9 (−3)	–	–	–	–	–	0.2±0.4
Hg	–	–	0±2	0±2	–	20.6	0.3±0.5
Pb	249±15	54±5	465±23	124±10	6.9±7.0	84.4	38.2±47.5
SO_4[a]	11.2±0.86	0.52±0.3	14.6±1.2	0.91±0.17	4.1±3.1	11.28	2.75±1.31
NO_3[a]	0.13±0.16	0.57±0.17	0.59±0.62	1.63±0.23	2,200±1,900	9.47	2.25±2.36
NH_4[a]	3.6±0.27	–	4.18±0.38	–	1.9±1.6	4.61	1.12±1.11
TC[a]	3.9±2.4	1.0±1.0	7.1±0.5	3.1±0.04	5.1±3.8	14.8	12.6±8.0
OC	2.05±1.0	0.62±1.0	–	–	4.0±3.1	11.61	10.08±7.35
EC	1.87±0.14	0.42±0.06	–	–	1.1±0.7	3.19	2.50±1.76

[a]TM total mass, TC, OC, EC total, organic and elemental carbon, respectively; data for these species as well as for SO_4, NO_3 and NH_4 in ($\mu g\ m^{-3}$)

[b]Average of samples taken during summer 1982 in Camden, New Jersey, Dzubay et al. (1988); also observed: Sc 1.4±0.2 (−2), I 1.04±0.3, Cs 2.3±1.4 (−2), Ce 0.61±0.1, Sm 4±0.5 (−2)

[c]At University of Houston Campus, autumn 1980, Johnson et al. (1984)

[d]Data taken in 2001–2002, Ito et al. (2004)

[e]Downtown Los Angeles, summer 1987, Chow et al. (1994)

[f]Calexico, southern California, 1992–1993, Chow et al. (2000)

Table 5.20 Elemental composition of urban aerosols (ng m⁻³) in Europe[a]

	Helsinki[b]		Budapest[c]		Ghent[d]		Huelva[e]	Ankara[f]
	<2.3 μm	2.3–15.7 μm	<2 μm	2–10 μm	<2 μm	2–10 μm	<10 μm	<2.2 μm
TM[a]	11.8±7.5	13±11	25	44	20	11.8	91.3	110
Na	170±100	300±230	56	338	162	440	2,100	61±65
Mg	85±100	130±120	63	518	<180	133	700	70±121
Al	59±52	520±530	125	1,413	23	114	2699.1	110±150
Si	–	–	385	3,820	72	280	7198.5	–
P	–	–	10	64	<10	17.8	359	–
S	833±600	60±47	1,791	885	1,490	350	2,567	2,330±3,266
Cl	43±62	240±270	23	181	400	470	2,400	24±44
K	85±100	200±180	155	459	116	97	1,200	140±190
Ca	71±50	480±420	213	2,785	52	290	4,800	95±120
Sc	–	–	0.025	0.26	–	–	1.0	0.11±0.3
Ti	0.83±0.72	27±26	9.7	96	3.4	11.1	155	–
V	5.0±3.7	1.4±1.1	1.8	3.5	11.0	3.8	15	3.9±3.6
Cr	–	–	3.2	5.3	<6	<7	6	3.2±2.6
Mn	3.3±2.4	8.6±7.1	7.6	34	5.3	8.1	30	4.9±7.5
Fe	96±55	520±450	339	2,004	102	240	2,200	100±120
Co	0.14±0.25	0.31±0.35	0.24	0.47	–	–	1	3.3±3.3
Ni	2.0±1.3	0.79±0.72	0.86	2.4	3.9	1.94	5	3.1±2.9
Cu	3.1±1.6	6.2±3.9	11	34	4.2	4.9	210	–
Zn	14±8.6	7.9±6.3	32	61	38	20	112	16±12
Ga	–	–	0.17	0.48	–	20	0.5	–
As	0.80±0.74	0.21±0.25	1.4	1.4	–	–	9	1.5±1.9
Se	–	–	0.52	0.58	0.59	–	2	0.48±0.46
Br	–	–	16	13	9.0	2.7	–	21±18
Rb	0.24±0.20	–	1.9	4.1	<1.1	<0.9	4	–
Sr	0.46±0.30	2.1±1.8	0.82	7.6	<0.8	1.43	11	–
Zr	–	–	1.5	3.0	<1.0	<1.3	6	–

(continued)

Table 5.20 (continued)

	Helsinki[b]		Budapest[c]		Ghent[d]		Huelva[e]	Ankara[f]
	<2.3 μm	2.3–15.7 μm	<2 μm	2–10 μm	<2 μm	2–10 μm	<10 μm	<2.2 μm
Mo	0.18±0.12	0.26±0.16	0.43	1.2	<1.7	<1.8	1	–
Ag	.012±0.009	0.013±0.012	–	–	–	–	–	–
Cd	0.12±0.13	0.014±0.014	–	–	–	–	1	0.11±0.11
In	–	–	0.0039	0.0085	0.0114	0.00	–	0.03±0.07
Sn	–	–	–	–	<16	<17	3	–
Sb	0.77±1.4	0.78±0.52	2.5	6.6	2.1	0.3	4	1.3±2.2
I	–	–	1.2	0.48	–	–	–	–
Cs	–	–	0.083	–	–	–	0.2	3.1±2.9
Ba	2.2±1.4	8.3±6.1	8.2	38	<9	5.1	35	–
La	–	–	0.12	0.79	–	–	–	0.14±0.26
Ce	–	–	0.39	1.8	–	–	–	0.06±0.13
Sm	–	–	0.013	0.12	–	–	–	0.02±0.04
W	–	–	0.23	0.31	–	–	–	–
Au	–	–	0.002	0.0031	–	–	–	–
Tl	0.017±0.024	0.007±0.007	–	–	–	–	0.3	–
Pb	5.8±5.5	2.0±1.3	52	42	37	12.7	83	71±62
Bi	0.024±0.019	0.014±0.015	–	–	–	–	1	–
Th	–	–	–	0.21	–	–	0.6	–
U	0.005±0.004	0.050±0.046	–	0.062	–	–	0.6	–
SO_4[a]	2.5±1.8	0.18±0.14	–	–	–	–	7.7	6.4±9.8
NO_3[a]	1.6±2.2	0.41±0.41	–	–	–	–	4.0	2.1±3.5
NH_4[a]	1.1±1.0	0.010±0.022	–	–	–	–	1.4	–
BC[a]	1.4±0.36	–	8.0	2.3	3.8	0.59	8.3(TC)	–

[a]M total mass, TC total carbon. Data for these species as well as for SO_4, NO_3 and NH_4 are in $\mu g \ m^{-3}$

[b]Average of data obtained in 1996/1997 on the roof of the Meteorological Institute, Helsinki (Finland), 2 km northeast of city center, Pakkanen et al. (2001)

[c]Median concentrations observed in Széna Square, downtown Budapest (Hungary) in 1996, Salma et al. (2001)

[d]At an urban residential site in Ghent (Belgium) for a 14 month period, 1993–1994 (Maenhaut et al. 1996b)

[e]From July 1999 to January 2001 in Huelva, a highly industrialized city in southern Spain, Querol et al. (2002)

[f]At Middle East Technical University between February and June 1993, Yatin et al. (2000)

Table 5.21 Elemental composition of urban aerosols (ng m^{-3}) in Asia[a]

	Beijing[b], summer		Beijing[b], winter		Dhaka[c]		HCMC[d]	Hong Kong[e]		Taipei[f]	Seoul[g]
	<2.5 μm	<10 μm	<2.5 μm	<10 μm	<2.5 μm	2.5–10 μm	<20 μm	<2.5 μm	<10 μm	<10 μm	<10 μm
TM[a]	75±46	150±57	182±120	292±123	22.3±15.4	42.9±37.1	77.7±24.2	50.9±7.9	83.5±16.7	108±57	72.5
Na	270±180	920±840	1,280±1,100	2,520±1,820	592±285	477±523	802±317	640±290	2,580±1,320	–	556
Mg	290±120	1,460±70	330±180	1,310±900	433±228	919±412	640±242	110±70	340±200	–	466
Al	680±350	3,360±1,770	990±630	5,540±3,180	502±331	2,148±1,228	2,760±1,461	180±110	700±300	3,054±1,986	1,400
Si	–	–	–	–	792±648	4,459±2,732	–	450±221	2,086±543	8,454±5,644	–
P	–	–	–	–	277±204	641±413	–	–	–	184±100	–
S	6,680±4,910	8,180±6,820	9,440±7,550	14,600±12,600	1,292±776	1,223±936	–	4,830±830	5,301±891	3,996±1,788	1,864
Cl	1,690±1,300	2,290±1,450	7,360±4,690	9,660±6,120	140±106	827±933	1,204±500	236±240	1,085±1,121	1,887±1,108	720
K	1,210±1,640	1,680±1,770	4,230±2,100	3,210±2,620	390±218	747±524	915±425	1,259±606	1,587±704	1,536±841	1,587
Ca	960±610	5,780±3,000	2,440±1,810	13,600±95,700	163±166	1,385±818	3,360±2,838	172±71	1,100±218	1,987±1,182	–
Sc	–	–	–	–	–	–	0.58±0.21	–	–	–	–
Ti	50±30	220±100	80±70	340±180	16.9±17.9	150±105	266±106	12.5±11.7	65.4±15.4	316±198	85.6
V	57±27	57±31	13±10	18±10	7.99±7.76	25.2±127	7.6±4.5	4.9±1.8	5.3±1.8	43±19	12.6
Cr	30±70	20±10	30±20	30±20	–	–	9.4±3.3	2.4±1.7	6.9±6.0	34±14	18.5
Mn	50±20	90±30	100±60	190±110	8.87±5.57	30.7±18.8	37.8±11.6	16.7±10.1	35.3±9.1	67±30	65.1
Fe	1,000±500	3,260±1,560	1,180±890	4,170±2,570	207±132	1,306±766	3,078±1,054	212±24	756±121	2,410±1,590	1,544
Co	0.54±0.48	1.64±0.96	2.85±1.47	5.13±2.95	–	–	1.23±0.46	–	–	20±15	3.82
Ni	70±70	50±30	80±50	130±80	3.2±2.3	4.6±3.5	1.39±1.21	2.1±2.3	2.1±0.8	20±8	37.4
Cu	70±80	60±40	70±50	120±70	5.1±4.1	13.2±43.4	–	13.0±7.5	16.0±3.6	57±35	48.7
Zn	270±120	420±260	730±500	1,140±850	272±247	375±367	203±100	145±68	212.6±63.3	348±238	270
As	20±40	30±30	60±40	80±60	–	–	1.47±0.60	3.9±3.4	7.3±5.2	1.1±3.7	–
Se	–	–	–	–	–	–	–	–	–	73±14	–
Br	–	–	–	–	10.6±7.87	17.4±28.5	9.7±6.0	8.5±5.4	12.5±8.6	147±61	–
Rb	–	–	–	–	–	–	4.1±1.4	–	–	–	–
Sr	10±10	50±40	30±20	110±60	–	–	–	–	–	–	13.7

(continued)

Table 5.21 (continued)

	Beijing[b], summer		Beijing[b], winter		Dhaka[c]		HCMC[d]	Hong Kong[e]		Taipei[f]	Seoul[g]
	<2.5 μm	<10 μm	<2.5 μm	<10 μm	<2.5 μm	2.5–10 μm	<20 μm	<2.5 μm	<10 μm	<10 μm	<10 μm
Cd	1.7±1.0	2.7±2.2	10.6±11.8	21.9±25.6	–	–	–	–	–	–	7.63
Sb	–	–			–	–	3.8±5.3	–	–	–	–
Cs	–	–			–	–	053±0.46	–	–	–	–
Ba	–	–			–	–	33±11	–	–	–	49.3
La	–	–			–	–	1.48±0.61	–	–	–	–
Ce	–	–			–	–	3.3±1.0	–	–	–	–
Sm	–	–			–	–	0.22±0.09	–	–	–	–
Pb	100±50	110±80	320±230	490±420	164±557	124±350	163±113	104±70	145±83	244±98	120
Th	–	–			–	–	0.56±0.19	–	–	–	–
U	–	–			–	–	0.18±0.08	–	–	–	–
SO₄[a]	19.7±16.2	25.4±20.4	29.9±23.4	46.0±39.1	–	–		12.8±5.5	15.9±12.1	–	–
NO₃[a]	13.2±10.3	410±410	19.3±13.7	27.3±19.9	–	–		2.3±0.9	5.3±3.1	–	–
NH₄[a]	9.75±6.78	10.8±7.75	20.3±10.4	18.9±12.1	–	–		3.2±1.1	3.3±2.4	–	–
OC[a]	11.2±3.8	–	37.5±11.2	–	–	–		9.4±2.0	12.0±2.0	–	–
EC	5.9±2.6	–	21.9±12.1	–	–	–		5.8±1.1	6.9±1.2	–	–

[a] TM total mass, OC, EC organic and elemental carbon, respectively; Data for these species as well as SO_4, NO_3 and NH_4 in $\mu g\ m^{-3}$

[b] Data collected in 2002/2003 in a residential area of Beijing, China; Sun et al. (2004)

[c] Data collected in 2001/2002 in a residential area in Dhaka, Bangladesh, Begum et al. (2004)

[d] Ho Chi Minh City, Vietnam, for the period 1992–1996, Hien et al. (1999)

[e] In residential areas of Hong Kong, China, during the winter 1999/2000 and 2000/2001, Chao and Wong (2002), Ho et al. (2003)

[f] Taipei, Taiwan, urban residential area, February–April 1993, Li (1994)

[g] At the Natural Science Building, Sejong University, Seoul, Korea, March–May 2001, Kim et al. (2003)

Table 5.22 Elemental composition of urban aerosols (ng m^{-3}) in the southern hemisphere[a]

	São Paulo[b]		Santiago[c]		Chillán[d]	Brisbane[e]	
	<2.5 μm	2.5–10 μm	<2.5 μm	2.5–10 μm	<10 μm	<2.5 μm	2.5–10 μm
TM[a]	30.2±16.1	46.1±38.1	34±13	66±24	82.5±48.7	7.6±5.4	10.4
Na	292±239	441±344	–	–	–	190±250	905
Mg	83±47	184±135	40±20	250±100	290±60	37.2	132
Al	437±282	1,521±1,212	180±70	1,960±710	1,420±160	40±80	187.2
Si	511±288	2,269±1,726	660±350	5,400±1,970	3,640±7,380	87±150	655.2
P	14.3±4.9	25±14	90±40	90±20	–	7.2±5.2	–
S	1,520±1,166	733±580	2,100±1,080	420±150	50±50	400±210	104
Cl	52±35	250±279	–	–	320±120	208±254	1,082
K	407±252	486±433	980±460	960±330	1,910±650	81±154	78
Ca	146±93	1,196±864	280±70	2,720±950	1,610±210	34±38	239.2
Ti	31±24	217±165	17±2	260±100	250±260	6.9±11	32.2
V	11.7±6.5	12±12	16±8	11±5	20±50	0.54±0.66	–
Cr	9.6±3.7	24±15	6.5±2.1	15±5	–	0.63±0.78	6.5
Mn	12.6±8.1	32±23	17±11	80±30	40±150	2.4±2.7	3.85
Fe	532±273	1,981±1,426	200±90	2,930±1,100	2,110±360	57±73	197.6
Co	–	–	–	–	–	0.39±0.57	–
Ni	3.9±2.8	5.8±6.2	3.3±0.9	2.5±2.4	10±80	0.63±1.0	–
Cu	19±11	44±44	26±16	40±20	180±50	5.3±8.2	–
Zn	126±107	189±234	220±180	90±50	300±120	21±21	8.3
As	–	–	–	–	20±50	–	–
Se	3.0±2.6	3.0±3.4	–	–	10±23	–	–
Br	14.3±6.2	24±14	50±20	21±6	10±20	20±17	4.2
Rb	4.6±2.5	7.6±4.9	–	–	–	–	–
Sr	2.4±1.0	8.5±6.2	–	–	30±180	–	–
Zr	4.1±2.1	12±15	–	–	–	–	–

(continued)

Table 5.22 (continued)

	São Paulo[b]		Santiago[c]		Chillán[d]	Brisbane[e]	
	<2.5 μm	2.5–10 μm	<2.5 μm	2.5–10 μm	<10 μm	<2.5 μm	2.5–10 μm
Mo	–	–	–	–	50±20	–	–
Ba	–	–	–	–	130±40	–	–
Pb	42±34	38±45	260±120	100±30	20±110	66±46	10.4
SO$_4$[a]	–	–	–	–	4.15±1.47	0.79	0.31
NO$_3$[a]	–	–	–	–	8.6±6.33	0.19	0.45
NH$_4$[a]	–	–	–	–	4.78±2.74	–	–
TC[a]	23.4±12	–	–	–	29.1±21.0	–	–
OC	15.8±8.3	–	–	–	23.12±18.56	–	–
EC	7.6±3.7	–	–	–	5.95±2.61	1.39	0.31

[a]*TM* total mass, *TC, OC, EC* total, organic and elemental carbon, respectively; Data for these species as well as for SO$_4$, NO$_3$, and NH$_4$ are in µg m^{-3}

[b]Samples collected between July and September 1997 in metropolitan São Paulo, Brazil, Castanho and Artaxo (2001); data for Na and Mg from Andrade et al. (1994)

[c]At Universidad de Santiago de Chile Planetarium January–February 1987; Rojas et al. (1990)

[d]Mean values of data obtained at six measurement sites in Chillán, Chile, September 2001 and April 2003, Celis et al. (2004)

[e]PM2.5 data with standard deviations from the inner city of Brisbane, Australia, April 1995–July 1999, Thomas and Morawska (2002); the other data collected at five sites within 15 km of the city center, September 1993–November 1995, Chan et al. (1997)

References

Adams, F., R. Dams, L. Guzman, J.W. Winchester, Atmos. Environ. **11**, 629–634 (1977)

Aiuppa, A., G. Dongarrà, M. Valenza, Degassing of trace volatile metals during the 2001 eruption of Etna, in *Volcanism and the Earth's Atmosphere*, ed. by A. Robock, C. Oppenheimer. Geophysical Monograph 137 (American Geophysical Union, Washington, DC, 2003), pp. 41–54

Amundsen, C.E., J.E. Hanssen, A. Semb, E. Steinnes, Atmos. Environ. **26A**, 1309–1324 (1992)

Andrade, F., C. Orsini, W. Maenhaut, Atmos. Environ. **28**, 2307–2315 (1994)

Arimoto, R., X.Y. Zhang, B.J. Huebert, C.H. Kang, D.L. Savoie, J.M. Prospero, S.K. Sage, C.A. Schloesslin, H.M. Khaing, S.N. Oh, J. Geophys. Res. **109**, D19S04 (2004). doi:10.1029/2003JD004323

Artaxo, P., H. Storms, F. Bruynseels, R. Van Grieken, W. Maenhaut, J. Geophys. Res. **93**, 1605–1615 (1988)

Begum, B.A., E. Kim, S.K. Biswas, P.K. Hopke, Atmos. Environ. **38**, 3025–3038 (2004)

Cao, J.J., J.C. Chow, J.G. Watson, F. Wu, Y.M. Han, Z.D. Jin, Z.X. Shen, Z.S. An, Atmos. Environ. **42**, 2261–2275 (2008)

Castanho, A.D.A., P. Artaxo, Atmos. Environ. **35**, 4889–4902 (2001)

Cawse, P.A., in *Pollutant Transport and Fate in Terrestrial Ecosystems*, ed. by P.J. Coughtry, M.H. Martin, M.H. Unsworth. Special Publication No. 6, British Ecological Society (Blackwell, Oxford, 1987), pp. 89–111

Celis, J.E., J.R. Morales, C.A. Zaror, J.C. Inzunza, Chemosphere **54**, 541–550 (2004)

Chan, Y.C., R.W. Simpson, G.H. McTainsh, P.D. Vowles, D.D. Cohen, G.M. Bailey, Atmos. Environ. **31**, 3773–3785 (1997)

Chao, C.Y., K.K. Wong, Atmos. Environ. **36**, 265–277 (2002)

Chow, J.C., J.G. Watson, E.M. Fujita, Z. Lu, D.R. Lawson, Atmos. Environ. **28**, 2061–2080 (1994)

Chow, J.C., J.G. Watson, M.C. Green, D.H. Lowenthal, B. Bates, W. Oslund, G. Torres, Atmos. Environ. **34**, 1833–1843 (2000)

Chow, J.C., J.G. Watson, L.L. Ashbaugh, K.L. Magliano, Atmos. Environ. **37**, 1317–1340 (2003)

Cornille, P., W. Maenhaut, J.M. Pacyna, Atmos. Environ. **24A** 1083–1093 (1990)

Crecelius, E.A., E.A. Lepel, J.C. Laul, L.A. Rancitelli, R.L. McKeever, Environ. Sci. Technol. **14**, 422–428 (1980)

Crowe, B.M., D.L. Finnegan, W.H. Zoller, W.V. Boynton, J. Geophys. Res. **92**, 13708–13714 (1987)

Cunningham, W.C., W.H. Zoller, J. Aerosol Sci. **12**, 367–384 (1981)

Desboeufs, K.V., A. Sofikitis, R. Losno, J.L. Colin, P. Ausset, Chemosphere **58**, 195–203 (2005)

Duce, R.A., R. Arimoto, B.J. Ray, C.K. Unni, P.J. Harder, J. Geophys. Res. **88**, 5321–5342 (1983)

Dzubay, T.G., R. Stevens, G.E. Gordon, I. Olmez, A.E. Sheffield, W.J. Courtney, Environ. Sci. Technol. **22**, 46–52 (1988)

Eltayeb, M.A.H., C.F. Xhofer, P.J. Van Espen, R.E. Van Grieken, Atmos. Environ. **27B** 67–76 (1993)

Gladney, E.S., J.A. Small, G.E. Gordon, W.H. Zoller, Atmos. Environ. **10**, 1071–1077 (1975)

Hien, P.D., N.T. Binh, Y. Truong, N.T. Ngo, Atmos. Environ. **33**, 3133–3142 (1999)

Hildemann, L.M., G.R. Markowski, G.R. Cass, Environ. Sci. Technol. **25**, 744–759 (1991)

Ho, K.F., S.C. Lee, C.K. Chan, J.C. Yu, J.C. Chow, X.H. Yao, Atmos. Environ. **37**, 31–39 (2003)

Huang, X., I. Olmez, N.K. Aras, Atmos. Environ. **28**, 1385–1391 (1994)

Huggins, F.E., G.P. Huffman, J.D. Robertson, J. Hazardous Mat. **74**, 1–23 (2000)

Ito, K., N. Xue, G. Thurston, Atmos. Environ. **38**, 5269–5282 (2004)

Jaenicke, R., L. Schütz, Aerosol physics and chemistry, in *Meterologie*, ed. by G. Fischer. Landolt-Börnstein, Neue Serie 4b (Springer, Berlin, 1988)

Johnson, D.L., B.L. Davis, T.G. Dzubay, H. Hasan, E.R. Crutcher, W.J. Courtney, J.M. Jaklevic, A.C. Thompson, Atmos. Environ. **18**, 1539–1553 (1984)

Kemp, K., Atmos. Environ. **27A**, 823–830 (1993)

Kim, K.-H., G.-H. Choi, C.-H. Kang, J.-H. Lee, J.Y. Kim, Y.H. Youn, S.R. Lee, Atmos. Environ. **37**, 753–765 (2003)

Klein, D.H. et al., Environ. Sci. Technol. **9**, 973–979 (1975)

Lepel, E.A., K.M. Stefansson, W.H. Zoller, J. Geophys. Res. **83**, 6213–6220 (1978)

Li, C.-S., Atmos. Environ. **28**, 3139–3144 (1994)

Maenhaut, W., W.H. Zoller, R.A. Duce, G.L. Hoffman, J. Geophys. Res. **84**, 2421–2431 (1979)

Maenhaut, W., P. Cornille, J.M. Pacyna, V. Vitols, Atmos. Environ. **23**, 2551–2569 (1989)

Maenhaut, W., G. Ducastel, C. Leck, D. Nilsson, J. Heintzenberg, Tellus **48B**, 300–321 (1996a)

Maenhaut, W., F. Francois, J. Cafmeyer, O. Okunade, Nucl. Instrum. Methods Phys. Res. **B 109/110**, 476–481 (1996b)

Matschullat, J., W. Maenhaut, F. Zimmermann, J. Fiebig, Atmos. Environ. **34**, 3213–3221 (2000)

Na, K., A.A. Sawant, D.R. Cocker III, Atmos. Environ. **38**, 2867–2877 (2004)

Pakkanen, T.A., K. Loukkola, C.H. Korhonen, M. Aurela, T. Mäkelä, R.E. Hillamo, P. Aarnio, T. Koskentalo, A. Kousa, W. Maenhaut, Atmos. Environ. **35**, 5381–5391 (2001)

Penkett, S.A., D.H.F. Atkins, M.H. Unsworth, Tellus **31**, 295–307 (1979)

Polissar, A.V., P.K. Hopke, W.C. Malm, J.F. Sisler, Atmos. Environ. **30**, 1147–1157 (1996)

Polissar, A.V., P.K. Hopke, R.L. Poirot, Environ. Sci. Technol. **35**, 4604–4621 (2001)

Querol, X., A. Alastuey, S. Rodriguez, F. Plana, E. Mantilla, C.R. Ruiz, Atmos. Environ. **35**, 845–858 (2001)

Querol, X., A. Alastuey, J. de la Rosa, A. Sánchez-de-la-Campa, F. Plana, C.R. Ruiz, Atmos. Environ. **36**, 3113–3125 (2002)

Rahn, K.A., R.D. Borys, G.E. Shaw, L. Schütz, R. Jaenicke, in *Saharan Dust*, ed. by C. Morales (Wiley, Chichester, 1979), pp. 49–60

Rojas, C.M., P. Artaxo, R. Van Grieken, Atmos. Environ. **24B**, 227–241 (1990)

Salma, I., W. Maenhaut, E. Zemplé-Papp, G. Záray, Atmos. Environ. **35**, 4367–4378 (2001)

Sun, Y., G. Zhuang, Y. Wang, L. Han, J. Guo, M. Dan, W. Zhang, Z. Wang, Z. Hao, Atmos. Environ. **38**, 5991–6004 (2004)

Thomas, S., L. Morawska, Atmos. Environ. **36**, 4277–4288 (2002)

Tuncel, G., N.K. Aras, W.H. Zoller, J. Geophys. Res. **94**, 13025–13038 (1989)

Wang, Y.F., K.-L. Huang, C.-T. Li, H.-H. Mi, J.-H. Luo, P.-J. Tsai, Atmos. Environ. **37**, 4637–4643 (2003)

Wangen, L.E., Environ. Sci. Technol. **15**, 1080–1088 (1981)

Yamasoe, M.A., P. Artaxo, A.H. Miguel, A.G. Allen, Atmos. Environ. **24**, 1641–1653 (2000)

Yatin, M., S. Tuncel, N.K. Aras, I. Olmez, S. Aygun, G. Tuncel, Atmos. Environ. **34**, 1305–1318 (2000)

5.4 Organic Components of the Tropospheric Aerosol

Introductory remarks:

The tropospheric aerosol harbors a great many organic compounds with solubilities in organic solvents or water ranging from very soluble to insoluble. Solvent-extraction followed by chromatographic separation techniques has served to identify a variety of organic compounds. Typically, about 50% of total organic carbon may be extracted, but only about 25% of this fraction can be resolved and identified in the chromatograms. Alves (2008) has reviewed our knowledge about solvent-extractable organic constituents of the tropospheric aerosol. The biosphere is a major source of particulate organics in the continental atmosphere. The following production mechanisms must be considered: (a) the direct injection of pollen from plants, spores from mosses, ferns and fungi, fragments of leaf litter, etc., identifiable by protein-staining techniques and/or analysis for cellulose, generates about 20% to the number concentration of all particles in the size range >0.2 μm (Jaenicke et al. 2007). Bacteria also are present, although they contribute less mass. (b) The combustion of biomass is a significant source of organic particles. This includes compounds derived from cellulose. (c) A secondary source of particulate organics is the gas-phase oxidation of volatile compounds emitted by the biosphere (as well as by anthropogenic sources). Oxidation often leads to products of low volatility that condense and associate with the aerosol; an example is pinonic acid arising from the oxidation of α-pinene. (d) In addition, some organic vapors, emitted from both the biosphere and from anthropogenic activities, have physicochemical properties such that they partly condense onto aerosol particles, whereby organic material is transferred from the gas phase to the particulate phase. (e) Macromolecular compounds with properties similar to humic acids are either incorporated with soil particles or are generated by chemical reactions directly on the aerosol particles (Zappoli et al. 1999).

Table 5.23a Annual mean chemical composition of fine particles in Los Angeles (μg m^{-3})[a]

	Total mass	NH$_4^+$	NO$_3^-$	SO$_4^{2-}$	EC	OC	Other
West Los Angeles	24.5	2.18	1.86	5.72	3.58	7.10	4.07
Rubidoux (Riverside)	42.1	5.00	10.2	6.02	3.16	6.27	11.4

[a]EC elemental carbon, OC organic carbon, Other: mainly minerals; Source of data: Rogge et al. (1993a)

Table 5.23b Composition of organic material in fine particles in Los Angeles (ng m^{-3})[a]

	A	B	C	D	E	F	G	H	I	K	L
West Los Angeles	3,300	2,790	227	54.1	256.1	199.0	32.0	91.7	11.9	3.2	35.0
Rubidoux (Riverside)	2,400	2,730	294	50.2	261.7	312.2	14.7	106.5	3.7	1.9	25.2

[a]Key: A Non-extractable, B unresolved, C unidentified, D n-alkanes, E n-alkanoic acids, F alphatic dicarboxylic acids, G diterpenoid acids and retene, H aromatic polycarboxylic acids, I polycyclic aromatic hydrocarbons, K N-containing compounds, L other organic compounds; Source of data: Rogge et al. (1993a)

Table 5.24 Characterization of primary biological particles over the continents[a]

Type	Size, mass, N (m^{-3})	Chemical characterization	Shape
Viruses[b]	10–500 nm	DNA/RNA, Protein envelope	Spherical, oval
Bacteria[b]	~ 1 μm, 30 fg, 10^4–10^6	DNA/RNA, proteins, lipids, phospholipids, cellulose,	Rods, spheres, spirals, cocci
Spores	>0.5 μm, 35 pg, 10^4–10^5	DNA/RNA, protoplasma, Sporoderm: cellulose, protein, high-polymer esters, carotinoids	Spherical, oval, fusiform, needles, elongated, helical
Pollen	>10 μm, 4 ng, 1–10^3	Protein, carbohydrates, lipids Sporoderm as for spores	Various species-characteristic forms
Plant debris	>~0.1 μm, 10^5–10^6	Cellulose, starch, waxes	Irregular

[a]Data compiled from: Matthias-Maser (1998), Burrows et al. (2009), Elbert et al. (2007), Winiwarter et al. (2009)

[b]Viruses and bacteria do not occur isolated in the atmosphere. They are released together with other materials from plant, soil and water surfaces

Table 5.25 Emission rates for and fractional concentrations of organic compounds in particles derived from automobile traffic[a]

Compound	Automobile emissions			Tire wear	Brake line	Road dust
	Diesel	Non-catalyst	Catalyst			
***n*-Alkanes**						
Nonadecane, C_{19}	533	59	9	41	2.0	14
Eicosane, C_{20}	991	104	13	58	2.4	15
Heneicosane, C_{21}	673	95	12	48	1.7	29
Docosane, C_{22}	364	80	11	63	1.4	36
Tricosane, C_{23}	241	55	8	71	3.2	35
Tetracosane, C_{24}	216	72	11	114	1.9	40
Pentacosane, C_{25}	334	114	18	175	5.7	91
Hexacosane, C_{26}	163	53	9	186	3.3	51
Heptacosane, C_{27}	72	26	7	227	2.1	101
Octacosane, C_{28}	32	7	2	269	2.0	47
Nonacosane, C_{29}	37	8	3	389	3.3	215
Triacontane, C_{30}	64	9	3	546	2.1	62
Hentriacontane, C_{31}	22	5	1	743	2.3	151
Dotriacontane, C_{32}	12	5	1	969	1.6	84
Tritriacontane, C_{33}	–	–	–	1,230	1.6	99
Tetratriacontane, C_{34}	–	–	–	1,556	0.8	43
Pentatriacontane, C_{35}	–	–	–	2,006	0.67	44
Hexatriacontane, C_{36}	–	–	–	2,254	0.42	31
Heptatriacontane, C_{37}	–	–	–	2,302	–	34
Octatriacontane, C_{38}	–	–	–	2,181	–	34
Nonatriacontane, C_{39}	–	–	–	1,428	–	21
Tetracontane, C_{40}	–	–	–	1,158	–	–
Hentetracontane, C_{41}	–	–	–	831	–	–
***n*-Alkanoic acids**						
Hexanoic, C_6	159	3.6	66	122	82	7
Heptanoic, C_7	99	0.8	8	3	35	29
Octanoic, C_8	63	2.1	51	31	61	58
Nonanoic, C_9	145	8.6	196	91	87	135
Decanoic, C_{10}	77	3.2	73	38	18	55
Undecanoic, C_{11}	270	8.8	77	187	41	145
Dodcanoic, C_{12}	72	5.5	41	137	13	105
Tridecanoic, C_{13}	21	0.6	3	12	3	34
Tetradecanoic, C_{14}	46	1.7	12	634	8	172
Pentadecanoic, C_{15}	18	0.4	3	86	3	64
Hexadecanoic, C_{16} (palmitic)	153	6.0	64	5,818	83	1,217
Heptadecanoic, C_{17}	38	0.5	2	151	4	61
Octadecanoic, C_{18} (stearic)	73	3.0	18	6,009	80	689
Nonadecanoic, C_{19}	2	0.6	0.5	17	0.8	27
Eicosanoic, C_{20}	–	0.3	–	127	1.8	119
Heneicosanoic, C_{21}	–	0.1	–	–	0.9	21
Docosanoic, C_{22}	–	–	–	–	2.1	70
Tricosanoic, C_{23}	–	–	–	–	1.9	24
Tetracosanoic, C_{24}	–	–	–	–	1.6	116

(continued)

Table 5.25   (continued)

Compound	Diesel	Non-catalyst	Catalyst	Tire wear	Brake line	Road dust
	Automobile emissions					
Pentacosanoic, C_{25}	–	–	–	–	0.3	22
Hexacosanoic, C_{26}	–	–	–	–	0.7	93
Heptacosanoic, C_{27}	–	–	–	–	0.4	21
Octacosanoic, C_{28}	–	–	–	–	0.6	124
n-Alkenoic and benzoic acids						
cis-9-Octadecenoic (oleic)	8	5	1.2	1,116	11	131
9,12-Octadienoic (linoleic)	–	–	–	164	–	129
Benzoic	171	4	100	75	32	114
4-Methylbenzoic	14	0.6	8	–	2.6	5
Substituted benzaldehydes						
Methylbenzaldehydes	9	26	5	8	9	–
Dimethylbenzaldehydes	7	56	11	–	–	–
Trimethylbenzaldehydes	–	17	4	–	–	–
4-Formylbenzaldehyde	2	19	5	–	–	–
3-Methoxybenzaldehyde	–	5	2	–	13	–
3,4-Dimethoxybenzaldehde	–	10.9	0.7	–	0.5	–
2-Hydroxybenzaldehyde	–	–	–	–	4.7	–
Polycyclic aromatic hydrocarbons						
Phenanthrene	12	17	0.9	12	1	4
Antracene	1.6	5	0.1	–	–	0.8
Methylphenanthrenes[b]	31	42	2	24	0.7	1.5
Dimethylphenanthrenes[b]	57	76	3	39	0.7	3
Fluoranthene	13	48	2	11	0.7	7
Pyrene	23	31	2.5	54	1	9
Benzacenaphtylene	3	17	0.8	–	–	0.2
2-Phenylnaphtalene	3.5	12	0.6	–	–	0.2
2-Benzylnaphtalene	–	4	0.7	–	–	–
Methylfluoranthenes, -pyrenes	13	107	4	24	0.9	2
Benzofluorenes[b]	2	31	0.7	1	0.3	0.4
Benzo[g,h,i]fluoranthene	7	25	1.3	6	0.3	1.3
Benz[a]anthracene	4	74	2	–	1.5	1
Cyclopenta[c,d]pyrene	1.4	49	1.7	–	–	–
Chrysene/triphenylene	10	56	4	8	2	8
Methylbenz[a]anthracenes[b]	3	143	3	19	2	1.3
Dimethylfluoranthenes[b]	15	161	4	19	–	–
Benzo[k]fluoranthene	3	41	2	–	0.6	6
Benzo [b]fluoranthene	3	38	3	–	0.4	4
Benzo[e]pyrene	3	46	2	5	0.8	3
Benzo[a]pyrene	1.3	44	2	4	0.7	2
Perylene	1	14	0.6	–	–	0.5
Indeno[1,2,3-c,d]pyrene	–	6	0.5	–	–	–
Indeno[1,2,3-c,d]fluoranthene	–	33	2	–	–	1
Benzo[g,h,i]perylene	1.6	145	5	–	3	2
Methylbenzofluoranthenes	–	6	0.5	–	–	–

(continued)

Table 5.25 (continued)

Compound	Automobile emissions			Tire wear	Brake line	Road dust
	Diesel	Non-catalyst	Catalyst			
Anthanthrene	–	9	0.1	–	–	–
Benzo[b]triphenylene	–	4	0.1	–	–	–
Dibenz[a,h]anthracene	–	8	0.3	–	–	–
Benzo[b]chrysene	–	2	0.02	–	–	–
Coronene	–	105	1	–	–	–
Polycyclic aromatic Ketones						
9H-Fluoren-9-one	65	113	24	–	0.8	0.7
2-Methylfluorene-9-one	17	8	2	–	–	–
9,10-Phenanthrenedione	63	32	4	–	0.3	0.4
9,10-Anthracenedione	24	24	4	–	0.5	1
Phenantrone/anthrone	21	19	3	–	0.5	–
9H-Xanthen-9-one	3	29	3	0.5	–	–
1H-Benz[de]anthracene-a-one	3	26	1.2	–	–	–
7H-Benz[de]anthracene-7-one	6	28	1.3	–	0.6	1.0
6H-Benzo[cd]pyren-6-one	1	20	0.8	–	0.4	–
Steranes, Triterpanes, Natural Resins[c]						
Regular steranes	189	37	17	74	1.6	56
Pentacyclic tripterpanes	272	58	26	684	3	114
Natural resins	–	–	–	9,513	10	7

[a]Source of data: Rogge et al. (1993b). Units: automobile emissions (μg km^{-1}), tire wear, brake line and road dust (μg g^{-1})

[b]Methylphenanthrenes: includes methylanthracenes; dimethylphenantrenes: includes dimehtylantracenes; benzofluorenes: benzo[a]fluorine and benzo[b]fluorine

[c]Steranes include: (20S and R)-5α(H),14β(H),17β(H),17β(H)-cholestanes, (20R)-5α(H), 14α(H), 17α(H)-cholestane, (20S and R)-5α(H),14β(H),17β(H)-ergostanes, (20S and R)-5α(H),14β(H), 17β(H)-sitostanes. Pentacyclic triterpanes: 22,29,30-trisnorneohopane, 17α(H),21β(H)-29-norhopane, 17α(H),21β(H)-hopane, (22S)-17α(H),21β(H)-30-homohopane, (22R)-17α(H), 21β(H)-30-homohopane, (22S)-17α(H),21β(H)-30-bishomohopane, (22R)-17α(H),21β(H)-30-bishomohopane. Natural resins include: dehydroabietic acid, abietic acid, 7-oxodehydroabietic acid, 13β-ethyl-13-methylpodocarp-8-en-15-oic acid, 13α-isopropyl-13-methylpodocarp-8-en-15-oic acid, 13β-isopropyl-13-methylpodocarp-8-15-oic acid

Table 5.26 Fractional concentrations of organic compounds in fine particles derived from products of plant leaf abrasion (μg g^{-1})[a]

Compound	Green Leaves	Dead	Compound	Green Leaves	Dead
n-Alkanes					
Nonadecane, C_{19}	9	5	Octacosane, C_{28}	234	329
Eicosane, C_{20}	13	5	Nonacosane, C_{29}	5,958	6,617
Heneicosane, C_{21}	14	9	Triacontane, C_{30}	434	445
Docosane, C_{22}	15	8	Hentriacontane, C_{31}	9,493	9,396
Tricosane, C_{23}	34	21	Dotriacontane, C_{32}	759	881
Tetracosane, C_{24}	45	30	Tritriacontane, C_{33}	4,636	5,480
Pentacosane, C_{25}	220	200	Tetratriacontane, C_{34}	90	92
Hexacosane, C_{26}	94	126	Pentatriacontane, C_{35}	360	651
Heptacosane, C_{27}	822	1,035	Hexatriacontane, C_{36}	49	13

(continued)

Table 5.26 (continued)

Compound	Green Leaves	Dead	Compound	Green Leaves	Dead
Iso- and Anteisoalkanes					
Isononacosane, C_{29}	18	21	Anteisodotriacontane, C_{32}	43	36
Isotriacontane, C_{30}	8	5	Isotritriacontane, C_{33}	123	140
Anteisotriacontane	20	19	Anteisotritriacontane, C_{33}	7	6
Isohentriacontane, C_{31}	66	67	Anteisotetratriacontane, C_{34}	45	28
Anteisohentriacontane, C_{31}	16	16			
n-Alkanoic acids					
Octanoic, C_8	143	250	Heneicosanoic, C_{21}	13	38
Nonanoic, C_9	445	597	Docosanoic, C_{22}	152	212
Decanoic, C_{10}	184	133	Tricosanoic, C_{23}	19	48
Undecanoic, C_{11}	597	403	Tetracosanoic, C_{24}	257	647
Dodcanoic, C_{12}	65	91	Pentacosanoic, C_{25}	24	101
Tridecanoic, C_{13}	7	10	Hexacosanoic, C_{26}	321	840
Tetradecanoic, C_{14}	148	237	Heptacosanoic, C_{27}	27	146
Pentadecanoic, C_{15}	15	25	Octacosanoic, C_{28}	431	1,610
Hexadecanoic, C_{16} (palmitic)	835	590	Nonacosanoic, C_{29}	33	266
Heptadecanoic, C_{17}	10	25	Triacontanoic, C_{30}	323	2,813
Octadecanoic, C_{18} (stearic)	210	201	Hentriacontanoic, C_{31}	16	171
Nonadecanoic, C_{19}	11	30	Dotriacontanoic, C_{32}	128	2,062
Eicosanoic, C_{20}	92	781			
n-Alkenoic and other acids					
cis-9-Octadecenoic, C_{18}	62	28	3,7-Dimethyl-6-octenoic	60	5
9,12-Octadecadienoic, C_{18}	111	72	2,6,10-Trimethylundecanoic acid methyl ester	171	14
9,12,15-Octadecatrienoic, C_{18}	150	40			
n-Alkanols and n-Alkanals					
Pentacosanol, C_{25}	1,378	1,094	Hexacosanal, C_{26}	2,279	736
Hexacosanol, C_{26}	8,224	2,069	Octacosanal, C_{28}	6,522	3,081
Octacosanol, C_{28}	15,484	7,294	Triacontanal, C_{30}	6,658	8,376
Tricontanol, C_{30}	1,060	1,249	Dotriacontanal, C_{32}	4,537	7,687
Dotricontanol, C_{32}	443	4,325			
Mono-, Sesqui- and Triterpenoids					
Eucalyptol	234	49	p-Menth-1-en-8-ol	133	30
Camphor	3,570	1,654	cis-Terpin hydrate	830	25
Borneol acetate	–	568	β-Ionone	95	14
Linalool	35	16	β-Cadinene	61	9
Citral (geranaldehyde)	462	1	Olean-12-en-3-ol	48	22
Citronellol	869	13	3-Oxo-olean-12-en-28-oic acid	1,544	380
Geranyl methylpropionate	32	–			
p-Menth-8-en-3-ol	87	–	3-Hydroxy-urs-12-en-28-oic acid	1,779	478
p-Menth-1-en-8-ol	59	20			
Polycyclic aromatic hydrocarbons					
Fluoranthene	1	0.5	Chrysene/triphenylene	3	2
Pyrene	2	1			

[a]Source of data: Rogge et al. (1993c)

Table 5.27 Fractional concentrations of organic compounds in fine particles derived from products of wood log combustion (mg kg^{-1})[a]

Compound	Pine	Oak	Compound	Pine	Oak
n-Alkanes					
Heneicosane, C_{21}	0.44	0.32	Octacosane, C_{28}	0.41	0.13
Docosane, C_{22}	0.45	0.28	Nonacosane, C_{29}	0.61	0.15
Tricosane, C_{23}	0.41	0.41	Triacontane, C_{30}	0.42	0.09
Tetracosane, C_{24}	0.29	0.26	Hentriacontane, C_{31}	0.48	0.08
Pentacosane, C_{25}	0.28	0.25	Dotriacontane, C_{32}	0.19	–
Hexacosane, C_{26}	0.36	0.20	Tritriacontane, C_{33}	0.13	0.03
Heptacosane, C_{27}	0.47	0.09	Tetratriacontane, C_{34}	0.12	0.02
n-Alkanoic acids					
Nonanoic, C_9	–	0.24	Eicosanoic, C_{20}	6.5	2.6
Decanoic, C_{10}	0.1	0.4	Heneicosanoic, C_{21}	5.2	1.5
Undecanoic, C_{11}	–	–	Docosanoic, C_{22}	8.0	4.7
Dodcanoic, C_{12}	1.9	1.7	Tricosanoic, C_{23}	1.6	2.2
Tridecanoic, C_{13}	–	–	Tetracosanoic, C_{24}	9,9	13.9
Tetradecanoic, C_{14}	1.7	4.9	Pentacosanoic, C_{25}	0.9	1.3
Pentadecanoic, C_{15}	0.9	1.6	Hexacosanoic, C_{26}	1.6	6.8
Hexadecanoic, C_{16} (palmitic)	13.9	21.5	Heptacosanoic, C_{27}	0.1	0.3
Heptadecanoic, C_{17}	1.6	2.6	Octacosanoic, C_{28}	0.2	0.6
Octadecanoic, C_{18} (stearic)	4.3	3.3	Nonacosanoic, C_{29}	–	0.1
Nonadecanoic, C_{19}	0.5	0.4	Triacontanoic, C_{30}	–	0.08
n-Alkenoic acids					
cis-9-Octadecanoic	7.6	1.2	9,12-Octadecadienoic acid	8.4	1.5
Resin Acids					
Abietic	42	–	*Iso*pimaric	27	–
Dehydroabietic	37	6	Sandaracopimaric	47	–
7-Oxodehydroabietic	3.3	–	13-Isopropyl-5a-podocarpa-	2	–
Pimaric	24	–	6,8,11,13-tetraen-16-oic		
8,15-Pimaradien-18-oic	4	–			
n-Dicarboxylic and other Acids					
Propanedioic (malonic), C_3	38.4	–	3-Hydroxybenzoic	4.3	7.5
Butanedioic (succinic), C_4	0.9	11.7	3,4-Dimethoxybenzoic	65	19
Methylbutanedioic, C_5	–	3.4	3,4,5-Trimethoxybenzoic	–	23
Pentanedioic (glutaric), C_5	6.7	5.4	3,4-Dimethoxyphenylacetic	6	–
Hexanedioic (adipic), C_6	0.6	1.8	4-Hydroxy-3-methoxyphenyl	83	15
2-Furancarboxylic	3	2	acetic		
Substituted benzenes					
1,2-Dimethoxybenzene	3	2	3,4-Dimethoxybenzaldehyde	23	5
1,3-Dimethoxybenzene	1	2	3-Methoxy-4-hydroxybenzal	29	2
1,4-Dimethoxybenzene	–	1.5	dehyde		
1,4-Dimethoxy-2-methyl	28	–	4-Hydroxy-3,5-dimethoxy	–	67
benzene			benzaldehyde		
3-Methoxybenzaldehyde	0.7	–	3,4,5-Trimethoxybenzaldehyde	–	62
Substituted phenols					
1,3-Benzenediol	3	4	3-Methyl-1,2-benzendiol	35	13
1,4-Benzendiol	62	23	4-Methyl-1,2-benzendiol	20	3

(continued)

Table 5.27   (continued)

Compound	Pine	Oak	Compound	Pine	Oak
2,6-Dimethoxy-4-(2-propenyl) Phenol	–	2	1-(3,5-Dimethoxy-4-hydroxyphenyl)propan-2-one	–	21
4-Propylbenzenediol	19	–			
2-Methoxyphenol	0.2	0.07	1-(3,4,5,-Trimethoxyphenyl) ethanone	–	36
2,6-Dimethoxyphenol	1	11			
2-Methoxy-4-methylphenol	0.8	0.04	1-(3,4,5,-Trimethoxyphenyl) propan-2-one	–	80
2-Methoxy-4-propylphenol	20	3			
2-Methoxy-4-propenylphenol	8	0.2	Divanillyl	22	2
2-Methoxy-4-(2-propenyl) phenol	1.5	–	Dihydro-3,4-divanillyl-2 (3*H*)-furanone	2.8	–
2-Methyl-5-(1-methylethyl) -2,5-cyclohexadiene-1, 4-dione	0.6	–	Dihydro-3,4-diveratryl-2 (3*H*)-furanone	2	–
1-(4-Methoxyphenyl) ethanone	5	4	Dihydrovanillylsyringyl-2 (3*H*)-furanone	5	–
3,4-Dimethoxyphenylacetone	10	6	Tetrahydro-3,4-divanillylfuran	23	0.6
1-(2,4-Dimethoxyphenyl) propan-2-one	9	2	Tetrahydro-3,4-diveratrylfuran	4	0.1
1-(4-Hydroxy-3-methoxyphenyl)ethanone	36	4	Tetrahydro-3,4-vanillyl-4-veratrilfuran	9	0.4
1-(4-Hydroxy-3-methoxy-phenyl)propan-2-one			Bisguaiacylsyringyl	–	3
1-(3,5-Dimethoxy-4-hydroxyphenyl)ethanone	–	56	Bis(3,4-dimethoxyphenyl) methane	6	0.8
			Bis(3,4,5-trimethoxyphenyl) ethane	–	0.8
			Disyringyl	–	7

Phytosterols

Compound	Pine	Oak	Compound	Pine	Oak
β-Sitosterol	46	10	Stigmast-4-en-3-one	3	1.3

Polycyclic aromatic hydrocarbons

Compound	Pine	Oak	Compound	Pine	Oak
Phenantrene	0.5	0.3	Benz[a]anthracene	0.6	0.2
Anthracene	0.05	0.06	Chrysene/triphenylene	1.0	2.8
Fluoranthene	1.2	0.4	Benzo[k]fluoranthene	0.5	0.3
Pyrene	1.6	0.5	Benzo[b]fluoranthene	0.5	0.2
Retene	0.7	0.1	Benzo[j]fluoranthene	0.3	0.1
Benzacenaphtalene	0.6	0.2	Benzo[e]pyrene	0.3	0.1
Methyl(fluoranthenes, pyrenes)	1.2	0.3	Benzo[a]pyrene	0.6	0.2
			Perylene	0.1	0.04
Benzo[a]fluorene, benzo[b] Fluorene	0.06	0.01	Indeno[1,2,3-c,d]pyrene	0.09	0.05
			Indeno[1,2,3-c,d]fluoranthene	0.4	0.2
Benzo[g,h,i,]fluoranthene	0.3	0.1	Benzo[g,h,i]perylene	0.3	0.1
Cyclopenta[c,d]pyrene	0.7	0.2	Anthanthrene	0.1	0.04
Benzo[]phenanthrene	0.2	0.03	Dibenz[a,h]anthracene	0.08	0.01

[a]Source of data: Rogge et al. (1998)

Table 5.28 Particulate concentrations (ng m^{-3}), average carbon preference index, CPI, and the most abundant carbon number, C_{max}, of solvent-extractable organic compound classes[a]

Location	n-alkanes	n-alkanoic acids	n-alkanols
Marine environment			
Enetewak Atoll, North Pacific (1)	0.02–0.16 CPI 2.4, C_{max} 29	0.63–4.25 CPI 3.9, C_{max} 16	0.07–0.21 CPI 12.5, C_{max} 28
Chichi-Jima, West Pacific (2)	0.1–14 CPI 2–14, C_{max} 29	2.4–60 C_{max} 16, 24	0.18–20 C_{max} 28
Ninety Mile Beach New Zealand (3)	0.02–0.27 CPI 4.7 C_{max} 29, 31	0.43–0.72 CPI 2.4, C_{max} 16, 24	0.05–0.49 CPI 14.7, C_{max} 26
South Pacific, near Coast of Peru (4)	0.28–0.61 CPI 5.8, C_{max} 29	0.08–0.26 CPI 19, C_{max} 29	0.2–0.41 CPI 3.7, C_{max} 28
South Atlantic, African coast (5)	0.002–4.2 CPI 5.8, C_{max} 29	0.01–1.5 CPI 11.9, C_{max} 16, 26	0.47–2.5 CPI 3.7, C_{max} 28
South Atlantic, 30–70° S (6)	30–2,800 CPI 1.8, C_{max} 29	7–26 CPI 19, C_{max} 19	1–1,200 CPI 46, C_{max} 16
Terceira Island, Azores (7)	0.3–6.2 CPI 1.6–11, C_{max} 29	0.42–86 CPI 3.2–7.6, C_{max} 16	0.75–18.6 CPI 16, C_{max} 26
Rural continental			
Western USA (8)	1–390 CPI 4.3, C_{max} 29	90–300 CPI 9.1, C_{max} 16	200–1,390 CPI 6.6, C_{max} 26
Central Africa (8)	110–1,700 CPI, C_{max} 29	80–960 C_{max} 16	230–2,200 C_{max} 30
SE Australia (9)	16–80 CPI 1.9, C_{max} 29	30–110 CPI 11, C_{max} 16	70–380 CPI 14, C_{max} 26
Amazonia, Brazil (10)	300–810 CPI 2.5, C_{max} 29	200–620 CPI 10.7, C_{max} 16	10–110 CPI 14.7, C_{max} 16, 28
Melpitz, Germany (11)	16–262 CPI 3.1, C_{max} 29	72–366 CPI 6.7, C_{max} 16	2.2–262 CPI 18.3, C_{max}, 26
Hyytiälä, Finland Boreal forest (11)	7.2–95 CPI 2.7, C_{max} 27	39–192 CPI 6.9, C_{max} 16	1–17 CPI 8.5, C_{max} 26
Pertouli, Greece (12)	350–3,486 CPI 1.9, C_{max} 29	124–3,851 CPI 8.1, C_{max} 16	31–179 CPI 9.2, C_{max} 26
Giesta, Portugal (13)	455 av. CPI 1.8, C_{max} 29	787 av. CPI 3.0, C_{max} 22	138 av. CPI 7.9, C_{max} 26
Urban			
Campo Goytacazes Brazil (14)	20–145 CPI 1.3, C_{max} 25	6.9–28 CPI -, C_{max} 16	3.0–6.2 CPI -, C_{max} 16
Los Angeles, California (8)	178–361 CPI 2.4, C_{max} 29	217–308 CPI 14.5, C_{max} 16	1,360–2,016 CPI 9.8, C_{max} 26, 28
Birmingham, UK (15)	109 av. CPI 1.2, C_{max} 19	76 av. CPI 3.4, C_{max} 16	120 av. CPI 4.4, C_{max} 26
Lisbon, Portugal (16)	427 av. CPI 1.4, C_{max} 29	104 av. CPI 8.9, C_{max} 16	905 av. CPI 5.0, C_{max} 16, 22
Kuala Lumpur, Malaysia (17)	335–1,009 CPI 1.4, C_{max} 29	3–7,400 CPI -, C_{max} 16	90–1,280 CPI 5.9, C_{max} 28

(continued)

Table 5.28 (continued)

Location	n-alkanes	n-alkanoic acids	n-alkanols
Hong Kong, China (18)	6–41 CPI 1.5, C_{max} 29	31–169 CPI 8.4, C_{max} 16	8–39 CPI 8.2, C_{max} 18
Nanjing, China (19)	30–265 CPI 3.5, C_{max} 29	127–408 CPI 7.6, C_{max} 18	27–220 CPI 16.9, C_{max} 28

[a]na not available; CPI (n-alkanes)=(Σ concentration of C_{odd})/(Σ concentration of C_{even}), CPI (alkanoic acids, n-alkanols)=(Σ concentration of C_{even})/(Σ concentration of C_{odd}). Sources of data : (1) Gagosian et al. (1982), (2) Kawamura et al. (2003), (3) Gagosian et al. (1987), (4) Simoneit et al. (1977), (5) Schneider and Gagosian (1985), (6) Simoneit et al. (1991a), (7) Alves et al. (2007), (8) Simoneit (1989), (9) Simoneit et al. (1990), (10) Simoneit et al. (1991b), (11) Alves et al. (2006), (12) Pio et al. (2001a), (13) Pio et al. (2001b), (14) Azevedo et al. (2002), (15) Harrad et al. (2003), (16) Alves et al. (2002), (17) Abas et al. (2004), (18) Zheng et al. (2000), (19) Wang and Kawamura (2005)

Table 5.29 Particulate concentrations of individual n-alkanes in urban regions (ng m^{-3})[a]

Alkane	Santiago Aug	Rio de Janeiro RT	Rio de Janeiro TF	London S	London W	Houston S	Houston W	Florina Annual	Hong Kong
C12	–	–	–	0.25	0.34	–	–	–	–
C13	–	–	–	0.30	0.45	–	–	–	–
C14	8.6	0.89	0.13	0.25	1.91	–	–	0.24	0.08
C15	14.6	0.61	0.98	0.48	2.45	–	–	0.83	0.15
C16	36.1	0.98	2.35	0.84	2.62	0.06	0.26	0.42	0.16
C17	21.9	1.60	2.04	2.07	4.74	0.14	0.77	2.62	0.16
C18	32.5	3.05	1.56	2.15	7.22	0.09	0.34	1.18	0.14
C19	10.0	8.10	1.50	2.68	7.86	0.28	0.86	0.80	0.15
C20	14.3	23.17	3.69	6.65	15.13	0.29	1.14	0.85	0.28
C21	9.8	30.02	5.3	7.53	17.50	0.21	1.77	1.24	0.41
C22	14.8	57.15	7.1	6.96	19.37	0.47	2.65	1.51	0.84
C23	14.7	70.35	6.58	9.79	22.08	0.61	3.70	2.35	1.60
C24	18.0	72.92	13.99	12.73	27.90	1.22	6.12	3.44	2.32
C25	15.0	72.13	5.54	18.20	32.60	1.19	5.31	4.98	3.02
C26	13.5	69.67	2.38	16.40	38.62	1.24	4.94	5.07	2.33
C27	12.6	38.73	3.13	15.62	30.27	0.78	4.28	7.34	3.25
C28	12.5	13.30	2.65	16.39	35.93	0.94	3.40	4.88	1.79
C29	10.2	20.63	15.10	23.72	24.42	1.03	8.43	9.89	3.50
C30	7.1	9.5	2.98	9.08	18.20	0.78	5.41	3.80	0.96
C31	7.2	10.09	4.44	–	–	1.10	9.89	7.84	1.87
C32	4.1	3.08	2.24	3.62	5.29	–	4.43	2.39	0.35
C33	3.2	2.51	2.62	–	–	0.47	9.49	2.84	0.35
C34	1.4	4.34	1.79	1.08	2.95	–	–	0.79	0.07
C35	0.8	1.93	1.53	–	–	–	–	1.64	0.04

[a]S summer, W winter; Sources of data. Santiago, Chile, Tsapakis et al. (2002); Rio de Janeiro, Brazil, Azevedo et al. (1999) (RT in a road tunnel, TF in the central park); downtown London, UK, Kendall et al. (2001); Houston, Texas (industrial), Fraser et al. (2002), Florina, Greece, Kalaitzoglou et al. (2004), Hong Kong, China, Zheng et al. (2000), annual averages

Table 5.30 Distribution of *n*-alkanes between the gas phase and the particulate phase (ng m^{-3})[a]

| | Athens, Greece | | | | Guangzhou, China | | | | Prato, Italy | |
| | Urban | | Coastal | | Ground level | | 25 m height | | Industrial | |
Compound	G	P	G	P	G	P	G	P	G	P
C12	–	–	–	–	2.20	0.55	1.97	1.27	–	–
C13	–	–	–	–	1.64	1.10	1.05	2.21	–	–
C14	0.83	–	–	–	1.91	1.27	1.16	1.87	–	–
C15	1.83	0.10	0.06	0.02	3.36	0.07	0.93	1.60	2.84	–
C16	3.89	0.15	0.27	0.02	5.02	0.49	1.40	1.91	7.68	0.45
C17	4.00	0.09	0.54	0.03	5.56	0.63	1.43	1.48	15.33	0.40
C18	5.07	0.10	0.63	0.03	8.45	1.53	2.63	1.44	21.70	0.40
C19	5.20	0.19	2.05	0.03	15.90	2.30	4.59	1.08	30.05	0.43
C20	3.92	0.31	2.65	0.04	27.55	5.02	6.85	1.13	32.30	0.95
C21	5.05	0.52	3.43	0.10	42.87	7.80	11.80	1.95	30.58	1.02
C22	3.86	0.76	3.99	0.17	42.92	10.00	15.02	1.71	24.05	2.85
C23	3.97	1.41	4.17	0.33	31.86	12.27	12.95	2.82	19.10	4.98
C24	2.70	1.50	2.31	0.47	13.18	17.33	6.88	4.15	12.81	8.02
C25	2.17	2.71	1.85	0.97	5.31	19.62	3.29	6.97	8.24	14.97
C26	1.44	2.35	1.56	1.15	1.88	15.17	1.53	7.38	0.45	17.70
C27	1.25	5.28	1.66	1.69	0.65	11.53	0.89	7.74	1.59	21.95
C28	1.13	2.19	1.36	0.86	–	7.87	0.40	5.98	1.14	14.72
C29	1.21	8.25	1.13	2.14	–	8.01	0.18	7.93	1.25	19.05
C30	0.98	1.85	0.88	0.76	–	4.11	–	4.03	0.91	8.98
C31	0.85	8.54	–	2.79	–	5.28	–	5.66	1.14	19.05
C32	0.41	1.42	–	0.66	–	1.39	–	1.72	0.82	5.90
C33	0.52	1.23	–	1.27	–	1.41	–	2.25	0.57	5.46
C34	–	–	–	–	–	0.91	–	0.92	0.43	1.71
C35	–	–	–	–	–	0.71	–	0.68	0.46	2.62

[a]Key: G gas phase, P particulate phase; source of data: Athens, July, Mandalakis et al. (2002); Guangzhou, July, Bi et al. (2003); Prato, March–November, Cincinelli et al. (2007)

Table 5.31 Concentrations of individual *n*-alkanoic acids in urban regions (ng m^{-3})[a]

| | Santiago | Los Angeles | Houston | Algiers | Ghent | Hong Kong | | Nanjing | |
Acid	Aug	Annual	Aug/Sept	S	S	Aug	Feb	July	Jan
C10	74.0	1.3	–	–	0.1	–	–	–	–
C11	26.1	3.8	–	–	–	–	–	–	–
C12	32.2	3.7	–	–	2.6	–	1.7	1.7	5.0
C13	13.0	3.3	0.32	–	–	–	0.5	–	–
C14	38.1	14.4	4.59	2.8	3.6	2.6	4.4	2.8	5.3
C15	23.1	4.3	1.63	0.4	1.9	1.6	1.3	1.6	3.0
C16	280.9	118.3	17.10	11.5	23.7	35.3	40.6	35.8	84.5
C17	22.9	3.4	1.74	2.7	1.5	1.0	1.2	1.5	2.7
C18	93.6	57.7	43.98	31.1	14.8	16.4	22.2	17.8	43.5
C19	6.8	0.8	1.01	1.5	0.4	1.6	1.6	0.7	1.8
C20	13.8	4.3	4.28	11.2	3.4	2.2	6.3	2.9	6.6
C21	4.8	1.7	2.38	7.5	0.6	0.3	2.2	0.9	3.1

(continued)

Table 5.31   (continued)

Acid	Santiago Aug	Los Angeles Annual	Houston Aug/Sept	Algiers S	Ghent S	Hong Kong Aug	Hong Kong Feb	Nanjing July	Nanjing Jan
C22	35.3	7.5	5.47	4.7	3.7	2.7	12.3	5.5	10.4
C23	8.7	2.0	2.27	1.4	1.1	0.7	5.7	3.6	7.5
C24	10.2	12.5	7.32	6.2	4.7	2.5	20.3	15.9	14.9
C25	5.5	1.4	0.90	1.4	1.6	1.1	3.5	3.8	4.2
C26	6.6	7.1	3.42	2.3	2.4	1.6	10.0	14.2	10.4
C27	1.3	0.7	2.50	7.3	0.4	0.3	1.9	5.5	2.5
C28	12.9	2.9	1.74	0.9	2.4	0.2	1.4	24.3	11.0
C29	7.6	0.5	0.41	0.1	–	2.0	6.6	7.4	2.3
C30	–	1.2	1.84	0.2	1.1	0.5	0.5	44.4	16.5
C31	–	–	0.23	–	–	1.9	1.9	2.4	1.7
C32	–	–	0.93	–	1.3	–	–	16.9	2.9

[a]Sources of data: Santiago, Chile, 1998, Tsapakis et al. (2002); Los Angeles, Calif., Rogge et al. (1993a); Houston, Texas, 2000, Yue and Fraser (2004); Algiers, May-Sept, Yassaa et al. (2001); Ghent, Belgium, June–Aug, Kubátova et al. (2002); Hong Kong, China, Zheng et al. (2000), Nanjing, China, Wang and Kawamura (2005)

Table 5.32   Gas/particulate phase partitioning of polycyclic aromatic hydrocarbons (ng m^{-3})[a]

Compound	Rome, Italy G	Rome, Italy P	Taichung, Taiwan G	Taichung, Taiwan P	Tampa Bay G	Tampa Bay P
Naphtalene	687±580	12±6	183±168	1.0±1.2	72	24
Methylnaphtalenes	478±200	11±6	–	–	–	–
Fluorene	18±8	0.9±0.5	69±71	2.1±3.3	6.2	0.27
Acenaphtene	57±20	2.2±1.0	41±29	0.9±0.9	4.1	0.2
Acenaphtylene	39±18	4.6±2.0	75±101	2.6±2.7	0.2	0.0
Phenantrene	71±22	7.2±1.9	59±102	1.5±2.1	13.3	2.4
Anthracene	5.6±1.9	0.5±0.2	61±108	1.1±1.7	0.5	0.0
Fluoroanthene	18±9	3.5±1.1	41±53	4.0±4.6	4.9	1.0
Pyrene	7.6±6.0	9.2±3.5	35±39	2.8±6.5	1.7	0.6
Cyclopenta[cd]pyrene	–	–	28±41	1.9±4.5	–	–
Benzo[a]anthracene	0.4±0.1	1.4±0.7	15±35	0.9±1.8	0.04	0.01
Chrysene	0.5±0.2	3.9±1.2	14±29	2.2±4.2	0.46	0.04
Benzo[b]fluoroanthene	0.7±0.3	6.8±2.5	4±8	1.5±3.5	0.06	0.02
Benzo[k]fluoroanthene	[b]	[b]	4±8	3.2±7	0.01	0.01
Benzo[a]fluoroanthene	–	1.4±0.6	–	–	–	–
Benzo[a]pyrene	0.3±0.1	2.4±1.0	4.6±13	1.5±4.1	0.01	0.0
Benzo[e]pyrene	0.5±0.2	2.8±0.9	11±18	3.4±7.7	–	–
Dibenz(ah)anthracene	–	–	1.1±2.4	5.2±15	0.01	0.0
Perylene	–	0.5±0.2	14±36	6.0±15	–	–
Benzo[b]chrysene	–	–	1.4±2.7	8.4±12	–	–
Benzo(ghi)perylene	0.5±0.2	2.4±1.0	2.7±7.7	2.0±3.1	0.04	0.04
Indeno[1,2,3-cd]pyrene	–	1.6±0.6	3.9±10	1.9±3.3	0.02	0.02
Anthanthrene	–	1.0±0.5	–	–	–	–
Coronene	–	1.1±0.5	1.7±3.5	8.4±20	–	–

[a]G gas-phase, P particulate phase. Sources of data: Rome, Italy, Possanzini et al. (2004); Taichung, Taiwan, Fang et al. (2004); Tampa Bay, Florida, Poor et al. (2004)
[b]Included in the value for Benzo[b]fluoroanthene

Table 5.33 Partitioning of polycyclic aromatic hydrocarbons between the gas phase and the particulate phase in Birmingham, UK (ng m^{-3})[a]

Compound	Urban Summer		Winter		Rural Summer		Queensway road tunnel	
	G	P	G	P	G	P	G	P
Naphtalene	1.73	0.14	12.6	0.69	0.31	0.03	116	0.9
Fluorene	6.79	0.21	12.6	1.06	1.96	0.06	151	16.3
Acenaphtene	3.94	0.29	11.9	1.60	0.75	0.06	99.2	14.7
Acenaphtylene	2.60	0.12	14.8	0.61	0.64	0.02	94.3	0.94
Phenantrene	3.59	0.25	23.0	1.08	0.54	0.06	307	25.6
Anthracene	0.45	0.16	4.1	0.39	0.14	0.03	41.7	9.41
Fluoroanthene	1.76	0.35	11.2	1.17	0.34	0.07	26.4	21.1
Pyrene	2.78	0.55	35.7	2.36	0.90	0.10	25.3	30.0
Benzo[a]anthracene	0.21	0.13	4.11	1.48	0.04	0.04	2.39	11.6
Chrysene	0.40	0.21	4.48	2.21	0.08	0.07	7.34	18.5
BNT[b]	0.18	0.15	0.70	0.62	0.07	0.05	6.44	9.46
Benzo[b]fluoroanthene	0.04	0.34	0.28	1.87	0.02	0.12	<1.06	11.6
Benzo[k]fluoroanthene	0.02	0.14	0.09	1.12	0.01	0.06	<0.58	5.40
Benzo[a]pyrene	0.02	0.23	0.08	0.73	<0.01	0.06	<0.93	12.7
Dibenz(a,h)anthracene	<0.01	0.07	0.05	0.78	<0.01	0.03	<1.32	4.38
Benzo(ghi)perylene	<0.01	0.76	0.06	1.91	<0.01	0.21	<0.66	35.2
Indeno[1,2,3-cd]pyrene	<0.01	0.42	0.01	1.95	<0.01	0.11	<5.64	21.5
Coronene	<0.01	0.27	<0.01	1.03	<0.01	0.06	<0.31	11.8

[a]G gas-phase, P particulate phase; source of data: Smith and Harrison (1996); standard deviations generally are of the same magnitude as the averages presented
[b]Benzo(b)naphto[2,1-d]thiophene

Table 5.34 Concentrations of polycyclic aromatic hydrocarbons in urban regions (ng m⁻³)[a]

Compound	Houston	Boston	Santiago	Porto Allegre	Brisbane	Hong Kong	Lahore	Guangzhou	London	Manchester	BanjaLuka	Heraklion	Naples
Naphtalene	600	529	–	–	3.01	–	0.39	–	–	–	3.7	–	–
Acenaphtylene	78	15	–	–	0.04	0.01	1.01	0.18	13	16	1.4	–	–
Fluorene	21	23	16	0.17	0.02	0.03	0.98	1.1	13	16	5.5	5.2	–
Acenaphtene	–	–	–	0.06	0.02	0.01	2.78	0.05	2.1	1.5	1.2	–	–
Phenantrene	44	66	28	0.64	0.88	0.32	0.97	15.3	76	36	13.9	20	0.9
Anthracene	1.8	2.3	3.1	0.27	0.24	0.29	4.99	1.3	5.0	2.0	0.6	3.3	0.1
Fluoroanthene	8.5	13	7.2	1.12	0.19	0.48	2.81	23.4	7.4	7.6	3.7	4.9	0.7
Pyrene	7.2	8.3	8.4	0.38	0.34	0.26	2.93	17.0	6.8	5.5	2.8	6.6	1.2
Benz[a]anthracene	0.5	1.5	8.5	1.28	0.30	0.35	5.39	1.5	0.8	0.8	0.5	1.1	0.9
Chrysene	0.7	1.6	28	0.80	0.22	0.53	8.64	6.8	1.5	1.4	0.8	3.1	1.4
Benzofluoroanthenes	0.8	2.0	36	1.80	0.33	1.06	14.4	1.6	2.1	1.0	1.5	3.3	5.5
Benzo[a]pyrene	0.3	1.0	11	1.09	0.19	0.65	9.32	0.79	0.6	1.2	0.8	1.2	0.9
Benzo[e]pyrene	0.3	0.7	20	–	–	–	–	1.5	–	–	0.9	1.7	5.6
Dibenz[ah]anthracene	–	–	–	0.54	0.56	0.01	3.85	1.0	0.3	1.0	0.1	0.1	0.3
Benzo[ghi]perylene	0.9	1.9	38	2.29	0.25	0.05	14.6	2.32	4.4	1.8	1.1	3.4	9.1
Indeno[1,2,3-cd]pyrene	0.5	1.2	9.7	0.80	0.24	0.49	12.3	1.66	–	–	1.1	2.5	–
Coronene	0.6	1.5	5.8	–	–	–	5.40	0.30	0.7	–	0.6	0.2	5.5

[a]Data sources: Houston, Texas, Boston, Massachusetts, 1991/1992, Lewis and Coutant (1999); Brisbane, Australia, June/July 2002, Lim et al. (2005); Santiago, Chile, Aug. 1998, Tsapakis et al. (2002); Hong Kong, summer 2001, particles only, Guo et al. (2003); Lahore, Pakistan, Aug/Sept 1992, Smith et al. (1996); Guangzhou, China, July 2001, Bi et al. (2003); Porto Allegre, Brazil, 2001/2002, particles only, Dallarosa et al. (2005); London, Manchester, UK, 1991/1992, Halsall et al. (1994); Banja Luka, Bosnia/Hercegowina, July 2008, Lammel et al. (2010); Heraklion, Greece, 200/2002, Tsapakis and Stephanou (2005); Naples, Italy, Sept 1996, particles only, Caricchia et al. (1999)

Table 5.35 Dicarboxylic acids in rural continental regions (ng m^{-3})[a, b]

Compound	Central Japan	Austrian Alps	Southeast USA	Nylsvley S. Africa	Waldstein Germany	Amazonia Brazil
Ethanedioic (oxalic)	3.9	9.9	–	193	68.6	120
Propanedioic, (malonic)	11.1	0.9	4.3	142	–	41
Methyl propanedioic	0.7	–	0.5	–	–	–
Butanedioic, (succinic)	21.1	2.5	12.0	58	10.4	7.6
Methyl butanedioic	5.3	–	1.6	–	–	0.8
Pentanedioic (glutaric)	4.7	1.8	3.5	8.8	1.6	1.5
Methyl pentanedioic	1.7	–	–	–	–	–
Hexanedioic (adipic)	2.3	2.9	1.9	7.9	1.9	0.8
Heptanedioic (pimelic)	1.7	–	1.0	–	1.6	0.2
Octanedioic (suberic)	1.5	1.1	3.2	–	1.4	0.5
Nonanedioic (azelaic)	1.2	1.0	6.7	–	3.2	1.7
Decanedioic (sebatic)	1.3	–	–	–	–	–
Phtalic	24.7	1.7	8.6	3.3	–	0.9

[a]Source of data: Graham et al. (2003), Limbeck and Puxbaum (1999, 2000), Plewka et al. (2006), Satsumabayashi et al. (1990), Zheng et al. (2002)
[b]Site descriptions: (1) Karuizawa mountain pass, Japan; (2) Sonnblick Observatory, 3,106 m a.s.l., Austria; (3) Centreville, Alabama; (4) Nylsvley Nature Reserve, savanna, South Africa; (5) Balbina, Amazonia, 100 km north of Manaus

Table 5.36 Concentrations of dicarboxylic and related acids in remote regions (ng m^{-3})[a]

Acid	Alert, Canada 1987–1988	Syowa Station, Antarctic Winter/Spring 1991		North Pacific Summer 1989
Dicarboxylic acids				
Ethanedioic (oxalic)	13.6±12	2.65±0.76	10.3	17.9±6.3
Propanedioic, (malonic)	2.46±3.3	0.34±0.82	2.69	3.8±1.58
Methyl propanedioic	0.13±0.12	0.05	0.24	0.21±0.08
Butanedioic, (succinic)	3.73±3.5	2.53±2.30	61.5	2.84±1.03
Methyl butanedioic	0.36±0.37	0.11±0.01	0.18	1.09±0.38
Pentanedioic (glutaric)	0.90±0.91	0.41±0.12	2.26	0.53±0.15
Methyl pentanedioic	0.016±0.04	–	1.81	–
Hexanedioic (adipic)	0.82±1.43	0.56±0.22	0.94	0.43±0.02
Heptanedioic (pimelic)	0.13±0.26	0.40±0.12	0.96	0.10±0.04
Octanedioic (suberic)	0.15±0.26	0.23±0.03	2.61	0.16±0.05
Nonanedioic (azelaic)	0.26±0.32	0.65±0.33	2.22	0.34±0.14
Decanedioic (sebatic)	<0.003–1.7	0.05±0.03	0.17	0.06±0.02
Undecanedioic, C$_{11}$	<0.003–1.1	0.09±0.04	0.28	–
Unsaturated dicarboxylic acids				
cis-Butenedioic (maleic)	0.19±0.14	0.25±0.09	0.96	0.22±0.11
trans-Butenedioic (fumaric)	0.14±0.19	0.09±0.1	0.23	0.31±0.10
Methyl butenedioic	0.03±0.06	0.01–0.06	–	–
Phtalic	1.5±1.6	1.16±0.37	2.61	0.58±0.16

(continued)

Table 5.36 (continued)

Acid	Alert, Canada 1987–1988	Syowa Station, Antarctic Winter/Spring 1991		North Pacific Summer 1989
Acids containing a carbonyl or hydroxyl group				
Oxoethanoic (glyoxylic)	1.7±2.4	0.29±0.06	0.38	0.72±0.46
2-Oxopropanoic (pyruvic)	0.13±0.13	0.23±0.07	0.78	<0.04–0.19
3-Oxopropanoic	0.07±0.05	0.06±0.01	1.03	0.09±0.05
4-Oxobutanoic	0.35±0.24	0.22±0.05	1.00	0.18±0.09
5-Oxopentanoic	0.02±0.02	0.01–0.02	0.09	–
6-Oxohexanoic	–	0.16±0.16	0.11	–
4-Oxoheptanoic	0.20±0.17	–	–	–
8-Oxononanoic	–	0.06±0.04	0.18	–
9-Oxononanoic	0.01±0.02	0.10±0.06	1.16	0.04±0.01
Oxopropanedioic	0.31±0.38	0.17±0.17	0.53	
4-Oxoheptanedioic	–	0.17±0.17	–	
Hydroxybutanedioic	0.03±0.04	0.06±0.06	0.70	0.15±0.03

[a]Source of data: Kawamura and Usukura (1993), Kawamura et al. (1996a, b)

Table 5.37 Dicarboxylic and related acids in urban regions (ng m^{-3})[a]

Acid	Los Angeles		Tokyo		Hong Kong	
	West	Rubidoux	June	Nov	Feb	Aug
Dicarboxylic acids						
Ethanedioic (oxalic)	–	–	357	186	249	15
Propanedioic, (malonic)	28	51	71	41	44.7	24.5
Methyl propanedioic	–	–	7.2	5.1	2.5	1.3
Butanedioic, (succinic)	55	84	73	47	45.8	15.2
Methyl butanedioic	12	20	13	11	5.5	1.3
Pentanedioic (glutaric)	28	39	23	18	16.5	5.3
2-Methyl pentanedioic	16	24	7.3	2.2	1.4	0.5
Hexanedioic (adipic)	15	24	26	14	17.4	19
2-Methyl hexanedioic	–	–	3.5	1.3	–	–
Heptanedioic (pimelic)	–	–	10	8.2	7.0	2.4
Octanedioic (suberic)	2.9	2.5	11	9.1	3.3	3.4
Nonanedioic (azelaic)	34.2	44.7	15	21	10.9	7.4
Decanedioic (sebatic)	–	–	4.9	7.3	–	–
Unsaturated and aromatic dicarboxylic acids						
cis-Butenedioic (maleic)	0.6	0.8	12	8.4	11.8	4.3
trans-Butenedioic (fumaric)	[b]	[b]	4.5	7.3	8.1	5.9
Methyl butenedioic	–	–	18	11	6.0	3.0
1,2-Benzenedioic (Phtalic)	61	53	33	24	86	50
1,3-Benzenedioic	2.1	2.1	–	–	69	38
1,4-Benzenedioic	1.3	0.9	–	–	71	32
Acids containing a carbonyl or hydroxyl group						
Oxoethanoic (glyoxylic)	–	–	44	42	37	23
2-Oxopropanoic (pyruvic)	–	–	50	45	4.8	3.1
3-Oxopropanoic	–	–	6.8	7.2	1.4	0.13
4-Oxobutanoic	–	–	16	17	3.8	–
Oxopropanedioic	–	–	3.5	3.7	6.3	2.2
4-Oxoheptanedioic	–	–	6.6	5.3	4.9	1.2
Hydroxybutanedioic	–	–	27	8.8	6.5	2.5

[a]Source of data: Los Angeles, Rogge et al. (1993a); Tokyo, Kawamura and Yasui (2005); Hong Kong (entrance of Shing Mum Tunnel), Wang et al. (2006)
[b]Included in the value for *cis*-butendioic acid

Table 5.38 Indicators (molecular markers) for sources of particulate organic material[a]

Compound class	Usual range of carbon number	Usual major homologue	Major source
n-Alkanes	C_{23}–C_{35} (odd/even)	C_{29} (C_{27})	Vascular plants
Branched hydrocarbons	C_{10}–C_{20} (CPI ≈ 1)	C_{19} (pristane)	Petroleum-derived[b]
n-Alkanoic acids	C_{24}–C_{36} (even/odd)	C_{26}	Vascular plants
	C_{10}–C_{20} (even/odd)	C_{16}	Vehicular emissions
n-Alkanols	C_{14}–C_{36} (even/odd)	C_{28}	Biogenic
Alkylcyclohexanes	C_{16}–C_{29}	C_{23}	Fossil fuel
Steranes, disteranes	C_{27}–C_{29}	C_{27}	Lubricants
Hopane, triterpenoids	C_{27}–C_{35}	C_{30}	Petroleum residue
Tricyclic terpanes	C_{19}–C_{29}	C_{23}	Petroleum-derived[b]
Polynuclear aromatic hydrocarbons	C_{10}–C_{24}	C_{16}	Combustion (fossil, fuels, biomass, etc.)
Wax esters	C_{32}–C_{40}	C_{40}	Higher plants
Phytosterols (e.g. β-sitosterol)	C_{27}–C_{30}	C_{29}	Vegetation, biomass burning
Levoglucosan[c]	$C_6H_{11}O_5$		Pyrolysis of cellulose
Diterpenoids[d]	C_{16}–C_{19}	C_{18}	Burning of conifers
Monosaccharides	$C_6H_{12}O_6$		Fungi, microbiota

[a]Source of data: Simoneit (1986, 2002)
[b]Not in gasoline
[c]1,6-Anhydro-β-D-glucopyranose (monosaccharide)
[d]These compounds include retene, pimaric acids, dehydroabietic acid, pimanthrene

Table 5.39 Particulate concentrations (ng m^{-3}) of terpene oxidation products[a]

Location	Nopinone	Pinonaldehyde	Pinonic acid	Pinic acid
Kejimkujik National Park, Nova Scotia (1)	–	0.2–1.0	0.1–0.8	0.5–0.6
Sierra Nevada, Calif. (2)	1.5–5.1	16–320	2.6–37	1.7–10
Yosemite Natl. Park, Calif. (3)	0.9	21	21	12
Duke Forest, North Carolina (4)	–	–	0.11–21	1.5–25
Eucalyptus forest, Portugal (5)	0–13	0.2–32	1.5–98[b]	0.4–83
Giesta, rural Portugal (6)	1.4	–	–	6.8
Aveiro, Portugal (7)	2.4	13.7	25.3	15.8
Hyytiälä, forest, Finland (8)	0.2	5.5	2.6	5.0
Melpitz, rural, Germany (8)	0.1	14	7.8	7.2
Coniferous forest, Germany (9)	–	2.7–13.7	2.5–3.5	3.2–9.5

[a]Sources of data: (1) Yu et al. (1999); (2) Cahill et al. (2006); (3) Engling et al. (2006); (4) Bhat and Fraser (2007); (5) Kavouras et al. (1999); (6) Pio et al. (2001b); (7) Alves et al. (2002); (8) Alves et al. (2006); (9) Plewka et al. (2006)
[b]Composed of cis-pinonic acid, 7–98 ng m^{-3}, trans-pinonic acid, 1.5–26 ng m^{-3}

References

Abas, M.R.B., N.A. Rahman, N.Y.M.J. Omar, M.J. Maah, A.A. Samah, D.R. Oros, A. Otto, B.R.T. Simoneit, Atmos. Environ. **38**, 4223–4241 (2004)

Alves, C.A., Ann. Brazilian Acad. Sci. **80**, 21–82 (2008)

Alves, C., A. Carvalho, C. Pio, J. Geophys. Res. **107**, 8345–8353 (2002)

Alves, C., C. Pio, A. Carvalho, C. Santos, Chemosphere **63**, 153–164 (2006)

Alves, C., T. Oliveira, C. Pio, A.J.D. Silvestre, P. Fialho, F. Barata, M. Legrand, Atmos. Environ. **41**, 1359–1373 (2007)

Azevedo, D.A., L.S. Moreira, D.S. de Siqueira, Atmos. Environ. **33**, 4987–5001 (1999)

Azevedo, D.A., C.Y.M. dos Santos, F.R.A. Neto, Atmos. Environ. **36**, 2383–2395 (2002)

Bhat, S., M.P. Fraser, Atmos. Environ. **41**, 2958–2966 (2007)

Bi, X., G. Sheng, P. Peng, Y. Chen, Z. Zhang, J. Fu, Atmos. Environ. **37**, 289–298 (2003)

Burrows, S.M., W. Elbert, M.G. Lawrence, U. Pöschl, Atmos. Chem. Phys. **9**, 9263–9280 (2009)

Cahill, T.M., V.Y. Seaman, M.J. Charles, R. Holzinger, A.H. Goldstein, J. Geophys. Res. **111**, D16312 (2006). doi:10.1029/2006JD007178

Caricchia, A.M., S. Chiavarini, M. Pezza, Atmos. Environ. **33**, 3731–3738 (1999)

Cincinelli, A., M. Del Bubba, T. Martinelli, A. Gambaro, L. Lepri, Chemosphere **68**, 472–478 (2007)

Dallarosa, J.B., J.G. Mônego, E.C. Teixeira, J.L. Stefens, F. Wiegand, Atmos. Environ. **39**, 1609–1625 (2005)

Elbert, W., P.E. Taylor, M.O. Andreae, U. Pöschl, Atmos. Chem. Phys. **7**, 4569–4588 (2007)

Engling, G., P. Herckes, S.M. Kreidenweis, W.C. Malm, J.L. Collett, Atmos. Environ. **40**, 2959–2972 (2006)

Fang, G.-C., Y.-S. Wu, M.-H. Chen, T.-T. Ho, S.H. Huang, J.-Y. Rau, Atmos. Environ. **38**, 3385–3391 (2004)

Fraser, M.P., Z.W. Yue, R.J. Tropp, S.D. Kohl, J.C. Chow, Atmos. Environ. **36**, 5751–5758 (2002)

Gagosian, R.B., O.C. Zafiriou, E.T. Peltzer, J.B. Alford, J. Geophys. Res. **87**, 11133–11144 (1982)

Gagosian, R.B., E.T. Peltzer, J.T. Merrill, Nature **325**, 800–803 (1987)

Graham, B., P. Guyon, P.E. Taylor, P. Artaxo, W. Maenhaut, M.M. Glovsky, R.C. Flagan, M.O. Andreae, J. Geophys. Res. **108**, 4766 (2003). doi:10.1029/2003JD003990

Guo, H., S.C. Lee, K.F. Ho, X.M. Wang, S.C. Zou, Atmos. Environ. **37**, 5307–5317 (2003)

Halsall, C.J., P.J. Coleman, B. J. Davis, V. Burnett, K.S. Waterhouse, P. Harding-Jones, K.C. Jones, Environ. Sci. Technol. **28**, 2380–2386 (1994)

Harrad, S., S. Hassoun, M.S. Callén Romero, R.M. Harrison, Atmos. Environ. **37**, 4985–4991 (2003)

Jaenicke, R., S. Matthias-Maser, S. Gruber, Environ. Chem. **4**, 217–220 (2007)

Kalaitzoglou, M., E. Terzi, C. Samara, Atmos. Environ. **38**, 2545–2560 (2004)

Kavouras, I.G., N. Mihalopoulos, E.G. Stephanou, Environ. Sci. Technol. **33**, 1028–1037 (1999)

Kawamura, K., K. Usukura, J. Oceanogr. **49**, 271–283 (1993)

Kawamura, K., O. Yasui, Atmos. Environ. **39**, 1945–1960 (2005)

Kawamura, K., H. Kasukabe, L.A. Barrie, Atmos. Environ. **30**, 1709–1722 (1996)

Kawamura, K., R. Sempéré, Y. Imai, J. Geophys. Res. **101**, 18721–18728 (1996)

Kawamura, K., Y. Ishimura, K. Yamazaki, Glob. Biogeochem. Cycles **17**, 1003 (2003). doi:10.1029/2001GB001810

Kendall, M., R.S. Hamilton, J. Watt, I.D. Williams, Atmos. Environ. **35**, 2483–2495 (2001)

Kubátova, A., R. Vermeylen, M. Claeys, J. Cafmeyer, W. Maenhaut, J. Geophys. Res. **107**, 8343 (2002). doi:10.1029/2001JD000556

Lammel, G., J. Klánová, P. Ilić, J. Kohoutek, B. Gasić, I. Kovacić, N. Lakić, R. Radić, Atmos. Environ. **44**, 5015–5021 (2010)

Lewis, R.G., R.W. Coutant, in *Gas and Particle Phase Measurements of Atmospheric Organic Compounds*, ed. by D.A. Lane (Gordon and Breach Science Publication, Amsterdam, 1999), pp. 201–232

Lim, M.C.H., G.A. Ayoko, L. Morawska, Atmos. Environ. **39**, 463–476 (2005)
Limbeck, A., H. Puxbaum, Atmos. Environ. **33**, 1847–1852 (1999)
Limbeck, A., H. Puxbaum, J. Geophys. Res. **105**, 19857–19867 (2000)
Mandalakis, M., M. Tsapakis, A. Tsoga, E.G. Stephanou, Atmos. Environ. **36**, 4023–4035 (2002)
Matthias-Maser, S., in *Atmospheric Particles*, ed. by R.M. Harrison, R. Van Grieken. IUPAC Series
    on Analytical and Physical Chemistry of Environmental Systems, vol. 5 (Wiley, Chichester,
    1998), pp. 349–368
Pio, C., C. Alves, A. Duarte, Atmos. Environ. **35**, 389–401 (2001a)
Pio, C., C. Alves, A. Duarte, Atmos. Environ. **35**, 1365–1375 (2001b)
Plewka, A., T. Gnauk, E. Brüggemann, H. Herrmann, Atmos. Environ. **40**, S103–S115 (2006)
Poor, N., R. Tremblay, H. Kay, V. Bhethanabotla, E. Swartz, M. Luther, S. Campbell, Atmos.
    Environ. **38**, 6005–6015 (2004)
Possanzini, M., V. Di Palo, P. Gigliucci, M.C. Tomasi Scianò, A. Cecinato, Atmos. Environ. **38**,
    1727–1734 (2004)
Rogge, W.F., M.A. Mazurek, L.M. Hildemann, G.R. Cass, B.R.T. Simoneit, Atmos. Environ. **27A**,
    1309–1330 (1993a)
Rogge, W.F., L.M. Hildemann, M.A. Mazurek, G.R. Cass, B.R.T. Simoneit, Envion. Sci. Technol.
    27, 636–651, 1892–1904 (1993b)
Rogge, W.F., L.M. Hildemann, M.A. Mazurek, G.R. Cass, B.R.T. Simoneit, Envion. Sci. Technol.
    27, 2700–2711 (1993c)
Rogge, W.F., L.M. Hildemann, M.A. Mazurek, G.R. Cass, B.R.T. Simoneit, Envion. Sci. Technol.
    32, 13–22 (1998)
Satsumabayashi, H., H. Kurita, Y. Yokouchi, H. Ueda, Atmos. Environ. **24A**, 1443–1450 (1990)
Schneider, J.K., R.B. Gagosian, J. Geophys. Res. **90**, 7889–7898 (1985)
Simoneit, B.R.T., Int. J. Environ. Anal. Chem. **23**, 207–237 (1986)
Simoneit, B.R.T., J. Atmos. Chem. **8**, 231–275 (1989)
Simoneit, B.R.T., Appl. Geohem. **17**, 129–162 (2002)
Simoneit, B.R.T., R. Chester, G. Eglinton, Nature **267**, 682–685 (1977)
Simoneit, B.R.T., J.N. Cardoso, N. Robinson, Chemosphere **21**, 1285–1301 (1990)
Simoneit, B.R.T., J.N. Cardoso, N. Robinson, Chemosphere **23**, 447–465 (1991a)
Simoneit, B.R.T., P.T. Crisp, M.A. Mazurek, L.J. Standley, Environ. Intern. **17**, 405–419 (1991b)
Smith, D.J.T., R.M. Harrison, Atmos. Environ. **30**, 2513–2525 (1996)
Smith, D.J.T., R.M. Harrison, L. Luhana, C.A. Pio, L.M. Castro, M.N. Tariq, S. Hayat, T. Quraishi,
    Atmos. Environ. **23**, 4031–4040 (1996)
Tsapakis, M., E.G. Stephanou, Environ. Pollut. **133**, 147–156 (2005)
Tsapakis, M., E. Lagoudaki, E.G. Stephanou, I. G. Kavouras, P. Koutrakis, P. Oyola, D. von Baer,
    Atmos. Environ. **36**, 3851–3863 (2002)
Wang, G., K. Kawamura, Environ. Sci. Technol. **39**, 7430–7438 (2005)
Wang, H., K. Kawamura, K.F. Ho, S.C. Lee, Environ. Sci. Technol. **40**, 6255–6260 (2006)
Winiwarter, W., H. Bauer, A. Caseiro, H. Puxbaum, Atmos. Environ. **43**, 1403–1409 (2009)
Yassaa, N., B.Y. Meklati, A. Cecinato, Atmos. Environ. **35**, 1843–1851 (2001)
Yu, J., R.J. Griffin, D.R. Cocker, R.C. Flagan, J.H. Seinfeld, Geophys. Res. Lett. **26**, 1145–1148
    (1999)
Yue, Z., M.P. Fraser, Atmos. Environ. **38**, 3253–3261 (2004)
Zappoli, S., A. Andracchio, S. Fuzzi, M.C. Facchini, A. Gelencsér, G. Kiss, Z. Krivácsy, A. Molnár,
    E. Mészáros, H.-C. Hansson, K. Rosman, Y. Zebühr, Atmos. Environ. **33**, 2733–2743 (1999)
Zheng, M., M. Fang, F. Wang, K.L. To, Atmos. Environ. **34**, 2691–2702 (2000)
Zheng, M., G.R. Cass, J.J. Schauer, E.S. Edgerton, Environ. Sci. Technol. **36**, 2361–2371 (2002)

5.5 Hygroscopicity

The ability of atmospheric particles to absorb water is due to the deliquescence of inorganic salts and the uptake of water by organic compounds. Whereas the former is well understood, the latter is not well characterized. The tables below present laboratory data for saline solutions and observed growth factors for marine and continental aerosols.

Radius of particle: The radius r of a drop of saline solution adjusts to satisfy the condition of balance between water activity a_w determined by Raoult's law (depression of water vapor increasing with concentration of solute) and the Kelvin effect (rise of vapor pressure with increasing surface curvature)

$$\%\,(\mathrm{rh})/100 = e/e_s = a_w \exp\left(2\sigma V_m / R_g Tr\right)$$

where e is the ambient vapor pressure of water, e_s is the vapor pressure over a plane surface of pure water, σ is the surface tension of the solution, $V_m = M_w/\rho_w$ is the partial molar volume of water, R_g is the gas constant, and T is absolute temperature. For particles with $r>0.1$ μm and relative humidities $\leq 90\%$, the Kelvin effect is small and can be neglected. However, under conditions of supersaturation of water vapor it becomes important and determines the particle size capable of undergoing cloud formation.

Water activity: The amount of water acquired at relative humidities above the deliquescence point of a salt is determined by the water activity a_w, which can be expressed in the form

$$a_w = \exp\left[-z\phi_s\left(bM_w/10^3\right)\right]$$

where z is the number of ions produced by dissociation of the salt, b (Mol kg^{-1}) is the molal concentration of salt (molality), and M_w (g Mol^{-1}) is the molar mass of water, and ϕ_s is the practical osmotic coefficient. For dilute solutions $\phi_s \approx 1$, for concentrated solutions $\phi_s < 1$.

Density: Although measured densities of salt solutions as a function of concentration are available, they frequently do not reach up to the saturation limit. In such cases the density can be estimated with the assumption of volume additivity. The general formulation leads to

$$\rho = m\left\{\sum_i (m_i/\rho_i)\right\}^{-1}$$

where the sum involves all components, the ρ_i and m_i denote densities and masses of the pure components, respectively, and $m = \Sigma m_i$. For a binary solution, the density can also be expressed in terms of the mass fraction w^*

$$\rho = \left\{(w^*/\rho_s) + (1-w^*)/\rho_w\right\}^{-1}$$

Here, $w^* = m_s/(m_s + m_w)$ and m_s and m_w denote the masses of salt and water, respectively.

Table 5.40 Properties of common salts at 25°C: molar mass, crystalline density, solubility in water, density of saturated solution, deliquescence relative humidity[a]

Solid phase	M (g Mol^{-1})	ρ_i (g cm^{-3})	w^b (%)	b (mol kg^{-1})	ρ^c (g cm^{-3})	rh (%)
NH$_4$Cl	53.492	1.519	28.3	7.33	1.077	77.1
NaCl	58.443	2.17	26.5	6.15	1.198	75.3
KCl	74.551	1.988	26.2	4.77	1.178	84.3
CaCl$_2$	110.984	2.15		See next line		
CaCl$_2$.6H$_2$O	219.075	1.71	44.8^d	7.32	1.450†	28.7
MgCl$_2$	95.211	2.325		See next line		
MgCl$_2$.6H$_2$O	203.302	1.56	35.9^d	5.88	1.324†	33.0
NH$_4$HSO$_4$	115.110	1.78	74.2	24.98	1.479*	40.0
(NH$_4$)$_2$SO$_4$	132.140	1.77	43.4	5.80	1.245	80.0
(NH$_4$)$_3$H(SO$_4$)$_2$	247.250	–	54.0	4.75	1.322	69
Na$_2$SO$_4^c$	142.043	2.70	38.5	4.40	1.208	84.2
Na$_2$SO$_4$.10H$_2$O	322.196	1.46	21.9^d	1.97	1.208	87
K$_2$SO$_4$	174.260	2.66	10.8	0.70	1.086	97
CaSO$_4$	136.141	2.96	0.205	0.015	0.999	–
MgSO$_4$	120.368	2.66		See next line		
MgSO$_4$.7H$_2$O	246.474	1.67	26.7^d	3.03	1.304	88
NH$_4$NO$_3$	80.043	1.72	68.0	26.50	1.320	62
NaNO$_3$	84.995	2.261	47.7	10.72	1.391	74.5
KNO$_3$	101.103	2.105	27.7	3.79	1.193	92.5
Ca(NO$_3$)$_2^c$	132.089	2.23	77.3^e	25.78	1.741*	~13
Ca(NO$_3$)$_2$.4H$_2$O	236.149	1.82	58.0^d	10.45	1.579	50.0
Mg(NO$_3$)$_2$	148.314	2.3		See next line		
Mg(NO$_3$)$_2$.6H$_2$O	256.406	1.46	41.6^d	4.80	1.388	52.9

[a]Sources of data: Gmelin (1939, 1957, 1966), Lide (2006), Seidell and Linke (1965), Robinson and Stokes (1970), Tang and Munkelwitz (1994), Washburn (1928)

[b]Percent mass fraction: $w = 100w^* = 100m_s/(m_s + m_w)$, where m_s and m_w are the masses of salt and water, respectively; $b = 10^3 w^*/M(1-w^*)$

[c]An asterisk indicates that the density was calculated using the principle of volume additivity, a dagger indicates that it was obtained by extrapolation

[d]The concentration refers to the anhydride

[e]The anhydrous compound is metastable at 25°C. Na$_2$SO$_4$ is stable at temperatures ≥ 32.4°C, Ca(NO$_3$)$_2$ is stable at temperatures ≥ 51.6°C

Table 5.41 Activity of water, a_w above aqueous solutions of several salts as a function of molality, b (Mol kg^{-1}), at 25°C[a]

b	NH$_4$Cl	NaCl	CaCl$_2$	MgCl$_2$	(NH$_4$)$_2$SO$_4$	NH$_4$HSO$_4$	Na$_2$SO$_4$	MgSO$_4$	NH$_4$NO$_3$	NaNO$_3$	Mg(NO$_3$)$_2$	Ca(NO$_3$)$_2$
0.1	0.9967	0.9966	0.9954	0.9954	0.9959	–	0.9957	0.9978	0.9967	0.9967	0.9954	0.9955
0.5	0.9839	0.9835	0.9755	0.9747	0.9819	0.982	0.9815	0.9906	0.9847	0.9844	0.9749	0.9780
1.0	0.9682	0.9669	0.9450	0.9419	0.9660	0.964	0.9659	0.9813	0.9708	0.9698	0.9436	0.9546
2.0	0.9366	0.9316	0.8618	0.8482	0.9349	0.928	0.9351	0.9531	0.9456	0.9422	0.8622	0.9021
3.0	0.9048	0.8932	0.7494	0.7219	0.9022	0.894	0.8984	0.9051	0.9228	0.9162	0.7579	0.8433
4.0	0.8727	0.8515	0.6239	0.5450	0.8670	0.858	0.8522	–	0.9021	0.8915	0.6430	0.7787
4.5	0.8568	0.8295	0.5602	0.5082	0.8490	0.840	–	–	0.8924	0.8795	0.5844	0.7451
5.0	0.8415	0.8068	0.4988	0.4388	0.8308	0.822	–	–	0.8831	0.8677	0.5262	0.7108
5.5	0.8263	0.7835	0.4425	–	0.8124	0.803	–	–	0.8741	0.8556	–	0.6789
6.0	0.8110	0.7598	0.3916	–	–	0.784	–	–	0.8652	0.8433	–	0.6432

[a]Calculated from the molal osmotic coefficients tabulated by Robinson and Stokes (1970) except activities for NH$_4$HSO$_4$ interpolated from the data of Tang and Munkelwitz (1977)

Following Tang and Munkelwitz (1993) the temperature dependence of deliquescence is given by the equation $d\ln a_w/dT = n\Delta H_s/R_g T^2$ where a_w is the deliquescence activity of water, ΔH_s is the integral heat of solution of the salt in water, and n is the solubility in moles of solute per mole of water. The equation is integrated by setting $n = A + BT + CT^2$, so that the temperature dependence of the deliquescence relative humidity can be expressed by

$$\ln(a_w/a^\circ{}_w) = (\Delta H_s/R_g)\left[A\left(\frac{1}{T}-\frac{1}{T^\circ}\right)\right] - B\ln\frac{T}{T^\circ} - C\left(T-T^\circ\right)$$

where $a^\circ{}_w$ and T° refer to 298.15 K.

Table 5.42 Deliquescence humidities at 298 K, heats of solution and solubility data for salts[a]

Salt	rh (%)	ΔH_s (kJ mol^{-1})	A	B	C
Na$_2$SO$_4$	84.2	−9.749	0.3754	−1.763 (−3)	2.424 (−6)
(NH$_4$)$_2$SO$_4$	79.9	6.318	0.1149	−4.489 (−4)	1.385 (−6)
NaCl	75.3	1.874	0.185	−5.310 (−4)	9.965 (−7)
KCl[b]	84.2	15.334	−0.1984	1.174 (−3)	−7.398 (−7)
NaNO$_3$	74.3	13.230	0.1868	−1.677 (−3)	5.714 (−6)
NH$_4$NO$_3$	61.8	16.255	4.298	−3.623 (−2)	7.853 (−5)

[a]Numbers in parentheses indicate powers of ten. Source of data: Tang and Munkelwitz (1993)
[b]The solubility data for KCl have been reevaluated

Table 5.43 Water activities for saturated salt solutions as a function of temperature[a]

T(°C)	NH$_4$Cl	NaCl	(NH$_4$)$_2$SO$_4$	NH$_4$NO$_3$	NaNO$_3$	Ca(NO$_3$)$_2$	Mg(NO$_3$)$_2$
10	0.791	0.750	0.734	0.695	0.720	0.571	0.591
15	0.778	0.753	0.773	0.675	0.729	0.595	0.586
20	0.785	0.756	0.800	0.652	0.735	0.590	0.573
25	0.783	0.758	0.812	0.622	0.737	0.559	0.552
30	–	0.758	0.810	–	0.735	0.507	0.525
35	–	0.757	0.795	–	0.730	0.442	0.494

[a]Calculated from the vapor pressures reported by Apelblat (1992, 1993), Apelblat and Korin (1998)

Table 5.44 Solubility, deliquescence and re-crystallization relative humidities and polynomial coefficients for calculating water activities and densities of solutions of important salts as a function of concentration (mass percent) at 25°C[a]

	Na$_2$SO$_4$		NH$_4$HSO$_4$	(NH$_4$)$_2$SO$_4$	(NH$_4$)$_3$(SO$_4$)$_2$	NaNO$_3$
Deliquescence and crystallization						
Sol. w(%)	38.5		76	43.3	54	47.9
Del. rh(%)	84		40	80	69	74
cryst. rh(%)	59–57		22–0.05	40–37	44–35	30–0.05

(continued)

Table 5.44 (continued)

	Na_2SO_4		NH_4HSO_4	$(NH_4)_2SO_4$	$(NH_4)_3(SO_4)_2$	$NaNO_3$
Polynomial coefficients						
Range $w(\%)$	0–40	40–67[b]	0–97	0–78	0–78	0–98
C1	-3.55(-3)	-1.99(-2)	-3.05(-3)	-2.715(-3)	-2.42(-3)	-5.52(-3)
C2	9.63(-5)	-1.92(-5)	-2.94(-5)	3.113(-5)	-4.615(-5)	1.286(-4)
C3	-2.97(-6)	1.47(-6)	-4.43(-7)	-2.336(-6)	-2.83(-69)	-3.496(-6)
C4	–	–	–	1.412(-8)	–	1.843(-8)
A1	8.871(-3)		5.87(-3)	5.92(-3)	5.66(-3)	6.512(-3)
A2	3.195(-5)		-1.89(-6)	-5.036(-6)	2.96(-6)	3.025(-5)
A3	2.28(-7)		1.763(-7)	1.024(-8)	6.68(-8)	1.437(-7)

[a]Source: Tang and Munkelwitz (1994); powers of ten are shown in parentheses. The polynomial expressions are: water activity, $a_w = 1.0 + \Sigma\, C_i\, w^i$; density, $\rho = \rho_w + \Sigma\, A_i\, w^i$
[b]For this concentration range $a_w = 1.557 + \Sigma\, C_i\, w^i$

Table 5.45 Average growth with relative humidity of marine aerosol particles: relative increase of mass, m/m_0, and particle radius, r/r_0, in three size ranges[a,b]

	$r<0.1$		$0.1<r<0.3$		$0.3<r<1.0$		$r>1.0$		Sea salt[c]	
rh (%)	m/m_0	r/r_0	m/m_0	r/r_0	m/m_0	r/r_0	m/m_0	r/r_0	m/m_0	r/r_0
30	0.233	1.105	0.204	1.093	0.157	1.073	0.033	1.024	0.05	1.036
40	0.290	1.128	0.255	1.114	0.194	1.089	0.077	1.053	0.07	1.047
50	0.385	1.164	0.308	1.135	0.245	1.110	0.131	1.088	0.09	1.065
60	0.503	1.206	0.412	1.174	0.345	1.149	0.266	1.166	0.16	1.108
65	0.617	1.244	0.485	1.200	0.431	1.181	–	–	0.24	1.155
70	0.829	1.309	0.614	1.243	0.604	1.240	0.655	1.346	0.40	1.239
75	1.082	1.379	0.822	1.307	1.010	1.360	2.000	1.754	1.85	1.728
80	1.345	1.445	1.040	1.368	1.606	1.505	2.410	1.847	2.50	1.878
85	1.679	1.521	1.455	1.471	2.009	1.601	4.268	2.182	3.27	2.029
90	2.168	1.620	1.973	1.582	2.803	1.733	5.645	2.376	4.71	2.264
95	3.333	1.817	3.068	1.776	–	–	6.262	2.513	–	–

[a]Source of data: Winkler and Junge (1972), Winkler (1988). The increase of radius is calculated from the measured mass increase with the assumption of volume additivity: $(r/r_0)^3 = 1 + (m/m_0)$ (ρ_0/ρ_w), where ρ_0 and ρ_w are the densities of the dry particle and of water, respectively. The influence of surface curvature (Kelvin effect), which becomes important for particles smaller than $0.1\ \mu m$, is neglected
[b]Size ranges (in μm) were separated by inertial impactors. The density of solid particles was assumed to be $\rho_0 = 1.5\ g\,cm^{-3}$ except for radii $r>1.0\ \mu m$: $\rho_0 = 2.2\ g\,cm^{-3}$. The latter assumption is justified by the observation that the growth curve approaches that of sea salt
[c]The density of sea salt is $\rho_0 = 2.25\ g\,cm^{-3}$

Table 5.46 Average growth with relative humidity of continental aerosol particles: relative increase of mass, m/m_0, and particle radius, r/r_0[a]

	Germany						Israel		Tennessee	Arizona
	$r<0.1$		$0.1<r<1.0$		$r>1.0$		$0.15<r<5.0$		$0.1<r<2.5$	
rh (%)	m/m_0	r/r_0	m/m_0	r/r_0	m/m_0	r/r_0	m/m_0	r/r_0	r/r_0	r/r_0
30	0.158	1.050	0.039	1.019	0.080	1.038	0.022	1.019	1.077	1.010
40	0.141	1.066	0.070	1.033	0.130	1.061	0.030	1.025	1.100	1.032
50	0.180	1.083	0.110	1.052	0.211	1.096	0.051	1.042	1.140	1.063
60	0.280	1.124	0.180	1.083	0.330	1.143	0.080	1.065	1.192	1.114
65	0.371	1.159	0.270	1.120	0.410	1.173	0.104	1.083	1.221	1.145

(continued)

Table 5.46 (continued)

rh (%)	Germany						Israel		Tennessee	Arizona
	$r<0.1$		$0.1<r<1.0$		$r>1.0$		$0.15<r<5.0$		$0.1<r<2.5$	
	m/m_0	r/r_0	m/m_0	r/r_0	m/m_0	r/r_0	m/m_0	r/r_0	r/r_0	r/r_0
70	0.491	1.202	0.320	1.139	0.500	1.205	0.132	1.103	1.265	1.183
75	0.664	1.259	0.440	1.184	0.611	1.242	0.170	1.129	1.319	1.241
80	0.870	1.321	0.585	1.233	0.760	1.289	0.280	1.199	1.371	1.319
85	1.100	1.384	0.775	1.293	0.980	1.352	0.500	1.319	1.497	1.439
90	1.570	1.497	1.130	1.392	1.310	1.437	0.801	1.297	1.652	1.559
92.5	1.948	1.577	1.485	1.478	1.730	1.532	1.130	1.578	1.749	–
95	2.312	1.647	2.230	1.632	2.450	1.672	1.680	1.749	–	–
m_s/m_0[b]	0.45		0.75		0.46		0.31		≥ 0.64	≥ 0.36
C[b]	–		28		10		–		25	4

[a]For the calculation of r/r_0 from m/m_0 see bottom of preceding table. Germany: Deuselbach, rural location (Winkler and Junge 1972; Winkler 1988), $\rho_0 = 1.5$ g cm^{-3} (assumed); Israel: Mizpeh Ramon, desert conditions (Hänel and Lehmann 1980), $\rho_0 = 2.59$ g cm^{-3}; Arizona: Grand Canyon National Park; Tennessee: Great Smokey Mountains National Park, elevation 808 m, rural; values of r/r_0 for Arizona and Tennessee were derived from measurements of the ratio of dry and wet optical scattering coefficients (Malm and Day 2001)
[b]m_s/m_0 is the ratio of water-soluble to total material, C (μg/m^3) is the average ambient concentration

Table 5.47 Hygroscopic growth factors GF$_{90}$ for 80 nm particles from various sources[a]

Ammonium sulfate	1.68	Fresh Diesel engine exhaust (soot)	<1.05
Ammonium hydrogen sulfate	1.72	Fresh gasoline engine exhaust	<1.05
Ammonium nitrate	1.77	Biomass burner exhaust[b]	1.0–1.65
Sodium chloride	2.34	Humic substances	1.08–1.20
Fresh mineral dust	<1.05	Organic acids[c]	1.0–1.7
		Secondary organic aerosol[d]	1.07–1.14

[a]Source: McFiggans et al. (2006); GF$_{90} = r_{90}/r_0$, where r_{90} is the radius at 90% relative humidity and r_0 is the dry radius; measurements by tandem differential mobility analyzer
[b]Low values from smoldering combustion, high values from high efficiency burners that leave primarily inorganic residue
[c]Refers primarily to multifunctional acids; monofunctional acids of low volatility are only slightly hygroscopic
[d]Produced in dry air by ozonolysis/photo-oxidation of toluene, α-, β-pinene, limonene

Table 5.48 Hygroscopic growth factors GF$_{90}$ and frequency of occurrence F for 50–80 nm particles in various environments[a]

Environment	Dominant mode			Second mode		
	GF$_{90}$	F (%)	Fraction (%)	GF$_{90}$	F (%)	Fraction (%)
Urban	1.15–1.43	≤ 100	11–90	1.00–1.12	≤ 100	10–89
Continental polluted	1.32–1.53	90–100	28–97	1.05–1.15	10–90	3–72
Free troposphere	1.40–1.61	76–100	75–90	1.14	26	15
Amazon rain forest	1.15–1.20	>97	87–95	~1.07	15–35	<12
Remote marine	1.42–1.75	100	>75	2.05–2.13	3–7	<7

[a]Summary of various measurements provided by McFiggans et al. (2006); GF$_{90} = r_{90}/r_0$, where r_{90} is the radius at 90% relative humidity and r_0 is the dry radius. The presence of at least two groups of particles with different hygroscopic growth behavior indicates externally mixed Aitken particles

Table 5.49 Hygroscopic growth factors GF_{90} and frequencies of occurrence F for 100–150 nm particles in various environments[a]

Environment	Dominant mode			Second mode		
	GF_{90}	F (%)	Fraction (%)	GF_{90}	F (%)	Fraction (%)
Urban	1.23–1.50	100	16–90	1.00–1.14	≤ 100	10–84
Continental polluted	1.41–1.64	100	37–100	1.03–1.18	10–90	2–63
Free troposphere	1.62	100	~85	<1.3	–	~15
Amazon rain forest	1.20–1.25	>93	90–96	~1.08	~25	<7
Remote marine	1.47–1.78	100	>80	2.06–2.14	13–40	<15

[a]Summary of various measurements provided by McFiggans et al. (2006); $GF_{90} = r_{90}/r_0$, where r_{90} is the radius at 90% relative humidity and r_0 is the dry radius. The presence of at least two groups of particles with different hygroscopic growth behavior indicates externally mixed particles

References

Apelblat, A., J. Chem. Thermodyn. **24**, 619–626 (1992), **25**, 63–71, 1513–1520 (1993)

Apelblat, A., E. Korin, J. Chem. Thermodyn. **30**, 59–71 (1998)

Gmelin, *Handbuch der Anorganischen Chemie,* 8th edn., System Number 21 (Na), 27 (Mg), 28 (Ca) (Verlag Chemie, Weinheim, 1939, 1957, 1966)

Hänel, G., M. Lehmann, Contrib. Atmos. Phys. **54**, 57–71 (1980)

Lide, D.R. (ed.), *Handbook of Chemistry and Physics*, 87th edn. (CRC Press, Boca Raton, 2006)

Malm, W.C., D.E. Day, Atmos. Environ. **35**, 2845–2860 (2001)

McFiggans, G., P. Artaxo, U. Baltensperger, H. Coe, M.C. Facchini, G. Feingold, S. Fuzzi, M. Gysel, A. Laaksonen, U. Lohmann, T.F. Mentel, D.M. Murphy, C.D. O'Dowd, J.R. Snider, E. Weingartner, Atmos. Chem. Phys. **6**, 2593–2649 (2006)

Robinson, R.A., R.H. Stokes, *Electrolyte Solutions*, 2nd edn. (Butterworth, London, 1970)

Seidell, A., W.F. Linke, *Solubilities of Inorganic and Metal-Organic Compounds* (American Chemical Society, Washington, DC, 1965)

Tang, I.N., H.R. Munkelwitz, J. Aerosol Sci. **8**, 321–330 (1977)

Tang, I.N., H.R. Munkelwitz, Atmos. Environ. **27A**, 467–473 (1993)

Tang, I., H.R. Munkelwitz, J. Geophys. Res. **99**, 18801–18808 (1994)

Washburn, E.W. (ed.), *International Critical Tables III* (McGraw-Hill, New York, 1928)

Winkler, P., *Physica Scripta* **37**, 223–230 (1988)

Winkler, P., C. Junge, J. Rech. Atmos. **6**, 617–638 (1972)

Chapter 6
Gas-Phase Photochemistry

Chemical reactions in the atmosphere result largely from the absorption of solar radiation in the visible and ultraviolet spectral regions provided the incoming photons (light quanta) carry sufficient energy for chemical processes to be initiated. Photochemical reactions always occur in two steps: the first is the excitation of a molecule by the absorption of radiation; the second is a reaction of the excited molecule. In atmospheric chemistry, the most important process to be considered is dissociation of the excited molecule. Other modes of energy dissipation, such as fluorescence or collisional energy transfer, usually do not lead to chemical changes so that they are of lesser interest. Any quantitative assessment of the photochemical activity of an atmospheric constituent requires knowledge of (a) the photon flux of solar radiation, (b) the absorption cross section of the species under consideration, and (c) the primary quantum yield of the photo-dissociation process. These parameters depend on the wavelength of radiation; absorption cross sections and quantum yields may additionally depend on temperature and/or pressure.

6.1 Solar Radiation

Solar radiation originates from the very thin outer layers of the sun that (with increasing distance from the center) are designated photosphere, chromosphere and corona. The spectrum consists of a continuum with superimposed line structure. It extends from the soft x-ray region at wavelengths $\lambda < 10$ nm to the far infrared at wavelengths $\lambda > 100$ μm with an intensity maximum occurring in the visible region. At wavelengths beyond about 350 nm the spectrum is that of blackbody at ~5,770 K, slightly modified by narrow absorption lines (Fraunhofer lines). At shorter wavelengths the solar flux decreases more strongly and, at the shortest wavelengths, departs significantly from that predicted by the blackbody curve. In addition, emission lines become more prominent. The strongest one is the hydrogen Lyman-alpha line at 121.6 nm.

P. Warneck and J. Williams, *The Atmospheric Chemist's Companion: Numerical Data for Use in the Atmospheric Sciences*, DOI 10.1007/978-94-007-2275-0_6, © Springer Science+Business Media B.V. 2012

As described in Table 6.1, the extent of attenuation in the terrestrial atmosphere divides the ultraviolet spectrum into characteristic wavelength intervals. In the lower atmosphere solar radiation becomes increasingly scattered. Both Rayleigh scattering by air molecules and Mie scattering by atmospheric particles are important, with Mie scattering occurring mainly in the troposphere. The degree of scattering increases with solar zenith angle (because of increasing slant path length) and with decreasing wavelength ($\sigma_{sc} \propto \lambda^{-n}$, with n = 4 for Rayleigh scattering and n = 0.5–2.5 for Mie scattering). In addition, radiation is reflected from Earth's surface. Accordingly, a volume of air receives radiation from all directions. This so-called actinic flux (because it would be determined by a chemical actinometer) must be considered when photochemical processes are involved. In the stratosphere the actinic flux is dominated by direct radiation. In the lower troposphere more than 50% of the actinic flux is due to diffuse sky radiation. Whereas the attenuation of radiation by absorption can be reasonably well calculated and the reflected radiation is determined by the albedo of the ground surface, an assessment of diffuse sky radiation caused by multiple scattering processes requires complex calculation procedures based on radiative transfer models.

The solar flux outside Earth's atmosphere is subject to natural variations, especially in the far ultraviolet spectral region where the variability increases with decreasing wavelength (for details see Chap. 10). Variations in the visible region of the solar spectral irradiance are essentially negligible. However, the eccentricity of Earth's orbit around the sun modulates all wavelength regions equally at ±3.3%.

Solar intensities presented in this Section cover the wavelength region 200–870 nm. Data in the far UV at wavelengths below 200 nm are treated in Chap. 10.

Table 6.1 Origin of solar emissions and regions of Earth's atmosphere thereby affected[a]

Wavelength range (nm)	Source region	Variation over 11 year cycle	Atmospheric absorption region (principal absorbers)
750–50,000 (IR)	Photosphere		Entire atmosphere (CO_2, H_2O, O_3)
400–750 (visible)		0.08%	Entire atmosphere (O_3 Chappuis band)
300–400 (UV)		0.1%	Troposphere (O_3, NO_2, HCHO, others)
200–300 (UV)		0.5–4%	Stratosphere (O_2, O_3 Hartley band)
175–200 (UV)	Upper photosphere	15%	Mesosphere (O_2, SRB)
120–175 (far UV)	Chromosphere	50%	Thermosphere (O_2, SC)
Lyman-α, 121.6		2 fold	Mesosphere (O_2, H_2O, NO ionization)
20–120 (EUV)	Transition region[b]	2–10 fold	Thermosphere (N_2, O_2, O ionization)
0–20 (X-rays)	Corona	10–10^3 fold	Mesosphere (N_2, O_2 ionization)

[a] Adapted from Lean (1991). Abbreviations: IR infrared, UV ultraviolet, EUV extreme ultraviolet, SC Schumann continuum, SRB Schumann-Runge bands
[b] The transition region is the region between chromosphere and corona

Table 6.2 Solar radiation intensities at wavelengths from 180 to 400 nm in 1 nm intervals[a]

λ_c (nm)	I_{av}	λ_c (nm)	I_{av}	λ_c (nm)	I_{av}	λ_c (nm)	I_{av}	λ_c (nm)	I_{av}
180.5	0.172	226.5	4.651	272.5	28.040	318.5	110.36	364.5	192.48
181.5	0.211	227.5	4.733	273.5	28.005	319.5	118.30	365.5	219.51
182.5	0.213	228.5	6.186	274.5	18.973	320.5	134.48	366.5	235.42
183.5	0.215	229.5	5.706	275.5	25.144	321.5	117.21	367.5	226.07
184.5	0.199	230.5	6.261	276.5	34.993	322.5	115.74	368.5	211.48
185.5	0.223	231.5	6.150	277.5	33.807	323.5	111.49	369.5	236.05
186.5	0.262	232.5	6.454	278.5	23.119	324.5	128.66	370.5	219.90
187.5	0.294	233.5	5.445	279.5	12.462	325.5	148.57	371.5	227.41
188.5	0.323	234.5	4.762	280.5	13.834	326.5	166.67	372.5	204.59
189.5	0.352	235.5	6.311	281.5	30.553	327.5	162.61	373.5	179.11
190.5	0.374	236.5	5.989	282.5	43.304	328.5	157.09	374.5	175.73
191.5	0.416	237.5	5.996	283.5	46.312	329.5	182.13	375.5	207.75
192.5	0.413	238.5	5.216	284.5	34.889	330.5	175.03	376.5	212.85
193.5	0.384	239.5	5.508	285.5	24.275	331.5	166.83	377.5	252.18
194.5	0.528	240.5	4.958	286.5	48.186	332.5	165.48	378.5	261.80
195.5	0.553	241.5	6.479	287.5	50.395	333.5	160.18	379.5	210.91
196.5	0.615	242.5	8.923	288.5	48.450	334.5	167.16	380.5	238.29
197.5	0.637	243.5	8.443	289.5	71.324	335.5	164.94	381.5	215.10
198.5	0.640	244.5	7.769	290.5	91.035	336.5	141.96	382.5	157.74
199.5	0.693	245.5	6.319	291.5	88.237	337.5	148.07	383.5	136.72
200.5	0.765	246.5	6.475	292.5	79.558	338.5	162.57	384.5	192.30
201.5	0.843	247.5	7.159	293.5	81.913	339.5	169.52	385.5	202.60
202.5	0.872	248.5	5.912	294.5	78,086	340.5	178.61	386.5	202.74
203.5	0.978	249.5	7.331	295.5	85.194	341.5	164.42	387.5	202.68
204.5	1.092	250.5	7.846	296.5	78.750	342.5	177.25	388.5	199.29
205.5	1.129	251.5	6.030	297.5	78.657	343.5	170.17	389.5	238.24
206.5	1.192	252.5	5.467	298.5	72.414	344.5	142.85	390.5	253.79
207.5	1.368	253.5	6.811	299.5	75.552	345.5	169.11	391.5	273.75
208.5	1.635	254.5	7.913	300.5	65.774	346.5	167.28	392.5	208.46
209.5	2.334	255.5	10.554	301.5	71.579	347.5	163.30	393.5	119.41
210.5	3.062	256.5	13.610	302.5	77.085	348.5	162.95	394.5	210.91
211.5	3.697	257.5	16.826	303.5	98.165	349.5	162.08	395.5	270.78
212.5	3.466	258.5	16.852	304.5	95.606	350.5	191.80	396.5	176.57
213.5	3.689	259.5	14.069	305.5	94.705	351.5	179.43	397.5	192.40
214.5	4.475	260.5	11.608	306.5	88.766	352.5	166.79	398.5	316.16
215.5	4.060	261.5	12.237	307.5	98.561	353.5	190.24	399.5	346.32
216.5	3.641	262.5	14.179	308.5	98.306	354.5	205.23	400.5	351.01
217.5	4.000	263.5	22.763	309.5	80.084	355.5	193.64	401.5	365.43
218.5	5.052	264.5	34.420	310.5	100.71	256.5	121.96	402.5	368.98
219.5	5.367	265.5	34.964	311.5	118.38	357.5	148.31	403.5	354.05
220.5	5.439	266.5	34.640	312.5	106.82	358.5	129.58	404.5	352.89
221.5	4.568	267.5	34.958	313.5	114.36	359.5	185.68	405.5	352.33
222.5	5.757	268.5	34.116	314.5	108.66	360.5	182.39	406.5	336.63
223.5	7.339	269.5	33.551	315.5	104.08	361.5	165.44	407.5	345.66
224.5	6.900	270.5	37.679	316.5	104.44	362.5	188.51	408.5	369.34
225.5	6.145	271.5	31.750	317.5	128.06	363.5	188.85	409.5	365.91

[a] Wavelengths λ_c at interval centers; irradiances in units of 10^{12} (photon cm^{-2} s^{-1} nm^{-1}) (converted from originally (mW m^{-2} nm^{-1})); Upper Atmosphere Research Satellite data validated by Spacelab ATLAS 2 measurements in April 1993. Source: Woods et al. (1996)

Table 6.3 Comparison of solar flux measurements in the near ultraviolet spectral region averaged over 5 nm intervals (units: 10^{12} photon cm^{-2} s^{-1} nm^{-1})[a]

	A	B	C	D	E	F	G	H
	Nov	July	May	Dec	Dec	Aug	Mar	Apr
$\Delta\lambda$ (nm)	1978	1980	1982	1983	1984	1985	1992	1993
180–185	0.16	0.19	0.17	–	0.17	–	0.22	0.20
185–190	0.24	0.27	0.24	–	0.26	–	0.31	0.29
190–195	0.36	0.39	0.40	–	0.41	–	0.44	0.42
195–200	0.57	0.59	0.61	–	0.63	–	0.66	0.63
200–205	0.80	0.87	0.84	0.82	0.94	0.94	0.95	0.91
205–210	1.37	1.45	1.39	1.45	1.59	1.53	1.59	1.53
210–215	3.28	3.52	3.36	3.27	3.83	3.56	3.73	3.68
215–220	3.98	4.42	4.06	4.35	4.52	4.26	4.48	4.42
220–225	5.54	6.19	5.74	5.74	6.56	5.72	6.05	6.00
225–230	4.94	5.60	5.45	5.12	5.83	5.04	5.53	5.48
230–235	5.23	5.93	5.42	5.43	5.92	5.28	5.86	5.81
235–240	5.16	5.93	5.67	5.46	5.93	5.46	5.85	5.80
240–245	6.79	7.07	7.35	6.98	7.22	7.10	7.35	7.31
245–250	5.95	6.42	6.49	6.32	6.69	6.48	6.70	6.64
250–255	6.13	6.61	6.64	6.40	6.95	6.61	6.89	6.81
255–260	12.9	13.8	14.3	13.8	14.7	13.9	14.5	14.4
260–265	17.7	19.0	19.4	19.6	20.3	19.5	19.0	19.0
265–270	32.3	32.7	34.2	35.9	36.1	35.3	34.7	34.4
270–275	27.5	27.5	28.7	30.2	30.1	30.3	29.1	28.9
275–280	25.2	25.4	26.6	27.4	26.8	26.9	26.2	25.9
280–285	32.0	33.3	34.5	37.4	36.2	37.2	34.1	33.8
285–290	46.5	47.6	50.8	53.7	54.1	52.4	48.5	48.5
290–295	84.0	79.8	84.5	88.2	88.4	87.8	83.8	83.8
295–300	74.8	73.0	78.1	81.4	80.4	82.0	77.9	78.1
300–305	82.3	73.6	77.8	89.7	90.1	84.6	82.1	81.6
305–310	94.1	81.9	87.5	100.3	97.1	101.2	92.6	92.1
310–315	112	95.6	99.1	116.6	119.8	120.0	109.8	109.8
315–320	112	–	–	119.1	–	120.4	113.1	113.1
320–325	119	–	–	126.6	–	131.0	121.5	121.5
325–330	165	–	–	171.6	–	170.6	163.4	163.4
330–335	159	–	–	171.4	–	168.4	166.9	166.9
335–340	159	–	–	163.3	–	162.8	157.5	157.5

[a]A: Nimbus 7 satellite solar UV backscatter measurements, reconstructed from data reported by Simon (1981) and Mentall et al. (1981), originally Heath (1980, unpublished); B: Rocket flight July 15, 1980, near solar maximum, Mount and Rottman (1983); C: Rocket flight May 17, 1982, Mount and Rottman (1983); D: Spacelab 1 (Space Shuttle Columbia) November-December 1983, Labs et al. (1987); E: Rocket flight December 1984, Mentall and Williams (1988); F: Spacelab 2 (Space Shuttle Challenger) August 1985, VanHoosier et al. (1988) as reported by Nicolet (1989); G, H: Upper Atmosphere Research Satellite validated by Space Shuttle ATLAS 1 and 2 measurements in March 1992 (near solar maximum) and April 1993, respectively, Woods et al. (1996). (*ATLAS* Atmospheric Laboratory for Applications and Science)

Table 6.4a Solar radiation intensities at wavelengths from 330 to 630 nm in 1 nm intervals[a]

λ_c (nm)	+0.0	+1.0	+2.0	+3.0	+4.0	+5.0	+6.0	+7.0	+8.0	+9.0
330.5	167.5	161.7	154.3	152.1	158.4	166.0	129.8	147.3	156.3	160.3
340.5	170.2	161.1	171.7	170.5	124.9	168.4	160.5	158.0	166.5	152.4
350.5	197.8	175.9	154.7	198.8	202.6	189.5	168.5	160.5	113.3	206.0
360.5	177.9	162.9	214.8	175.5	186.4	232.8	230.8	225.0	202.2	248.0
370.5	200.9	244.8	199.9	157.8	165.7	216.1	209.1	245.7	255.9	191.2
380.5	247.3	210.9	141.3	132.2	199.0	185.3	208.6	188.6	178.6	241.0
390.5	240.8	275.9	188.9	97.1	219.1	274.8	129.9	208.3	308.9	333.2
400.5	332.9	363.6	365.9	337.2	326.6	341.7	332.7	317.4	375.7	352.1
410.5	310.8	377.4	372.5	366.4	363.3	363.5	387.3	350.8	355.6	360.1
420.5	373.0	382.4	337.3	365.6	378.7	363.9	365.4	338.5	343.2	319.8
430.5	246.6	367.1	359.2	378.6	366.2	378.6	425.0	398.9	346.8	404.9
440.5	380.7	430.3	442.2	427.3	442.6	409.5	426.2	469.0	446.6	459.8
450.5	487.4	480.5	443.3	450.9	453.9	467.6	478.5	484.8	456.1	465.9
460.5	474.1	478.6	491.0	477.2	463.2	479.7	452.3	475.4	471.5	471.5
470.5	445.8	480.2	486.7	475.8	491.1	482.8	470.4	500.0	485.1	502.3
480.5	493.5	507.8	492.6	492.6	481.5	448.5	399.0	450.3	471.9	484.2
490.5	496.8	470.4	471.3	470.3	513.6	481.7	505.4	506.7	469.5	496.6
500.5	469.1	458.7	480.4	491.5	475.9	508.4	501.3	488.2	492.5	492.7
510.5	501.6	515.5	483.0	482.4	486.7	494.4	435.0	450.7	432.8	479.4
520.5	481.1	501.7	480.8	500.5	518.3	511.9	444.7	486.8	506.0	512.6
530.5	522.6	526.6	475.8	517.8	501.3	537.8	506.7	510.6	517.5	498.9
540.5	482.7	514.1	499.8	515.5	516.4	523.4	518.3	506.6	515.8	525.6
550.5	517.4	520.8	514.8	525.8	531.2	531.9	511.5	519.5	503.8	510.6
560.5	521.4	517.0	525.3	529.3	528.3	513.3	523.0	540.5	519.4	534.7
570.5	509.5	525.9	546.7	543.1	541.4	531.6	537.2	541.3	520.7	534.7
580.5	538.6	543.9	550.7	546.9	548.8	527.0	541.8	548.0	519.6	479.6
590.5	540.4	533.6	540.8	538.1	532.1	533.0	543.5	536.9	530.9	536.9
600.5	529.0	531.4	522.6	544.4	542.0	538.9	538.6	538.9	535.2	536.3
610.5	524.6	538.7	526.9	521.0	531.1	532.0	500.6	531.9	538.0	533.6
620.5	542.9	530.0	538.1	524.2	521.9	515.2	536.5	537.3	538.2	532.7

[a]Wavelengths λ_c at interval centers; irradiances in units of 10^{12} (photon $cm^{-2}\,s^{-1}\,nm^{-1}$) (converted from originally ($\mu W\,cm^{-2}\,\mathring{A}^{-1}$)); Source: Neckel and Labs (1984)

Table 6.4b Solar radiation intensities at wavelengths from 630 to 870 nm in 2 nm intervals[a]

λ_c (nm)	I_{av}	λ_c (nm)	I_{av}	λ_c (nm)	I_{av}	λ_c (nm)	I_{av}	λ_c (nm)	I_{av}
631.0	521.9	679.0	513.4	727.0	493.7	775.0	464.7	823.0	445.8
633.0	527.4	681.0	512.9	729.0	485.2	777.0	469.0	825.0	447.7
635.0	530.6	683.0	509.9	731.0	489.1	779.0	466.7	827.0	449.2
637.0	531.7	685.0	503.1	733.0	487.5	781.0	467.9	829.0	448.6
639.0	532.4	687.0	508.7	735.0	485.4	783.0	464.7	831.0	447.6
641.0	522.1	689.0	508.1	737.0	486.0	785.0	467.5	833.0	434.0
643.0	526.0	691.0	505.1	739.0	476.6	787.0	467.5	835.0	443.0
645.0	529.6	693.0	506.6	741.0	470.4	789.0	467.5	837.0	443.7
647.0	523.4	695.0	503.8	743.0	482.1	791.0	462.3	839.0	440.5
649.0	510.3	697.0	498.2	745.0	480.8	793.0	457.5	841.0	442.8
651.0	527.6	699.0	502.8	747.0	483.6	795.0	455.0	843.0	436.7
653.0	527.0	701.0	490.5	749.0	480.0	797.0	463.4	845.0	439.8
655.0	506.5	703.0	492.6	751.0	478.2	799.0	457.7	847.0	437.5
657.0	459.1	705.0	503.6	753.0	478.4	801.0	461.7	849.0	415.4
659.0	515.2	707.0	499.7	755.0	478.1	803.0	457.6	851.0	430.1
661.0	524.1	709.0	495.4	757.0	476.7	805.0	453.1	853.0	418.2
663.0	520.3	711.0	497.2	759.0	474.9	807.0	456.2	855.0	377.9
665.0	523.6	713.0	494.3	761.0	475.0	809.0	447.2	857.0	436.6
667.0	516.8	715.0	493.1	763.0	477.8	811.0	456.0	859.0	431.6
669.0	522.0	717.0	489.8	765.0	471.4	813.0	457.6	861.0	432.6
671.0	513.4	719.0	481.8	767.0	458.7	815.0	455.4	863.0	434.4
673.0	516.7	721.0	484.2	769.0	466.9	817.0	455.3	865.0	422.8
675.0	514.5	723.0	491.7	771.0	468.5	819.0	439.5	867.0	384.5
677.0	515.3	725.0	493.8	773.0	471.2	821.0	447.6	869.0	423.5

[a] Wavelengths λ_c at interval centers, irradiances in units of 10^{12} (photon cm^{-2} s^{-1} nm^{-1}) (originally (μW cm^{-2} Å^{-1})); Source: Neckel and Labs (1984)

Comments: Center-of-disc spectral intensities were measured at the Jungfraujoch (Swiss Alps, 3,600 m a.s.l.) and corrected for (wavelength-dependent) limb darkening and telluric absorptions. These data have served as a standard for many years despite indications by some other studies that at wavelengths <500 nm the intensities are slightly too low. Recent measurements have confirmed this effect (see the following Table 6.5 for a comparison).

Table 6.5 Comparison of solar irradiances measured in the wavelength range 330–670 nm[a]

Δλ (nm)	A Intensity[a]	B	C	Δλ (nm)	A Intensity[a]	B	C
330–335	158.8	162.7	–	500–505	475.1	481.4	485.4
335–340	151.9	160.2	–	505–510	496.6	501.3	511.5
340–345	159.7	170.3	–	510–515	493.8	493.0	492.8
345–350	161.1	169.9	174.1	515–520	458.4	459.5	459.5
350–355	186.0	197.0	193.1	520–525	496.5	502.1	494.2
355–360	203.1	178.0	193.9	525–530	492.4	497.4	490.5
360–365	183.5	194.9	197.4	530–535	508.8	513.1	512.9
365–370	227.7	242.5	236.1	535–540	514.3	518.2	512.4
370–375	193.8	206.7	215.5	540–545	505.7	510.7	512.0
375–380	223.6	241.4	231.8	545–550	517.9	520.1	522.0
380–385	186.1	197.0	188.4	550–555	522.0	523.2	527.7
385–390	200.4	212.2	211.2	555–560	515.5	513.9	516.9
390–395	204.4	215.0	216.6	560–565	524.3	519.9	515.2
395–400	251.0	263.3	209.6	565–570	526.2	523.1	517.8
400–405	345.3	359.9	375.2	570–575	533.3	530.9	527.2
405–410	343.9	358.8	355.2	575–580	533.1	528.2	527.7
410–415	358.1	373.4	359.2	580–585	545.8	538.1	539.5
415–420	463.5	381.3	372.2	585–590	523.2	517.9	524.0
420–425	367.4	384.1	378.6	590–595	537.0	532.1	530.6
425–430	346.1	358.3	360.3	595–600	536.2	532.4	536.1
430–435	343.5	357.1	341.9	600–605	533.9	527.8	532.3
435–440	390.8	404.1	395.6	605–610	537.6	528.8	534.3
440–445	424.6	438.6	435.7	610–615	528.5	516.8	524.5
445–450	442.2	454.8	450.3	615–620	527.2	514.2	517.9
450–455	463.2	475.4	466.5	620–625	531.9	516.4	527.2
455–460	470.6	480.2	486.4	625–630	532.0	516.5	530.3
460–465	476.8	486.8	491.4	630–635	525.8	510.4	526.1
465–470	470.1	481.5	483.5	635–640	531.8	515.7	530.0
470–475	475.9	483.6	492.8	640–645	525.2	509.7	523.2
475–480	488.1	496.4	510.7	645–650	519.4	504.6	521.2
480–485	493.6	504.7	508.6	650–655	523.1	514.4	525.7
485–490	450.8	461.6	466.9	655–660	491.0	478.3	486.4
490–495	484.5	493.6	497.9	660–665	522.5	510.9	520.4
495–500	492.0	498.6	498.0	665–670	520.2	508.1	518.6

[a]Averaged intensities in 5 nm intervals: Units: 10^{12} (photon cm^{-2} s^{-1} nm^{-1}). A: Neckel and Labs (1984), Ground-based measurements at the Jungfraujoch (Swiss Alps, 3,600 m a.s.l.); B: Burlov-Vasiljev et al. (1995), ground-based measurements at the high altitude observatory at Terskol Peak (Caucasus, 3,100 m a.s.l.), data presented in 5 nm intervals; C: Thuillier et al. (1998) Observations from Space Shuttle Atlantis (ATLAS 1 mission) in March 1992, 1 nm intervals (approximately)

References

Burlov-Vasiljev, K.A., A. Gurtovenko, Y.B. Matvejev, Solar Phys. **157**, 51–73 (1995)

Labs, D., H. Neckel, P.C. Simon, G. Thuillier, Solar Phys. **107**, 203–219 (1987)

Lean, J., Rev. Geophys. **29**, 505–535 (1991)

Mentall, J.E., D.E. Williams, J. Geophys. Res. **93**, 735–746 (1988)

Mentall, J.E., J.E. Frederick, J.R. Herman, J. Geophys. Res. **86**, 9881–9884 (1981)

Mount, G.H., G.J. Rottman, J. Geophys. Res. **88**, 5403–5410, 6807–6811 (1983)

Neckel, H., D. Labs, Solar Phys. **90**, 205–258 (1984)

Nicolet, M., Planet. Space Sci. **37**, 1249–1289 (1989)

Simon, P.C., Solar Phys. **74**, 273–291 (1981)

Thuillier, G., M. Hersé, P.C. Simon, D. Labs, H. Mandel, D. Gillotay, T. Foujols, Solar Phys. **177**, 41–61 (1998)

VanHoosier, M.E., J.-D.F. Bratoe, G.E. Brueckner, D.K. Prinz, Astrophys. Lett. Comm. **27**, 163–168 (1988)

Woods, T.N., D.K. Prinz, G.J. Rottman, J. London, P.C. Crane, R.P. Cebula, E. Hilsenrath, G.E. Brueckner, M.D. Andrews, O.R. White, M.E. VanHoosier, L.E. Floyd, L.C. Herring, B.G. Knapp, C.K. Pankratz, P.A. Reiser, J. Geophys. Res. **101** D6, 9541–9569 (1996)

6.2 Photodissociation Coefficients

The absorption cross section (σ_s) is defined by the Beer-Lambert law written in the form

$$\sigma_s(\lambda) = (1/n_s l) \ln(I_0(\lambda)/I_t(\lambda))$$

where l is the absorption path length, n_s is the number concentration of the absorbing species s assumed to be constant along the absorption path, and $I_0(\lambda)$ and $I_t(\lambda)$ are the incident and transmitted light intensities at wavelength λ, respectively. The units adopted for the absorption cross section: (cm^2 molecule^{-1}) are associated with the units for number concentration (molecule cm^{-3}) and path length (cm). Owing to space limitations, the unit for absorption cross sections often is simplified to (cm^2) (with the entity omitted).

The primary quantum yield of a photochemical process is the probability of a molecule to enter into a specific chemical reaction following excitation by absorption of a photon. The dissociation quantum yield ϕ_i (molecule photon^{-1}) is the probability for decomposition of the exited molecule along a specific product channel i. Other possible primary processes such as isomerization, fluorescence or collisional energy transfer must be assigned the corresponding quantum yields. The sum of quantum yields for all primary processes is equal to unity. This follows directly from the energy balance. Discrete band spectra indicate transitions to bound upper energy levels that may or may not undergo predissociation by interacting with another unstable state. Conversely, a continuous spectrum usually indicates a transition to an unstable, that is dissociating, upper state. If this is the only allowed photodissociation process, the dissociation quantum yield is close to unity.

These quantities are combined with the actinic radiation flux to derive photodissociation coefficients for atmospheric constituents. The photodissociation coefficient, j, (sometimes called the photodissociation frequency) is defined by the integral over the effective wavelength region

$$j_i = \int_{\Delta\lambda} \phi_i(\lambda)\sigma_s(\lambda)I_a(\lambda)d\lambda \quad [s^{-1}]$$

where ϕ_i is the associated primary quantum yield, $\sigma_s(\lambda)$ is the absorption cross section of the dissociating molecule, and $I_a(\lambda)$ is the actinic radiation flux, with units (photon cm^{-2} s^{-1} nm^{-1}) if $\sigma_s(\lambda)$ is given in (cm^2 $molecule^{-1}$). The photodissociation rate $-dn_s/dt$ (molecule cm^{-3} s^{-1}) refers to the loss of photo-active species n_s. The rate of product formation must take the stoichiometry of the process into account. For example, the rate of OH formation associated with the process $H_2O_2 + h\nu \rightarrow OH + OH$ is $dn_{OH}/dt = -2dn_{H_2O_2}/dt = 2j_{H_2O_2}n_{H_2O_2}$.

Table 6.6 presents an overview of absorption spectra and photochemical processes in the atmosphere. Individual data are treated subsequently.

Table 6.6 Overview on absorption spectra and photochemical products of gas-phase atmospheric constituents and their significance to atmospheric chemistry[a]

Constituent	Absorption spectrum Type and approximate long wavelength limit (nm)	Dissociation limit (nm)	Photochemistry Main primary photochemical process	Quantum yield	Significance in the atmosphere
N_2	Lyman–Birge–Hopfield bands	145	127	Predissociation		Not significant, the absorption is too weak
O_2	Herzberg bands	275				Night-sky emission, high altitude
	Herzberg continuum	240	240	$O_2 \rightarrow O(^3P)+O(^3P)$	$\phi \approx 1.0$	O_3 formation in the stratosphere
	Schumann–Runge bands	200		Predissociation	$\phi \approx 1.0$	Important in the mesosphere
	Schumann continuum	175	175	$O_2 \rightarrow O(^3P)+O(^1D)$	$\phi \approx 1.0$	O_2 dissociation in upper atmosphere
O_3	Hartley bands	320	310	$O_3 \rightarrow O_2\,(^1\Delta_g)+O(^1D)$	$\phi \approx 0.9$	Important in the stratosphere
	Huggins bands	360	410	$O_3 \rightarrow O_2\,(^3\Sigma_g^-)+O(^3P)$	$\phi \approx 1.0$	Important in the troposphere
	Chappuis bands	850	1,180	$O_3 \rightarrow O_2\,(^3\Sigma_g^-)+O(^3P)$	$\phi \approx 1.0$	Entire atmosphere
CO_2	Continuum overlapped by bands	200	226 / 167	(a) $CO_2 \rightarrow CO+O(^3P)$ (b) $CO_2 \rightarrow CO+O(^1D)$	$\phi \approx 1.0$	(a) Mesosphere and stratosphere, 175–200 nm; (b) shielded by O_2
CO	4th positive band system	155	111	Fluorescence-quenching		Not significant
CH_4	Continuum, some band structure	160	277	Dissociation		Not significant, shielded by O_2
H_2	Lyman bands	111	275	Fluorescence-quenching		Not significant
H_2O	Continuum, some band structure	200	243 / 176	$H_2O \rightarrow OH+H$ $H_2O \rightarrow H_2+O(^1D)$	$\phi \approx 1.0$ Small	Mesosphere and stratosphere, 175–200 nm
H_2O_2	Continuum	350	557 / 324	$H_2O_2 \rightarrow OH+OH$ $H_2O_2 \rightarrow HO_2+H$	$\phi \approx 1.0$	Stratosphere and mesosphere
HO_2	Continuum	270	437	$HO_2 \rightarrow O+HO$ (presumably)		Possibly in the mesosphere
N_2O	Continuum, some band structure	240	742 / 341	(a) $N_2O \rightarrow N_2+O(^3P)$ (b) $N_2O \rightarrow N_2+O(^1D)$	Spin-forbidden	Stratosphere >185 nm
NH_3	Bands+continuum	235	281	Dissociation	$\phi \approx 1.0$	Not significant

Species	Spectral region			Process	ϕ	Atmospheric significance
NO	Resonance bands	230	191	(a) Predissociation, mainly the bands 0–0 and 0–1		(a) Mesosphere and upper stratosphere
	Continuum+bands	135	134	(b) Photoionization		(b) Lyman α, upper mesosphere
NO_2	Complex band system	700	398	(a) $NO_2 \to NO + O(^3P)$	$\phi \approx 1.0$	(a) Very important throughout the entire atmosphere
	Continuum<250 nm		244	(b) $NO_2 \to NO + O(^1D)$		(b) Possibly in the mesosphere
NO_3	Diffuse bands	690	587	$NO_3 \to NO_2 + O$	$\phi \approx 1.0$	Important
			1,139	$NO_3 \to NO + O_2$	$\phi\ 0{-}0.5$	$586 < \lambda < 640$ nm
$HONO$	Diffuse bands	395	578	$HNO_2 \to NO + OH$	$\phi \approx 0.9$	Important in the entire atmosphere
			361	$HNO_2 \to H + NO_2$	$\phi \approx 0.1$	
HNO_3	Continuum	330	604	$HNO_3 \to NO_2 + OH$	$\phi \approx 1.0$	Mainly in the stratosphere
N_2O_5	Continuum	410	1,252	$N_2O_5 \to NO_2 + NO_3$	$\phi \approx 1.0$	Mainly in the stratosphere
			298	$N_2O_5 \to NO_3 + NO + O$		
HO_2NO_2	Continuum	330	1,191	$HO_2NO_2 \to HO_2 + NO_2$	$\phi \approx 0.66$	Stratosphere
			731	$HO_2NO_2 \to OH + NO_3$	$\phi \approx 0.33$	
Cl_2	Continuum, bands	>500	493	$Cl_2 \to Cl + Cl$	$\phi = 1.0$	Entire atmosphere
HCl	Continuum	220	277	$HCl \to H + Cl$		Possibly in the mesosphere, > 175 nm
$HOCl$	Continuum	420	500	$HOCl \to OH + Cl$	$\phi = 1.0$	Entire atmosphere
CH_3Cl	Continuum	220	347	$CH_3Cl \to CH_3 + Cl$	$\phi \approx 1.0$	Stratosphere
CH_3Br	Continuum	300	417	$CH_3Br \to CH_3 + Br$	$\phi \approx 1.0$	Not competitive
CH_3I	Continuum	370	501	$CH_3I \to CH_3 + I$	$\phi \approx 1.0$	Important
CF_4	Continuum, bands	103	220	$CF_4 \to CF_3 + F$ (presumably)		Only known loss process, occurring in the upper atmosphere
$CFCl_3$	Continuum	230	375	$CFCl_3 \to CFCl_2 + Cl$		Important in the stratosphere, in the O_2 absorption window
				$CFCl_3 \to CFCl + 2Cl$		
CF_2Cl_2	Continuum	220	354	$CF_2Cl_2 \to CF_2Cl + Cl$		Important in the stratosphere in the O_2 absorption window
				$CF_2Cl_2 \to CFCl + 2Cl$		
CCl_4	Continuum	240	407	$CCl_4 \to CCl_3 + Cl$		Important in the stratosphere in the O_2 absorption window
				$CCl_4 \to CCl_2 + 2Cl$		

(continued)

Table 6.6 (continued)

Constituent	Absorption spectrum		Photochemistry		Significance in the atmosphere
	Type and approximate long wavelength limit (nm)	Dissociation limit (nm)	Main primary photochemical process	Quantum yield	
HF	Continuum	162	211	$HF \rightarrow H+F$		Shielded by O_2
$ClONO_2$	Continuum	430	1,065 735 425	$ClONO_2 \rightarrow Cl+NO_3$ $ClONO_2 \rightarrow ClO+NO_2$ $ClONO_2 \rightarrow ClONO+O$	$0.4 < \phi < 1.0$ $0 < \phi < 0.6$ $\varphi = 0.17$ (at $\lambda = 220$ nm)	Stratosphere
ClONO	Continuum	400	880	$ClONO \rightarrow Cl+NO_2$ (expected)		Not established
Cl_2	Continuum	470	495	$Cl_2 \rightarrow Cl+Cl$	$\phi \approx 1$	Entire atmosphere
ClOOCl	Continuum	450	1,615	$ClOOCl \rightarrow ClOO+Cl$	$\phi \approx 1$	Polar stratosphere
PAN^b	Continuum	350	1,086	$PAN \rightarrow CH_3CO_3+NO_2$ $PAN \rightarrow CH_3CO_2+NO_3$	$\phi \approx 0.75$ $\phi \approx 0.25$	Significance not established
C_2–C_5 alkanes	Continuum	170		Photodecomposition		Not significant
C_2–C_5 alkenes	Continuum	205		Photodecomposition		Not significant
C_2H_2	Continuum, weak structure	237	230	$C_2H_2 \rightarrow C_2H+H$		Significance not established
HCHO	Bands	360	324	$HCHO \rightarrow H+HCO$ $HCHO \rightarrow H_2+CO$	ϕ depends on λ and p	Important in the entire atmosphere
CH_3CHO	Quasi-continuum, weak structure	340	337 320	$CH_3CHO \rightarrow CH_3+CHO$ $CH_3CHO \rightarrow CH_3CO+H$	ϕ depends on λ and p	Significant in the troposphere
CH_3COCH_3	Quasi-continuum, weak structure	350	338	$CH_3COCH_3 \rightarrow CH_3CO+CH_3$	ϕ depends on λ and p	Significant in the troposphere

CH₃OOH	Continuum	350		CH₃OOH → CH₃O + OH	$\phi \approx 1.0$	Possibly stratosphere
SO₂	Very weak bands, Stronger bands	390, 340	647	Excited state quenching and reaction		Not significant because of collisional quenching
	Continuum + bands	220	220	SO₂ → SO + O		Stratosphere
OCS	Continuum	300	388	OCS → CO + S(^{3}P)	$\phi \approx 0.27$	Stratosphere
			286	OCS → CO + S(^{1}D)	$\phi \approx 0.67$	
CH₃SH	Continuum	280	311	CH₃SH → CH₃S + H	$\phi \approx 0.9$	Not significant
CS₂	Two strongly structured bands	370	281	CS₂ → CS + S(^{3}P)		Not significant
			223	CS₂ → CS + S(^{1}D)		Not significant
CH₃SSCH₃	Continuum	355	437	CH₃SSCH₃ → 2 CH₃S	$\phi = 1.65$ (at $\lambda = 248$ nm)	Significant in the troposphere

[a]Source of data: Warneck (2000), with changes and additions (with permission of Elsevier)
[b]PAN = peroxyacetyl nitrate, CH₃C(O)OONO₂

Table 6.7 Photodissociation coefficients for important atmospheric processes[a]

Reaction	$\lambda_{threshold}$ or $\Delta\lambda$ (nm)	Ground level	25 km altitude
		Photolysis frequency (s^{-1})	
$O_2 + h\nu \rightarrow O\,(^3P) + O\,(^3P)$	245	4.3 (−26)	1.2 (−11)
$O_3 + h\nu \rightarrow O_2 + O\,(^3P)$	310–850	4.2 (−4)	5.0 (−4)
$O_3 + h\nu \rightarrow O_2\,(a\,^1\Delta_g) + O\,(^1D)$	310	5.1 (−5)	1.1 (−4)
$H_2O + h\nu \rightarrow H + OH$	210	0	6.6 (−11)
$H_2O_2 + h\nu \rightarrow OH + OH$	355	7.7 (−6)	1.2 (−5)
$CH_3OOH + h\nu \rightarrow CH_3O + OH$	360	5.7 (−6)	9.9 (−6)
$NO_2 + h\nu \rightarrow NO + O\,(^3P)$	420	8.8 (−3)	1.2 (−2)
$NO_3 + h\nu \rightarrow NO_2 + O\,(^3P)$	410–640	2.8 (−1)	3.0 (−1)
$NO_3 + h\nu \rightarrow NO + O_2$	586–640	2.5 (−2)	2.7 (−2)
$N_2O_5 + h\nu \rightarrow NO_2 + NO_3$	385	5.0 (−5)	4.7 (−5)
$N_2O + h\nu \rightarrow N_2 + O\,(^1D)$	240	1.8 (−23)	2.5 (−8)
$HONO + h\nu \rightarrow NO + OH$	400	1.9 (−3)	2.8 (−3)
$HNO_3 + h\nu \rightarrow NO_2 + OH$	335	8.2 (−7)	5.6 (−6)
$HO_2NO_2 + h\nu \rightarrow HO_2 + NO_2$	330	3.7 (−6)	1.1 (−5)
$HO_2NO_2 + h\nu \rightarrow OH + NO_3$	334	1.8 (−6)	5.7 (−6)
$HCHO + h\nu \rightarrow H_2 + CO$	360	4.4 (−5)	7.6 (−5)
$HCHO + h\nu \rightarrow H + HCO$	340	3.3 (−5)	6.2 (−5)
$CH_3CHO + h\nu \rightarrow CH_3 + HCO$	325	6.4 (−6)	1.3 (−5)
$CH_3COCH_3 + h\nu \rightarrow CH_3 + CH_3CO$	335	8.4 (−7)	1.9 (−6)
$CH_3COCHO + h\nu \rightarrow CH_3CO + CO + H$	465	2.9 (−4)	3.8 (−4)
$CH_3ONO_2 + h\nu \rightarrow CH_3O + NO_2$	330	1.3 (−6)	2.5 (−6)
$C_2H_5ONO_2 + h\nu \rightarrow C_2H_5O + NO_2$	330	1.8 (−6)	1.5 (−5)
$C_3H_7ONO_2 + h\nu \rightarrow C_3H_7O + NO_2$	330	3.4 (−6)	1.9 (−5)
$CH_3OONO_2 + h\nu \rightarrow CH_3O_2 + NO_2$	325	6.4 (−6)	1.8 (−5)
$CH_3CO_3NO_2 + h\nu \rightarrow CH_3CO_3 + NO_2$	300	7.2 (−8)	3.5 (−6)
$CH_3SSCH_3 + h\nu \rightarrow CH_3S + CH_3S$	400	6.3 (−5)	1.3 (−4)
$COS + h\nu \rightarrow CO + S$	260	1.7 (−21)	2.8 (−8)
$Cl_2 + h\nu \rightarrow Cl + Cl$	450	2.4 (−3)	3.5 (−3)
$ClO + h\nu \rightarrow Cl + O$	305	4.1 (−5)	1.3 (−4)
$OClO + h\nu \rightarrow ClO + O$	450	9.9 (−2)	1.4 (−1)
$ClOOCl + h\nu \rightarrow ClOO + Cl$	360	1.2 (−3)	2.0 (−3)
$HOCl + h\nu \rightarrow Cl + OH$	420	2.6 (−4)	4.1 (−4)
$ClONO_2 + h\nu \rightarrow Cl + NO_3$	400	4.8 (−5)	6.2 (−5)
$ClNO_2 + h\nu \rightarrow Cl + NO_2$	372	3.9 (−4)	6.5 (−4)
$CCl_4 + h\nu \rightarrow CCl_3 + Cl$	250	2.1 (−21)	8.5 (−7)
$CFCl_3 + h\nu \rightarrow CFCl_2 + Cl$	250	4.3 (−22)	5.1 (−7)
$CF_2Cl_2 + h\nu \rightarrow CF_2Cl + Cl$	226	1.7 (−23)	4.6 (−8)
$CFCl_2CF_2Cl + h\nu \rightarrow CF_2ClCF_2 + Cl$	230	2.6 (−23)	8.8 (−8)
$CH_3Cl + h\nu \rightarrow CH_3 + Cl$	220	2.3 (−24)	1.3 (−8)
$CH_3CCl_3 + h\nu \rightarrow Cl + products$	240	5.1 (−23)	7.2 (−7)
$CHF_2Cl + h\nu \rightarrow Cl + products$	205	3.0 (−27)	5.7 (−10)
$Br_2 + h\nu \rightarrow Br + Br$	600	3.0 (−2)	3.6 (−2)
$BrCl + h\nu \rightarrow Br + Cl$	570	4.6 (−3)	5.9 (−3)
$HOBr + h\nu \rightarrow Br + OH$	480	6.1 (−4)	9.1 (−4)
$BrO + h\nu \rightarrow Br + O$	375	3.8 (−3)	5.9 (−3)
$BrONO_2 + h\nu \rightarrow Br + NO_3$	390	1.1 (−3)	1.6 (−3)
$CH_3Br + h\nu \rightarrow Br + CH_3$	260	4.0 (−21)	1.1 (−6)
$CH_3I + h\nu \rightarrow I + CH_3$	365	1.4 (−6)	2.5 (−5)

[a] Powers of ten in parentheses; calculated with overhead sun, surface albedo 0.03; from the summary of Jacobson (2005), data for CH_3I from Rattigan et al. (1997)

Table 6.8 Stratospheric photodissociation rates (s^{-1}) as a function of altitude[a]

Altitude (km)	20	25	30	40	50	60
N_2O	3.08 (−9)	2.76 (−8)	1.07 (−7)	3.87 (−7)	6.15 (−7)	7.09 (−7)
$CFCl_3$	6.46 (−8)	5.75 (−7)	2.21 (−6)	7.69 (−6)	1.15 (−5)	1.37 (−5)
CF_2Cl_2	4.34 (−9)	4.17 (−8)	1.75 (−7)	7.96 (−7)	1.58 (−6)	2.14 (−6)

[a] Powers of ten in parentheses; calculated by Mérienne et al. (1990) for an overhead sun

Table 6.9 Photodissociation coefficients for different wavelength regions[a]

Constituent or process	Wavelength region (nm)	j (s^{-1})	Constituent or process	Wavelength region (nm)	j (s^{-1})
O_3	<310	8.7 (−3)	HO_2NO_2	<300	6.3 (−4)
	>310	4.6 (−4)		>300	1.5 (−4)
CO_2	Ly α	2.2 (−8)	HCl	>175	1.2 (−6)
	>175	1.9 (−9)	HOCl	<300	9.2 (−5)
H_2O	Ly α	4.2 (−6)		>300	3.7 (−4)
	>175	8.6 (−7)	CH_3Cl	>175	1.0 (−7)
H_2O_2	<300	1.0 (−4)	$CFCl_3$	>175	1.0 (−7)
	>300	1.3 (−5)	CF_2Cl_2	>175	6.3 (−7)
N_2O	>175	7.3 (−7)	CCl_4	>175	3.5 (−5)
NO_2	175–240	5.9 (−5)	ClO	<300	6.5 (−3)
	240–307	2.3 (−4)		>300	5.5 (−4)
	>310	8.0 (−3)	$ClONO_2$	<310	2.0 (−5)
$NO_3 \rightarrow NO_2+O$	>400	1 (−1)		>310	9.5 (−4)
$NO_3 \rightarrow NO+O_2$		4 (−2)	ClONO	<310	2.0 (−5)
N_2O_5	<310	5.8 (−4)		>310	9.5 (−4)
	>310	3.6 (−5)	HCHO → H	<300	4.5 (−5)
HNO_3	<200	7.1 (−5)	+ HCO	>300	1.1 (−5)
	200–307	7.2 (−5)	HCHO → H_2	<300	4.0 (−5)
	>307	2.0 (−6)	+ CO	>300	1.1 (−5)
HNO_2	<310	3.4 (−4)	CH_3OOH	>300	1.2 (−5)
	>310	2.5 (−3)			

[a] Powers of ten in parentheses; from Warneck (2000) with permission of Elsevier, originally Nicolet (1978) calculated with solar radiation intensities incident on Earth's atmosphere

References

Jacobson, M.Z., *Fundamentals of Atmospheric Modeling*, 2nd edn. (Cambridge University Press, Cambridge, 2005)

Mérienne, M.F., B. Coquart, A. Jenouvrier, Planet. Space Sci. **38**, 617–625 (1990)

Nicolet, M., *Etude des reactions chimiques de l'ozone dans la stratosphère* (Institut Royal Météorologique de Belgique, Belgium, 1978)

Rattigan, O.V., D.E. Shallcross, R.A. Cox, J. Chem. Soc. Faraday Trans. **93**, 2839–2846 (1997)

Warneck, P., *Chemistry of the Natural Atmosphere*, 2nd edn. (Academic Press, San Diego, 2000), Copyright Elsevier

6.3   Absorption Cross Sections and Primary Quantum Yields

Absorption cross sections and quantum yields for many gaseous species of interest in atmospheric chemistry have been critically reviewed by Atkinson et al. (2004, 2006, 2007) and by Sander et al. (2006). These publications have served as a guide in selecting the data presented below. For definitions of absorption cross sections and quantum yields see the introduction to the preceding Sect. 6.2 on photodissociation coefficients.

Table 6.10   Absorption cross sections of molecular oxygen in the Herzberg continuum[a]

λ (nm)	$10^{24}\sigma_0$ (cm^2)	$10^{27}\alpha$ (cm^2 hPa^{-1})	λ (nm)	$10^{24}\sigma_0$ (cm^2)	$10^{27}\alpha$ (cm^2 hPa^{-1})	λ (nm)	$10^{24}\sigma_0$ (cm^2)	$10^{27}\alpha$ (cm^2 hPa^{-1})
194.6	1.97	18.2	215	5.36	7.32	229	2.50	3.87
200.8	5.63	12.5	217	4.92	6.83	231	2.13	3.47
205	6.90	10.3	219	4.45	6.40	233	1.98	3.10
207	6.66	9.68	221	4.09	5.78	235	1.67	2.76
209	6.35	9.08	223	3.73	5.22	237	1.24	2.48
211	5.95	8.63	225	3.30	4.72	239	1.01	2.18
213	5.63	8.03	227	2.84	4.37	240	0.92	2.02

[a] Source of data: Yoshino et al. (1988, 1992). The absorption cross sections are pressure-dependent: $\sigma(\lambda) = \sigma_0(\lambda) + \alpha(\lambda)P$, where $\alpha(\lambda)$ is a cross section involving two oxygen molecules. Values for $\sigma_0(\lambda)$ and $\alpha(\lambda)$ are presented above

Comments: At wavelength below 200 nm the Herzberg continuum underlies the Schumann-Runge band system and is effective only in windows between rotational lines. In this region the intensity of the Herzberg continuum declines to low values near 190 nm. The quantum yield for the process $O_2 + h\nu \rightarrow 2\,O\,(^3P)$ is $\phi = 1.0$ for the entire wavelength region 175–242 nm. Yoshino et al. (1992) have fitted the observed wavelength dependence of cross sections of the Herzberg continuum to a fourth order polynomial with the following parameters: $y(\lambda\,[\text{nm}]) = 10^{24}\sigma_0$, $a_0 = -2.3837947 \times 10^4$, $a_1 = 4.1973085 \times 10^2$, $a_2 = -2.7640139$, $a_3 = 8.0723193 \times 10^{-3}$, $a_4 = 8.8255447 \times 10^{-6}$.

Table 6.11a Absorption cross sections of ozone in the Hartley band[a]

λ (nm)	$10^{18}\,\sigma$ (cm^2 molecule^{-1})		λ (nm)	$10^{20}\,\sigma$ (cm^2 molecule^{-1})		λ (nm)	$10^{22}\,\sigma$ (cm^2 molecule^{-1})	
	298 K	226 K		298 K	226 K		298 K	226 K
185	0.6537	0.6437	230	4.476	4.506	275	5.913	5.871
186	0.6187	0.6259	232	5.181	5.230	276	5.450	5.379
188	0.5659	0.5655	234	5.891	5.923	278	4.668	4.617
190	0.5114	0.5163	235	6.318	6.349	280	4.001	3.983
192	0.4606	0.4595	236	6.722	6.772	282	3.250	3.220
194	0.4066	0.4088	238	7.486	7.532	284	2.712	2.644
195	0.3864	0.3827	240	8.314	8.367	285	2.465	2.398
196	0.3673	0.3642	242	8.971	9.016	286	2.238	2.195
198	0.3349	0.3333	244	9.717	9.752	288	1.750	1.687
200	0.3154	0.3145	245	9.932	10.28	290	1.418	1.365
202	0.3179	0.3156	246	10.33	10.42	292	1.111	1.043
204	0.3365	0.3400	248	10.71	10.79	294	0.8711	0.8148
205	0.3385	0.3623	250	11.24	11.34	295	0.7753	0.7270
206	0.3855	0.3887	252	11.55	11.65	296	0.6727	0.6202
208	0.4640	0.4684	254	11.59	11.69	298	0.5124	0.4714
210	0.5716	0.5806	255	11.61	11.45	300	0.3964	0.3616
212	0.7194	0.7312	256	11.54	11.58	301	0.3463	0.3128
214	0.9096	0.9255	258	11.24	11.30	302	0.3073	0.2796
215	1.023	1.041	260	10.80	10.86	303	0.2650	0.2325
216	1.081	1.169	262	10.57	10.64	304	0.2401	0.2156
218	1.439	1.464	264	10.06	10.13	305	0.2015	0.1771
220	1.785	1.799	265	9.657	9.612	306	0.1808	0.1603
222	2.200	2.217	266	9.485	9.467	307	0.1565	0.1367
224	2.684	2.688	268	8.754	8.722	308	0.1364	0.1202
225	2.943	2.963	270	7.980	7.961	309	0.1243	0.1064
226	3.226	3.239	272	7.147	7.107	310	0.1020	0.08637
228	3.829	3.857	274	6.140	6.013	311	0.0926	0.07925

[a]The temperature dependence of cross sections in the Hartley band is slight and shows up mainly in the long wavelength tail. Source of data: Molina and Molina (1986)

Table 6.11b Absorption cross sections of ozone in the wavelength region of the Huggins bands[a]

λ (nm)	$10^{20}\,\sigma$ (cm² molecule⁻¹) 295 K	218 K	λ (nm)	$10^{21}\,\sigma$ (cm² molecule⁻¹) 295 K	218 K	λ (nm)	$10^{21}\,\sigma$ (cm² molecule⁻¹) 295 K	218 K
312.0	7.868	6.416	322.0 p	2.423	2.158	330.8 v	6.379	4.524
312.5	7.221	5.792	322.5 v	21.70	17.15	331.0	8.406	7.088
313.0	6.712	5.495	322.6 p	24.13	20.98	331.1 p	8.529	7.269
313.5	6.968	5.952	323.0	19.60	14.32	331.5	6.239	4.356
314.0	6.291	5.073	323.5	13.93	8.248	332.0	4.182	2.259
314.5	5.364	4.041	323.9 v	11.77	6.562	332.5 v	3.083	1.433
315.0	5.096	3.973	324.5 s	12.83	8.769	333.0 s	3.416	2.072
315.1	5.015	3.911	324.8 p	18.23	16.43	333.5	4.465	3.240
315.5 p	5.461	4.675	325.0	17.28	14.57	333.7 p	5.888	5.022
316.0	4.664	3.636	325.3 v	14.19	10.47	334.0	5.319	3.785
316.5	4.180	3.141	325.5 p	15.09	12.16	334.5	3.516	2.008
316.8 v	3.907	2.880	326.0	11.06	6.997	335.0	2.391	1.208
317.0 p	4.128	3.317	326.5	8.575	4.677	335.6 s	1.876	8.112
317.4 v	3.910	3.061	326.8 v	7.702	4.087	336.0	1.770	7.504
317.6 p	4.219	3.614	327.0	8.383	5.557	336.5 p	2.702	1.810
318.0	3.699	2.948	327.2 p	9.634	7.280	336.8 v	2.513	1.536
318.5	3.082	2.185	327.5 v	8.839	6.129	337.0	2.072	1.283
319.1 v	2.660	1.836	327.9 p	13.09	11.97	337.3 p	3.855	3.392
319.5 p	3.205	2.644	328.0	12.99	11.40	337.5	3.251	2.597
319.8 v	2.959	2.318	328.5	9.406	5.991	338.0	2.072	1.283
320.0	3.250	2.835	329.0	6.419	3.167	338.5	1.405	0.6684
320.5 s	2.540	1.862	329.4 v	5.137	2.413	339.0 v	1.322	0.5069
321.0	1.996	1.268	329.6 s	5.048	2.528	339.5 p	1.655	0.6423
321.1 v	1.915	1.239	329.9 v	4.637	2.458	339.7 v	1.555	0.6129
321.4 p	1.923	1.322	330.0	4.695	2.634	340.0	2.031	1.403
321.6 v	1.837	1.233	330.3 p	7.376	6.043	340.2 p	2.211	1.667

λ (nm)	$10^{21}\,\sigma$ (cm² molecule⁻¹) 295 K	220 K	λ (nm)	$10^{21}\,\sigma$ (cm² molecule⁻¹) 295 K	220 K	λ (nm)	$10^{21}\,\sigma$ (cm² molecule⁻¹) 295 K	220 K
340.5	1.747	1.134	345.0	0.685	0.355	349.1 v	0.231	0.070
341.0	1.095	0.587	345.5	0.606	0.247	349.5 p	0.374	0.102
341.5	0.755	0.325	346.0	0.592	0.184	350.0	0.294	0.080
341.8 v	0.692	0.298	346.5	0.493	0.141	350.5	0.258	0.088
342.0	0.749	0.300	346.7 v	0.440	0.116	350.7 v	0.235	0.084
342.1 p	0.781	0.297	346.9 v	0.408	0.140	350.9 v	0.220	0.088
342.6 v	0.670	0.219	347.0	0.426	0.190	351.0	0.230	0.101
343.0	0.907	0.327	347.2 p	0.518	0.356	351.4 p	0.399	0.321
343.2 p	0.952	0.445	347.5	0.455	0.298	352.0	0.291	0.186
343.5 v	0.882	0.423	348.0	0.316	0.159	352.5	0.295	0.132
344.0 p	1.371	1.179	348.5	0.269	0.103	353.0	0.222	0.081
344.5	0.954	0.683	349.0	0.249	0.074	354.0	0.127	0.042

[a] Peaks (p), valleys (v) and shoulders (s) are indicated. Source of data (upper part): Malicet et al. (1995), numerical data were taken from the compilation of Keller-Rudek and Moortgat (2010); (lower part) Cacciani et al. (1989)

Table 6.11c Quantum yields for the process $O_3 + h\nu \rightarrow O_2 + O$ (^1D) at 298 and 223 K[a]

λ (nm)	ϕ_{298}	ϕ_{223}	λ (nm)	ϕ_{298}	ϕ_{223}	λ (nm)	ϕ_{298}	ϕ_{223}	λ (nm)	ϕ_{298}	ϕ_{223}
305	0.900	–	311	0.310	0.23	317	0.222	0.08	323	0.113	0.05
306	0.884	0.86	312	0.310	0.16	318	0.206	0.08	324	0.101	0.06
307	0.862	0.84	313	0.265	0.13	319	0.187	0.08	325	0.092	0.08
308	0.793	0.74	314	0.246	0.09	320	0.166	0.07	326	0.086	0.08
309	0.671	0.61	315	0.239	0.09	321	0.146	0.07	327	0.082	0.08
310	0.523	0.39	316	0.233	0.09	322	0.128	0.07	328	0.080	0.07

[a] In the wavelength region $220 < \lambda < 305$ nm, ϕ (O ^1D) $= 0.90$, ϕ (O ^3P) $= 0.10$; in the wavelength region $330 < \lambda < 370$, ϕ (O ^1D) $= 0.08$, ϕ (O ^3P) $= 0.92$ (independent of temperature in both regions). Source of data: Talukdar et al. (1998)

Matsumi et al. (2002) have proposed the following formula for O (^1D) quantum yields in the wavelength region 306–328 nm:

$$\phi(\lambda, T) = A_1[1 / (1+q)]\exp(-[(x_1 - \lambda)/\omega_1]^4) + A_2[q / (1+q)](T / 300)^2$$
$$\times \exp(-[(x_2 - \lambda)/\omega_2]^2) + A_3(T / 300)^{1.5} \exp(-[(x_3 - \lambda)/\omega_3]^2) + C$$

with $q = \exp(-825.518/R_g T)$ and the constants $A_1 = 0.8036$, $A_2 = 8.9061$, $A_3 = 0.1192$, $x_1 = 304.225$, $x_2 = 314.957$, $x_3 = 310.737$, $\omega_1 = 5.576$, $\omega_2 = 6.601$, $\omega_3 = 2.187$, $C = 0.08$.

Table 6.12 Absorption cross sections of ozone in the Chappuis band[a]

λ (nm)	$10^{22}\,\sigma$ (cm^2)	λ (nm)	$10^{22}\,\sigma$ (cm^2)	λ (nm)	$10^{22}\,\sigma$ (cm^2)	λ (nm)	$10^{22}\,\sigma$ (cm^2)	λ (nm)	$10^{22}\,\sigma$ (cm^2)
377.5*	0.044	490	8.135	590	44.34	680	13.64	800	1.554
390	0.069	500	11.96	595	47.33	690	11.19	810	1.846
395	0.103	510	15.55	600	51.18	700	8.618	815	2.226
400	0.110	520	18.15	603	51.90	710	7.268	820	2.074
410	0.270	530	26.67	605	51.38	720	6.174	825	1.427
420	0.369	540	29.26	610	47.58	730	4.743	830	0.991
430	0.651	550	33.56	620	40.32	740	4.274	838	0.749
440	1.368	560	39.56	630	35.40	750	4.276	853	1.417
450	1.899	570	46.56	640	29.49	760	2.838	877	0.368
460	3.732	575	47.98	650	24.94	770	2.569	898	0.620
470	4.174	580	46.17	660	20.92	780	3.214	933	0.161
480	7.601	585	44.35	670	16.85	790	2.064	944	0.407

[a] The asterisk indicates the minimum of absorption between the visible and near ultraviolet spectral regions. Absorption maxima occur at 575 and 603 nm. The cross sections are independent of temperature. The quantum yield for $O_3 + h\nu \rightarrow O_2(X^3\Sigma_g^-) + O$ (^3P) is unity. Source of data: Brion et al. (1998) (1 nm averages), numerical values taken from the compilation of Keller-Rudek and Moortgat (2010); above 830 nm Anderson et al. (1993) (spectral resolution 3.1 nm)

Table 6.13a Absorption cross sections of nitrogen dioxide, NO_2 (298 K)[a]

λ (nm)	$10^{19}\,\sigma$ (cm^2)	λ (nm)	$10^{19}\,\sigma$ (cm^2)	λ (nm)	$10^{19}\,\sigma$ (cm^2)	λ (nm)	$10^{19}\,\sigma$ (cm^2)	λ (nm)	$10^{19}\,\sigma$ (cm^2)
205	3.581	295	1.064	385	5.942	475	3.849	565	0.870
210	4.451	300	1.299	390	6.200	480	3.345	570	0.848
215	4.887	305	1.604	395	5.920	485	2.518	575	0.471
220	4.672	310	1.882	400	6.385	490	3.075	580	0.447
225	3.904	315	2.174	405	5.768	495	2.930	585	0.469
230	2.765	320	2.542	410	6.153	500	1.82	590	0.539
235	1.653	325	2.879	415	5.892	505	2.43	595	0.408
240	0.830	330	3.188	420	5.950	510	2.31	600	0.395
245	0.375	335	3.587	425	5.670	515	1.60	605	0.185
250	0.146	340	4.020	430	5.405	520	1.61	610	0.254
255	0.109	345	4.175	435	5.555	525	1.79	615	0.353
260	0.154	350	4.616	440	4.842	530	1.53	620	0.257
265	0.218	355	4.982	445	4.882	535	1.06	625	0.196
270	0.292	360	5.077	450	4.958	540	1.08	630	0.121
275	0.406	365	5.501	455	4.123	545	1.27	635	0.133
280	0.527	370	5.607	460	4.299	550	1.10	640	0.153
285	0.682	375	5.888	465	4.086	555	0.797	645	0.192
290	0.864	380	5.924	470	3.357	560	0.605	650	0.135

[a] Cross sections represent $\lambda \pm 2.5$ nm averages; the temperature dependence is minimal. Sources of data: Jenouvrier et al. (1996), Mérienne et al. (1995), Vandaele et al. (1998)

Comments: Nitrogen dioxide features a band system with complex structure ranging from ~275 to 900 nm. To display full details requires very high spectral resolution, too detailed for a complete listing of numerical values. Photodissociation begins near 420 nm, ~20 nm above the nominal threshold at 398 nm. A second absorption region below 250 nm consists of a continuum overlapped by well-defined bands. In this region, photodissociation leads at least partly to the formation of O (^{1}D) (threshold 244 nm).

Table 6.13b Quantum yield for the process $NO_2 + h\nu \rightarrow NO + O$ (^{3}P) at 298 and 248 K[a]

λ (nm)	ϕ_{298}	ϕ_{248}	λ (nm)	ϕ_{298}	ϕ_{248}	λ (nm)	ϕ_{298}	ϕ_{248}	λ (nm)	ϕ_{298}	ϕ_{248}
300–398	1.00	1.00	403	0.53	0.44	408	0.22	0.14	413	0.09	0.06
399	0.95	0.94	404	0.44	0.34	409	0.18	0.12	414	0.08	0.04
400	0.88	0.86	405	0.37	0.28	410	0.15	0.10	415	0.06	0.03
401	0.75	0.69	406	0.30	0.22	411	0.13	0.08	416	0.05	0.02
402	0.62	0.56	407	0.26	0.18	412	0.11	0.07	417	0.04	0.02

[a] From the critical assessment of data by Troe (2000)

Table 6.14a Absorption cross sections of nitrogen trioxide, NO_3 (298 K)[a]

λ (nm)	$10^{19}\,\sigma$ (cm^2)	λ (nm)	$10^{19}\,\sigma$ (cm^2)	λ (nm)	$10^{19}\,\sigma$ (cm^2)	λ (nm)	$10^{19}\,\sigma$ (cm^2)	λ (nm)	$10^{19}\,\sigma$ (cm^2)
410	0.1	488	10.3	566	30.9	613	28.6	652	6.6
412	0.5	490	11.2	568	30.9	614	27.7	653	7.7
414	0.2	492	10.8	570	30.3	615	24.5	654	8.9
416	0.7	494	11.0	572	29.8	616	22.7	655	10.6
418	0.5	496	13.1	574	31.0	617	22.9	656	14.4
420	0.9	498	13.0	576	35.5	618	25.9	657	18.6
422	1.0	500	12.3	578	35.9	619	27.8	658	26.3
424	1.0	502	12.0	580	36.3	620	35.5	659	44.2
426	1.5	504	13.7	582	35.6	621	56.9	660	80.9
428	1.3	506	14.5	583	31.8	622	110.5	661	157.2
430	1.8	508	13.8	584	30.6	623	159.9	662	228.0
432	1.6	510	16.4	585	31.4	624	130.8	663	189.4
434	2.0	512	19.2	586	36.0	625	91.0	664	122.6
436	1.6	514	17.1	587	45.2	626	79.3	665	80.5
438	2.3	516	16.9	588	54.7	627	81.7	666	53.9
440	2.1	518	15.6	589	66.5	628	80.0	667	33.0
442	2.3	520	18.2	590	64.7	629	75.8	668	20.6
444	2.1	522	20.9	591	59.1	630	73.4	669	13.6
446	2.6	524	17.8	592	55.5	631	52.5	670	10.3
448	2.6	526	17.7	593	49.7	632	35.8	671	8.6
450	3.1	528	22.8	594	45.5	633	23.6	672	8.2
452	3.6	530	24.2	595	46.6	634	17.8	673	7.0
454	3.9	532	21.9	596	50.2	635	15.6	674	5.6
456	3.8	534	22.1	597	47.3	636	18.3	675	5.2
458	4.0	536	27.9	598	39.8	637	22.5	676	5.3
460	4.3	538	25.4	599	33.7	638	22.0	677	6.4
462	4.3	540	22.8	600	30.0	639	17.1	678	8.1
464	5.2	542	20.4	601	31.1	640	13.4	679	8.5
466	5.9	544	18.5	602	36.0	641	10.9	680	7.5
468	6.1	546	26.3	603	41.3	642	10.0	681	5.8
470	6.4	548	32.4	604	47.4	643	10.5	682	4.3
472	7.0	550	26.9	605	47.3	644	10.3	683	3.3
474	6.7	552	26.8	606	36.0	645	9.3	684	2.8
476	8.5	554	30.2	607	26.1	646	8.1	685	1.9
478	7.9	556	35.4	608	20.1	647	7.6	686	1.7
480	7.6	558	38.1	609	18.6	648	6.7	687	1.3
482	7.7	560	36.0	610	19.2	649	5.9	688	1.3
484	8.4	562	31.5	611	20.7	650	5.4	689	1.3
486	9.9	564	29.5	612	24.2	651	6.0	690	1.1

[a]Source of data: Sander (1986), abbreviated

Comment: The band maxima increase significantly with decreasing temperature. The following formula was suggested by Orphal et al. (2003) to account for the effect:

$$\sigma(T)/\sigma(298\text{K}) = \left\{1 - \exp(-A/T) - 2\exp(-B/T)\right\} \Big/$$
$$\left\{1 - \exp(-A/298) - 2\exp(-B/298)\right\}$$

where A = 1096.4 and B = 529.5 are the vibrational energies divided by the Boltzmann constant.

Table 6.14b Quantum yields for the photodissociation of NO_3 at 298 and 230 K[a]

	298 K		230 K			298 K		230 K	
λ (nm)	ϕ_1	ϕ_2	ϕ_1	ϕ_2	λ (nm)	ϕ_1	ϕ_2	ϕ_1	ϕ_2
585	0.983	0.0	0.996	0.0	615	0.147	0.166	0.058	0.110
590	0.793	0.190	0.696	0.300	620	0.090	0.131	0.030	0.072
595	0.608	0.359	0.453	0.359	625	0.049	0.099	0.014	0.045
600	0.472	0.291	0.307	0.346	630	0.026	0.065	0.006	0.025
605	0.323	0.264	0.176	0.253	635	0.015	0.037	0.003	0.012
610	0.226	0.236	0.105	0.193	640	0.007	0.020	0.001	0.005

[a] For the two processes: $NO_3 + h\nu \rightarrow NO_2\ (^2A_1) + O\ (^3P)\ (\phi_1)$ and $NO_3 + h\nu \rightarrow NO\ (^2\Pi) + O_2\ (X^3\Sigma_g^-)$ (ϕ_2). Source of data: Johnston et al. (1996), abbreviated

Table 6.15 Absorption cross sections of N_2O (298 K)[a]

λ (nm)	$10^{20}\ \sigma$ (cm^2)	λ (nm)	$10^{20}\ \sigma$ (cm^2)	λ (nm)	$10^{20}\ \sigma$ (cm^2)	λ (nm)	$10^{21}\ \sigma$ (cm^2)	λ (nm)	$10^{22}\ \sigma$ (cm^2)
165	5.61	182	14.7	193	8.95	204	23.0	220	9.22
170	8.30	183	14.6	194	8.11	205	19.5	222	5.88
173	11.3	184	14.4	195	7.57	206	16.5	224	3.75
174	11.9	185	14.3	196	6.82	207	13.8	226	2.39
175	12.6	186	13.6	197	6.10	208	11.6	228	1.51
176	13.4	187	13.1	198	5.35	209	9.80	230	0.955
177	14.0	188	12.5	199	4.70	210	7.55	232	0.605
178	13.9	189	11.7	200	4.09	212	5.18	234	0.360
179	14.4	190	11.1	201	3.58	214	3.42	236	0.240
180	14.6	191	10.4	202	3.09	216	2.23	238	0.152
181	14.6	192	9.75	203	2.67	218	1.42	240	0.101

[a] Source of data: Hubrich and Stuhl (1980), Selwyn et al. (1977). Temperature dependence (for the range: 173–240 nm, 194–302 K): $\ln \sigma(\lambda, T) = \Sigma A_n \lambda^n + (T-300) \exp(\Sigma B_m \lambda^m)$, n = 0–4: $A_0 = 68.21023$, $A_1 = -4.071805$, $A_2 = 4.301146 \times 10^{-2}$, $A_3 = -1.777846 \times 10^{-4}$, $A_4 = 2.520672 \times 10^{-7}$; m = 0–3: $B_0 = 123.4014$, $B_1 = -2.116255$, $B_2 = 1.111572 \times 10^{-2}$, $B_3 = -1.881058 \times 10^{-5}$. The photodissociation process is: $N_2O + h\nu \rightarrow N_2 + O\ (^1D)$, $\phi = 1.0$

Table 6.16 Absorption cross sections of N_2O_5 (298 K)[a]

λ (nm)	$10^{19}\,\sigma$ (cm^2)	λ (nm)	$10^{20}\,\sigma$ (cm^2)	λ (nm)	$10^{21}\,\sigma$ (cm^2)	λ (nm)	$10^{21}\,\sigma$ (cm^2)	λ (nm)	$10^{22}\,\sigma$ (cm^2)
170	332	290	6.52	322	10.50	348	2.34	374	5.45
180	225	295	5.02	324	9.30	350	2.10	376	4.84
190	149	300	3.81	326	8.26	352	1.88	378	4.31
200	90.0	302	3.40	328	7.35	354	1.67	380	3.83
210	38.0	304	3.03	330	6.54	356	1.49	382	3.41
220	16.5	306	2.70	332	5.82	358	1.33	384	3.05
230	8.38	308	2.40	334	5.18	360	1.20	386	2.73
240	5.71	310	2.13	336	4.62	362	1.07	388	2.42
250	3.86	312	1.90	338	4.12	364	0.958	390	2.15
260	2.52	314	1.68	340	3.68	366	0.852	392	1.93
270	1.62	316	1.49	342	3.28	368	0.763	394	1.72
280	1.05	318	1.33	344	2.93	370	0.685	396	1.50
285	8.34	320	1.18	346	2.62	372	0.613	398	1.34

[a] Source of data: Harwood et al. (1998), 210–398 nm, Osborne et al. (2000), 170–200 nm

Comments: The maximum of absorption occurs at $\lambda \sim 160$ nm. In the wavelength region covered the absorption cross section is a smoothly varying function of λ. At wavelengths <280 nm the temperature dependence is insignificant. Yao et al. (1982) have provided a simple formula for the temperature effect:

$$\ln(10^{19}\sigma) = 0.432537 + (4728.48 - 17.1269\lambda)/\,T$$

covering the wavelength region 280–380 nm and the temperature interval 225–300 K. Harwood et al. (1993) have presented a different algorithm for calculating the temperature dependence, although their data are in good agreement with those of Yao et al. (1982).

Quantum yield: $N_2O_5 + h\nu \rightarrow NO_3 + NO_2$ (ϕ_1), $N_2O_5 + h\nu \rightarrow NO_3 + NO + O$ (ϕ_2), $\lambda >$ 300 nm: $\phi_1 \approx 1$, $\lambda < 300$ nm: ϕ_1 decreases toward ϕ_1 0.7–0.8 at 248 nm while ϕ_2 increases from $\phi_2 = 0.15$ at 289 nm toward 0.72 at 248 nm.

Table 6.17 Absorption cross sections of nitrous acid, HNO_2 (298 K)[a]

λ (nm)	$10^{20}\,\sigma$ (cm²)	λ (nm)	$10^{20}\,\sigma$ (cm²)	λ (nm)	$10^{20}\,\sigma$ (cm²)	λ (nm)	$10^{21}\,\sigma$ (cm²)	λ (nm)	$10^{22}\,\sigma$ (cm²)
296	0.326	321	5.96	339.5	9.96	355	27.64	372	7.96
297	0.565	322	4.05	340	7.79	355.5	16.40	373	6.30
298	0.517	323	4.56	340.5	8.51	356	11.13	374	4.57
299	0.429	324	5.89	341	16.13	357	9.45	375	3.55
300	0.617	325	4.05	341.5	31.52	357.5	10.08	376	3.36
301	0.690	326	2.65	342	29.40	358	9.84	377	3.66
302	0.579	326.5	3.55	342.5	18.47	359	8.37	378	4.33
303	0.925	327	6.44	343	11.43	360	6.87	379	5.66
304	1.04	327.5	10.26	343.5	8.29	361	6.05	380	7.21
305	1.57	328	9.22	344	7.59	362	5.98	381	9.13
306	1.29	328.5	6.38	345	8.77	363	7.39	382	12.44
307	0.916	329	5.20	346	9.64	364	11.49	383	17.03
308	1.45	330	9.92	347	7.80	364.5	12.71	384	19.47
309	2.01	330.5	15.06	348	6.63	365	12.82	385	16.09
310	1.51	331	14.32	349	6.00	365.5	13.19	386	10.52
311	2.07	331.5	9.88	350	9.06	366	14.84	387	6.59
312	2.42	332	6.94	350.5	14.95	366.5	18.43	388	4.30
313	2.25	333	6.31	351	16.94	367	25.08	389	2.81
314	3.35	334	8.35	351.5	14.07	367.5	35.18	390	1.71
315	2.54	335	7.71	352	12.42	368	43.56	391	0.992
316	1.61	336	5.33	352.5	12.81	368.5	41.37	392	0.731
317	3.21	337	4.23	353	16.34	369	31.45	393	0.597
318	4.49	338	9.38	353.5	28.49	369.5	21.72	394	0.528
319	3.19	338.5	16.52	354	48.73	370	15.05	395	0.403
320	4.66	339	14.32	354.5	44.34	371	9.49	396	0.237

[a]Source of data: Stutz et al. (2000). The photodissociation process $HNO_2 + h\nu \rightarrow NO + OH$ occurs with quantum yield $\phi \approx 1.0$

Table 6.18 Absorption cross sections of H_2O_2, HO_2NO_2 and HNO_3[a]

λ (nm)	H_2O_2 $10^{20}\,\sigma$ (cm²)	HO_2NO_2 $10^{20}\,\sigma$ (cm²)	HNO_3 $10^{20}\,\sigma$ (cm²)	$10^3\,B$ (K⁻¹)	λ (nm)	H_2O_2 $10^{20}\,\sigma$ (cm²)	HO_2NO_2 $10^{20}\,\sigma$ (cm²)	HNO_3 $10^{20}\,\sigma$ (cm²)	$10^3\,B$ (K⁻¹)
190	67.2	1,010	1,360	0	235	15.0	68.0	3.75	1.93
195	56.3	816	1,016	0	240	12.4	57.9	2.58	1.97
200	47.5	563	588	1.66	245	10.2	49.7	2.11	1.68
205	40.8	367	288	1.75	250	8.3	41.1	1.97	1.34
210	35.7	239	104	1.97	255	6.7	34.9	1.95	1.16
215	30.7	161	36.5	2.17	260	5.3	28.4	1.91	1.14
220	25.8	118	14.9	2.15	265	4.2	22.9	1.80	1.20
225	21.7	93.2	8.81	1.90	270	3.3	18.0	1.62	1.45
230	18.2	78.8	5.78	1.80	275	2.6	13.3	1.38	1.60

(continued)

Table 6.18 (continued)

λ (nm)	H_2O_2 $10^{20}\,\sigma$ (cm^2)	HO_2NO_2 $10^{20}\,\sigma$ (cm^2)	HNO_3 $10^{20}\,\sigma$ (cm^2)	10^3 B (K^{-1})	λ (nm)	H_2O_2 $10^{20}\,\sigma$ (cm^2)	HO_2NO_2 $10^{20}\,\sigma$ (cm^2)	HNO_3 $10^{20}\,\sigma$ (cm^2)	10^3 B (K^{-1})
280	2.0	9.3	1.12	1.78	320	0.22	0.24	0.020	6.45
285	1.5	6.2	0.858	1.99	325	0.16	0.15	0.0095	7.35
290	1.2	3.9	0.615	2.27	330	0.13	0.09	0.0043	9.75
295	0.90	2.4	0.412	2.61	335	0.10	–	0.0022	10.1
300	0.68	1.4	0.263	3.10	340	0.07	–	0.0010	11.8
305	0.51	0.85	0.150	3.64	345	0.05	–	0.0006	11.2
310	0.39	0.53	0.081	4.23	350	0.04	–	0.0004	9.30
315	0.29	0.39	0.041	5.20					

[a] Absorption cross sections at 298 K; the temperature dependence for HNO_3 is expressed by $\sigma(\lambda, T) = \sigma(\lambda, 298)\exp(B(\lambda)(T\text{-}298))$, with $B(\lambda)$ shown next to the cross sections. Source of data: Sander et al. (2006) (review)

Quantum yields (for the predominant photodissociation processes)

(a) $H_2O_2 + h\nu \rightarrow 2\ OH$ (ϕ_1), $H_2O_2 + h\nu \rightarrow H + HO_2$ (ϕ_2), $\lambda > 230$ nm: $\phi_1 \approx 1$, $\phi_2 = 0$; $\lambda = 193$ nm: $\phi_2 = 0.15$ $(\phi_1 = 0.85$ by difference).

(b) $HO_2NO_2 + h\nu \rightarrow HO_2 + NO_2$ (ϕ_1), $HO_2NO_2 + h\nu \rightarrow OH + NO_3$ (ϕ_2), $\lambda > 248$ nm: $\phi_1 \approx 0.9$, $\phi_2 = 0.08$; $\lambda = 193$ nm: $\phi_1 \approx 0.56$, $\phi_2 = 0.35$ (Jimenez et al. 2005);

(c) $HNO_3 + h\nu \rightarrow OH + NO_2$ (ϕ_1), $HNO_3 + h\nu \rightarrow O + NO_3$ (ϕ_2), $\lambda = 248$ nm: $\phi_1 = 0.95$, $\phi_2 = 0.03$; $\lambda = 222$ nm: $\phi_1 = 0.90$; $\lambda = 193$ nm: $\phi_1 = 0.33$, $\phi_2 = 0.67$ (Sander et al. 2006).

Table 6.19a Absorption cross sections of formaldehyde[a]

λ (nm)	$10^{20}\,\sigma$ (cm^2)	λ (nm)	$10^{20}\,\sigma$ (cm^2)	λ (nm)	$10^{20}\,\sigma$ (cm^2)	λ (nm)	$10^{20}\,\sigma$ (cm^2)
240	0.078	255	0.450	270	0.963	285	4.050
241	0.078	256	0.628	271	1.941	286	2.095
242	0.123	257	0.443	272	1.430	287	1.153
243	0.159	258	0.307	273	0.811	288	3.169
244	0.110	259	0.617	274	0.658	289	3.225
245	0.131	260	0.605	275	2.143	290	1.173
246	0.163	261	0.659	276	2–584	291	1.836
247	0.151	262	0.603	277	1.573	292	0.797
248	0.234	263	1.077	278	1.035	293	3.128
249	0.318	264	0.947	279	2.451	294	7.154
250	0.257	265	0.531	280	23.38	295	4.054
251	0.204	266	0.539	281	1.562	296	2.474
252	0.337	267	1.360	282	0.973	297	1.467
253	0.289	268	1.243	283	0.722	298	4.217
254	0.342	269	0.991	284	4.265	299	3.175

(continued)

Table 6.19a (continued)

λ (nm)	$10^{20}\,\sigma\,(cm^2)$	λ (nm)	$10^{20}\,\sigma\,(cm^2)$	λ (nm)	$10^{20}\,\sigma\,(cm^2)$	λ (nm)	$10^{20}\,\sigma\,(cm^2)$
300	0.964	319	0.978	338	1.919	357	0.0345
301	1.625	320	1.194	339	5.381	358	0.0186
302	0.854	321	1.598	340	3.151	359	0.0111
303	3.021	322	0.722	341	0.978	360	0.0087
304	7.219	323	0.328	342	0.509	361	0.0100
305	4.752	324	0.858	343	1.922	362	0.0211
306	4.292	325	1.578	344	1.268	363	0.0141
307	1.781	326	6.876	345	0.437	364	0.0094
308	1.385	327	4.370	346	0.119	365	0.0088
309	3.252	328	1.220	347	0.044	366	0.0085
310	1.737	329	3.120	348	0.075	367	0.0091
311	0.462	330	3.865	349	0.038	368	0.0142
312	1.188	331	1.412	350	0.036	369	0.0297
313	0.964	332	0.347	351	0.089	370	0.0635
314	5.637	333	0.214	352	0.729	371	0.0571
315	5.565	334	0.159	353	2.275	372	0.0198
316	2.561	335	0.097	354	1.645	373	0.0113
317	5.777	336	0.126	355	0.696	374	0.0091
318	3.151	337	0.383	356	0.148	375	0.0087

[a]Temperature 298 K; Source of data Meller and Moortgat (2000)

Product channels are: $HCHO + h\nu \rightarrow H + CHO$ (ϕ_1), $HCHO + h\nu \rightarrow H_2 + CO$ (ϕ_2). Quantum yields vary with wavelength. They are independent of temperature and pressure at wavelengths $\lambda < 330$ nm, but show Stern-Volmer pressure dependence at $\lambda > 330$ nm.

Table 6.19b Quantum yields of HCHO in air at 1 atm (101,325 Pa) and 298 K[a]

λ	ϕ_1	ϕ_2	λ	ϕ_1	ϕ_2	λ	ϕ_1	ϕ_2	λ	ϕ_1	ϕ_2
250	0.31	0.49	280	0.56	0.34	310	0.75	0.25	340	0	0.66
255	0.30	0.50	285	0.64	0.30	315	0.72	0.28	345	0	0.51
260	0.31	0.49	290	0.68	0.29	320	0.61	0.40	350	0	0.38
265	0.32	0.48	295	0.75	0.26	325	0.46	0.64	355	0	0.23
270	0.38	0.45	300	0.77	0.23	330	0.30	0.67	360	0	0.07
275	0.45	0.39	305	0.78	0.22	335	0.14	0.68	365	0	~0

[a]Wavelength λ (nm); quantum yields derived by interpolation of data reported by Moortgat et al. (1983) and earlier studies cited therein

The observed pressure dependence is $(\phi_2(\lambda, n, T))^{-1} = (\phi_{CO}(\lambda, n, T))^{-1} = 1 + \alpha(\lambda, T)\,n$, where n (molecule cm^{-3}) is the number density of air; $\alpha(\lambda, T)$ (10^{-19} cm^3 molecule^{-1}) is the quenching coefficient:

$$
\begin{array}{ccc}
 & \alpha(298K) & \alpha(220K) \\
\lambda = 339\,[nm] & 0.26 \pm 0.10 & 0.39 \pm 0.07 \\
\lambda = 353\,[nm] & 1.12 \pm 0.17 & 2.47 \pm 0.59
\end{array}
$$

Table 6.20 Absorption cross sections of acetaldehyde, CH_3CHO at 298 K[a]

λ (nm)	$10^{20}\,\sigma$ (cm^2)	λ (nm)	$10^{20}\,\sigma$ (cm^2)	λ (nm)	$10^{20}\,\sigma$ (cm^2)	λ (nm)	$10^{20}\,\sigma$ (cm^2)	λ (nm)	$10^{20}\,\sigma$ (cm^2)
210	0.049	295	4.27	309	3.14	323	1.24	337	0.222
220	0.059	296	4.24	310	2.93	324	1.09	338	0.205
230	0.151	297	4.38	311	2.76	325	1.14	339	0.219
240	0.469	298	4.41	312	2.53	326	1.07	340	0.150
250	1.13	299	4.26	313	2.47	327	0.858	341	0.074
260	2.22	300	4.16	314	2.44	328	0.747	342	0.042
270	3.42	301	3.99	315	2.20	329	0.707	343	0.031
280	4.50	302	3.86	316	2.04	330	0.688	344	0.026
285	4.49	303	3.72	317	2.07	331	0.588	345	0.021
290	4.89	304	3.48	318	1.98	332	0.530	346	0.019
291	4.78	305	3.42	319	1.87	333	0.398	347	0.015
292	4.68	306	3.42	320	1.72	334	0.363	348	0.016
293	4.53	307	3.36	321	1.48	335	0.350	349	0.010
294	4.33	308	3.33	322	1.40	336	0.238	350	0.008

[a]Source of data: Martinez et al. (1992)

Quantum yields: The main processes are (1) $CH_3CHO + h\nu \rightarrow CH_3 + CHO$ (ϕ_1), which is active in the entire wavelength region, and (2) $CH_3CHO + h\nu \rightarrow CH_4 + CO$ (ϕ_2), which occurs at wavelengths below $\lambda \approx 290$ nm so that it is inactive in the troposphere.

Table 6.20a Acetaldehyde quantum yields in air at 1 atm (101,325 Pa) and 295 K[a]

λ (nm)	290	295	300	305	310	315	320	325	330
ϕ_{1atm}	0.63	0.57	0.50	0.42	0.32	0.20	0.09	0.04	0.01

[a]Source of data: Moortgat et al. (2010); quantum yields increase with decreasing pressure in accordance with the Stern-Volmer relation and may be estimated for any reduced pressure p(atm) from $\phi_1 = (1 + kp)^{-1}$ with k (atm^{-1}) = $(1 - \phi_{1atm})/\phi_{1atm}$

Table 6.21 Absorption cross sections of acetone, CH_3COCH_3, at 295 and 230 K[a]

λ (nm)	$10^{20}\,\sigma$ (cm^2) 295 K	230 K	λ (nm)	$10^{20}\,\sigma$ (cm^2) 295 K	230 K	λ (nm)	$10^{20}\,\sigma$ (cm^2) 295 K	230 K	λ (nm)	$10^{21}\,\sigma$ (cm^2) 295 K	230 K
215	0.167	0.169	255	3.15	3.17	291	3.95	3.64	299	2.82	2.49
220	0.246	0.250	260	3.81	3.81	292	3.82	3.50	300	2.67	2.35
225	0.380	0.390	265	4.41	4.37	293	3.71	3.39	301	2.58	2.26
230	0.584	0.599	270	4.79	4.71	294	3.57	3.25	302	2.45	2.14
235	0.885	0.906	275	4.94	4.78	295	3.42	3.10	303	2.30	2.01
240	1.30	1.33	280	4.91	4.706	296	3.26	2.93	304	2.18	1.89
245	1.83	1.86	285	4.54	4.26	297	3.11	2.76	305	2.05	1.77
250	2.47	2.50	290	4.06	3.75	298	2.98	2.65	306	1.89	1.62

(continued)

Table 6.21 (continued)

	$10^{20}\ \sigma\,(cm^2)$			$10^{20}\ \sigma\,(cm^2)$			$10^{20}\ \sigma\,(cm^2)$			$10^{21}\ \sigma\,(cm^2)$	
λ (nm)	295 K	230 K	λ (nm)	295 K	230 K	λ (nm)	295 K	230 K	λ (nm)	295 K	230 K
307	1.75	1.48	318	0.598	0.451	329	0.0913	0.0423	340	0.0912	0.0516
308	1.61	1.35	319	0.523	0.385	330	0.0740	0.0331	341	0.0729	0.0418
309	1.49	1.23	320	0.455	0.324	331	0.0586	0.0258	342	0.0583	0.0331
310	1.36	1.17	321	0.411	0.280	332	0.465	0.204	343	0.0494	0.0275
311	1.24	1.00	322	0.348	0.227	333	0.375	0.165	344	0.0365	0.0194
312	1.14	0.916	323	0.294	0.183	334	0.311	0.139	345	0.0301	0.0156
313	1.06	0.846	324	0.248	0.147	335	0.248	0.114	346	0.0235	0.0124
314	0.944	0.749	325	0.210	0.118	336	0.199	0.0953	347	0.0158	0.0089
315	0.837	0.658	326	0.174	0.0929	337	0.162	0.0812	348	0.0111	0.0069
316	0.760	0.591	327	0.141	0.0718	338	0.135	0.0709	349	0.0107	0.0069
317	0.684	0.525	328	0.113	0.0549	339	0.113	0.0618			

[a] Source of data: Gierczak et al. (1998)

Quantum yields: The main processes are (1) $CH_3COCH_3 + h\nu \rightarrow CH_3 + CH_3CO$ (ϕ_1) and (2) $CH_3COCH_3 + h\nu \rightarrow 2CH_3 + CO$ (ϕ_2); under tropospheric conditions, i.e. $\lambda > 290$ nm, $\phi_2 = \phi_{CO} \leq 0.05$, so that it may be neglected to a first approximation. Warneck (2001) has derived an expression for the dependence of ϕ_1 on wavelength (nm) and air number density n_M (molecule cm^{-3}): $\phi_1 = \{(1/\Phi s) + 7.145 \times 10^{-7}\ n_M\ exp$ $(-8780.6/\lambda)\}^{-1}$, where $\Phi s = 0.113 + 0.887\ [1 + exp\ (\lambda-307.5)/3.0)]^{-1}$, $(T \approx 295$ K). Blitz et al. (2004), who have studied the temperature dependence of the quantum yield, prefer different formulae.

Table 6.22 Absorption cross sections of methyl iodide[a]

λ (nm)	$10^{20}\ \sigma$ (cm^2)	$10^3\ B$ (K^{-1})	ϕ_2	λ (nm)	$10^{20}\ \sigma$ (cm^2)	$10^3\ B$ (K^{-1})	ϕ_2	λ (nm)	$10^{21}\ \sigma$ (cm^2)	$10^3\ B$ (K^{-1})	ϕ_2
235	18.66	0.67	0.69	280	26.28	2.43	0.69	325	1.12	6.79	0.47
240	35.18	0.61	0.71	285	14.18	3.74	0.62	330	0.64	7.82	0.72
245	60.56	0.34	0.73	290	7.351	4.98	0.57	335	0.36	9.34	0.94
250	88.46	0.08	0.75	295	3.736	6.38	0.52	340	0.21	10.95	
255	106.4	0.08	0.76	300	1.946	6.97	0.47	345	0.11	13.58	
260	107.2	−0.10	0.78	305	1.029	6.84	0.43	350	0.059	16.83	
265	92.03	−0.12	0.79	310	0.579	6.78	0.41	355	0.032	18.91	
270	68.14	0.54	0.76	315	0.334	6.75	0.40	360	0.019	17.28	
275	44.76	1.33	0.72	320	0.192	6.53	0.41	365	0.009	23.63	

[a] Absorption cross sections at 298 K; the temperature dependence is expressed by $\ln\ \sigma(\lambda,\ T) = \ln\ \sigma(\lambda,\ 298) + B(\lambda)\ (T - 298)$, with $B(\lambda)$ shown next to the cross sections. Source of data: Rattigan et al. (1997). ϕ_2 is the quantum yield for the formation of I $(^2P_{1/2})$ atoms

Quantum yields: Photodissociation produces iodine atoms in the $^2P_{3/2}$ ground state as well as in the metastable $^2P_{1/2}$ state, which lies ~ 0.943 eV above the ground state: $CH_3I + h\nu \rightarrow CH_3 + I(^2P_{3/2})$ (ϕ_1), $CH_3I + h\nu \rightarrow CH_3 + I(^2P_{1/2})$ (ϕ_2). The total quantum yield is unity ($\phi_1 + \phi_2 = 1$). Values of ϕ_2 listed in Table 6.22 were derived by interpolation of quantum yields measured in the wavelength range 222–333.5 nm by Kang et al. (1996), Ogorzalek et al. (1989) and Uma and Das (1994). The strong increase of ϕ_2 at wavelengths approaching the onset of absorption suggests that in this wavelength region $\phi_2 \approx 1$.

Table 6.23 Absorption cross sections of carbonyl sulfide, OCS, at 295 and 225 K[a]

λ (nm)	$10^{20}\,\sigma\,(cm^2)$ 295 K	225 K	λ (nm)	$10^{20}\,\sigma\,(cm^2)$ 295 K	225 K	λ (nm)	$10^{21}\,\sigma\,(cm^2)$ 295 K	225 K	λ (nm)	$10^{22}\,\sigma\,(cm^2)$ 295 K	225 K
185	27.8	21.2	214	22.2	22.2	246	31.8	20.5	276	1.25	0.561
186	18.8	12.5	215	23.9	23.1	248	22.8	14.0	277	1.01	0.396
187	10.5	7.02	216	26.1	25.3	250	16.1	9.74	278	0.721	0.281
188	7.73	5.09	218	28.3	26.9	252	11.1	6.13	279	0.831	0.250
189	4.39	2.76	220	31.9	30.2	254	7.72	4.06	280	0.615	0.238
190	3.41	2.35	222	31.5	29.4	255	6.42	3.32	281	0.429	0.144
191	2.37	1.62	224	31.7	28.9	256	5.38	2.71	282	0.318	0.116
192	1.97	1.58	225	31.2	28.4	258	3.65	1.71	283	0.315	0.131
194	1.81	1.64	226	30.6	27.9	260	2.41	1.11	284	0.298	0.099
195	1.92	1.81	228	27.8	24.6	262	1.62	0.730	285	0.181	0.061
196	2.16	2.10	230	24.6	21.5	264	1.13	0.493	286	0.141	0.062
198	2.85	2.82	232	21.0	17.6	265	0.12	0.403	287	0.154	0.068
200	3.84	3.83	234	16.9	13.8	266	0.77	0.339	288	0.128	0.048
202	5.13	5.15	235	15.1	12.0	268	0.52	0.214	289	0.076	0.034
204	7.02	7.06	236	13.6	10.7	270	0.369	0.152	290	0.057	0.031
205	7.96	8.02	238	10.5	7.96	271	0.297	0.116	291	0.062	0.036
206	9.39	9.48	240	8.00	5.78	272	0.250	0.106	292	0.061	0.024
208	12.0	12.1	242	6.02	4.20	273	0.222	0.0921	293	0.032	0.015
210	15.0	15.1	244	4.35	2.88	274	0.168	0.0630	294	0.019	0.011
212	18.1	18.1	245	3.76	2.46	275	0.136	0.0526	295	0.029	0.015

[a] Source of data: Molina et al. (1981). Carbonyl sulfide features a broad absorption continuum with superimposed band structure at wavelengths $\lambda > 275$ nm and at the absorption maximum near 225 nm

Quantum yields: Photodissociation processes are (1) $OCS + h\nu \rightarrow CO + S$ (^3P) (φ_1) and (2) $OCS + h\nu \rightarrow CO + S$ (^1D) (φ_2). The second process is dominant; $\varphi_1 + \varphi_2 = 1.0$ (Zhao et al. 1995), $\varphi_1 \approx 0.05$ at 220 nm (Nan et al. 1993).

Table 6.24 Absorption cross sections of carbon tetrachloride, CCl_4, at 295 and 210 K[a]

λ (nm)	$10^{18}\,\sigma\,(cm^2)$ 295 K	210 K	λ (nm)	$10^{18}\,\sigma\,(cm^2)$ 295 K	210 K	λ (nm)	$10^{18}\,\sigma\,(cm^2)$ 295 K	210 K	λ (nm)	$10^{18}\,\sigma\,(cm^2)$ 295 K	210 K
174	9.90	9.90	180	7.20	7.20	186	3.10	3.10	192	0.992	0.992
176	10.10	10.10	182	5.90	5.90	188	1.98	1.98	194	0.767	0.767
178	9.75	9.75	184	4.40	4.40	190	1.469	1.469	196	0.695	0.695

(continued)

Table 6.24 (continued)

λ (nm)	10^{19} σ (cm^2) 295 K	210 K	λ (nm)	10^{19} σ (cm^2) 295 K	210 K	λ (nm)	10^{20} σ (cm^2) 295 K	210 K	λ (nm)	10^{20} σ (cm^2) 295 K	210 K
198	6.80	6.80	212	4.10	3.48	226	7.60	4.45	240	0.83	0.342
200	6.60	6.60	214	3.45	2.79	228	5.65	3.16	242	0.59	0.234
202	6.38	6.38	216	2.78	2.17	230	4.28	2.27	244	0.413	0.258
204	6.10	6.01	218	2.21	1.63	232	3.04	1.52	246	0.29	0.108
206	5.70	5.44	220	1.75	1.25	234	2.20	1.05	248	0.21	0.0762
208	5.25	4.83	222	1.36	0.900	236	1.60	0.723	250	0.148	0.0528
210	4.69	4.15	224	1.02	0.640	238	1.16	0.50			

[a] Data source: Simon et al. (1988)

Comments: Simon et al. (1988) have approximated wavelength and temperature dependence of the cross section by means of the expression $\log_{10} \sigma\,(\lambda,\,T) = \Sigma\,A_n\lambda^n + (T\text{-}273)\,\Sigma\,B_n\lambda^n$ ($n = 0\text{-}4$), covering the wavelength region 194–250 nm and the temperature range 210–300 K, with the following parameters

$A_0 = -37.104$, $A_1 = -5.8218 \times 10^{-1}$, $A_2 = 9.9974 \times 10^{-3}$, $A_3 = -4.6765 \times 10^{-5}$, $A_4 = 6.8501 \times 10^{-8}$; $B_0 = 1.0739$, $B_1 = -1.6275 \times 10^{-2}$, $B_2 = 8.8141 \times 10^{-5}$, $B_3 = -1.9811 \times 10^{-7}$, $B_4 = 1.5022 \times 10^{-10}$.

Quantum yields: The principal photolytic processes are (1) $CCl_4 + h\nu \rightarrow CCl_3 + Cl$ (φ_1) and (2) $CCl_4 + h\nu \rightarrow CCl_2 + 2\,Cl$ (φ_2), with $\varphi_1 + \varphi_2 \approx 1.0$. At the long-wavelength limit $\varphi_2 \approx 0$, but φ_2 increases at the expense of φ_1 until near 165 nm both processes occur with essentially equal probability (Rebbert and Ausloos 1977).

Quantum yields for $CFCl_3$ and CF_2Cl_2 behave similarly (Rebbert and Ausloos 1975) (absorption cross sections for these two compounds are given in the next table).

Table 6.25 Absorption cross sections of $CFCl_3$ and CF_2Cl_2 at 295 and 210 K[a]

λ (nm)	10^{19} σ (cm^2 molecule^{-1}) CFCl$_3$ 295 K	210 K	CF$_2$Cl$_2$ 295 K	210 K	λ (nm)	10^{20} σ (cm^2 molecule^{-1}) CFCl$_3$ 295 K	210 K	CF$_2$Cl$_2$ 295 K	210 K
174	31.30	31.30	16.20	16.20	204	37.4	30.0	0.344	0.169
176	32.40	32.40	18.10	18.10	206	28.0	21.6	0.209	0.989
178	32.35	32.35	18.70	18.70	208	19.7	14.9	0.127	0.557
180	31.40	31.40	17.90	17.90	210	14.8	9.94	0.759	0.324
182	29.60	29.60	16.00	16.00	212	10.5	6.63	0.454	0.183
184	27.20	27.20	13.40	13.40	214	7.56	4.31	0.271	0.106
186	24.30	23.00	10.70	10.70	216	5.38	2.78	0.158	0.0577
188	21.30	20.20	8.28	7.93	218	3.79	1.77	0.100	0.0330
190	17.90	17.05	6.32	5.29	220	2.64	1.13	0.060	0.0180
192	15.40	14.12	4.55	3.58	222	1.82	0.714	0.036	0.0105
194	12.43	11.51	3.15	2.28	224	1.24	0.454	0.022	0.0060
196	9.91	9.05	2.11	1.44	226	0.842	0.291	0.013	0.0034
198	7.80	7.18	1.39	0.882	228	0.565	0.188		
200	6.45	5.58	0.889	0.511	230	0.375	0.123		
202	5.00	4.20	0.551	0.297					

[a] Source of data: Simon et al. (1988). For quantum yields see preceding entry

Comments: Mérienne et al. (1990) have studied the spectra of $CFCl_3$ and CF_2Cl_2 in the wavelength region 200–238 nm and 200–231 nm, respectively, and found good agreement with the data of Simon et al. (1988) except at long wavelengths where deviations up to 15% occurred. Simon et al. (1988) have used a formula for wavelength and temperature dependence of the cross sections similar to that for carbon tetrachloride, but the published parameters contain numerical errors. Mérienne et al. (1990) have used the expression $\ln \sigma(\lambda, T) = \Sigma A_n(200 - \lambda)^n + (T - 296) \times \Sigma B_n(200 - \lambda)^n$ $(n = 0 - 3, 220 < T \text{ (K)} < 296)$, which represents their data quite accurately. $CFCl_3$: $A_0 = -41.925548$, $A_1 = -1.142857 \times 10^{-1}$, $A_2 = -3.12034 \times 10^{-3}$, $A_3 = 3.6699 \times 10^{-5}$; $B_0 = 3.58977 \times 10^{-4}$, $B_1 = 3.02973 \times 10^{-4}$, $B_2 = -1.13 \times 10^{-8}$, $B_3 = 0$; CF_2Cl_2: $A_0 = -43.8954569$, $A_1 = -2.403597 \times 10^{-1}$, $A_2 = -4.2619 \times 10^{-4}$, $A_3 = 9.8743 \times 10^{-6}$; $B_0 = 4.8438 \times 10^{-3}$, $B_1 = 4.96145 \times 10^{-4}$, $B_2 = -5.6953 \times 10^{-6}$, $B_3 = 0$.

Table 6.26a Absorption cross sections of chlorine nitrate, $ClONO_2$, at 296 and 220 K[a]

λ (nm)	$10^{18} \sigma$ (cm²) 296 K	220 K	λ (nm)	$10^{19} \sigma$ (cm²) 296 K	220 K	λ (nm)	$10^{21} \sigma$ (cm²) 296 K	220 K	λ (nm)	$10^{21} \sigma$ (cm²) 296 K	220 K
200	2.82	2.70	260	3.38	2.91	320	8.31	5.78	380	1.210	1.044
205	2.84	2.78	265	2.65	2.26	325	6.13	4.26	385	1.060	0.907
210	3.14	3.15	270	2.05	1.73	330	4.66	3.29	390	0.909	0.773
215	3.42	3.47	275	1.58	1.32	335	3.61	2.67	395	0.760	0.633
220	3.32	3.35	280	1.19	0.983	340	3.02	2.30	400	0.638	0.519
225	2.77	2.74	285	0.881	0.718	345	2.58	2.04	405	0.541	0.430
230	2.08	2.00	290	0.641	0.515	350	2.29	1.86	410	0.444	0.341
235	1.98	1.86	295	0.438	0.345	355	2.08	1.73	415	0.345	0.270
240	1.05	0.967	300	0.313	0.240	360	2.00	1.69	420	0.316	0.226
245	0.764	0.691	305	0.224	0.167	365	1.80	1.53	425	0.232	0.164
250	0.560	0.497	310	0.160	0.116	370	1.59	1.36	430	0.189	0.131
255	0.432	0.377	315	0.115	0.081	375	1.41	1.21			

[a]Source of data : Burkholder et al. (1994); the authors presented wavelength-dependent parameters for the calculation of absorption coefficients from a second order polynomial in the temperature range 220–396 K

Quantum yields: The major processes at wavelengths above 220 nm are:

(1) $ClONO_2 + h\nu \rightarrow Cl + NO_3$ (φ_1); (2) $ClONO_2 + h\nu \rightarrow ClO + NO_2$ (φ_2).

Table 6.26b Quantum yields for the photolysis of $ClONO2$[a]

λ	ϕ_1	ϕ_2	λ	ϕ_1	ϕ_2	λ	ϕ_1	ϕ_2	λ	ϕ_1	ϕ_2
240	0.48	0.52	270	0.57	0.43	300	0.68	0.32	330	0.80	0.20
250	0.50	0.50	280	0.61	0.39	310	0.72	0.28	340	0.85	0.15
260	0.54	0.46	290	0.64	0.36	320	0.76	0.24	350	0.90	0.10

[a] Wavelength λ (nm); quantum yields derived by interpolation of data summarized by Sander et al. (2006). At 193 and 220 nm the production of oxygen atoms is observed possibly indicating the occurrence of a third photolytic process

Table 6.27 Absorption cross sections of the chlorine oxide dimer, ClOOCl, at ~210 K[a]

λ (nm)	$10^{20}\,\sigma$ (cm^2)	λ (nm)	$10^{20}\,\sigma$ (cm^2)	λ (nm)	$10^{20}\,\sigma$ (cm^2)	λ (nm)	$10^{20}\,\sigma$ (cm^2)	λ (nm)	$10^{20}\,\sigma$ (cm^2)
190	565	235	510	280	175	325	20	370	3.2
195	470	240	636	285	142	330	16	380	2.1
200	379	245	680	290	114	335	13	390	1.3
205	297	250	637	295	91	340	12	400	0.85
210	229	255	531	300	71	345	9.5	410	0.54
215	188	260	414	305	55	350	7.7	420	0.35
220	188	265	321	310	42	355	6.3	430	0.23
225	287	270	256	315	32	360	5.0	440	0.15
230	362	275	210	320	25	365	3.9	450	0.09

[a] Source of data: DeMore and Tschuikow-Roux (1990); $\lambda > 360$ nm, Sander et al. (2006). The predominant photolytic process ClOOCl + $h\nu \rightarrow$ Cl + ClOO occurs with essentially unity quantum yield ($\phi_{Cl} \approx 1$)

Table 6.28 Absorption cross sections of hypochlorous acid, HOCl, at 298 K[a]

λ (nm)	$10^{19}\,\sigma$ (cm^2)	λ (nm)	$10^{19}\,\sigma$ (cm^2)	λ (nm)	$10^{19}\,\sigma$ (cm^2)	λ (nm)	$10^{20}\,\sigma$ (cm^2)	λ (nm)	$10^{20}\,\sigma$ (cm^2)
200	0.72	245	1.94	290	0.513	335	2.81	380	0.708
205	0.56	250	1.73	295	0.561	340	2.22	385	0.602
210	0.55	255	1.40	300	0.600	345	1.76	390	0.491
215	0.71	260	1.09	305	0.612	350	1.43	395	0.384
220	1.02	265	0.821	310	0.597	355	1.20	400	0.288
225	1.39	270	0.625	315	0.555	360	1.06	405	0.182
230	1.75	275	0.510	320	0.495	365	0.968	410	0.144
235	1.99	280	0.464	325	0.424	370	0.888	415	0.097
240	2.07	285	0.473	330	0.350	375	0.804	420	0.063

[a] Source of data: Burkholder (1993), Barnes et al. (1998). The dominant photolytic process is HOCl + $h\nu \rightarrow$ OH + Cl occurring with unity quantum yield at wavelengths $\lambda > 200$ nm (as reviewed by Sander et al. (2006)

Table 6.29 Absorption cross sections of chlorine, Cl$_2$, at 295 K[a]

λ (nm)	$10^{20}\,\sigma$ (cm^2)	λ (nm)	$10^{20}\,\sigma$ (cm^2)	λ (nm)	$10^{20}\,\sigma$ (cm^2)	λ (nm)	$10^{20}\,\sigma$ (cm^2)	λ (nm)	$10^{21}\,\sigma$ (cm^2)
260	0.917	320	23.71	380	5.00	440	0.546	500	0.283
270	0.824	330	25.55	390	2.94	450	0.387	510	0.142
280	2.58	340	23.51	400	1.84	460	0.258	520	0.0681
290	6.22	350	18.77	410	1.28	470	0.162	530	0.0313
300	11.92	360	13.22	420	0.956	480	0.0957	540	0.0137
310	18.50	370	8.41	430	0.732	490	0.0534	550	0.0058

[a] Source of data: Maric et al. (1993). The absorption band narrows somewhat with decreasing temperature. Photodissociation Cl$_2$ + $h\nu \rightarrow$ Cl ($^2P_{3/2}$) + Cl ($^2P_{3/2}$) occurs with a quantum yield of unity

Table 6.30 Absorption cross sections of mixed halogen molecules BrCl, ICl, IBr[a]

	$10^{19}\,\sigma$ (cm² molecule⁻¹)				$10^{19}\,\sigma$ (cm² molecule⁻¹)				$10^{20}\,\sigma$ (cm² molecule⁻¹)		
λ (nm)	BrCl	ICl	IBr	λ (nm)	BrCl	ICl	IBr	λ (nm)	BrCl	ICl	IBr
280	0.0653	1.29	2.11	390	3.487	1.38	0.696	500	3.68	35.4	121.7
290	0.0357	0.64	1.68	400	2.895	1.90	1.20	510	2.59	29.0	115.9
300	0.0504	0.32	1.24	410	2.251	2.50	2.05	520	1.74	21.1	103.1
310	0.147	<0.1	0.795	420	1.780	2.85	3.17	530	1.13	15.6	85.8
320	0.408	<0.1	0.539	430	1.463	3.20	4.48	540	0.700	11.0	67.5
330	0.925	<0.1	0.336	440	1.255	3.54	5.87	550	0.419	7.3	52.3
340	1.721	<0.1	0.214	450	1.098	3.88	7.19	560	0.243	5.5	36.6
350	2.668	<0.1	0.145	460	0.952	4.17	8.52	570	0.136	4.2	27.2
360	3.503	0.23	0.152	470	0.802	4.20	9.85	580	0.074	3.4	19.9
370	3.961	0.46	0.237	480	0.647	4.18	11.11	590	0.039	2.9	14.6
380	3.926	0.874	0.417	490	0.499	4.00	11.99	600	0.020	2.1	11.3

[a]Source of data: BrCl, Maric et al. (1994); ICl, Jenkin et al. (1990); IBr, Seery and Britton (1964)

Table 6.31 Absorption cross sections of bromine, Br₂, at 298 K[a]

λ (nm)	$10^{21}\,\sigma$ (cm²)	λ (nm)	$10^{20}\,\sigma$ (cm²)	λ (nm)	$10^{20}\,\sigma$ (cm²)	λ (nm)	$10^{20}\,\sigma$ (cm²)	λ (nm)	$10^{20}\,\sigma$ (cm²)
270	1.61	355	5.63	430	60.1	505	29.0	580	3.50
280	0.728	360	8.66	435	57.1	510	26.2	585	2.98
290	0.299	365	12.7	440	54.0	515	23.4	590	2.52
295	0.187	370	17.8	445	51.2	520	20.6	595	2.11
300	0.122	375	23.9	450	48.8	525	18.0	600	1.76
305	0.100	380	30.7	455	46.8	530	15.7	605	1.45
310	0.135	385	37.9	460	45.2	535	13.6	610	1.19
315	0.274	390	45.1	465	44.0	540	11.7	615	0.958
320	0.626	395	51.8	470	42.8	545	10.1	620	0.767
325	1.41	400	57.4	475	41.6	550	8.68	625	0.607
330	3.00	405	61.6	480	40.3	555	7.47	630	0.475
335	6.02	410	64.2	485	38.6	560	6.43	635	0.368
340	11.4	415	65.1	490	36.6	565	5.54	640	0.282
345	20.5	420	64.5	495	34.3	570	4.77	645	0.214
350	34.9	425	62.8	500	31.8	575	4.09	650	0.161

[a]Source of data: Maric et al. (1994); the spectrum consists of a broad continuum, with some overlapping band structure at wavelengths >510 nm. A second, much weaker continuum occurs in the region 200–270 nm. Photodissociation $Br_2 + h\nu \rightarrow Br\,(^2P_{3/2}) + Br\,(^2P_{3/2})$ features a constant quantum yield in the range 480–680 nm, which is assumed to be unity. The formation of excited $Br\,(^2P_{1/2})$ atoms becomes possible – and has been observed – at wavelengths below 505 nm

Table 6.32 Absorption cross sections of iodine, I_2, at 295 K[a]

λ (nm)	$10^{20}\,\sigma$ (cm²)	λ (nm)	$10^{20}\,\sigma$ (cm²)	λ (nm)	$10^{20}\,\sigma$ (cm²)	λ (nm)	$10^{20}\,\sigma$ (cm²)	λ (nm)	$10^{20}\,\sigma$ (cm²)
210	418	310	18.1	410	4.43	510	277	610	40.8
220	302	320	12.2	420	5.96	520	309	620	30.5
230	225	330	7.79	430	13.5	530	326	630	28.0
240	169	340	4.71	440	20.3	540	306	640	23.6
250	128	350	2.58	450	33.3	550	265	650	21.6
260	97.1	360	1.24	460	57.1	560	191	660	19.0
270	72.9	370	0.659	470	89.7	570	130	670	17.7
280	54.4	380	1.14	480	131	580	92.7	680	14.9
290	38.9	390	0.925	490	179	590	65.8	690	12.8
300	27.4	400	2.93	500	228	600	47.4	700	10.3

[a]Averages over 5 nm intervals; source of data: Saiz-Lopez et al. (2004). The absorption spectrum is continuous below 500 nm, pressure-dependent band structure occurs at longer wavelengths. The data were taken at sufficiently high bath gas pressures to approximate atmospheric conditions. Under these conditions, the quantum yield for photodissociation: $I_2 + h\nu \rightarrow I\,(^2P_{3/2}) + I\,(^2P_{3/2})$ is unity. At wavelength below 530 nm a fraction of the iodine atoms is formed in the excited $I\,(^3P_{1/2})$ state, which in the atmosphere are rapidly quenched by collisions

Table 6.33 Absorption cross sections of iodine nitrate, $IONO_2$, at 298 K[a]

λ (nm)	$10^{18}\,\sigma$ (cm²)	λ (nm)	$10^{18}\,\sigma$ (cm²)	λ (nm)	$10^{18}\,\sigma$ (cm²)	λ (nm)	$10^{18}\,\sigma$ (cm²)	λ (nm)	$10^{18}\,\sigma$ (cm²)
245	12.1	280	7.41	315	4.41	350	3.34	385	1.53
250	11.7	285	6.91	320	4.04	355	3.16	390	1.30
255	10.6	290	6.31	325	3.96	360	2.94	395	1.03
260	9.46	295	5.77	330	3.80	365	2.70	400	0.78
265	8.80	300	5.25	335	3.74	370	2.42	405	0.605
270	7.97	305	4.95	340	3.60	375	2.13	410	0.496
275	7.72	310	4.62	345	3.48	380	1.84	415	0.416

[a]Source of data: Mössinger et al. (2002), cross sections represent 5 nm averages

Table 6.34 Absorption cross sections of bromine nitrate, BrONO$_2$, at 296 and 220 K[a]

λ (nm)	$10^{19}\,\sigma\,(cm^2)$ 298 K	230 K	λ (nm)	$10^{19}\,\sigma\,(cm^2)$ 298 K	230 K	λ (nm)	$10^{20}\,\sigma\,(cm^2)$ 298 K	230 K	λ (nm)	$10^{21}\,\sigma\,(cm^2)$ 298 K	230 K
200	68.0	55.3	275	3.05	2.83	350	7.01	7.12	425	13.8	14.3
205	52.0	44.7	280	2.79	2.62	355	6.52	6.62	430	12.9	13.6
210	36.1	34.5	285	2.56	2.43	360	5.99	6.07	435	12.0	12.9
215	29.2	29.4	290	2.32	2.25	365	5.43	5.51	440	11.1	12.0
220	25.6	26.5	295	2.08	2.06	370	4.89	4.94	445	10.3	11.2
225	23.0	24.1	300	186	1.88	375	4.35	4.40	450	9.28	10.1
230	20.5	21.5	305	1.65	1.70	380	3.85	3.84	455	8.31	8.93
235	17.5	18.2	310	1.45	1.52	385	3.37	3.34	460	7.42	7.85
240	14.0	14.3	315	1.27	1.34	390	2.97	2.91	465	6.52	6.64
245	10.6	10.6	320	1.13	1.18	395	2.59	2.52	470	5.66	4.92
250	7.97	7.72	325	1.02	1.04	400	2.28	2.21	475	4.61	4.31
255	6.00	5.70	330	0.932	0.950	405	2.01	1.96	480	3.92	3.29
260	4.71	4.40	335	0.862	0.879	410	1.81	1.76	485	3.97	2.40
265	3.89	3.61	340	0.806	0.818	415	1.65	1.63	490	2.49	1.67
270	3.38	3.13	345	0.757	0.766	420	1.50	1.51	495	2.07	1.05

[a]Data source: Burkholder et al. (1995), Deters et al. (1998). At wavelength λ > 300 nm, the photo-dissociation process BrONO$_2$ + hν → Br + NO$_3$ (φ_1) is dominant ($\varphi_1 \approx 1$); at 248 nm φ_1 is reduced to $\varphi_1 = 0.28$, indicating the occurrence of additional processes (Harwood et al. 1998)

Table 6.35 Absorption cross sections of hypobromous acid, HOBr, and hypoiodous acid, HOI[a]

λ (nm)	$10^{20}\,\sigma\,(cm^2)$ HOBr	HOI	λ (nm)	$10^{20}\,\sigma\,(cm^2)$ HOBr	HOI	λ (nm)	$10^{20}\,\sigma\,(cm^2)$ HOBr	HOI	λ (nm)	$10^{21}\,\sigma\,(cm^2)$ HOBr	HOI
250	4.15		325	10.5	27.2	400	2.43	32.2	475	16.2	5.25
255	6.19		330	10.8	32.9	405	1.80	33.2	480	13.0	2.96
260	10.5		335	11.3	37.0	410	1.36	32.7	485	9.93	1.61
265	14.6		340	11.9	38.5	415	1.08	30.7	490	7.23	0.86
270	18.7		345	12.3	37.7	420	0.967	27.5	495	5.02	
275	22.1		350	12.4	34.7	425	0.998	23.5	500	3.33	
280	24.3	0.077	355	12.1	30.4	430	1.15	19.2	505	2.12	
285	25.0	0.226	360	11.5	25.8	435	1.40	15.0	510	1.29	
290	24.0	0.589	365	10.5	22.1	440	1.68	11.3	515	0.76	
295	21.9	1.37	370	9.32	19.8	445	1.96	8.13	520	0.42	
300	19.1	2.86	375	7.99	19.4	450	2.18	5.63	525	0.23	
305	16.2	5.41	380	6.65	20.7	455	2.29	3.76	530	0.12	
310	13.6	9.26	285	5.38	23.3	460	2.28	2.42	535	0.059	
315	11.8	14.5	390	4.22	26.6	465	2.14	1.50	540	0.029	
320	10.8	20.7	395	3.23	29.8	470	1.91	0.904	545	0.013	

[a]Source of data: HOBr (298 K) Ingham et al. (1998); HOI (295 K) Bauer et al. (1998)

Quantum yields:

1. The photodissociation process $HOBr + h\nu \rightarrow Br + OH$ (φ_1) is expected to occur with unity quantum yield ($\varphi_1 \approx 1$), similar to photodissociation of HOCl.
2. The process $HOI + h\nu \rightarrow I + OH$ (φ_2) has been found to occur with a quantum yield $\varphi_2 = 1$ at 355 nm (Bauer et al. (1998), which may be assumed to apply to the entire wavelength region.

References

Anderson, S.M., P. Hupalo, K. Mauersberger, Geophys. Res Lett. **20**, 1579–1582 (1993)

Atkinson, R., D.L. Baulch, R.A. Cox, J.N. Crowley, R.F. Hampson Jr., R.G. Hynes, M.E. Jenkin, M.J. Rossi, J. Troe, Atmos. Chem. Phys. **4**, 1461–1738 (2004)

Atkinson, R., D.L. Baulch, R.A. Cox, J.N. Crowley, R.F. Hampson Jr., R.G. Hynes, M.E. Jenkin, M.J. Rossi, J. Troe, Atmos. Chem. Phys. **6**, 3625–4055 (2006)

Atkinson, R., D.L. Baulch, R.A. Cox, J.N. Crowley, R.F. Hampson Jr., R.G. Hynes, M.E. Jenkin, M.J. Rossi, J. Troe, Atmos. Chem. Phys. **7**, 981–1191 (2007)

Barnes, R.J., A. Sinha, H.A. Michelsen, J. Phys. Chem. A **102**, 8855–8859 (1998)

Bauer, D., T. Ingham, S.A. Carl, G.K. Moortgat, J.N. Crowley, J. Phys. Chem. A **102**, 2857–2864 (1998)

Blitz, M.A., D.E. Heard, M.J. Pilling, S.R. Arnold, M.P. Chipperfield, Geophys. Res. Lett. **31**, L06111, 1–5 (2004). doi:10.1029/2003GL018793

Brion, J., A. Chakir, J. Charbonnier, D. Daumont, C. Parisse, J. Malicet, J. Atmos. Chem. **30**, 29–299 (1998)

Burkholder, J.B., J. Geophys. Res. **98**, 2963–2974 (1993)

Burkholder, J.B., R.K. Talukdar, A.R. Ravishankara, Geophys. Res. Lett. **21**, 585–588 (1994)

Burkholder, J.B., A.R. Ravishankara, S. Solomon, J. Geophys. Res. **100**, 16793–16800 (1995)

Cacciani, M., A. di Sarra, G. Fiocco, A.A. Amuroso, J. Geophys. Res. **94**, 8485–8490 (1989)

DeMore, W.B., E. Tschuikow-Roux, J. Phys. Chem. **94**, 5856–5860 (1990)

Deters, B., J.P. Burrows, J. Orphal, J. Geophys. Res. **103**, 3563–3570 (1998)

Gierczak, T., J.B. Burkholder, S. Bauerle, A.R. Ravishankara, Chem. Phys. **231**, 229–244 (1998)

Harwood, M.H., R.L. Jones, R.A. Cox, E. Lutman, O.V. Rattigan, J. Photochem. Photobiol. A **73**, 167–175 (1993)

Harwood, M.H., J.B. Burkholder, A.R. Ravishankara, J. Chem. Phys. A **102**, 1309–1317 (1998)

Hubrich, C., F. Stuhl, J. Photochem. **12**, 93–107 (1980)

Ingham, T., D. Bauer, J. Landgraf, J.N. Crowley, J. Chem. Phys. A **104**, 3293–3298 (1998)

Jenkin, M.E., R.A. Cox, A. Mellouki, G. Le Bras, G. Poulet, J. Phys. Chem. **94**, 2927–2934 (1990)

Jenouvrier, A., B. Coquart, M.F. Mérienne, J. Atmos. Chem. **25**, 21–32 (1996)

Jimenez, E., T. Gierczak, H. Stark, J.B. Burkholder, A.R. Ravishankara, Phys. Chem. Chem. Phys. **7**, 342–348 (2005)

Johnston, H.S., H.F. Davis, Y.T. Lee, J. Phys. Chem. **100**, 4713–4723 (1996)

Kang, W.K., K. Jung, D.-C. Kim, K.-H. Jung, J. Chem. Phys. **104**, 5815–5820 (1996)

Keller-Rudek, H., G.K. Moortgat, *MPI-Mainz-UV–VIS-Spectral-Atlas of Gaseous Molecules* (2010). www.atmosphere.mpg.de/spectral-atlas-mainz

Malicet, J., D. Daumont, J. Charbonnier, C. Parisse, A. Chakir, J. Brion, J. Atmos. Chem. **21**, 263–273 (1995)

Maric, D., J.P. Burrows, R. Meller, G.K. Moortgat, J. Photochem. Photobiol. A **70**, 205–214 (1993)

Maric, D., J.P. Burrows, G.K. Moortgat, J. Photochem. Photobiol. A **83**, 179–192 (1994)

Martinez, R.D., A.A. Buitrago, N.W. Howell, C.H. Hearn, J.A. Joens, Atmos. Environ. **26A**, 685–792 (1992)

Matsumi, Y., F.J. Comes, G. Hancock, A. Hofzumahaus, A.J. Hynes, M. Kawasaki, A.R. Ravishankara, J. Geophys., Res. **107**, D3 4124, (2002). doi:10.1029/2001JD000510

Meller, R.E., G.K. Moortgat, J. Geophys. Res. **105**, 7089–7101 (2000)

Mérienne, M.F., B. Coquart, A. Jenouvrier, Planet. Space Sci. **38**, 617–625 (1990)

Mérienne, M.F., A. Jenouvrier, B. Coquart, J. Atmos. Chem. **20**, 281–297 (1995)

Molina, L.T., M.J. Molina, J. Geophys. Res. **91**, 14501–14508 (1986)

Molina, L.T., J.J. Lamb, M.J. Molina, Geophys. Res. Lett. **8**, 1008–1011 (1981)

Moortgat, G.K., W. Seiler, P. Warneck, J. Chem. Phys. **78**, 1185–1190 (1983)

Moortgat, G.K., H. Meyrahn, P. Warneck, Chem. Phys. Chem. **11**, 3896–3908 (2010)

Mössinger, J.C., D.M. Rowley, R.A. Cox, Atmos. Chem. Phys. **2**, 227–234 (2002)

Nan, G., I. Burak, P.L. Houston, Chem. Phys. Lett. **209**, 383–389 (1993)

Ogorzalek, L.R., H.-P. Haerri, G.E. Hall, P.L. Houston, J. Chem. Phys. **90**, 4222–4236 (1989)

Orphal, J.S., E. Fellows, J.-M. Flaud, J. Geophys. Res. **108**, 4077 (2003). doi: 10.1029/2002JD002489

Osborne, B.A., G. Marston, L. Kaminski, N.C. Jones, J.M. Gingell, N. Mason, I.C. Walker, J. Delwiche, M.-J. Hubin-Franskin, J. Quant. Spectros. Radiat. Transfer **64**, 67–74 (2000)

Rattigan, O.V., D.E. Shallcross, R.A. Cox, J. Chem. Soc. Faraday Trans. **93**, 2839–2846 (1997)

Rebbert, R.E., P. Ausloos, J. Photochem. **4**, 419–434 (1975)

Rebbert, R.E., P. Ausloos, J. Photochem. **6**, 265–276 (1977)

Saiz-Lopez, A., R.W. Saunders, D.M. Joseph, S.H. Ashworth, J.M.C. Plane, Atmos. Chem. Phys. **4**, 1443–1450 (2004)

Sander, S.P., J. Phys. Chem. **90**, 4135–4142 (1986)

Sander, S.P., R.R. Friedl, D.M. Golden, M.J. Kurylo, G.K. Moortgat, H. Keller-Rudek, P.H Wine, A.R. Ravishankara, C.E. Kolb, M.J. Molina, B.J. Finlayson-Pitts, R.E. Huie, V.L. Orkin, NASA Chemical Kinetics and Photochemical Data for Use in Atmospheric Studies, Evaluation Number 15, JPL Publication 06–2 (Jet Propulsion Laboratory, California Institute of Technology, Pasadena, 2006)

Seery, D.J., D. Britton, J. Phys. Chem. **68**, 2263–2266 (1964)

Selwyn, G., J. Podolske, H.S. Johnston, Geophys. Res. Lett. **4**, 427–430 (1977)

Simon, P.C., D. Gillotay, N. Vanlaethem-Meuree, J. Wisemberg, J. Atmos. Chem. **7**, 107–135 (1988)

Stutz, J., E.S. Kim, U. Platt, P. Bruno, C. Perrino, A. Febo, J. Geophys. Res. **105**, 14585–14592 (2000)

Talukdar, R.K., C.A. Longfellow, M.K. Gilles, A.R. Ravishankara, Geophys. Res. Lett. **25**, 143–146 (1998)

Troe, J., Z. Phys. Chem. **214**, 573–581 (2000)

Uma, S., P.K. Das, Can J. Chem. **72**, 865–869 (1994)

Vandaele, A.C., C. Hermans, P.C. Simon, M. Carleer, R. Colin, S. Fally, M.F. Merienne, A. Jenouvrier, B. Coquart, J. Quant. Spectros. Radiat. Transfer **59**, 171–184 (1998)

Warneck, P., Atmos. Environ. **35**, 5773–5777 (2001)

Yao, F., I. Wilson, H. Johnston, J. Phys. Chem. **86**, 3611–3615 (1982)

Yoshino, K., A.S.C. Cheung, J.R. Esmond, W.H. Parkinson, D.E. Freeman, S.L. Guberman, A. Jenouvrier, B. Coquart, M.F. Merienne, Planet. Space Sci. **36**, 1469–1475 (1988)

Yoshino, K., J.R. Esmond, A.S.C. Cheung, D.E. Freeman, W.H. Parkinson, Planet. Space Sci. **40**, 185–192 (1992)

Zhao, Z., R.E. Stickel, P.H. Wine, Geophys. Res. Lett. **22**, 615–618 (1995)

Chapter 7
Rate Coefficients for Gas-Phase Reactions

7.1 General Introduction

The description of atmospheric gas-phase chemistry generally requires a consideration of large sets of elementary chemical reactions. The reactions are *elementary*
in the sense that they cannot or need not be further broken up into sub-processes.
Elementary reactions are classified according to the number of atoms or molecules
participating in reactive molecular encounters: as bimolecular, when new products
arise from the collision of two reactants, and as termolecular when the reaction
requires three reactants. In particular, the association of two reactants requires a
third collision partner to remove excess energy in order to stabilize the newly formed
molecule. In the atmosphere the third partner is primarily nitrogen or oxygen.

Bimolecular reactions: The convention for the rate expression of a bimolecular
(second order) reaction such as

$$A + A \rightarrow B + C \tag{7.1}$$

is
$$-\tfrac{1}{2}d[A]/dt = d[B]/dt = d[C]/dt = k[A]^2 \tag{7.2}$$

The temperature dependence of the rate coefficient over not too wide a temperature
range is usually expressed by the normal Arrhenius form $k_{bim} = A \exp(-B/T)$ with
$B = E_a/R_g$, where E_a is the activation energy, A is a pre-exponential factor, and R_g is the
gas constant, and T is the absolute temperature. For a wider temperature range, the
modified expression $k_{bim} = A (T/T^\sigma)^n \exp(-E_a/R_g T)$, where $T^\sigma = 300$ K, often gives a better fit to the experimental data. In some cases the experimental data reveal a negative
activation energy suggesting that the reaction proceeds via the addition of reactants to
form an intermediate species, which subsequently breaks up into the final products.

Association reactions: The rates of termolecular association (recombination) reactions and the reverse dissociation processes

$$A + B + M \rightleftharpoons AB + M \tag{7.3}$$

P. Warneck and J. Williams, *The Atmospheric Chemist's Companion: Numerical Data
for Use in the Atmospheric Sciences*, DOI 10.1007/978-94-007-2275-0_7,
© Springer Science+Business Media B.V. 2012

depend on the concentration and nature of a third body M, which in the atmosphere is that of air as the carrier gas (mainly molecular nitrogen and oxygen). The reactions can be represented by two consecutive processes. The association reaction first forms an unstable AB* molecule capable of re-dissociation, and the second stabilizes the AB* molecule:

$$A + B \rightleftharpoons AB^* \quad k_1, k_{-1} \tag{7.3a}$$

$$AB^* + M \rightarrow AB + M \quad k_2 \tag{7.3b}$$

The corresponding rate expression is

$$d[AB]/dt = \{k_1 k_2 [M]/(k_{-1} + k_2[M])\}[A][B] = k[A][B] \tag{7.4}$$

where k is the experimentally defined rate coefficient. Re-dissociation and stabilization of AB* compete with each other so that at low pressures, when $k_{-1} \gg k_2 [M]$, the reaction will follow a third order rate law with an effective rate coefficient $k_0 = k_1 k_2/k_{-1}$, whereas at high pressures, when $k_{-1} \ll k_2 [M]$, the reaction changes to second order with an effective rate coefficient $k_\infty = k_1$. This makes the pseudo-second order rate coefficient k pressure-dependent. A similar procedure is applicable to the reverse of reaction (7.3), the dissociation of the substance AB:

$$AB + M \rightleftharpoons AB^* + M \quad k_{-2}, k_2 \tag{7.3c}$$

$$AB^* \rightarrow A + B \quad k_{-1} \tag{7.3d}$$

In this case the rate expression is

$$-d[AB]/dt = \{k_{-1} k_{-2} [M]/(k_{-1} + k_2[M])\}[AB] = k[AB] \tag{7.5}$$

Dissociation reactions follow formally a first order rate law, but the experimentally observed rate coefficient k is pressure-dependent. In the low pressure limit $k_0 = k_{-2}[M]$ and at high pressure $k_\infty = k_{-1} k_{-2}/k_2$. Since detailed balancing applies, the equilibrium constant K_{eq} is given by the ratio of rate coefficients for recombination and dissociation reactions, $K_{eq} = k_1 k_2/k_{-1} k_{-2}$ at fixed T and [M].

Although in the general case the dependence of k on [M] is complex, for small molecules in the atmosphere it can be approximated by a combination of the three parameters k_0, k_∞ and F. The first two correspond to the low and high pressure limits, and F is a form factor that describes the transition region, the so-called fall-off region.

$$k = \frac{k_0 k_\infty [M]}{k_\infty + k_0 [M]} F \tag{7.6}$$

The temperature dependence of dissociation reactions is expressed by an Arrhenius form similar to that for bimolecular reactions. Association reactions, in contrast, do not require an activation energy. Rather, the excess energy has to be dissipated. The temperature dependence of recombination reactions is weak and if an Arrhenius expression is applied the activation energy is negative. Frequently, the temperature dependence of the rate coefficient k for an association/recombination reaction is written in the form

$$k_{ter} = k^{300}(T/T^{\ominus})^{-m} \qquad (7.7)$$

where k^{300} refers to the value of k at $T^{\ominus} = 300$ K. Values for m range from 0.5 to 5.

The broadening factor F is essentially independent of temperature in the range encountered in the atmosphere. In its simplest form deduced from statistical unimolecular rate theory it may be approximated by

$$\log_{10}F = (\log_{10}F_C)/\{1+[\log_{10}(k_0[M]/k_{\infty})/C]^2\}$$

where $C = (0.75 - 1.27 \log_{10}F_c)$. This representation is used in the evaluation and compilation of rate coefficients by Atkinson et al. (2004). Earlier assessments (Atkinson et al. 1997) chose $C = 1$. Sander et al. (2006) in an independent evaluation adopted instead the simpler procedure of setting $F_c = 0.6$ and $C = 1$ throughout. Both representations are useful for atmospheric applications but may lead to slightly different values for k_0 and k_{∞}.

Combined expression for rate coefficients: For the purpose of presenting rate coefficients for atmospheric reactions, the above expressions for the temperature dependence of bimolecular, termolecular and dissociation reactions are in some of the following tables combined in the form

$$k = A^*(T/T^{\ominus})^{\alpha}\exp(-B/T)$$

and the parameters A^*, α and B are listed together with the rate coefficient at 298 K. In this representation the values of both α and B may be either positive or negative, or zero.

Units of rate coefficients: The units involve concentration (either chemical amount per volume or number of entities per volume) and time. In atmospheric chemistry, number concentrations are commonly employed to express gas-phase concentrations in the form (molecule cm^{-3}) or (atom cm^{-3}). Accordingly, the units are (s^{-1}) for a first order reaction rate coefficient, (cm^3 molecule^{-1} s^{-1}) for a bimolecular rate coefficient, and (cm^6 molecule^{-2} s^{-1}) for a termolecular rate coefficient. These units are used in the following tables. If chemical amount concentrations (mol m^{-3}) are required, the appropriate rate coefficients are obtained by multiplying bimolecular rate coefficients given in (cm^3 molecule^{-1} s^{-1}) with the factor $10^{-6}N_A$, and termolecular rate coefficients given in (cm^6 molecule^{-2} s^{-1}) with the factor $10^{-12}N_A^2$, where $N_A = 6.022 \times 10^{23}$ is Avogadro's constant.

References

Atkinson, R., D.L. Baulch, R.A. Cox, R.F. Hampson Jr., J.A. Kerr, M.J. Rossi, J. Troe, J. Phys. Chem. Ref. Data **26**, 1329–1499 (1997)

Atkinson, R., D.L. Baulch, R.A. Cox, J.N. Crowley, R.F. Hampson Jr., R.G. Hynes, M.E. Jenkin, M.J. Rossi, J. Troe, Atmos. Chem. Phys. **4**, 1461–1738 (2004)

Sander, S.P., R.R. Friedl, D.M. Golden, M.J. Kurylo, G.K. Moortgat, H. Keller-Rudek, P.H. Wine, A.R. Ravishankara, C.E. Kolb, M.J. Molina, B.J. Finlayson-Pitts, R.E. Huie, V.L. Orkin, NASA Chemical Kinetics and Photochemical Data for Use in Atmospheric Studies, Evaluation Number 15, JPL Publication 06–2, Jet Propulsion Laboratory, California Institute of Technology, Pasadena (2006)

7.2　Reactions of Excited Oxygen Species

Table 7.1　Rate coefficients and Arrhenius parameters for reactions of ^{1}D oxygen atoms[a]

Reaction	k_{298}	A	B
$O(^1D) + N_2 \rightarrow O + N_2$	3.1 (−11)	2.15 (−11)	−110
$O(^1D) + O_2 \rightarrow O + O_2$	4.0 (−11)	3.2 (−11)	−67
$O(^1D) + O_3 \rightarrow O_2 + O_2$	1.2 (−10)	1.2 (−10)	0
$\rightarrow O_2 + O + O$	1.2 (−10)	1.2 (−10)	0
$O(^1D) + H_2 \rightarrow OH + H$	1.1 (−10)	1.1 (−10)	0
$O(^1D) + H_2O \rightarrow OH + OH$	2.0 (−10)	1.63 (−10)	−60
$O(^1D) + N_2O \rightarrow N_2 + O_2$	5.0 (−11)	4.7 (−11)	−20
$\rightarrow NO + NO$	6.7 (−11)	6.7 (−11)	−20
$O(^1D) + NH_3 \rightarrow OH + NH_2$	2.5 (−10)	2.5 (−10)	0
$O(^1D) + CO_2 \rightarrow O + CO_2$	1.1 (−10)	7.5 (−11)	−115
$O(^1D) + CH_4 \rightarrow OH + CH_3$	1.05 (−10)	1.05 (−10)	0
$\rightarrow HCHO + H_2$	7.5 (−12)	7.5 (−12)	0
$\rightarrow CH_3O/CH_2OH + H$	3.45 (−11)	3.45 (−11)	0
$O(^1D) + HF \rightarrow$ products[b]	1.5 (−11)	5.0 (−11)	0
$O(^1D) + HCl \rightarrow$ products	1.5 (−10)	1.5 (−10)	0
$O(^1D) + HBr \rightarrow$ products	1.5 (−10)	1.5 (−10)	0
$O(^1D) + COCl_2 \rightarrow$ products	2.4 (−10)	2.2 (−10)	−30
$O(^1D) + COClF \rightarrow$ products	1.9 (−10)	1.9 (−10)	0
$O(^1D) + COF_2 \rightarrow$ products	7.4 (−11)	7.4 (−11)	0
$O(^1D) + CCl_4 \rightarrow$ products	3.3 (−10)	3.3 (−10)	0
$O(^1D) + CCl_3F \rightarrow$ products	2.3 (−10)	2.3 (−10)	0
$O(^1D) + CCl_2F_2 \rightarrow$ products	1.4 (−10)	1.4 (−10)	0
$O(^1D) + CClF_3 \rightarrow$ products	8.7 (−11)	8.7 (−11)	0
$O(^1D) + CF_4 \rightarrow CF_4 + O$	2.0 (−14)[c]	–	–
$O(^1D) + CH_3F \rightarrow$ products	1.5 (−10)	1.5 (−10)	0
$O(^1D) + CH_2F_2 \rightarrow$ products	5.1 (−11)	5.1 (−11)	0
$O(^1D) + CHF_3 \rightarrow$ products	9.1 (−12)	9.1 (−12)	0
$O(^1D) + CHCl_2F \rightarrow$ products	1.9 (−10)	1.9 (−10)	0
$O(^1D) + CHClF_2 \rightarrow$ products	1.0 (−10)	1.0 (−10)	0
$O(^1D) + CCl_3CF_3 \rightarrow$ products	2.0 (−10)	2.0 (−10)	0
$O(^1D) + CCl_2FCClF_2 \rightarrow$ products	2.0 (−10)	2.0 (−10)	0
$O(^1D) + CCl_2FCF_3 \rightarrow$ products	1.0 (−10)	1.0 (−10)	0
$O(^1D) + CClF_2CClF_2 \rightarrow$ products	1.3 (−10)	1.3 (−10)	0
$O(^1D) + CClF_2CF_3 \rightarrow$ products	5.0 (−11)	5.0 (−11)	0
$O(^1D) + C_2F_6 \rightarrow$ products	1.5 (−13)[b]	–	–
$O(^1D) + CH_3CH_2F \rightarrow$ products	2.6 (−10)	2.6 (−10)	0

(continued)

Table 7.1 (continued)

Reaction	k_{298}	A	B
$O(^1D)+CH_3CHF_2 \rightarrow products$	2.0 (−10)	2.0 (−10)	0
$O(^1D)+CH_3CF_3 \rightarrow products$	1.0 (−10)	1.0 (−10)	0
$O(^1D)+CH_3CClF_2 \rightarrow products$	2.2 (−10)	2.2 (−10)	0
$O(^1D)+CH_3CCl_2F \rightarrow products$	2.6 (−10)	2.6 (−10)	0
$O(^1D)+CH_2ClCClF_2 \rightarrow products$	1.6 (−10)	1.6 (−10)	0
$O(^1D)+CH_2ClCF_3 \rightarrow products$	1.2 (−10)	1.2 (−10)	0
$O(^1D)+CH_2FCF_3 \rightarrow products$	4.9 (−11)	4.9 (−11)	0
$O(^1D)+CHCl_2CF_3 \rightarrow products$	2.0 (−10)	2.0 (−10)	0
$O(^1D)+CHClFCF_3 \rightarrow products$	8.6 (−11)	8.6 (−11)	0
$O(^1D)+CHF_2CF_3 \rightarrow products$	1.2 (−10)	1.2 (−10)	0
$O(^1D)+CH_3Br \rightarrow products$	1.8 (−10)	1.8 (−10)	0
$O(^1D)+CH_2Br_2 \rightarrow products$	2.7 (−10)	2.7 (−10)	0
$O(^1D)+CHBr_3 \rightarrow products$	6.6 (−10)	6.6 (−10)	0
$O(^1D)+CHF_2Br \rightarrow products$	1.75 (−10)	2.2 (−10)	−70
$O(^1D)+CF_3Br \rightarrow products$	1.0 (−10)	1.0 (−10)	0
$O(^1D)+CClF_2Br \rightarrow products$	1.5 (−10)	1.5 (−10)	0
$O(^1D)+CF_2Br_2 \rightarrow products$	2.2 (−10)	2.2 (−10)	0
$O(^1D)+CF_2BrCF_2Br \rightarrow products$	1.6 (−10)	1.6 (−10)	0
$O(^1D)+SF_6 \rightarrow products$	1.8 (−14)[c]	–	–

[a]Powers of ten are shown in parentheses. The notation $O(^1D)$ refers to an oxygen atom excited to the first electronically (metastable) excited state, $^1D_2$ (life time 110 s), whereas O denotes an oxygen atom in its ^3P ground state; $k=A \exp(-B/T)$ (cm^3 molecule^{-1} s^{-1}); Source of data: Sander et al. (2006)
[b]The products are OH+F (30%) and O+HF (70%)
[c]The value may represent an upper limit

Table 7.2 Rate parameters for reactions of excited oxygen molecules[a]

	k_{298}	A	B	T-range
$O_2(^1\Delta_g)+O_2 \rightarrow O_2+O_2$	1.6 (−18)	3.0 (−18)	200	100–450
$O_2(^1\Delta_g)+O_3 \rightarrow 2O_2+O$	3.8 (−15)	5.2 (−11)	2,840	280–360
$O_2(^1\Delta_g)+N_2 \rightarrow O_2+N_2$	<1.4 (−19)			
$O_2(^1\Delta_g)+H_2O \rightarrow O_2+H_2O$	5.0 (−18)			
$O_2(^1\Delta_g)+CO_2 \rightarrow O_2+CO_2$	<2.0 (−20)			
$O_2(^1\Sigma_g^+)+O \rightarrow O_2+O$	8.0 (−14)			
$O_2(^1\Sigma_g^+)+O_2 \rightarrow O_2+O_2$	4.1 (−17)			
$O_2(^1\Sigma_g^+)+O_3 \rightarrow products$	2.2 (−11)	2.2 (−11)	0	295–360
$O_2(^1\Sigma_g^+)+N_2 \rightarrow O_2+N_2$	2.1 (−15)	2.1 (−15)	0	200–350
$O_2(^1\Sigma_g^+)+H_2O \rightarrow O_2+H_2O$	4.6 (−12)			
$O_2(^1\Sigma_g^+)+CO_2 \rightarrow O_2+CO_2$	4.1 (−13)			
$O_2(^1\Sigma_g^+)+N_2O \rightarrow products$[b]	9.0 (−14)			

[a]Powers of ten are shown in parentheses. The notations $O_2(^1\Delta_g)$ and $O_2(^1\Sigma_g^+)$ refer to oxygen molecules in the first and second electronically excited states, respectively, whereas O_2 denotes an oxygen molecule in its $O_2(^3\Sigma_g^-)$ ground state (which may become vibrationally excited); $k=A \exp(-B/T)$ (cm^3 molecule^{-1} s^{-1}); Source of data: Atkinson et al. (2004), Sander et al. (2006)
[b]The fraction of NO+NO$_2$ occurring as products is $<3.0 \times 10^{-4}$

References

Atkinson, R., D.L. Baulch, R.A. Cox, J.N. Crowley, R.F. Hampson Jr., R.G. Hynes, M.E. Jenkin, M.J. Rossi, J. Troe, Atmos. Chem. Phys. **4**, 1461–1738 (2004)

Sander, S.P., R.R. Friedl, D.M. Golden, M.J. Kurylo, G.K. Moortgat, H. Keller-Rudek, P.H. Wine, A.R. Ravishankara, C.E. Kolb, M.J. Molina, B.J. Finlayson-Pitts, R.E. Huie, V.L. Orkin, NASA Chemical Kinetics and Photochemical Data for Use in Atmospheric Studies, Evaluation Number 15, JPL Publication 06–2, Jet Propulsion Laboratory, California Institute of Technology, Pasadena (2006)

7.3 Reactions of Inorganic Species (Containing Hydrogen, Oxygen, Nitrogen, Sulfur)

Table 7.3 Rate coefficients and Arrhenius parameters for reactions in the oxygen/hydrogen system[a]

Reaction	k_{298}	$A*$	α	B
$O+O_2+M \rightarrow O_3+M$	5.6 (−34) [N_2] (k_0)	6.0 (−34) [N_2]	−2.6	0
	6.0 (−34) [O_2] (k_0)	5.6 (−34) [O_2]	−2.6	0
$O+O_3 \rightarrow O_2+O_2$	8.0 (−15)	8.0 (−12)	0	2,060
$O+OH \rightarrow O_2+H$	3.5 (−11)	2.4 (−11)	0	−110
$O+HO_2 \rightarrow OH+O_2$	5.8 (−11)	2.7 (−11)	0	224
$O+H_2O_2 \rightarrow OH+HO_2$	1.7 (−15)	1.4 (−12)	0	2,000
$H+O_2+N_2 \rightarrow HO_2+N_2$	5.4 (−32) (k_0)	5.4 (−32)	−1.8	0
$H+O_3 \rightarrow OH+O_2$	2.8 (−11)	1.4 (−10)	0	480
$H+HO_2 \rightarrow OH+OH$	7.2 (−11)	7.2 (−11)	0	0
$\rightarrow H_2O+O$	2.4 (−12)	2.4 (−12)	0	0
$\rightarrow H_2+O_2$	5.6 (−12)	5.6 (−12)	0	0
$OH+OH \rightarrow H_2O+O$	1.5 (−12)	6.2 (−14)	2.6	−945
$OH+OH+N_2 \rightarrow H_2O_2+N_2$	6.9 (−31) (k_0)	6.9 (−31)	−0.8	0
	2.6 (−11) (k_∞)	$F_c=0.5$	0	
$OH+HO_2 \rightarrow H_2O+O_2$	1.1 (−10)	4.8 (−11)	0	−250
$OH+H_2 \rightarrow H_2O+O_2$	6.7 (−15)	7.7 (−12)	0	2,100
$OH+H_2O_2 \rightarrow H_2O+HO_2$	1.7 (−12)	2.9 (−12)	0	160
$OH+O_3 \rightarrow HO_2+O_2$	7.3 (−14)	1.7 (−12)	0	940
$HO_2+HO_2 \rightarrow H_2O_2+O_2$	1.6 (−12)	2.2 (−13)	0	−600
$HO_2+HO_2+M \rightarrow H_2O_2+O_2+M$	5.2 (−32) [N_2]	1.9 (−33) [N_2]	0	−980
	4.5 (−32) [O_2]			
$HO_2+O_3 \rightarrow OH+2\,O_2$	2.0 (−15)	2.0 (−16)	4.57	−693

[a]Powers of ten in parentheses, $k=A*(T/T^{\ominus})^{\alpha} \exp(-B/T)$ (cm³ molecule⁻¹ s⁻¹); Source: Atkinson et al. (2004)

Table 7.4 Rate coefficients and Arrhenius parameters for reactions of nitrogen species[a]

Reaction	k_{298}	$A*$	α	B
$O + NO + N_2 \rightarrow NO_2 + N_2$	1.0 (−31) (k_0)	1.0 (−31)	−1.6	0
	3.0 (−11) (k_∞)	$F_c = 0.85$	0.3	0
$O + NO_2 \rightarrow NO + O_2$	1.0 (−11)	5.5 (−12)	0	−188
$O + NO_2 + N_2 \rightarrow NO_3 + N_2$	1.3 (−31) (k_0)	1.3 (−31)	−1.5	0
	2.3 (−11) (k_∞)	$F_c = 0.6$	0.24	0
$O + NO_3 \rightarrow O_2 + NO_2$	1.7 (−11)			
$O + HO_2NO_2 \rightarrow$ products	8.6 (−16)	7.8 (−11)	0	3,400
$N + O_2 \rightarrow NO + O$	8.5 (−17)	1.5 (−11)	0	3,600
$N + NO \rightarrow N_2 + O$	3.0 (−11)	2.1 (−11)	0	−100
$N + NO_2 \rightarrow N_2O + O$	1.2 (−11)	5.8 (−12)	0	−220
$2NO + O_2 \rightarrow 2NO_2$	2.0 (−38)	3.3 (−39)	0	−530
$NO + O_3 \rightarrow NO_2 + O_2$	1.8 (−14)	1.4 (−12)	0	1,310
$NO_2 + O_3 \rightarrow NO_3 + O_2$	3.5 (−17)	1.4 (−13)	0	2,470
$NO + NO_3 \rightarrow 2\,NO_2$	2.6 (−11)	1.8 (−11)	0	−110
$NO + NO_2 + N_2 \rightarrow N_2O_3 + N_2$	3.1 (−34) (k_0)	3.1 (−34)	−7.7	0
	7.9 (−12) (k_∞)	$F_c = 0.6$	1.4	0
$N_2O_3 + N_2 \rightarrow NO + NO_2 + N_2$	1.6 (−14) (k_0)	1.9 (−7)	−8.7	4,880
	3.6 (8) (k_∞) (s^{-1})	4.7 (15)	0.4	4,880
		$F_c = 0.6$		
$NO_2 + NO_2 + N_2 \rightarrow N_2O_4 + N_2$	1.4 (−33) (k_0)	1.4 (−33)	−3.8	0
	1.0 (−12) (k_∞) (s^{-1})	$F_c = 0.4$	0	0
$N_2O_4 + N_2 \rightarrow NO_2 + NO_2 + N_2$	6.1 (−15) (k_0)	1.3 (−5)	−3.8	6,400
	4.4 (6) (k_∞) (s^{-1})	1.15 (16)	0	6,460
		$F_c = 0.4$		
$NO_3 + NO_2 + N_2 \rightarrow N_2O_5 + N_2$	3.6 (−30) (k_0)	3.6 (−30)	−4.1	0
	1.9 (−12) (k_∞)	$F_c = 0.35$	0.2	0
$N_2O_5 + N_2 \rightarrow NO_3 + NO_2 + N_2$	1.2 (−19) [N$_2$] (k_0)	1.3 (−3)	−3.5	11,000
	6.9 (−2) (k_∞) (s^{-1})	9.7 (14)	0.1	11,080
		$F_c = 0.35$		
$N_2O_5 + H_2O \rightarrow 2\,HNO_3$	2.5 (−22)			
$N_2O_5 + 2\,H_2O \rightarrow 2\,HNO_3 + H_2O$	1.8 (−39)			
$OH + NO + N_2 \rightarrow HNO_2 + N_2$	7.4 (−31) (k_0)	7.4 (−31)	−2.4	0
	3.3 (−11) (k_∞)	$F_c = 0.81$	−0.3	0
$OH + NO_2 + N_2 \rightarrow HNO_3 + N_2$	3.3 (−30) (k_0)	3.3 (−30)	−3.0	0
	4.1 (−11) (k_∞)	$F_c = 0.4$	0	0
$OH + HNO_2 \rightarrow H_2O + NO_2$	6.0 (−12)	2.5 (−12)	0	−260
$OH + HNO_3 \rightarrow H_2O + NO_3$[b]	1.1 (−13)	(a) 2.4 (−14)	0	−460
$OH + HNO_3 + N_2 \rightarrow$	5.7 (−32) (k_0)	(b) 6.5 (−34)	0	−1,335
$\qquad H_2O + NO_3 + N_2$[b]	4.3 (−14) (k_∞)	(b) 2.7 (−17)	0	−2,199
$OH + HO_2NO_2 \rightarrow$ products	4.7 (−12)	1.9 (−12)	0	−270
$OH + NO_3 \rightarrow HO_2 + NO_2$	2.1 (−11)			
$HO_2 + NO \rightarrow OH + NO_2$	8.8 (−12)	3.6 (−12)	0	−270
$HO_2 + NO_2 + N_2 \rightarrow HO_2NO_2 + N_2$	1.8 (−31) (k_0)	1.8 (−31)	−3.2	0
	4.7 (−12) (k_∞)	$F_c = 0.6$	0	0

(continued)

Table 7.4 (continued)

Reaction	k_{298}	$A*$	α	B
$HO_2NO_2 + N_2 \rightarrow HO_2 + NO_2 + N_2$	1.3 (−20)	4.1 (−5)	0	10,650
	0.25 (k_∞) (s⁻¹)	4.8 (15)	0	11,170
		$F_c = 0.6$		
$HO_2 + NO_3 \rightarrow$ products[c]	4.0 (−12)			
$OH + NH_3 \rightarrow H_2O + NH_2$	1.6 (−13)	3.5 (−12)	0	925
$NH_2 + O_2 \rightarrow$ products	<6 (−21)			
$NH_2 + O_3 \rightarrow NH_2O + O_2$	1.7 (−13)	4.9 (−12)	0	1,000
$NH_2 + NO \rightarrow$ products[d]	1.6 (−11)	1.6 (−11)	−1.4	0
$NH_2 + NO_2 \rightarrow$ products[e]	2.0 (−11)	2.0 (−11)	−1.3	0
$OH + HCN \rightarrow$ products[f]	3.0 (−14) (k_∞)	1.2 (−13)	0	400
$OH + CH_3CN \rightarrow$ products[f]	2.2 (−14)	8.1 (−13)	0	1,080

[a]Powers of ten in parentheses, $k = A*(T/T^\ominus)^\alpha \exp(-B/T)$ (cm³ molecule⁻¹ s⁻¹); Source: Atkinson et al. (2004)
[b]The reaction proceeds directly and via an intermediate addition product. At 298 K and 1013.25 hPa the sum of the two rate coefficients $k_a + k_b = 1.6$ (−13) (cm³ molecule⁻¹ s⁻¹)
[c]The major products are $NO_2 + OH + O_2$
[d]The major products are $N_2 + H_2O$
[e]The products are $N_2O + H_2O$ (25%) and $NH_2O + NO$ (75%)
[f]The reaction with HCN proceeds by OH radical addition, the reaction with CH_3CN proceeds by both abstraction of H from the CH_3 group and OH radical addition

Table 7.5 Rate coefficients and Arrhenius parameters for reactions involving sulfur species[a]

Reaction	k_{298}	$A*$	α	B
$O + H_2S \rightarrow OH + HS$	2.2 (−14)	9.2 (−12)	0	1,800
$O + HS \rightarrow SO + H$	1.6 (−10)			
$O + OCS \rightarrow SO + CO$	1.2 (−14)	1.6 (−11)	0	2,150
$O + CS_2 \rightarrow SO + CS$	3.7 (−12)	3.3 (−11)	0	650
$O + CS \rightarrow CO + S$	2.1 (−11)	2.7 (−10)	0	760
$O + SO_2 + M \rightarrow SO_3 + M$	1.4 (−33) [N_2] (k_0)	4.0 (−32)	0	1,000
$O + CH_3SCH_3 \rightarrow CH_3SO + CH_3$	5.0 (−11)	1.3 (−11)	0	−409
$O + CH_3SSCH_3 \rightarrow CH_3SO + CH_3S$	1.5 (−10)	6.5 (−11)	0	−250
$O + CH_3SOCH_3 \rightarrow$ products	8.8 (−12)	2.0 (−12)	0	−440
$S + O_2 \rightarrow SO + O$	2.1 (−12)	2.1 (−12)	0	0
$S + O_3 \rightarrow SO + O_2$	1.2 (−11)			
$SO + O_2 \rightarrow SO_2 + O$	7.6 (−17)	1.6 (−13)	0	2,280
$SO + O_3 \rightarrow SO_2 + O_2$	8.9 (−14)	4.5 (−12)		1,170
$SO + NO_2 \rightarrow SO_2 + NO$	1.4 (−11)			
$HS + O_2 \rightarrow OH + SO$	<4 (−19)			
$HS + O_3 \rightarrow HSO + O_2$	3.7 (−12)	9.5 (−12)	0	280
$HS + NO_2 \rightarrow HSO + NO$	6.7 (−11)	2.9 (−11)	0	−240
$HSO + O_2 \rightarrow$ products	≤2 (−17)			
$HSO + O_3 \rightarrow$ products[b]	1.1 (−13)			
$HSO + NO_2 \rightarrow$ products[c]	9.6 (−12)			
$CS + O_2 \rightarrow$ products	2.9 (−19)			
$CS + O_3 \rightarrow OCS + O_2$	3.0 (−16)			
$CS + NO_2 \rightarrow OCS + NO$	1.4 (−11)			

(continued)

Table 7.5 (continued)

Reaction	k_{298}	$A*$	α	B
$OH + S \rightarrow H + SO$	6.6 (−11)			
$OH + SO \rightarrow H + SO_2$	8.3 (−11)	2.7 (−11)	0	−335
$OH + H_2S \rightarrow H_2O + HS$	4.7 (−12)	6.1 (−12)	0	80
$OH + OCS \rightarrow SH + CO_2$	2.0 (−15)	1.1 (−13)	0	1,200
$OH + CS_2 + M \rightarrow HOCS_2 + M$	8.0 (−31) [N_2] (k_0)	8.0 (−31)	0	0
	8.0 (−12) (k_∞)	$F_c = 0.8$	0	0
$HOCS_2 + M \rightarrow HO + CS_2 + M$	4.8 (−14) [N_2] (k_0)	1.6 (−6)	0	5,160
	4.8 (5) (s^{-1})	1.6 (13)	$F_c = 0.8$	5,160
$HOCS_2 + O_2 \rightarrow$ products[d]	2.8 (−14)	2.8 (−14)	0	0
$OH + SO_2 + M \rightarrow HOSO_2 + M$	4.5 (−31) [N_2] (k_0)	4.5 (−31)	−3.9	0
	1.3 (−12) (k_∞)	1.3 (−12)	−0.7	0
$HOSO_2 + O_2 \rightarrow HO_2 + SO_3$	4.3 (−13)	1.3 (−12)	0	330
$SO_3 + H_2O \rightarrow H_2SO_4$	5.7 (4) (s^{-1}) (50% r. h.)			
$SO_3 + NH_3 \rightarrow SO_3 \cdot NH_3$	2.0 (−11) (10^5 Pa)			
$OH + CH_3SH \rightarrow H_2O + CH_3S$	3.3 (−11)	9.9 (−12)	0	−356
$OH + CH_3SCH_3 \rightarrow CH_3SCH_2 + H_2O$	4.8 (−12)	1.13 (−11)	0	253
$\xrightarrow{O_2} CH_3S(OH)CH_3$	1.7 (−12) (10^5 Pa)	See note[e]		
$OH + CH_3SSCH_3 \rightarrow$ products	2.3 (−10)	7.0 (−11)	0	−350
$OH + CH_3S(O)CH_3 \rightarrow$ products	8.9 (−11)	6.1 (−12)	0	−800
$NO_3 + CH_3SH \rightarrow CH_3S + HNO_3$	9.2 (−13)	9.2 (−13)	0	0
$NO_3 + CH_3SCH_3 \rightarrow$ $CH_3SCH_2 + HNO_3$	1.1 (−12)	1.9 (−13)	0	−520
$NO_3 + CH_3SSCH_3 \rightarrow$ $CH_3S + CH_3SO + NO_2$	7.0 (−13)	7.0 (−13)	0	0
$CH_2SH + O_2 \rightarrow$ products	6.6 (−12)			
$CH_2SH + O_3 \rightarrow$ products	3.5 (−11)			
$CH_2SH + NO \rightarrow$ products	1.5 (−11)			
$CH_2SH + NO_2 \rightarrow$ products	4.4 (−11)			
$CH_3SCH_2 + O_2 \rightarrow CH_3SCH_2O_2$	5.7 (−12)			
$CH_3SCH_2O_2 + NO \rightarrow CH_3S +$ $HCHO + NO_2$	1.2 (−11)	4.9 (−12)	0	−260
$CH_3SCH_2O_2 + NO_2 + M \rightarrow$ $CH_3SCH_2O_2NO_2 + M$	9.0 (−12) (10^5 Pa)			
$CH_3S + O_2 + M \rightarrow CH_3SOO + M$	See note[f]	1.2 (−16)	0	−1,580
$CH_3SOO + M \rightarrow CH_3S + O_2 + M$	See note[f]	3.5 (−10) (s^{-1})	0	3,560
$CH_3S + O_3 \rightarrow$ products	4.9 (−12)	1.15 (−12)	0	−430
$CH_3S + NO + N_2 \rightarrow$ $CH_3SO + NO + N_2$	3.3 (−19) [N_2] (k_0)	3.3 (−19)	−4	0
	4.0 (−11) (k_∞)	$F_c = 0.54$	0	0
$CH_3S + NO_2 \rightarrow CH_3SO + NO$	6.0 (−11)	3.0 (−11)	0	−210
$CH_3SO + O_3 \rightarrow$ products	6.0 (−13)			
$CH_3SO + NO_2 \rightarrow$ products	1.2 (−11)			
$CH_3SOO + O_3 \rightarrow$ products	<8 (−13)			
$CH_3SOO + NO \rightarrow$ products	1.1 (−11)			
$CH_3SOO + NO_2 \rightarrow$ products	2.2 (−11)			
$CH_3SS + O_3 \rightarrow$ products	4.6 (−13)			

(continued)

Table 7.5 (continued)

Reaction	k_{298}	$A*$	α	B
$CH_3SS + NO_2 \rightarrow$ products	1.8 (−11)			
$CH_3SSO + NO_2 \rightarrow$ products	4.5 (−12)			

[a]Powers of ten in parentheses, $k = A*(T/T^{\ominus})^{\alpha} \exp(-B/T)$ (cm^3 molecule^{-1} s^{-1}); Source: Atkinson et al. (2004)
[b]The major products are $HS + 2O_2$ (55%) and $HSO_2 + O_2$ (45%)
[c]The major reaction products presumably are $HSO_2 + NO$
[d]The most likely pathway is addition of O_2 to the carbon atom followed by O transfer to the sulfur atom bearing the OH group, with breakup leading to $HOSO + OCS$; the HOSO radical reacts further with oxygen to form $HO_2 + SO_2$
[e]This channel proceeds only in the presence of O_2, but it is independent of other third bodies. The T dependence in the range 240–320 K at pressures near 1×10^5 Pa is described by $k = 1.0^{-39}$ [O_2] $\exp(5820/T)/\{1 + 5.0^{-30}$ [O_2] $\exp(6280/T)\}$ (cm^3 molecule^{-1} s^{-1})
[f]The data refer to $M = 1.07 \times 10^4$ Pa He and a temperature range 210–250 K. At ground level in the atmosphere and 298 K, forward and reverse reactions would result in 33% CH_3S being present as CH_3SOO

Reference

Atkinson, R., D.L. Baulch, R.A. Cox, J.N. Crowley, R.F. Hampson Jr., R.G. Hynes, M.E. Jenkin, M.J. Rossi, J. Troe, J. Atmos. Chem. Phys. **4**, 1461–1738 (2004)

7.4   Reactions and Oxidation of Organic Compounds

Introductory remarks on the oxidation of hydrocarbons: The oxidation of hydrocarbons in the atmosphere is mainly initiated by reaction with OH radicals, and less frequently by reaction with NO_3 radicals (only at night because NO_3 undergoes photolysis during the day). Alkenes react also with ozone. Although oxidation mechanisms follow a common scheme, significant differences occur in the behavior of alkanes, alkenes, and aromatic compounds, so that they must be considered separately.

Alkanes: The reactions of OH (and NO_3) with alkanes proceed by abstraction of a hydrogen atom. The resulting alkyl radical combines rapidly with oxygen to produce a peroxy radical:

$$RH \xrightarrow{+OH/-H_2O} R \cdot \xrightarrow{+O_2} ROO \cdot$$

Generally, three types of alkyl radicals and alkyl peroxy radicals may be formed, depending on whether the hydrogen atom undergoing abstraction is a primary, secondary or tertiary H-atom. In the atmosphere, peroxy radicals react with NO, NO_2, and HO_2, unless they react with other alkyl peroxy radicals (the dominant peroxy radical in the atmosphere is methyl peroxy except under polluted urban conditions). The reactions are summarized as follows:

$$
\begin{aligned}
\text{(a}_1\text{)} \qquad & \text{ROO·} + \text{NO} \;\rightarrow \text{RO·} + \text{NO}_2 \quad \text{(dominant)} \\
\text{(a}_2\text{)} \qquad & \qquad\qquad\quad \rightarrow \text{RONO}_2 \qquad \text{(minor)} \\
\text{(b)} \qquad & \text{ROO·} + \text{NO}_2 \;\rightleftharpoons \text{ROONO}_2 \\
\text{(c)} \qquad & \text{ROO·} + \text{HO}_2 \;\rightarrow \text{ROOH} + \text{O}_2 \\
\text{(d}_1\text{)} \qquad & \text{ROO·} + \text{CH}_3\text{OO·} \;\rightarrow \text{RO·} + \text{CH}_3\text{O·} + \text{O}_2 \\
\text{(d}_2\text{)} \qquad & \qquad\qquad\qquad\quad \rightarrow \text{alcohol} + \text{aldehyde/ketone} + \text{O}_2
\end{aligned}
$$

Reactions (a) and (b) are most important under NO_x-rich conditions, reaction (c) becomes prominent when the concentration of NO_x is low. The peroxynitrate formed in reaction (b) is stable only at low temperatures (in the upper troposphere and lower stratosphere), normally it reverts to the peroxy radical and NO_2 as shown. Reactions (a$_1$) and (d$_1$) generate alkoxy radicals, RO·, that enter into further reactions.

Alkoxy radicals undergo four types of reactions: decomposition, isomerization, reaction with oxygen, and reaction with NO or NO_2. Examples are summarized as follows:

$$
\begin{aligned}
\text{R}_1\text{R}_2\text{CHO·} &\rightarrow \text{R}_1\text{CHO} + \text{R}_2 & \text{decomposition} \\
\text{R}_1\text{CH(O·)(CH}_2)_2\text{CH}_2\text{R}_2 &\rightarrow \text{R}_1\text{CH(OH)(CH}_2)_2\dot{\text{C}}\,\text{HR}_2 & \text{isomerization} \\
\text{R}_1\text{R}_2\text{CHO·} + \text{O}_2 &\rightarrow \text{R}_1\text{R}_2\text{CHO} + \text{HO}_2 & \text{reaction with oxygen} \\
\text{R}_1\text{R}_2\text{CHO·} + \text{NO} &\rightarrow \text{R}_1\text{R}_2\text{CHONO} & \text{reaction with NO} \\
\text{R}_1\text{R}_2\text{CHO·} + \text{NO}_2 &\rightarrow \text{R}_1\text{R}_2\text{CHONO}_2 & \text{reaction with NO}_2
\end{aligned}
$$

Here, R_1 and R_2 represent an alkyl moiety or a hydrogen atom. Decomposition leads to the formation of an aldehyde and a new, smaller alkyl radical, which combines with oxygen and re-enters into the oxidation reactions (a–d). Isomerization results from the internal abstraction of a hydrogen atom during intermediate ring formation. The example shown, 1,5-hydrogen shift, is energetically favored. The new moiety formed attaches oxygen and the resulting hydroxy-alkyl peroxy radical enters the reaction scheme (a–d). Decomposition and isomerization are important only for the larger alkanes starting with butane. The reaction with oxygen generates aldehydes or ketones in addition to a hydrogen peroxy radical. The reactions with NO and NO_2, which produce stable nitrites or nitrates, are of minor importance compared with the other reactions.

Alkenes: OH radicals react with alkenes mainly by addition to either carbon atom of the >C=C< double bond. Hydrogen abstraction from alkyl substituent groups usually is a minor process. The hydroxy-alkyl radical formed by OH addition combines rapidly with oxygen to produce the corresponding hydroxy-alkyl peroxy radical, which subsequently enters into one of the reactions (a–d). The radicals R_1CH(OH)CH$_2$O· and R_1CH(O·)CH$_2$OH arising from reactions of 1-alkenes decompose mostly to produce aldehydes, but in the case of 1-pentene and longer alkenes isomerization has also been observed to occur. Conjugated alkadienes admit the formation of several different hydroxy-alkoxy radicals. In the important case of isoprene, H_2C=C(CH$_3$)CH=CH$_2$, the dominant process is decomposition of the hydroxy-alkoxy radicals to form HCHO + methyl vinyl ketone, CH_3C(O)CH=CH$_2$, and HCHO + methacrolein, H_2C=C(CH$_3$)CHO, depending onto which of the two double bonds the OH radical attaches.

The reaction of ozone with alkenes also proceeds by addition of O_3 to the double bond. The energy-rich ozonide thus formed is unstable and decomposes rapidly toward an aldehyde (or ketone) and an unstable bi-radical, $R_1R_2\dot{C}OO$ (Criegee-intermediate), which either decomposes or is stabilized by collisions. Decomposition results in a variety of products. Noteworthy is the production of OH radicals in addition to carbon monoxide and carbon dioxide. A significant fraction of the Criegee-radical is stabilized and reacts with other atmospheric constituents, notably water vapor, which is abundant in the troposphere. This reaction forms hydroxy-alkyl hydroperoxides as intermediates that decompose to yield acids. Although the reactions of alkenes with ozone follow a common scheme, they are complex and individual product yields have proven difficult to determine.

Aromatic compounds: The reaction with OH radicals proceeds either by addition to the aromatic ring or by interaction with elements of a side chain. In the latter case, the oxidation scheme is similar to that of alkanes, if H-atom abstraction occurs, or alkenes if double bonds are present (OH-addition). The addition of OH to the aromatic ring produces a substituted hydroxy-cyclo-hexadienyl radical (several isomers), which reacts further with oxygen either to stabilize OH addition or to cause ring opening and fragmentation. Figure 7.1 illustrates prominent oxidation pathways for toluene as an example. About 7.5% of the reaction leads to oxidation of the methyl group to produce benzaldehyde. The methyl-hydroxy-cyclo-hexadienyl radical formed by addition to the aromatic ring reacts with oxygen to produce cresols with about 35% yield. Other reaction products such as formaldehyde, glyoxal, methyl glyoxal, and 1,4-dicarbonyl compounds, result from ring cleavage in a mechanism, which is still incompletely understood.

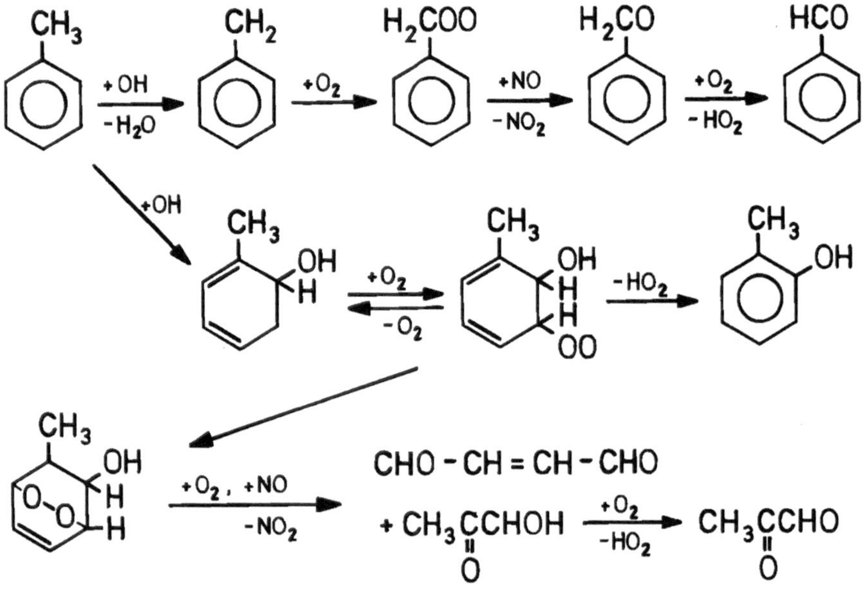

Fig. 7.1 Proposed reaction mechanism for the oxidation of toluene (Source: Warneck (2000) *Chemistry of the Natural Atmosphere*, Copyright Elsevier, reproduced with kind permission)

Table 7.6 Rate coefficients and Arrhenius parameters for reactions in the oxidation of simple organic compounds[a]

Reaction	k_{298}	$A*$	α	B
Reactions involving oxygen atoms				
$O + CH_3 \rightarrow products$[b]	1.3 (−10)			
$O + CH_4 \rightarrow OH + CH_3$	5.0 (−18)	8.3 (−12)	1.56	4,270
$O + C_2H_6 \rightarrow OH + C_2H_5$	4.7 (−16)	8.5 (−12)	1.5	2,920
$O + C_3H_8 \rightarrow OH + C_3H_7$	7.9 (−15)	5.8 (−13)	3.5	1,280
$O + C_2H_4 \rightarrow products$[c]	8.1 (−13)	8.1 (−13)	2.08	0
$O + C_3H_6 \rightarrow products$[c]	4.8 (−12)	1.2 (−12)	2.15	−400
$O + C_2H_2 \rightarrow products$[c]	1.4 (−13)	3.0 (−11)	0	1,600
$O + HCHO \rightarrow OH + HCO$	1.6 (−13)	3.4 (−11)	0	1,600
$O + CH_3CHO \rightarrow CH_3CO + OH$	4.5 (−13)	1.8 (−11)	0	1,100
$O + HCN \rightarrow products$	1.5 (−17)	1.0 (−11)	0	4,000
Oxidation of methane				
$OH + CH_4 \rightarrow CH_3 + H_2O$	6.4 (−15)	1.85 (−12)	0	1,690
$CH_3 + O_2 + N_2 \rightarrow CH_3O_2 + N_2$	1.0 (−30) (k_0)	1.0 (−30)	−3.3	0
	1.8 (−12) (k_∞)	$F_c = 0.27$	1.1	0
$CH_3 + O_3 \rightarrow products$	2.3 (−12)	4.7 (−12)	0	210
$CH_3O_2 + NO \rightarrow CH_3O + NO_2$	7.7 (−12)	2.95 (−12)	0	−285
$CH_3O_2 + NO_2 + N_2 \rightarrow$	2.5 (−30) (k_0)	2.5 (−30)	−5.5	0
$\quad CH_3O_2NO_2 + N_2$	1.8 (−11) (k_∞)	$F_c = 0.36$	0	0
$CH_3O_2NO_2 + N_2 \rightarrow$	6.8 (−19) N$_2$ (k_0)	9.0 (−5)	0	9,690
$\quad CH_3O_2 + NO_2 + N_2$	4.5 (k_∞) (s^{-1})	1.1 (16)	$F_c = 0.6$	10,560
$CH_3O_2 + HO_2 \rightarrow CH_3OOH + O_2$	5.2 (−12)	3.8 (−13)	0	−780
$OH + CH_3OOH \rightarrow products$[d]	5.5 (−12)	2.9 (−12)	0	−190
$CH_3O_2 + CH_3O_2$	3.5 (−13) (overall)	1.0 (13)	0	−365
$\quad \rightarrow HCHO + CH_3OH + O_2$	2.2 (−13)			
$\quad \rightarrow 2\ CH_3O + O_2$	1.3 (−13)	7.4 (−13)	0	520
$\quad \rightarrow CH_3OOCH_3 + O_2$	<3 (−14)			
$CH_3O + O_2 \rightarrow HCHO + HO_2$	1.9 (−15)	7.2 (−14)	0	1,080
$CH_3O + NO \rightarrow HCHO + HNO$	2.3 (−12)	2.3 (−12)	−0.7	0
$CH_3O + NO_2 \rightarrow HCHO + HNO_2$	2.0 (−13)	9.6 (−12)	0	1,150
$CH_3O + NO + N_2 \rightarrow CH_3ONO + N_2$	2.6 (−29) (k_0)	2.6 (−29)	2.8	0
	3.3 (−11) (k_∞)	$F_c = \exp(-T/900)$	0.6	0
$CH_3O + NO_2 + N_2 \rightarrow CH_3ONO_2 + N_2$	8.1 (−29) (k_0)	8.1 (−29)	−4.5	0
	2.1 (−11) (k_∞)	$F_c = 0.44$	0	0
Oxidation of formaldehyde (methanal)				
$OH + HCHO \rightarrow HCO + H_2O$	8.5 (−12)	5.4 (−12)	0	−135
$HCO + O_2 \rightarrow CO + HO_2$	5.1 (−12)	5.1 (−12)	0	0
$OH + CO \rightarrow CO_2 + H$	1.3 (−13)(1 + 6.0 × 10^4 (Pa))		0	0
$HO_2 + HCHO \rightarrow HOCH_2OO$	7.9 (−14)	9.5 (−15)	0	−625
$HOCH_2OO \rightarrow HCHO + HO_2$	1.5 (2)	2.4 (12)	0	7,000
Oxidation of ethane				
$OH + C_2H_6 \rightarrow H_2O + C_2H_5$	2.4 (−13)	6.9 (−12)	0	1,000
$C_2H_5 + O_2 + N_2 \rightarrow C_2H_5O_2 + N_2$	5.9 (−29) (k_0)	5.9 (−29)	−3.8	0
	7.8 (−12) (k_∞)	$F_c = 0.54$ (298 K)[e]	0	0

(continued)

Table 7.6 (continued)

Reaction	k_{298}	$A*$	α	B
$C_2H_5+O_2 \rightarrow C_2H_4+HO_2$	3.8 (−15) ($p=1.01325 \times 10^5$ (Pa))			
$C_2H_5O_2+NO \rightarrow C_2H_5O+NO_2$	9.2 (−12)	2.6 (−12)	0	−380
$C_2H_5O_2+NO_2+N_2 \rightarrow$	1.3 (−29) (k_0)	1.3 (−29)	−6.2	0
$\rightarrow C_2H_5O_2NO_2+N_2$	8.8 (−12) (k_∞)	$F_c=0.31$	0	0
$C_2H_5O_2NO_2+N_2 \rightarrow$	1.4 (−17) (k_0)	4.8 (−4)	0	9,825
$\rightarrow C_2H_5O_2+NO_2+N_2$	5.4 (k_∞) (s^{-1})	8.8 (15)	0	10,440
		$F_c=0.31$		
$C_2H_5O_2+HO_2 \rightarrow C_2H_5OOH+O_2$	7.7 (−12)	3.8 (−13)	0	−900
$C_2H_5O_2+C_2H_5O_2$	6.4 (−14) (overall)	6.4 (−14)	0	0
$\rightarrow C_2H_5OH+CH_3CHO+O_2$	2.4 (−14)			
$\rightarrow 2\ C_2H_5O+O_2$	4.0 (−14)			
$C_2H_5O+O_2 \rightarrow CH_3CHO+HO_2$	8.1 (−15)	2.4 (−14)	0	325
$C_2H_5O+NO+N_2 \rightarrow C_2H_5ONO+N_2$	2.2 (−28) (k_0)			
	4.4 (−11) (k_∞)	$F_c=0.6$		
$C_2H_5O+NO_2+N_2 \rightarrow C_2H_5ONO_2+N_2$	2.8 (−11) (k_∞)			
Oxidation of acetaldehyde (ethanal)				
$OH+CH_3CHO \rightarrow CH_3CO+H_2O$	1.5 (−11)	4.4 (−12)	0	−365
$CH_3CO+O_2+M \rightarrow CH_3C(O)O_2+M$	5.1 (−12) (k_∞)			
$CH_3C(O)O_2+CH_3O_2$	1.1 (−11) (overall)	2.0 (−12)	0	−500
$\rightarrow CH_3O+CH_3+CO_2+O_2$	9.9 (−12)			
$\rightarrow CH_3COOH+HCHO+O_2$	1.1 (−12)			
$CH_3C(O)O_2+C_2H_5O_2$	1.6 (−11) (overall)	4.4 (−13)	0	−1,070
$\rightarrow C_2H_5O+CH_3+CO_2+O_2$				
$\rightarrow CH_3CHO+CH_3C(O)OH$				
$+O_2$				
$CH_3C(O)O_2+CH_3C(O)O_2$	1.6 (−11)	2.9 (−12)	0	−500
$\rightarrow 2\ CH_3+2\ CO_2+O_2$				
$CH_3C(O)O_2+NO \rightarrow CH_3+CO_2+NO_2$	2.0 (−11)	7.5 (−12)	0	−290
$CH_3C(O)O_2+NO_2+N_2$	2.7 (−28) (k_0)	2.7 (−28)	−7.1	0
$\rightarrow CH_3C(O)O_2NO_2+N_2$	1.2 (−11) (k_∞)	1.2 (−11)	−0.9	0
		$F_c=0.31$		
$CH_3C(O)O_2NO_2+N_2$	1.1 (−20)	4.9 (−3)	0	12,100
$\rightarrow CH_3C(O)O_2+NO_2+N_2$	3.8 (−4)	5.4 (16)	0	13,830
		$F_c=0.3$		
$CH_3C(O)O_2+HO_2$	1.4 (−11) (overall)	5.2 (−13)	0	−983
$\rightarrow CH_3C(O)OOH+O_2$	1.1 (−11)			
$\rightarrow CH_3C(O)OH+O_3$	2.8 (−12)			
Oxidation of propane				
$OH+CH_3CH_2CH_3$	1.1 (−12) (overall)	7.6 (−12)	0	585
$\rightarrow CH_3CH_2CH_2+H_2O$	3.0 (−13)			
$\rightarrow CH_3CHCH_3+H_2O$	8.0 (−13)			
$CH_3CH_2CH_2+O_2 \rightarrow CH_3CH_2CH_2O_2$	8.0 (−12) (k_∞)	8.0 (−12)	0	0
$CH_3CHCH_3+O_2 \rightarrow CH_3CH(O_2)CH_3$	1.1 (−11) (k_∞)	1.1 (−11)	0	0

(continued)

Table 7.6 (continued)

Reaction	k_{298}	$A*$	α	B
$CH_3CH_2CH_2O_2 + NO$ $\rightarrow CH_3CH_2CH_2O + NO_2$	9.4 (−12)	2.9 (−12)	0	−350
$CH_3CH(O_2)CH_3 + NO$ $\rightarrow CH_3CH(O)CH_3 + NO_2$	9.0 (−12)	2.7 (12)	0	−360
$CH_3CH_2CH_2O + O_2$ $\rightarrow CH_3CH_2CHO + HO_2$	1.0 (−14)	2.6 (−14)	0	253
$CH_3CH(O)CH_3 + O_2$ $\rightarrow CH_3COCH_3 + HO_2$	7.0 (−15)	1.5 (−14)	0	230
$CH_3CH_2CH_2O + NO_2$ $\rightarrow CH_3CH_2CH_2ONO_2$	3.6 (−11) (k_∞)			
$CH_3CH(O)CH_3 + NO_2$ $\rightarrow (CH_3)_2CHONO_2$	3.4 (−11) (k_∞)			
$OH + CH_3CH_2CHO \rightarrow products$	2.0 (−11)	5.1 (−12)	0	−405
Oxidation of acetone (propanone)				
$OH + CH_3COCH_3$ $\rightarrow CH_3COCH_2 + H_2O$	1.7 (−13)	8.8 (−12)	0	1,320
$CH_3COCH_2 + O_2 \rightarrow CH_3COCH_2O_2$	1.5 (−12) (k_∞)			
$CH_3COCH_2O_2 + NO$ $\rightarrow CH_3COCH_2O + NO_2$	8.0 (−12)			
$CH_3COCH_2O_2 + HO_2$ $\rightarrow CH_3COCH_2OOH + O_2$	9.0 (−12)			
$CH_3COCH_2O_2 + CH_3COCH_2O_2$ $\rightarrow CH_3COCH_2OH +$ $CH_3COCHO + O_2$	8.0 (−12) (overall) 3.0 (−12)			
$\rightarrow 2\ CH_3COCH_2O + O_2$	5.0 (−12)			
$CH_3COCH_2O \rightarrow CH_3CO + H_2CO$	dominant			
Oxidation of ethene[f]				
$OH + C_2H_4 + N_2 \rightarrow HOC_2H_4 + N_2$	7.0 (−29) (k_0) 9.0 (−12) (k_∞)	7.0 (−29) 9.0 (−12)	−3.1 $F_c = 0.48$	0 0
$HOC_2H_4 + O_2 \rightarrow HOC_2H_4O_2$	3.0 (−12)			
$HOC_2H_4O_2 + NO \rightarrow HOC_2H_4O + NO_2$	9.0 (−12) (overall)			
$HOC_2H_4O \rightarrow HCHO + CH_2OH$	78% ($p \approx 1.0 \times 10^5$ (Pa))			
$HOC_2H_4O + O_2 \rightarrow HOCH_2CH_2O + HO_2$	22% ($p \approx 1.0 \times 10^5$ (Pa))			
$CH_2OH + O_2 \rightarrow HCHO + HO_2$	9.7 (−12)			
$HOC_2H_4O_2 + HO_2$ $\rightarrow HOC_2H_4OOH + O_2$	1.5 (−11)			
Oxidation of acetylene (ethyne)[g]				
$OH + C_2H_2 + N_2 \rightarrow C_2H_2OH + N_2$	5.0 (−30) (k_0) 1.8 (−12) (k_∞)	5.0 (−30) 3.8 (−11)	−1.5 $F_c = 0.35$	0 910
$C_2H_2OH \rightarrow CH_2CHO$				
$CH_2CHO + O_2 \rightarrow OHCCHO + OH$	60%			

(continued)

Table 7.6 (continued)

Reaction	k_{298}	$A*$	α	B
$C_2H_2OH+O_2 \rightarrow HOC_2H_2O_2$				
$HOC_2H_2O_2 \rightarrow HCOOH+HCO$	40%			
$HCO+O_2 \rightarrow CO+HO_2$	5.1 (−12)			
Oxidation of methanol				
$OH+CH_3OH \rightarrow$	9.3 (−13) (overall)	3.1 (−12)	0	360
$\rightarrow CH_2OH+H_2O$	7.9 (−13)			
$\rightarrow CH_3O+H_2O$	1.4 (−13)			
$CH_2OH+O_2 \rightarrow HCHO+HO_2$	9.7 (−12)			
$CH_3O+O_2 \rightarrow HCHO+HO_2$	1.9 (−15)	7.2 (−14)	0	1,080
Oxidation of ethanol				
$OH+C_2H_5OH \rightarrow$	3.2 (−12) (overall)	4.1 (−12)	0	70
$\rightarrow CH_3CHOH$	2.9 (−12)			
$\rightarrow CH_2CH_2OH+H_2O$	1.6 (−13)			
$\rightarrow CH_3CH_2O$	1.6 (−13)			
$CH_3CHOH+O_2 \rightarrow CH_3CHO+HO_2$	1.9 (−11)			

[a]Powers of ten in parentheses, $k=A*(T/T^{\ominus})^{\alpha} \exp(-B/T)$ (cm^3 molecule^{-1} s^{-1}), temperature range ~200–350 K; Data taken largely from Atkinson et al. (2006)
[b]The products at 298 K are HCHO+H (83%) and CO+H$_2$+H (17%)
[c]These reactions occur by addition of the oxygen atom to the alkene or alkyne
[d]The products are CH$_3$OO+H$_2$O (65%) and CH$_2$OOH+H$_2$O (35%) over the temperature range 220–430 K; CH$_2$OOH is expected to disintegrate to form HCHO and OH
[e]The temperature dependence is given as $F_c=0.58 \exp(-T/1250)+0.42 \exp(-T/183)$
[f]These data are largely from Calvert et al. (2000)
[g]The mechanism is taken from Atkinson (1994). Also expected to occur is the reaction HOC$_2$H$_2$O$_2$+NO$\rightarrow$HOC$_2$H$_2$O+NO$_2$ followed by HOC$_2$H$_2$O+O$_2\rightarrow$OHCCHO+HO$_2$

Table 7.7 Rate coefficients and Arrhenius parameters for reactions of OH radicals with alkanes[a]

Species	k_{298}	A	n	B (K)	T-range (K)
Methane	6.4 (−15)	1.85 (−20)	2.82	987	195–1,234
Methane-d_1	5.28 (−15)	3.19 (−18)	2	1,322	249–422
Methane-d_2	3.4 (−15)	2.18 (−12)	0	1,926	270–354
Methane-d_3	1.95 (−15)	1.46 (−12)	0	1,972	270–354
Methane-d_4	9.16 (−16)	5.7 (−18)	2	1,882	244–800
Ethane	2.48 (−13)	1.49 (−17)	2	499	180–1,225
Propane	1.09 (−12)	1.65 (−17)	2	87	190–1,220
n-Butane	2.36 (−12)	1.81 (−17)	2	−114	231–1,146
2-Methylpropane	2.12 (−12)	1.17 (−17)	2	−213	213–1,146
n-Pentane	3.80 (−12)	2.52 (−17)	2	−158	224–753
2-Methylbutane	3.6				
2,2-Dimethylpropane	8.25 (−13)	1.86 (−17)	2	207	287–901
n-Hexane	5.20 (−12)	2.54 (−14)	1	112	292–962
2-Methylpentane	5.2 (−12)				
3-Methylpentane	5.2 (−12)				

(continued)

Table 7.7 (continued)

Species	k_{298}	A	n	B (K)	T-range (K)
2,2-Dimethylbutane	2.23 (−12)	3.37 (−11)	0	809	245–328
2,3-Dimethylbutane	5.78 (−12)	1.66 (−17)	2	−407	247–1,220
n-Heptane	6.76 (−12)	1.95 (−17)	2	−406	299–1,086
2,2-Dimethylpentane	3.4 (−12)				
2,4-Dimethylpentane	4.77 (−12)				
2,2,3-Trimethylbutane	3.81 (−12)	9.20 (−18)	2	−459	243–753
n-Octane	8.11 (−12)	2.72 (−17)	2	−361	299–1,078
2,2,4-Trimethylpentane	3.34 (−12)	2.35 (−17)	2	−140	297–1,186
2,3,4-Trimethylpentane	6.6 (−12)				
2,2,3,3-Tetramethylbutane	9.72 (−13)	1.99 (−17)	2	178	290–1,180
n-Nonane	9.70 (−12)	2.72 (−17)	2	−436	295–1,097
2-Methyloctane	1.0 (−11)				
4-Methyloctane	9.7 (−12)				
2,3,5-Trimethylhexane	7.9 (−12)				
3,3-Diethylpentane	4.8 (−12)				
n-Decane	1.10 (−11)	3.17 (−17)	2	−406	296–1,109
3,3-Diethylhexane	6.92 (−12)				
n-Undecane	1.23 (−11)				
n-Dodecane	1.32 (−11)				
n-Tridecane	1.51 (−11)				
n-Tetradecane	1.79 (−11)				
n-Pentadecane	2.07 (−11)				
n-Hexadecane	2.32 (−11)				
Cyclopropane	8.15 (−14)	4.21 (−18)	2	454	200–459
Isopropylcyclopropane	2.61 (−12)				
Cyclobutane	2.03 (−12)	2.1 (−17)	2	−25	272–366
Cyclopentane	4.97 (−12)	2.73 (−17)	2	−214	273–1,194
Cyclohexane	6.97 (−12)	3.26 (−17)	2	−262	292–497
Methylcyclohexane	9.64 (−12)				
n-Butylcyclohexane	1.47 (−11)				
Cycloheptane	1.24 (−11)	3.99 (−17)	2	−373	298–388
Cyclooctane	1.33 (−11)	5.91 (−17)	2	−276	298–387
Bicyclo[2.2.1]heptane	5.12 (−12)				
Bicyclo[2.2.2]octane	1.37 (−11)				
Bicyclo[3.3.0]octane	1.03 (−11)				
1,1,3-Trimethylcyclohexane	8.70 (−12)				
cis-Bicyclo[4.3.0]nonane	1.60 (−11)				
trans-Bicyclo[4.3.0]nonane	1.65 (−11)				
cis-Bicyclo[4.4.0]decane	1.86 (−11)				
trans-bicyclo[4.4.0]decane	1.90 (−11)				
Tricyclo[5.2.1.0^{2,6}]decane	1.06 (−11)				
Tricyclo[3.3.1.1^{3,7}]decane	2.15 (−11)				
trans-Pinane	1.24 (−11)				
Tricyclene	2.66 (−12)				
Quadricyclane	1.70 (−12)				

[a]Powers of ten are shown in parentheses, $k = A\,T^n\exp(-B/T)$ (cm^3 $molecule^{-1}$ s^{-1}); Sources: Atkinson and Arey (2003), Atkinson (1997)

Table 7.8 Rate coefficients and Arrhenius parameters for reactions of OH radicals with alkenes, alkadienes and terpenes;[a] supplemental k_{298} values are shown in Table 7.8a

	k_{298}	A	B	T-range
Ethene	8.52 (−12)	1.96 (−12)	−438	291–425
Propene	2.63 (−11)	4.85 (−12)	−504	293–467
1-Butene	3.14 (−11)	6.55 (−12)	−467	295–424
cis-2-Butene	5.64 (−11)	1.10 (−11)	−487	295–425
trans-2-Butene	6.40 (−11)	1.01 (−11)	−550	295–425
2-Methylpropene	5.14 (−11)	9.47 (−12)	−504	295–425
1-Pentene	3.14 (−11)			
cis-2-Pentene	6.5 (−11)			
trans-2-Pentene	6.7 (−11)			
2-Methyl-1-butene	6.1 (−11)			
3-Methyl-1-butene	3.18 (−11)	5.32 (−12)	−533	295–423
2-Methyl-2-butene	8.69 (−11)	1.92 (−11)	−450	295–426
1-Hexene	3.7 (−11)			
Propadiene	9.82 (−12)	7.66 (−12)	−74	295–478
1,2-Butadiene	2.6 (−11)			
1,3-Butadiene	6.66 (−11)	1.48 (−11)	−448	295–483
2-Methyl-1,3-butadiene	1.00 (−10)	2.7 (−11)	−390	249–422
3-Methyl-1,2-butadiene	5.7 (−11)			
α-Pinene	5.23 (−11)	1.21 (−11)	−436	294–364
β-Pinene	7.43 (−11)	1.55 (−11)	−467	294–364
Limonene	1.64 (−10)	4.28 (−11)	−401	294–363

[a]Powers of ten are shown in parentheses; $k = A \exp(−B/T)$ (cm^3 molecule^{-1} s^{-1}). Source: Atkinson and Arey (2003)

Comments: Reactions of OH radicals with alkenes, alkadienes, and terpenes occur largely by addition to the double bond; hydrogen abstraction from side chains makes a minor contribution to the overall rate coefficient. At pressures >10^4 Pa, the rates for hydrogen addition are in the high pressure limit.

Table 7.8a Rate coefficients at 298 K (cm^3 molecule^{-1} s^{-1}) for additional gas-phase reactions of OH radicals with alkenes, alkadienes and terpenes[a]

Species	k_{298}	Species	k_{298}
2-Methyl-1-pentene	6.3 (−11)	cis-1,3-Pentadiene	1.01 (−10)
2-Methyl-2-pentene	8.9 (−11)	1,4-Pentadiene	5.3 (−11)
trans-4-Methyl-2-pentene	6.1 (−11)	trans-1,3-Hexadiene	1.12 (−10)
3,3-Dimethyl-1-butene	2.8 (−11)	trans-1,4-Hexadiene	9.1 (−11)
2,3-Dimethyl-2-butene	1.1 (−10)	1,5-Hexadiene	6.2 (−11)
1-Heptene	4.0 (−11)	cis- and trans-2,4-Hexadiene	1.34 (−10)
trans-2-Heptene	6.8 (−11)	2-Methyl-1,3-pentadiene	1.36 (−10)
2,3-Dimethyl-2-pentene	1.03 (−10)	2-Methyl-1,4-pentadiene	7.9 (−11)
trans-4,4-Dimethyl-2-pentene	5.5 (−11)	4-Methyl-1,3-pentadiene	1.31 (−10)
trans-4-Octene	6.9 (−11)	2,3-Dimethyl-1,3-butadiene	1.22 (−10)
1,2-Pentadiene	3.55 (−11)	2-Methyl-1,5-hexadiene	9.6 (−11)

(continued)

Table 7.8a (continued)

Species	k_{298}	Species	k_{298}
2,5-Dimethyl-1,5-hexadiene	1.20 (−10)	Ocimene (cis and trans)	2.52 (−10)
2,5-Dimethyl-2,4-hexadiene	2.10 (−10)	Camphene	5.3 (−11)
cis-1,3,5-Hexatriene	1.10 (−10)	2-Carene	8.0 (−11)
trans-1,3,5-Hexatriene	1.11 (−10)	3-Carene	8.8 (−11)
Cyclopentene	6.7 (−11)	α-Phellandrene	3.13 (−10)
Cyclohexene	6.77 (−11)	β-Phellandrene	1.68 (−10)
Cycloheptene	7.4 (−11)	Sabinene	1.17 (−10)
1,3-Cyclohexadiene	1.64 (−10)	α-Terpinene	3.63 (−10)
1,4-Cyclohexadiene	9.95 (−11)	γ-Terpinene	1.77 (−10)
1,3-Cycloheptadiene	1.39 (−10)	Terpinolene	2.25 (−10)
1,3,5-Heptatriene	9.7 (−11)	α-Cedrene	6.7 (−11)
1-Methyl-1-cyclohexene	9.4 (−11)	α-Copaene	9.0 (−11)
Bicyclo[2.2.1]-2-heptene	4.9 (−11)	β-Caryophyllene	1.97 (−10)
Bicyclo[2.2.1]-2,5-heptadiene	1.2 (−10)	α-Humulene	2.93 (−10)
Bicyclo[2.2.2]-2-octene	4.1 (−11)	Longifolene	4.7 (−11)
Myrcene	2.15 (−10)		

[a]Powers of ten are shown in parentheses; From Atkinson and Arey (2003)

Table 7.9 Rate coefficients at 298 K (cm^3 molecule^{-1} s^{-1}) for gas-phase reactions of OH radicals with aromatic compounds[a]

Species	k_{298}	Species	k_{298}
Benzene[b]	1.22 (−12)	Indene	7.8 (−11)
Toluene[b]	5.63 (−12)	Tetralin	3.4 (−11)
Ethylbenzene	7.0 (−12)	Styrene	5.8 (−11)
ortho-Xylene	1.36 (−11)	α-Methylstyrene	5.1 (−11)
meta-Xylene[c]	2.31 (−11)	β-Methylstyrene	5.7 (−11)
para-Xylene[c]	1.43 (−11)	β,β-Dimethylstyrene	3.2 (−11)
n-Propylbenzene	5.8 (−12)	Biphenyl	7.1 (−12)
Isopropylbenzene	6.3 (−12)	Phenol[b]	2.7 (−11)
ortho-Ethyltoluene	1.19 (−11)	ortho-Cresol[b]	8.2 (−11)
meta-Ethyltoluene	1.86 (−11)	meta-Cresol	6.8 (−11)
para-Ethyltoluene	1.18 (−11)	para-Cresol	5.0 (−11)
1,2,3-Trimethylbenzene	3.27 (−11)	Nitrobenzene[b]	1.4 (−13)
1,2,4-Trimethylbenzene	3.25 (−11)	Naphthalene[b]	2.3 (−11)
1,3,5-Trimethylbenzene	5.67 (−11)	Fluorene	1.4 (−11)
tert-Butylbenzene	4.5 (−12)	Acenaphthene	1.0 (−10)
4-Isopropyltoluene	1.45 (−11)	Phenanthrene	1.8 (−11)
Hexamethylbenzene	1.13 (−10)	Dibenzo-p-dioxin	1.4 (−11)
Indan	1.9 (−11)	Dibenzofuran[b]	3.8 (−12)

[a]Powers of ten are shown in parentheses; Source of data: Atkinson and Arey (2003), Calvert et al. (2002)

[b]The parameters for the temperature dependence described by $k = A \exp(-B/T)$ are for benzene $A = 2.33 \times 10^{-12}$, $B = 193$; for toluene $A = 1.18 \times 10^{-12}$, $B = -338$; for phenol $A = 1.73 \times 10^{-12}$, $B = -840$; for ortho-cresol $A = 1.7 \times 10^{-12}$, $B = -950$; for nitrobenzene $A = 6.2 \times 10^{-13}$, $B = 444$; for naphthalene $A = 1.56 \times 10^{-11}$, $B = -117$; for dibenzofuran $A = 1.48 \times 10^{-11}$, $B = 404$;

[c]The reactions of OH with xylenes and trimethylbenzenes are essentially independent of temperature in the range 250–315 K and 296–335 K, respectively

Table 7.10 Rate coefficients and Arrhenius parameters for reactions of OH radicals with organic carbonyl compounds;[a] Supplemental k_{298} data are shown in Table 7.10a

	k_{298}	A	B	T-range
Aldehydes				
Formaldehyde (methanal)	8.5 (−12)	5.4 (−12)	−135	200–300
Acetaldehyde (ethanal)	1.5 (−11)	4.4 (−12)	−365	200–350
Propanal	2.0 (−11)	5.1 (−12)	−405	240–370
Butanal	2.4 (−11)	6.0 (−12)	−410	260–420
2-Methylpropanal	2.6 (−11)	7.3 (−12)	−390	240–370
Metacrolein (2-methylpropenal)	2.9 (−11)	8.0 (−12)	−380	230–370
Pentanal	2.8 (−11)	9.9 (−12)	−310	245–410
2,2-Dimethylpropanal	2.8 (−11)	4.3 (−12)	−560	245–425
Ketones				
Acetone (propanone)	1.7 (−13)	See footnote[b]		200–400
2-Butanone	1.2 (−12)	1.3 (−12)	25	240–300
Methyl vinyl ketone	2.0 (−11)	2.6 (−12)	−610	230–380
3-Methyl-2-butanone	2.9 (−12)	1.58 (−12)	−193	250–370
4-Methyl-2-pentanone	1.3 (−11)	7.9 (−13)	−834	250–370
5-Methyl-2-hexanone	1.0 (−11)	1.33 (−12)	−649	260–370
Ketene[c]	3.3 (−11)	4.5 (−12)	−599	200–420

[a]Powers of ten are shown in parentheses, $k = A\ \exp(-B/T)$ (cm^3 molecule^{-1} s^{-1}); Source: Atkinson and Arey (2003)
[b]The temperature dependence is best described by the bi-exponential expression 8.8×10^{-12} $\exp(-1320/T) + 1.7 \times 10^{-14} \exp(420/T)$
[c]Ketene ($CH_2{=}CO$), although not a ketone, is included here. The data of Brown et al. (1989) were used to derive the temperature dependence, but note that their k_{298} value is larger than $k_{298} = (1.2 \pm 0.3) \times 10^{-11}$ derived by Oehlers et al. (1992)

Table 7.10a Rate coefficients at 298 K (cm^3 molecule^{-1} s^{-1}) for additional gas-phase reactions of OH radicals with organic carbonyl compounds[a]

Species	k_{298}	Species	k_{298}
Aldehydes			
2-Methylbutanal	3.3 (−11)	Benzaldehyde	1.2 (−11)
3-Methylbutanal	2.7 (−11)	*ortho*-Tolualdehyde	1.8 (−11)
Hexanal	3.0 (−11)	*meta*-Tolualdehyde	1.7 (−11)
2-Methylpentanal	3.3 (−11)	*para*-Tolualdehyde	1.3 (−11)
3-Methylpentanal	2.9 (−11)	Glyoxal	1.1 (−11)
4-Methylpentanal	2.6 (−11)	Methylglyoxal	1.5 (−11)
3,3-Dimethylbutanal	2.1 (−11)	Pinonaldehyde	4.4 (−11)
2-Ethylbutanal	4.0 (−11)	Caronaldehyde	4.8 (−11)
Heptanal	3.0 (−11)	3-Isopropenyl-6-oxo-heptanal	1.1 (−10)
Ketones			
2-Pentanone	4.4 (−12)	3-Methyl-2-pentanone	6.9 (−12)
3-Pentanone	2.0 (−12)	3,3-Dimethyl-2-butanone	1.2 (−12)
2-Hexanone	9.1 (−12)	2-Heptanone	1.1 (−11)
3-Hexanone	6.9 (−12)	2,4-Dimethyl-3-pentanone	5.0 (−12)

(continued)

Table 7.10a (continued)

Species	k_{298}	Species	k_{298}
2-Octanone	1.1 (−11)	cyclo-Hexanone	6.4 (−12)
2-Nonanone	1.2 (−12)	Camphenilone	4.8 (−12)
2,6-Dimethyl-4-heptanone	2.6 (−11)	Nopinone	1.5 (−11)
2-Decanone	1.3 (−11)	Sabonaketone	4.9 (−12)
cyclo-Butanone	8.7 (−13)	Camphor	4.3 (−12)
cyclo-Pentanone	2.9 (−12)		

[a]Powers of ten are shown in parentheses; Source: Atkinson and Arey (2003)

Table 7.11 Rate coefficients and Arrhenius parameters for gas-phase reactions of OH radicals with alcohols and ethers[a]

	k_{298}	A	n	B	T-range
Alcohols					
Methanol	9.4 (−13)	6.0 (−18)	2	−170	235–1,205
Ethanol	3.2 (−12)	6.1 (−18)	2	−530	227–600
1-Propanol	5.8 (−12)	4.6 (−12)	0	−70	263–372
2-Propanol	5.1 (−12)	4.0 (−18)	2	−792	253–587
1-Butanol	8.5 (−12)	5.3 (−12)	0	−140	263–372
2-Butanol	8.7 (−12)				
2-Methyl-1-propanol	9.3 (−12)				
2-Methyl-2-propanol	1.1 (−12)	4.1 (−18)	2	−321	240–440
1-Pentanol	1.1 (−11)				
2-Pentanol	1.2 (−11)				
3-Pentanol	1.3 (−11)				
cyclo-Pentanol	1.1 (−11)				
3-Methyl-1-butanol	1.3 (−11)				
3-Methyl-2-butanol	1.2 (−11)				
2,2-Dimethyl-1-propanol	5.5 (−12)				
1-Hexanol	1.5 (−11)				
2-Hexanol	1.2 (−11)				
cyclo-Hexanol	1.9 (−11)				
1-Heptanol	1.4 (−11)				
1-Octanol	1.4 (−11)				
Ethers					
Dimethyl ether	2.8 (−12)	1.14 (−17)	2	−303	230–650
Diethyl ether	1.31 (−11)	8.91 (−18)	2	−837	230–440
Methyl tert-butyl ether	2.94 (−12)	6.54 (−18)	2	−483	230–440
Ethyl tert-butyl ether	8.7 (−12)	6.70 (−18)	2	−800	235–370
tert-Amyl methyl ether	6.1 (−12)	7.56 (−18)	2	−656	230–370

[a]Powers of ten are shown in parentheses, $k = AT^n \exp(-B/T)$ (cm^3 molecule^{-1} s^{-1}); Source: Atkinson and Arey (2003)

Table 7.12 Rate coefficients and Arrhenius parameters for gas-phase reactions of OH radicals with organic acids[a]

	k_{298}	A	B	T-range
$OH + HCOOH \rightarrow products$	4.5 (−13)	4.5 (−13)	0	290–450
$OH + CH_3COOH \rightarrow CH_3CO_2 + H_2O$	8.0 (−13)			
$OH + C_2H_5COOH \rightarrow products$	1.2 (−12)	1.2 (−12)	0	290–450

[a]Powers of ten are shown in parentheses, $k = A \exp(-B/T)$ (cm^3 molecule^{-1} s^{-1}); Source: Atkinson et al. (2006)

Table 7.13 Rate coefficients at 298 K (cm^3 molecule^{-1} s^{-1}) for gas-phase reactions of OH radicals with organic nitrates[a]

Species	k_{298}	Species	k_{298}
Methyl nitrate[b]	2.3 (−14)	3-Methyl-1-butyl nitrate	2.3 (−12)
Ethyl nitrate[b]	1.8 (−13)	3-Methyl-2-butyl nitrate	1.7 (−12)
1-Propyl nitrate	5.8 (−13)	2,2-Dimethyl-1-propyl nitrate	7.9 (−13)
2-Propyl nitrate[b]	2.9 (−13)	2-Hexyl nitrate	3.0 (−12)
1-Butyl nitrate	1.6 (−12)	3-Hexyl nitrate	2.5 (−12)
2-Butyl nitrate	8.6 (−13)	2-Methyl-2-pentyl nitrate	1.6 (−12)
2-Methyl-1-propyl nitrate	1.5 (−12)	3-Methyl-2-pentyl nitrate	2.8 (−12)
1-Pentyl nitrate	3.0 (−12)	cyclo-Hexyl nitrate	3.1 (−12)
2-Pentyl nitrate	1.7 (−12)	3-Heptyl nitrate	3.4 (−12)
3-Pentyl nitrate	1.0 (−12)	3-Octyl nitrate	3.6 (−12)
2-Methyl-1-butyl nitrate	2.3 (−12)		

[a]Powers of ten are shown in parentheses; Source: Atkinson and Arey (2003)
[b]The parameters for the temperature dependence, described by $k = A \exp(-B/T)$, are for methyl nitrate $A = 4.0 \times 10^{-13}$, $B = 845$, for ethyl nitrate $A = 6.7 \times 10^{-13}$, $B = 395$, and for 2-propyl nitrate $A = 6.2 \times 10^{-13}$, $B = 230$
Note that these compound also photolyse in the troposphere

Table 7.14 Rate coefficients and Arrhenius parameters for gas-phase reactions of NO_3 radicals with some alkanes, alkenes, alkynes, aldehydes and alcohols;[a] supplemental k_{298} data are shown in Table 7.14a

	k_{298}	A	B	T-range
n-Butane	4.6 (−17)	2.8 (−12)	3,280	290–430
2-Methylpropane	1.06 (−16)	3.05 (−12)	3,060	290–430
n-Pentane	8.7 (−17)			
2-Methylbutane	1.62 (−16)	3.0 (−12)	2,927	290–525
Ethene	2.1 (−16)	3.3 (−12)	2,880	290–430
Propene	9.5 (−15)	4.6 (−13)	1,155	290–430
1-Butene	1.35 (−14)	3.14 (−13)	938	230–440
cis-2-Butene	3.5 (−13)			
trans-2-Butene[b]	3.9 (−13)	1.22 (−18) $\times T^2$	−382	204–378
2-Methyl-1,3-butadiene (isoprene)	7.0 (−13)	3.15 (−12)	450	250–380
α-Pinene	6.2 (−12)	1.2 (−12)	−490	260–385
Ethyne (acetylene)	5.1 (−17)	4.9 (−13)	2,742	295–520

(continued)

Table 7.14   (continued)

	k_{298}	A	B	T-range
Propyne	2.7 (−16)	1.5 (−11)	3,284	295–475
1-Butyne	4.5 (−16)	3.2 (−11)	3,320	295–475
2-Butyne	6.7 (−14)			
Formaldehyde (methanal)	5.6 (−16)			
Acetaldehyde (ethanal)	2.7 (−15)	1.4 (−12)	1,860	265–375
Propanal	6.5 (−15)			
Butanal	1.1 (−14)	1.7 (−12)	1,500	265–325
2-Methylpropanal	1.1 (−14)	3.6 (−12)	1,724	265–335
Methanol	1.3 (−16)	9.4 (−13)	2,650	260–370
Ethanol	<2 (−15)			
2-Propanol	1.4 (−15)			
2-Butanol	2.1 (−15)			
3-Hydroxy-3-methyl-1-butene	1.2 (−14)	4.6 (−14)	400	260–40

[a] Powers of ten are shown in parentheses, $k = A\exp(-B/T)$ (cm^3 molecule^{-1} s^{-1}); Source: Atkinson and Arey (2003), data for alkynes from Wayne et al. (1991)
[b] The rate coefficient for the reaction with *trans*-2-butene does not follow the normal Arrhenius expression

Table 7.14a   Further rate coefficients at 298 K (cm^3 molecule^{-1} s^{-1}) for gas-phase reactions of NO$_3$ radicals with hydrocarbons and oxygen-containing organic compounds[a]

Species	k_{298}	Species	k_{298}
Alkanes			
n-Hexane	1.1 (−16)	*n*-Octane	1.9 (−16)
2-Methylpentane	1.8 (−16)	2,2,4-Trimethylpentane	9.0 (−17)
3-Methylpentane	2.2 (−16)	2,2,3,3-Tetramethylbutane	<5 (−17)
2,3-Dimethylbutane	4.4 (−16)	*n*-Nonane	2.3 (−16)
n-Heptane	1.5 (−16)	*n*-Decane	2.8 (−16)
2,4-Dimethylpentane	1.5 (−16)	*cyclo*-Hexane	1.4 (−16)
2,2,3-Trimethylbutane	2.4 (−16)		
Alkenes			
2-Methylpropene	3.4 (−13)	Ocimene (*cis*- and *trans*-)	2.2 (−11)
1-Pentene	1.5 (−14)	*cyclo*-Pentene	4.2 (−13)
2-Methyl-2-butene	9.4 (−12)	*cyclo*-Hexene	5.1 (−13)
1-Hexene	1.8 (−14)	1,3-*cyclo*-Hexadiene	1.15 (−11)
2,3-Dimethyl-2-butene	5.7 (−11)	1,4-*cyclo*-Hexadiene	6.6 (−13)
2-Ethyl-1-butene	4.5 (−13)	*cyclo*-Heptene	5.1 (−13)
1-Heptene	2.0 (−14)	1,3-*cyclo*-Heptadiene	6.5 (−12)
1,3-Butadiene	1.0 (−13)	1,3,5-*cyclo*-Heptatriene	1.2 (−12)
2,3-Dimethyl −1,3-butadiene	1.4 (−12)	1-Methyl-1-*cyclo*-hexene	1.0 (−11)
cis-1,3-Pentadiene	1.4 (−12)	Methylene-*cyclo*-propane	1.5 (−14)
trans-1,3-Pentadiene	1.6 (−12)	Methylene-*cyclo*-butane	4.2 (−13)
trans-2, *trans*-4-Hexadiene	1.6 (−11)	Methylene-*cyclo*-pentane	1.5 (−12)
Myrcene	1.1 (−11)	Methylene-*cyclo*-hexane	5.4 (−13)

(continued)

Table 7.14a (continued)

Species	k_{298}	Species	k_{298}
Methylene-*cyclo*-heptane	1.0 (−12)	β-Pinene	2.5 (−12)
3-Methylene-*cylo*-hexene	5.7 (−12)	Sabinene	1.0 (−11)
1,2-Dimethyl-1-*cyclo*-hexene	5.2 (−11)	α-Terpinene	1.4 (−10)
2,3-Dimethyl-1-*cyclo*-hexene	1.5 (−11)	γ-Terpinene	2.9 (−11)
bicyclo[2.2.1]-2-Heptene	2.5 (−13)	Terpinolene	9.7 (−11)
bicyclo[2.2.1]-2,5-Hetadiene	1.0 (−12)	α-Cedrene	8.2 (−12)
bicyclo[2.2.2]-2-Octene	1.5 (−13)	α-Copaene	1.6 (−11)
Camphene	6.6 (−13)	β–Caryophyllene	1.9 (−11)
2-Carene	1.9 (−11)	α-Humulene	3.9 (−11)
3-Carene	9.1 (−12)	Longifolene	6.8 (−13)
Limonene	1.2 (−11)	Isolongifolien	3.9 (−12)
α-Phellandrene	7.3 (−11)	Alloisolongifolene	1.4 (−12)
β-Phellandrene	8.0 (−12)	α-Neoclovene	8.2 (−12)
Aromatic compounds			
Benzene	<3 (−17)	4-Isopropyltoluene	1.0 (−15)
Toluene	7.0 (−17)	Indan	7.3 (−15)
Ethylbenzene	<6 (−16)	Tetralin	8.5 (−15)
ortho-Xylene	4.1 (−16)	Indene	4.1 (−12)
meta-Xylene	2.6 (−16)	Styrene	1.5 (−12)
para-Xylene	5.0 (−16)	Phenol	3.6 (−12)
4-Ethyltoluene	8.6 (−16)	*ortho*-Cresol	1.4 (−11)
1,2,3-Trimethylbenzene	1.9 (−15)	*meta*-Cresol	9.7 (−12)
1,2,4-Trimethylbenzene	1.8 (−15)	*para*-Cresol	1.1 (−11)
1,3,5-Trimethylbenzene	8.8 (−16)		
Carbonyl compounds			
Pentanal	1.5 (−14)	Nonanal	2.0 (−14)
2-Methylbutanal	2.7 (−14)	Decanal	2.2 (−14)
3-Methylbutanal	1.9 (−14)	Benzaldehyde	2.4 (−15)
2,2-Dimethylpropanal	2.4 (−14)	Methyl vinyl ketone	<6 (−16)
Hexanal	1.6 (−14)	Methacrolein	3.4 (−15)
2-Methylpentanal	2.7 (−14)	Pinonaldehyde	2.0 (−14)
3-Methylpentanal	2.4 (−14)	Caronaldehyde	2.5 (−14)
4-Methylpentanal	1.7 (−14)	3-Isopropenyl-6-oxo-heptanal	2.6 (−13)
2-Ethylbutanal	4.5 (−14)	Acetone	<3 (−17)
3,3-Dimethylbutanal	1.8 (−14)	Nopinone	<2 (−15)
Heptanal	1.9 (−14)	Sabinaketone	<6 (−16)
Octanal	1.7 (−14)	Camphor	<3 (−16)
Miscellaneous			
Furan	1.4 (−12)	2,3-Dihydro-benzofuran	1.1 (−13)
3-Methylfuran	1.9 (−11)	1,4-Benzodioxan	5.9 (−11)

[a]Powers of ten are shown in parentheses; From Atkinson and Arey (2003), Atkinson (1994). Reactions of NO_3 radicals with halogenated hydrocarbons are shown in Table 7.28

Table 7.15   Rate coefficients and Arrhenius parameters for important gas-phase reactions of ozone with alkenes, alkadienes, terpenes and unsaturated carbonyl compounds;[a] for supplemental k_{298} data see Table 7.15a

	k_{298}	A	B	T-range
Ethene	1.59 (−18)	9.14 (−15)	2,580	178–362
Propene	1.01 (−17)	5.51 (−15)	1,878	235–362
1-Butene	9.64 (−18)	3.36 (−15)	1,774	225–362
cis-2-Butene	1.25 (−16)	3.22 (−15)	968	225–364
trans-2-Butene	1.90 (−16)	6.64 (−15)	1,059	225–363
2-Methylpropene	1.13 (−17)	2.70 (−15)	1,632	225–363
1-Pentene	1.06 (−17)	2.13 (−15)	1,580	240–324
cis-2-Pentene	1.30 (−16)	3.70 (−15)	1,002	278–353
trans-2-Pentene	1.60 (−16)	7.10 (−15)	1,132	278–353
2-Methyl-1-butene	1.40 (−17)	4.90 (−15)	1,741	278–353
3-Methyl-1-butene	9.5 (−18) (293 K)			
2-Methyl-2-butene	4.03 (−16)	6.51 (−15)	829	227–363
1-Hexene	1.13 (−17)	1.62 (−15)	1,480	240–324
cis-2-Hexene	1.10 (−16)	3.20 (−15)	1,017	278–353
trans-2-Hexene	1.50 (−16)	7.6 (−15)	1,163	278–353
2,3-Dimethyl-2-butene	1.13 (−15)	3.03 (−15)	294	227–363
1-Heptene	1.20 (−17)	4.2 (−15)	1,756	278–353
Propadiene	1.85 (−19)	1.54 (−15)	2,690	226–325
1,3-Butadiene	6.30 (−18)	1.34 (−14)	2,283	231–324
1,3-Pentadiene	4.30 (−17)	2.10 (−15)	1,158	240–324
1,4-Pentadiene	1.45 (−17)			
2-Methyl-1,3-butadiene[b]	1.27 (−17)	1.03 (−14)	1,995	242–353
2,3-Dimethyl-1,3-butadiene	2.60 (−17)	6.90 (−15)	1,668	240–324
cis-Ocimene	5.4 (−16)			
cyclo-Pentene	5.7 (−16)	1.80 (−15)	350	240–324
cyclo-Hexene	8.1 (−17)	2.87 (−15)	1,063	240–324
cyclo-Heptene	2.5 (−16)	1.30 (−15)	494	240–324
1-Methyl-1-cyclohexene	1.6 (−16)	5.25 (−15)	1,040	240–324
4-Methyl-1-cyclohexene	8.9 (−17)	2.16 (−15)	952	240–324
α-Pinene	8.4 (−17)	5.0 (−16)	530	276–363
β-Pinene	1.5 (−17)	1.2 (−15)	1,300	296–363
Limonene	2.1 (−16)	2.95 (−15)	783	296–363
Acrolein	2.8 (−19)			
Methacrolein	1.2 (−18)	1.4 (−15)	2,100	240–324
Methyl vinyl ketone	5.2 (−18)	8.5 (−16)	1,520	240–324

[a]Powers of ten shown in parentheses, $k = A \exp(-B/T)$ (cm^3 molecule^{-1} s^{-1}); From Atkinson and Arey (2003)
[b]Commonly called isoprene

Comments: Alkanes, alkylbenzenes and saturated aldehydes and ketones do not react with ozone at measureable rates. Rate coefficients for reactions of ozone with alkenes are fairly small; yet these reactions are important in the atmosphere at prevalent ambient concentrations of ozone (especially at night when concentrations of OH radicals are negligible). The rate coefficients for alkenes increase regularly with the number of alkyl groups attached to the C=C double bond. At 298 K, with increasing substitution around the >C=C< bond are CH_2=CH–, 1.0×10^{-17}; CH_2=C<, 1.1×10^{-17}; *cis* –CH=CH–, 1.2×10^{-16}; *trans* –CH=CH–, 1.9×10^{-16}; –CH=CH<, 4.0×10^{-16}; and >C=C<, 1.1×10^{-15}. For non-cyclic dienes and non-strained six-member cyclodienes with non-conjugated >C=C< bonds, the overall ozone reaction rate coefficients are approximately the sum of those for the individual >C=C< bonds. There is, however, a large effect of ring strain in cyclo-alkenes, which introduces a much larger uncertainty in estimating the rate coefficient for reactions that have not been studied.

Table 7.15a Rate coefficients at 298 K (cm^3 molecule^{-1} s^{-1}) for additional gas-phase reactions of ozone with alkenes, alkadienes and terpenes[a]

Species	k	Species	k
cis-3-Hexene	1.4 (−16) 295 K	1,3-Cyclo-hexadiene	1.2 (−15)
trans-3-Hexene	1.6 (−16) 290 K	1,4-Cyclo-hexadiene	4.6 (−17)
2-Methyl-1-pentene	1.6 (−17)	Cyclo-heptene	2.5 (−16)
3-Methyl-1-pentene	3.8 (−18) 286 K	1,3-Cyclo-heptadiene	1.5 (−16)
4-Methyl-1-pentene	1.0 (−17)	1,3,5-Cyclo-heptatriene	5.2 (−17) 294 K
cis-3-Methyl-2-pentene	4.5 (−16)	1-Methyl-cyclo-hexene	1.6 (−16)
trans-3-Methyl-2-pentene	5.6 (−16)	4-Methyl-cyclo-hexene	8.9 (−17)
2,3-Dimethyl-1-butene	1.0 (−17) 285 K	1,2-Dimethyl-1-cyclohexene	2.1 (−16)
3,3-Dimethyl-1-butene	3.9 (−18) 285 K	Bicyclo[2.2.1]-2-heptene	1.6 (−15)
2-Ethyl-1-butene	1.3 (−17)	Bicyclo[2.2.1]-2, 5-heptadiene	3.6 (−15)
2,2,3-Trimethyl-1-butene	7.8 (−18) 294 K	Bicyclo[2.2.2]-2-octene	7.1 (−17)
1-Octene	1.4 (−17)	3,7-Dimethyl-1,3, 6-octatriene	5.4 (−16) 296 K
cis-4-Octene	9.0 (−17) 293 K	Indene	1.7 (−16)
trans-4-Octene	1.3 (−16) 290 K	Styrene	1.7 (−17)
trans-2,5-Dimethyl-3-hexene	3.8 (−17) 291 K	Myrcene	4.7 (−16) 296 K
trans-2,2-Dimethyl-3-hexene	4.0 (−17) 295 K	2-Carene	2.3 (−16)
2,4,4-Trimethyl-2-pentene	1.4 (−16) 297 K	3-Carene	3.7 (−17)
1-Decene	9.3 (−18)	Sabinene	8.3 (−17)
cis-5-Decene	1.1 (−16) 293 K	Camphene	0.9 (−18)
3,4-Diethyl-2-hexene	4.0 (−18) 293 K	α-Phellandrene	3.0 (−15)
cis-2, *trans*-4-Hexadiene	3.1 (−16)	β-Phellandrene	4.7 (−17)
trans-2, *trans*-4-Hexadiene	3.7 (−16)	α-Terpinene	2.1 (−14)
2-Methyl-1,4-pentadiene	1.3 (−17)	γ-Terpinene	1.4 (−16)
2-Methyl-1,3-pentadiene	8.0 (−17)	Terpinolene	1.9 (−15)
2,4-Dimethyl-1,3-butadiene	8.0 (−17)	α-Cedrene	2.8 (−17)
2,5-Dimethyl-1,5-hexadiene	1.4 (−17)	α-Copaene	1.6 (−16)
1,3,5-Hexatriene (*cis* + *trans*)	2.6 (−17) 294 K	β-Caryophyllene	1.16 (−14)
1-Methyl-1-cyclopentene	6.7 (−16)	α-Humulene	1.17 (−14)

[a]Powers of ten shown in parentheses; Source: Atkinson and Arey (2003); temperatures differing from 298 K are indicated

Table 7.16 Yield of OH radical formation at 298 K from gas-phase reactions of ozone with alkenes[a]

Alkene	OH yield	Alkene	OH yield
Ethene	0.12	1-Methyl-*cyclo*-hexene	0.90
Propene	0.33	1,2-Dimethyl-1-*cyclo*-hexene	$1.02 \pm 0.16^*$
1-Butene	0.41	Camphene	≤ 0.18
cis-2-Butene	0.41	3-Carene	1.06
trans-2-Butene	0.64	Limonene	0.86
2-Methylpropene	0.84	Myrcene	1.15
1-Pentene	0.37	*cis*- and *trans*-Ocimene	0.63
2-Methyl-2-butene	$0.83, 0.93 \pm 0.14^*$	β-Phellandrene	0.14
1-Hexene	0.32	α-Pinene	$0.85, 0.76 \pm 0.11^*$
2,3-Dimehtyl-1-butene	0.50	β-Pinene	0.35
2,3-Dimethyl-2-butene	$0.80 \pm 0.12^*$	Sabinene	$0.26, 0.33 \pm 0.06^*$
1-Heptene	0.27	Terpinolene	1.03
1-Octene	0.18	α-Cedrene	0.67
1,3-Butadiene	0.08	α-Copaene	0.38
2-Methyl-2,3-butadiene	0.27	β-Caryophyllene	0.06
cyclo-Pentene	0.61	α-Humulene	0.22
cyclo-Hexene	0.68		

[a] Source: Atkinson (1997). The yields were determined from the amounts of cyclohexanone and cyclohexanol formed in the presence of cyclohexane as an OH scavenger. These data are uncertain by a factor of 1.5

Values marked by an asterisk were obtained from the yield of butanone in the presence of 2-butanol as scavenger. For a critical discussion of the techniques involved see Calvert et al. (2000)

Table 7.17 Rate coefficients and Arrhenius parameters for gas-phase reactions of alkyl peroxyl radicals with nitric oxide[a]

	k_{298}	A	B	T-range
$CH_3OO + NO \rightarrow CH_3O + NO_2$	7.7 (−12)	2.95 (−12)	−285	200–430
$C_2H_5OO + NO \rightarrow$ products	9.2 (−12)	2.6 (−12)	−380	200–410
$HOCH_2OO + NO \rightarrow$ products	5.6 (−12)			
$HOCH_2CH_2OO + NO$	9.0 (−12)			
$\rightarrow HOCH_2CH_2O + NO_2$				
$CH_3CH_2CH_2OO + NO \rightarrow$ products	9.4 (−12)	2.9 (−12)	−350	200–410
$(CH_3)_2CHOO + NO \rightarrow$ products	9.0 (−12)	2.7 (−12)	−360	200–410
$CH_3COCH_2OO + NO$	8.0 (−12)			
$\rightarrow CH_3COCH_2O + NO_2$				
$CH_3C(O)OO + NO \rightarrow CH_3C(O)O + NO_2$	2.0 (−11)	7.5 (−12)	−290	200–350
$C_2H_5C(O)OO + NO \rightarrow C_2H_5C(O)O + NO_2$	2.1 (−11)	6.7 (−12)	−340	220–410
$CH_3OCH_2OO + NO \rightarrow$ products	9.1 (−12)			
$(CH_3)_3COO + NO \rightarrow$ products	4.0 (−12)			
$(CH_3)_3CH_2OO + NO \rightarrow$ products	4.7 (−12)			
$HO(CH_3)_2CCH_2OO + NO \rightarrow$ products	4.9 (−12)			
$CH_3CH(OO)(CH_2)_2CH_3 + NO \rightarrow$ products	8.0 (−12)			
$cyclo\text{-}C_5H_9OO + NO \rightarrow$ products	1.1 (−11)			
$(CH_3)_2C(CH_3)_2CH_2OO + NO \rightarrow$ products	1.8 (−12)			
$CH_2 \rightarrow CHCH_2OO + NO \rightarrow$ products	1.0 (−11)			

[a] Powers of ten shown in parentheses, $k = A \exp(-B/T)$ (cm³ molecule⁻¹ s⁻¹); From Atkinson (1997), Atkinson et al. (2006), Wallington et al. (1997)

Table 7.18 Yields of $RONO_2$ observed in reactions of alkyl peroxyl radicals with nitric oxide[a]

Precursor	Alkyl radical	$k_b/(k_a+k_b)$[b]
Ethane	Ethyl	≤0.014
Propane	1-Propyl	0.020±0.009, 0.019±0.010
	1-,2-Propyl	0.036±0.005, 0.04
	2-Propyl	0.049±0.030
n-Butane	1-Butyl	0.041±0.012
	1-,2-Butyl	0.077±0.009
	2-Butyl	0.083±0.020
2-Methyl propane	2-Methyl-1-propyl	0.075±0.014
	tert-Butyl	0.180±0.080
n-Pentane	1-Pentyl	0.061±0.018
	2-Pentyl	0.129±0.014
	2-,3-pentyl	0.14±0.05, 1.125±0.03, 0.129±0.019
	3-Pentyl	0.118±0.014
2-Methyl butane	2-Methyl-1-butyl	0.040±0.005
	2-Methyl-2-butyl	0.044±0.002, 0.0560±0.0017
	2-Methyl-3-butyl	0.109±0.003, 0.074±0.002,
		0.150±0.004
	2-Methyl-3-butyl	0.043±0.008
neo-Pentane	neo-Pentyl	0.0507±0.0056, 0.0513±0.0053
n-Hexane	1-Hexyl	0.121±0.036
	2-Hexyl	0.220±0.034
	2-, 3-Hexyl	0.223±0.035, 0.37
	3-Hexyl	0.219±0.029
2-Methyl pentane	2-Methyl-2-pentyl	0.0350±0.0096
3-Methyl pentane	3-Methyl-2-pentyl	0.162±0.009, 0.140±0.014
n-Heptane	1-Heptyl	0.195±0.059
	2-Heptyl	0.324±0.044
	3-Heptyl	0.312±0.041
	4-Heptyl	0.290±0.037
	2-, 3-, 4-Heptyl	0.309±0.05, 0.287±0.016,
		0.220±0.017
n-Octane	1-Octyl	0.36±0.14
	2-Octyl	0.353±0.027
	3-Octyl	0.343±0.031
	4-Octyl	0.324±0.032
	2-, 3-, 4-Octyl	0.37±0.12, ≤ 0.43±0.10,
		0.332±0.034
cyclo-Pentane	cyclo-Pentyl	0.045±0.015
cyclo-Hexane	cyclo-Hexyl	0.090±0.044, 0.160±0.015,
		0.074±0.014
cyclo-Heptane	cyclo-Heptyl	0.050±0.011
Propene	2-Hydroxy-1-propyl	0.015±0.008
	1-Hydroxy-2-propyl	0.017±0.009
2-Methyl-propene	2-Methyl-1-hydroxy-2-propyl	≤0.057±0.022

(continued)

Table 7.18 (continued)

Precursor	Alkyl radical	$k_b/(k_a+k_b)^b$
2-Butanone	2-Keto-3-butyl	0.0021 ± 0.0010
Toluene	Benzyl	0.12, 0.09
2-Methyl-1,3-butadiene (isoprene)		$0.08–0.13, \leq 0.10\pm0.06$

[a]Source: Lightfoot et al. (1992). The data are derived from laboratory experiments and refer to room temperature and 1 atm (1.01325×10^5 Pa) total pressure
[b]Ratio of the rate coefficient k_b to the sum of both rate coefficients for the two channels of the reaction (a) $RO_2+NO \rightarrow RO+NO_2$ and (b) $RO_2+NO\ (+M) \rightarrow RONO_2\ (+M)$. If more than one value is shown, the data are from different independent studies

Comment: The relative importance of channel (b) increases with increasing size of the radical from <0.5% for R=methyl to 35% for R=octyl. Whereas the rate coefficient for channel (a) is independent of pressure (p> 150 Pa), the rate coefficient for channel (b) does depend on pressure (at low pressures) and decreases with increasing temperature as expected for an association reaction.

Table 7.19 Rate coefficients and Arrhenius parameters for gas-phase reactions of HO_2 with alkyl peroxyl radicals and substituted alkyl peroxyl radicals[a]

	k_{298}	A	B	T-range
CH_3OO	5.2 (−12)	3.8 (−12)	−780	225–580
C_2H_5OO	7.7 (−12)	3.8 (−13)	−2,300	200–500
$(CH_3)_3CCH_2OO$	1.5 (−11)	1.4 (−13)	−1,380	250–360
$cyclo$-C_5H_9OO	1.8 (−11)	2.1 (−13)	−1,323	250–360
$cyclo$-$C_6H_{11}OO$	1.7 (−11)	2.6 (−13)	−1,245	250–360
$C_6H_5CH_2OO$	1.0 (−11)	3.8 (−13)	−980	275–450
$CH_3C(O)OO^b$	1.4 (−11)	5.2 (−13)	−983	250–400
$C_2H_5C(O)OO$	1.0 (−11)			
CH_3COCH_2OO	9.0 (−12)			
$HOCH_2OO$	1.2 (−11)	5.6 (−15)	−2,300	275–335
$HOCH_2CH_2OO$	1.0 (−11)			
$(CH_3)_2C(OH)CH_2OO$	1.3 (−11)			
$(CH_3)CH(OH)CH(CH_3)OO$	1.5 (−11)			
$(CH_3)_2C(OH)C(CH_3)_2OO$	2 (−11)			
CH_2ClOO^c	5.2 (−12)	3.3 (−13)	−820	260–590
$CHCl_2OO$	5.8 (−12)	5.6 (−13)	−700	285–440
CCl_3OO	5.1 (−12)	4.8 (−13)	−706	298–375
CF_2ClOO	3.4 (−12)			
$CFCl_2CH_2OO$	9.1 (−12)			
CF_2ClCH_2OO	6.8 (−12)			
CF_3CFHOO	3.8 (−12)	1.8 (−13)	−910	210–360
CF_3CF_2OO	1.2 (−12)			
CH_2BrOO	6.7 (−12)			

[a]Powers of ten in parentheses, $k=A\ \exp(−B/T)$ (cm^3 molecule^{-1} s^{-1}); Source: Atkinson (1997), Atkinson et al. (2006), Lesclaux (1997). In the majority of cases the reactions lead to the formation of hydroperoxides
[b]See Table 7.6 for the distribution of products
[c]Two product channels are observed leading to the products CH_2ClOOH (20%), and $HC(O)Cl$ (80%)

Table 7.20 Rate coefficients and Arrhenius parameters for self-reactions of alkyl peroxyl radicals[a]

	k_{298}	$k_a/(k_a+k_b)$[b]	A	B	T-range
Alkyl peroxyl radicals					
CH_3OO	3.7 (−13)	0.33	9.5 (−14)	−410	250–650
C_2H_5OO	7.0 (−14)	0.63	7.0 (−14)	0	230–460
$CH_3CH_2CH_2OO$	3 (−13)				
$CH_3CH(OO)CH_3$	1.1 (−15)	0.56	1.7 (−12)	2,200	215–260
$(CH_3)_3CH_2OO$	1.2 (−12)	0.39–0.43	1.7 (−15)	−1,960	250–375
cyclo-C_5H_9OO	4.5 (−14)		2.9 (−13)	555	245–375
cyclo-$C_6H_{11}OO$	4.2 (−14)	0.3	7.7 (−14)	184	250–275
$(CH_3)_3COO$	3.0 (−17)	1	4.1 (−11)	4,200	295–420
Acyl peroxyl radicals					
$CH_3C(O)OO$	1.4 (−11)	1	2.3 (−12)	−530	250–370
$C_2H_5C(O)OO$	1.2 (−11)	1			
$C_6H_5C(O)OO$	1.4 (−11)	1			
Substituted alkyl peroxyl radicals					
$HOCH_2OO$	6.2 (−12)	0.89	5.7 (−14)	−745	275–325
$HOCH_2CH_2OO$	2.3 (−12)	0.50	8.0 (−14)	−1,000	298–470
$HOC(CH_3)_2CH_2OO$	4.8 (−12)	0.59	1.4 (−14)	−1,740	305–400
CH_3OCH_2OO	2.1 (−12)				
$(CH_3)_2COC(CH_3)_2CH_2OO$	2.7 (−12)				
CH_3SCH_2OO	7.9 (−12)				
$CH_3C(O)CH_2OO$	8.0 (−12)	0.75			
$(CH_3)_2C(OH)C(OO)(CH_3)_2$	5.0 (−15)	1	3.0 (−13)	1,220	
$CH_2{\rightarrow}CHCH_2OO$	7.0 (−13)	0.61			
$C_6H_5CH_2OO$	7.7 (−12)	0.40	2.8 (−14)	−1,680	275–450
$CH_3CH(OH)CH(OO)CH_3$	6.6 (−13)	<0.3	8.4 (−15)	−1,300	298–470
$CH_3CHBrCH(OO)CH_3$	7.3 (−13)	~0.5	1.3 (−14)	−1,200	275–370

[a]Powers of ten in parentheses; $k=A\exp(-B/T)$ (cm^3 molecule^{-1} s^{-1}); Source: Lesclaux (1997)
[b]Ratio of the rate coefficient k_a to the sum of both rate coefficients for the two channels of the reaction (a) $RO_2+RO_2\rightarrow RO+RO+O_2$ and (b) $RO_2+RO_2\rightarrow ROH+R_{-H}O+O_2$

Table 7.21 Rate coefficients at 298 K (cm^3 molecule^{-1} s^{-1}) for reactions of CH_3OO with other peroxy radicals in the gas-phase[a]

	k_{298}	k_a/k[b]
$CH_3OO+C_2H_5OO$	2.0 (−13)	0.48
$CH_3OO+CH_3C(O)OO$	9.5 (−12)	0.66
$CH_3OO+CH_3C(O)CH_2OO$	3.8 (−12)	0.54
$CH_3OO+(CH_3)_3CH_2OO$	1.5 (−12)	0.36
$CH_3OO+C_6H_5CH_2OO$	<2 (−12)	0.37
$CH_3OO+cyclo\text{-}C_6H_{11}OO$	9.0 (−14)	0.32
$CH_3OO+(CH_3)_3COO$	3.1 (−15)	0.66
CH_3OO+CH_2ClOO	2.5 (−12)	0.66
$CH_3OO+CCl_3OO$	6.6 (−12)	0.66

[a]Powers of ten in parentheses; Source of data: Lesclaux (1997)
[b]Ratio of the rate coefficient k_a for the non-terminating channel (a) $CH_3O_2+RO_2\rightarrow CH_3O+RO+O_2$ to the sum of rate coefficients of all three channels (a) and (b) $CH_3O_2+RO_2\rightarrow CH_3OH+R_{-H}O+O_2$ and (c) $CH_3O_2+RO_2\rightarrow HCHO+ROH+O_2$

Table 7.22 Rate coefficients and Arrhenius parameters for reactions of alkoxyl radicals with molecular oxygen[a]

	k_{298}	A	B	T-range
CH_3O	1.9 (−15)	7.2 (−14)	1,080	290–610
C_2H_5O	8.1 (−15)	2.4 (−14)	325	295–355
$CH_3CH_2CH_2O$	1.0 (−14)	2.6 (−14)	253	220–380
$(CH_3)_2CHO$	7.0 (−15)	1.5 (−14)	230	210–370
$CH_3(CH_2)_2CH_2O$	1.4 (−14)			
$C_2H_5CH(O)CH_3$	8.7 (−15)			
$C_2H_5CH(O)C_2H_5$	9.6 (−15)			

[a]Powers of ten in parentheses, $k = A \exp(−B/T)$ (cm^3 molecule^{-1} s^{-1}); the first five entries are from Atkinson et al. (2006), values for 2-butoxyl and 3-pentoxyl are averages of data listed by Devolder (2003)

Table 7.23 Experimental rate coefficients and Arrhenius parameters for the decomposition and isomerization of alkoxyl radicals[a]

	k_{298}/k_{O2}	k_{298}	A	B	T-range
Decomposition					
$CH_3CH_2O \rightarrow CH_3 + HCHO$	6.4 (14)	5	1.1 (13)	8,660	390–470
$(CH_3)_2CHO \rightarrow CH_3 + CH_3CHO$	1.2 (17)	8.5 (2)	1.2 (14)	7,650	330–405
$C_2H_5CH(O)CH_3 \rightarrow CH_3CHO + C_2H_5$	2.3 (18)	2.0 (4)	2.0 (14)	6,860	300–500
$(CH_3)_2CHCH_2O \rightarrow (CH_3)_2CH + HCHO$	5.0 (18)	4.0 (4)			
$(CH_3)_3CO \rightarrow CH_3COCH_3 + CH_3$	–	2.5 (3)	1.0 (14)	7,320	325–385
$C_3H_7CH(O)CH_3 \rightarrow C_3H_7 + CH_3CHO$	1.1 (18)	9.0 (3)			
$(C_2H_5)_2CHO \rightarrow C_2H_5 + C_2H_5CHO$	3.3 (18)	2.6 (4)			
Isomerization					
$C_3H_7CH_2O \cdot \rightarrow \dot{C}H_2(CH_2)_2CH_2OH$	1.6 (19)	2.3 (5)	2.4 (11)	4,130	295–400
$C_3H_7CH(O \cdot)CH_3 \rightarrow CH_2(CH_2)_2CHOHCH_3$	3.1 (19)	2.5 (5)			
$C_4H_9CH(O \cdot)CH_3 \rightarrow CH_3\dot{C}H(CH_2)_2CHOHCH_3$	~4 (20)	~3 (6)			
$C_2H_5CH(O \cdot)C_3H_7 \rightarrow C_2H_5CHOH(CH_2)_2\dot{C}H_2$	~4 (19)	~3 (5)			

[a]Powers of ten in parentheses, $k = A \exp(−B/T)$ (s^{-1}); the data refer to atmospheric pressure (1 atm = 1.01325×10^5 Pa). Source of data: Atkinson et al. (1995, 2006), Devolder (2003), Peeters et al. (2004)

Comments: The rate coefficients of decomposition and isomerization reactions of alkoxy radicals are pressure dependent. The fall-off behavior has been assessed from theoretical studies, which show that at atmospheric pressure the reactions have not quite reached the high pressure limit. Most data have been obtained relative to the reactions with oxygen (see previous table). Recent theoretical studies based on *ab-initio* and density functional theory have provided rate coefficients in fairly good agreement with experimental data (see following table). Experimental data for pentoxyl and hexoxyl radicals are still tentative owing to the small data base.

Table 7.24 Calculated rate coefficients and Arrhenius parameters for the decomposition and isomerization of alkoxyl radicals[a]

Reaction	k_{298}	$k_{\infty 298}$	A_∞	E_∞	Ref.[b]
Decomposition					
Ethoxyl → CH_3 + HCHO	3.4	6.4	1.1 (14)	75.5	(a)
	1.5	2.5	3.0 (13)	75.1	(b)
1-Propoxyl → C_2H_5 + HCHO	2.5 (2)	3.5 (2)	1.2 (14)	65.6	(a)
	2.4 (2)	3.4 (2)	4.2 (13)	63.7	(b)
2-Propoxyl → CH_3CHO + CH_3	1.3 (2)	1.8 (2)	2.0 (14)	68.6	(a)
	3.5 (2)	5.6 (2)	7.1 (13)	63.8	(b)
1-Butoxyl → C_3H_7 + HCHO	1.4 (2)	1.5 (2)	1.1 (14)	67.8	(a)
	3.9 (1)	1.7 (2)	4.5 (13)	65.6	(b)
2-Butoxyl → CH_3CHO + C_2H_5	3.5 (4)	5.0 (4)	1.0 (14)	53.1	(a)
	2.7 (4)	3.9 (4)	4.2 (13)	51.8	(b)
2-Butoxyl → CH_3 + C_2H_5CHO	3.1 (1)	3.3 (1)	1.1 (14)	71.5	(a)
	4.6 (1)	2.0 (2)	4.5 (13)	65.1	(b)
2-Methyl-1-propoxyl → $(CH_3)_2$CH$_2$ + HCHO	5.7 (4)	8.7 (4)	1.2 (14)	51.9	(a)
tert-Butoxyl → $(CH_3)_2$CO + CH_3	3.0 (3)	4.0 (3)	2.9 (14)	61.9	(a)
1-Pentoxyl → C_4H_9 + HCHO	2.6	1.8 (2)	4.3 (13)	65.2	(b)
2-Pentoxyl → C_3H_7CHO + CH_3	4.7 (1)	4.7 (1)	1.1 (14)	70.7	(a)
	5.1 (1)	2.6 (2)	7.1 (13)	65.7	(b)
2-Pentoxyl → C_3H_7 + CH_3CHO	2.2 (4)	2.6 (4)	9.8 (13)	54.8	(a)
	1.0 (4)	2.4 (4)	8.3 (13)	54.8	(b)
3-Pentoxyl → C_2H_5 + C_2H_5CHO	3.4 (4)	4.2 (4)	2.2 (14)	55.6	(a)
	3.3 (4)	4.6 (4)	9.1 (13)	53.4	(b)
Isomerization					
1-Butoxyl → HOCH$_2$(CH$_2$)$_2$CH$_2$	1.6 (5)	1.8 (5)	2.5 (12)	40.6	(a)
	1.1 (5)	2.0 (5)	2.9 (12)	42.1	(b)
1-Pentoxyl → HOCH$_2$(CH$_2$)$_2$CHCH$_3$	2.2 (6)	2.6 (6)	1.7 (12)	33.1	(a)
	2.0 (6)	2.7 (6)	2.0 (12)	33.7	(b)
2-Pentoxyl → CH_3CHOH(CH$_2$)$_2$CH	3.3 (5)	3.7 (5)	2.4 (12)	38.9	(a)
	5.0 (5)	6.6 (5)	2.8 (12)	38.0	(b)
1-Hexoxyl → HOCH$_2$(CH$_2$)$_2$CHC$_2$H$_5$	2.9 (6)	3.3 (6)	1.4 (12)	32.2	(a)
2-Hexoxyl → CH_3CHOH(CH$_2$)$_2$CHCH$_3$	5.0 (6)	5.8 (6)	1.5 (12)	31.0	(a)
3-Hexoxyl → C_2H_5CHOH(CH$_2$)$_2$CH$_2$	1.7 (5)	1.8 (5)	2.3 (12)	40.2	(a)

[a]Powers of ten are shown in parentheses; $k_\infty = A_\infty \exp(-E_\infty/R_g T)$ (s^{-1}) is the rate coefficient at infinite pressure, E_∞ (kJ mol^{-1}) is the activation energy, R_g is the gas constant, and T is temperature; the data refer to atmospheric pressure (1 atm = 1.01325×10^5 Pa)
[b](a) Méreau et al. (2000, 2003) (b) Somnitz and Zellner (2000); the data were derived from *ab initio* calculations of energy barriers, density functional theory, and RRKM statistical calculations of the pressure dependence

References

Atkinson, R., Gas-phase tropospheric chemistry of organic compounds, J. Phys. Chem. Ref. Data, Monogr. No. 2, 1–216 (1994)

Atkinson, R., J. Phys. Chem. Ref. Data **26**, 215–290 (1997)

Atkinson, R., J. Arey, Chem. Rev. **103**, 4605–4638 (2003)

Atkinson, R., E.S.C. Kwok, J. Arey, S. Aschmann, Faraday Discuss. **100**, 23–37 (1995)

Atkinson, R., D.L. Baulch, R.A. Cox, J.N. Crowley, R.F. Hampson, R.G., Hynes, M.E. Jenkin, M.J. Rossi, J. Troe, Atmos. Chem. Phys. **6**, 3625–4055 (2006)

Brown, A.C., C.E. Canosa-Mas, A.D. Parr, R.P. Wayne, Chem. Phys. Lett. **161**, 491–496 (1989)

Calvert, J.G., R. Atkinson, J.A. Kerr, S. Madronich, G.K. Moortgat, T.J. Wallington, G. Yarwood, *The Mechanisms of Atmospheric Oxidation of the Alkenes* (Oxford University Press, Oxford, 2000)

Calvert, J.G., R. Atkinson, K.H. Becker, R.M. Kamens, J.H. Seinfeld, T.J. Wallington, G. Yarwood, *The Mechanisms of Atmospheric Oxidation of Aromatic Hydrocarbons* (Oxford University Press, New York, 2002)

Devolder, P., J. Photochem. Photobiol. A **157**, 137–147 (2003)

Lesclaux, R., in *Peroxy Radicals*, ed. by Z.B. Alfassi (Wiley, Chichester, 1997), pp. 81–112

Lightfoot, P.D., R.A. Cox, J.N. Crowley, M. Destriau, G.D. Hayman, M.E. Jenkin, G.K. Moortgat, F. Zabel, Atmos. Environ. **26A**, 1805–1964 (1992)

Méreau, R., M.-T. Rayez, F. Caralp, J.-C. Rayez, Phys. Chem. Chem. Phys. **2**, 3765–3772 (2000); Phys. Chem. Chem. Phys. **5**, 4828–4833 (2003)

Oehlers, C., F. Temps, H.G. Wagner, M. Wolf, Ber. Bunsenges. Phys. Chem. **96**, 171–175 (1992)

Peeters, J., G. Fantechi, L. Vereecken, J. Atmos. Chem. **48**, 59–80 (2004)

Somnitz, H., R. Zellner, Phys. Chem. Chem. Phys. **2**, 4319–4325 (2000)

Wallington, T.J., O.J. Nielsen, J. Sehested, in: *Peroxy Radicals*, ed. by Z.B. Alfassi (Wiley, Chichester, 1997), pp. 113–172

Warneck, P., *Chemistry of the Natural Atmosphere*, 2nd edn. (Academic Press, San Diego, 2000)

Wayne, R.P., I. Barnes, P. Biggs, J.P. Burrows, C.E. Canosa-Mas, J. Hjorth, G. Le Bras, G.K. Moortgat, D. Perner, G. Poulet, G. Restrelli, H. Sidebotttom, Atmos. Environ. **25A**, 1–203 (1991)

7.5   Reactions of Halogen Compounds

Table 7.25   Rate parameters for reactions of (inorganic) halogen species[a]

Reaction	k_{298}	$A*$	α	B
Fluorine species				
$O + FO \rightarrow O_2 + F$	2.7 (–11)	2.7 (–11)	0	0
$O + FO2 \rightarrow F + O2$	5.0 (–11)	5.0 (–11)	0	0
$F + O_3 \rightarrow FO + O_2$	1.0 (–11)	2.2 (–11)	0	230
$F + O_2 + N_2 \rightarrow FO_2 + N_2$	5.8 (–33) (k_0)	5.8 (–33)	–1.7	0
	1.2 (–10) (k_∞)	1.2 (–10)	F_c=0.5	
$FO_2 + N_2 \rightarrow F + O_2 + N_2$	1.5 (–17) (k_0 (s⁻¹))	8.4 (–9)	–1.25	5,990
	3.1 (5) (k_∞ (s⁻¹))	1.7 (14)	0.45	5,990
	F_c=0.5			
$F + HNO_3 \rightarrow HF + NO_3$	2.3 (–11)	6.0 (–12)	0	–400
$F + H_2 \rightarrow HF + H$	2.4 (–11)	1.1 (–10)	0	450
$F + H_2O \rightarrow HF + OH$	1.4 (–11)	1.4 (–11)	0	0
$F + CH_4 \rightarrow HF + CH_3$	6.7 (–11)	1.6 (–10)	0	260
$FO + NO \rightarrow NO2 + F$	2.2 (–11)	8.2 (–12)	0	–300
$FO + FO \rightarrow 2\ F + O_2$	1.0 (–11)	1.0 (–11)	0	0
$FO_2 + NO \rightarrow FNO + O_2$	7.5 (–13)	7.5 (–12)	0	690
$FO_2 + NO_2 \rightarrow$ products	4.0 (–14)	3.8 (–11)	0	2,040
$CF_3 + O_2 + M \rightarrow CF_3O_2 + M$	3.0 (–29) (k_0)	3.0 (–29)	4.0	0
	3.0 (–12) (k_∞)	3.0 (12)	1.0	0

(continued)

Table 7.25 (continued)

Reaction	k_{298}	$A*$	α	B
$CF_3O_2 + NO \rightarrow CF_3O + NO_2$	1.6 (−11)	5.4 (−12)	0	−320
$CF_3O_2 + NO_2 + M \rightarrow CF_3NO_2 + M$	1.5 (−29) (k_0)	1.5 (−29)	2.2	0
	9.6 (−12) (k_∞)	9.6 (−12)	1	0
$CF_3O + O_3 \rightarrow CF_3O_2 + O_2$	1.8 (−14)	2.0 (−12)	0	1,400
$CF_3O + NO \rightarrow CF_2O + FNO$	5.4 (−11)	3.7 (−11)	0	−110
$CF_3O + CH_4 \rightarrow CH_3 + CF_3OH$	2.2 (−14)	2.6 (−12)	0	1,420
$CF_3O + C_2H_6 \rightarrow C_2H_5 + CF_3OH$	1.3 (−12)	4.9 (−12)	0	400
Chlorine species				
$O + HOCl \rightarrow HO + ClO$	1.7 (−13)	1.7 (−13)	0	0
$O + HCl \rightarrow OH + Cl$	1.5 (−16)	1.0 (−11)	0	3,300
$O + ClO \rightarrow Cl + O_2$	3.7 (−11)	2.5 (−11)	0	−110
$O + OClO \rightarrow ClO + O_2$	1.0 (−13)	2.4 (−11)	0	960
$O + OClO + N_2 \rightarrow ClO_3 + N_2$	1.8 (−31) (k_0)	1.8 (−31)	−1	0
	2.8 (−11) (k_∞)	2.8 (−11)	$F_c = 0.5$	
$O + Cl_2O \rightarrow ClO + ClO$	4.5 (−12)	2.7 (−11)	0	530
$O + ClONO_2 \rightarrow products$	2.2 (−13)	4.5 (−12)	0	900
$O_3 + OClO \rightarrow products$	3.0 (−19)	2.1 (−12)	0	4,700
$OH + Cl_2 \rightarrow HOCl + Cl$	6.7 (−14)	1.4 (−12)	0	900
$OH + ClO \rightarrow Cl + HO_2$	1.8 (−11)	7.4 (−12)	0	−270
$\rightarrow HCl + O_2$	1.2 (−12)	6.0 (−13)	0	−230
$OH + OClO \rightarrow HOCl + O_2$	6.8 (−12)	4.5 (−13)	0	−800
$OH + HCl \rightarrow H_2O + Cl$	8.0 (−13)	1.8 (−12)	0	240
$OH + HOCl \rightarrow H_2O + ClO$	5.0 (−13)	3.0 (−12)	0	500
$OH + CH_3OCl \rightarrow products$	7.1 (−13)	2.5 (−12)	0	370
$OH + ClNO_2 \rightarrow HOCl + NO_2$	3.6 (−14)	2.4 (−12)	0	1,250
$OH + ClONO_2 \rightarrow products$	4.0 (−13)	1.2 (−12)	0	330
$Cl + O_2 + M \rightarrow ClOO + M$	1.4 (−33) [N_2] (k_0)	1.4 (−33)	−3.9	0
	1.6 (−33) [O_2] (k_0)	1.6 (−33)	−2.9	0
$ClOO + N_2 \rightarrow Cl + O_2 + N_2$	6.3 (−13) (k_0 (s^{-1}))	4.1 (−10)	0	1,820
$Cl + O_3 \rightarrow ClO + O_2$	1.2 (−11)	2.3 (−11)	0	200
$Cl + NO_3 \rightarrow ClO + NO_2$	2.4 (−11)	2.4 (−11)	0	0
$Cl + ClOO \rightarrow ClO + ClO$	1.2 (−11)	1.2 (−11)	0	0
$\rightarrow Cl_2 + O_2$	2.3 (−10)	2.3 (−10)	0	0
$Cl + OClO \rightarrow ClO + ClO$	5.7 (−11)	3.2 (−11)	0	−170
$Cl + Cl_2O \rightarrow Cl_2 + ClO$	9.6 (−11)	6.2 (−11)	0	−130
$Cl + Cl_2O_2 \rightarrow Cl_2 + ClOO$	1.0 (−10)	1.0 (−10)	0	0
$Cl + ClONO_2 \rightarrow Cl_2 + NO_3$	1.0 (−11)	6.2 (−12)	0	−130
$Cl + ClNO \rightarrow NO + Cl_2$	8.1 (−11)	5.8 (−11)	0	−100
$Cl + H_2 \rightarrow HCl + H$	1.7 (−14)	3.9 (−11)	0	2,310
$Cl + HO_2 \rightarrow HCl + O_2$	3.4 (−11)			
$\rightarrow OH + ClO$	9.3 (−12)	6.3 (−11)	0	570
	4.3 (−11) overall	4.3 (−11)	0	0
$Cl + H_2O_2 \rightarrow HCl + HO_2$	4.1 (−13)	1.1 (−11)	0	980
$Cl + HOCl \rightarrow products$	1.6 (−12)	2.5 (−12)	0	130
$Cl + CH_4 \rightarrow HCl + CH_3$	1.0 (−13)	7.3 (−12)	0	1,280
$Cl + CH_3D \rightarrow products$	6.8 (−14)	7.0 (−12)	0	1,380
$Cl + C_2H_6 \rightarrow HCl + C_2H_5$	5.7 (−11)	7.2 (−11)	0	70

(continued)

Table 7.25 (continued)

Reaction	k_{298}	$A*$	α	B
$Cl+HCHO \rightarrow HCl+HCO$	7.3 (−11)	8.1 (−11)	0	30
$Cl+CH_3O_2 \rightarrow products$	1.6 (−10)			
$Cl+C_2H_5O_2 \rightarrow ClO+C_2H_5O$	7.4 (−11)			
$\rightarrow HCl+C_2H_4O_2$	7.7 (−11)			
$Cl+CH_3OOH \rightarrow products$	5.7 (−11)			
$Cl+CH_3ONO_2 \rightarrow products$	2.3 (−13)	1.3 (−11)	0	1,200
$ClO+HO_2 \rightarrow products^b$	6.9 (−12)	2.2 (−12)	0	−340
$ClO+NO \rightarrow Cl+NO_2$	1.7 (−11)	6.2 (−12)	0	−295
$ClO+NO_2+N_2 \rightarrow ClONO_2+N_2$	1.6 (−31) (k_0)	1.6 (−31)	−3.4	0
	7.0 (−11) (k_∞)	7.0 (−11)	$F_c=0.5$	0
$ClO+CH_3O_2 \rightarrow products$	2.2 (−12)	3.3 (−12)	0	115
$ClO+NO_3 \rightarrow ClOO+NO_2$	3.4 (−13)			
$\rightarrow OClO+NO_2$	1.2 (−13)			
	4.6 (−13) overall	4.6 (−13)	0	0
$ClO+ClO \rightarrow Cl_2+O_2$	4.8 (−15)	1.0 (−12)	0	1,590
$\rightarrow Cl+ClOO$	8.0 (−15)	3.0 (−11)	0	2,450
$\rightarrow Cl+OClO$	3.5 (−15)	3.5 (−13)	0	1,370
$ClO+ClO+N_2 \rightarrow Cl_2O_2+N_2$	2.0 (−32) (k_0)	2.0 (−32)	−4	0
	1.0 (−11) (k_∞)	1.0 (−11)	$F_c=0.45$	0
$Cl_2O_2+N_2 \rightarrow ClO+ClO$	2.3 (−18) $(k_0 (s^{-1}))$	3.7 (−7)	0	7,690
	1.1 (3) $(k_\infty (s^{-1}))$	1.8 (14)	$F_c=0.45$	7,690
$ClO+OClO+N_2 \rightarrow Cl_2O_3+N_2$	6.2 (−32) (k_0)	6.2 (−32)	−4.7	0
	2.4 (−11) (k_∞)	2.4 (−11)	$F_c=0.6$	0
$Cl_2O_3+N_2 \rightarrow ClO+OClO+N_2$	4.1 (−16) $(k_0 (s^{-1}))$	1.4 (−10)	0	3,810
	1.6 (5) $(k_\infty (s^{-1}))$	2.5 (12)	0	4,940
$OClO+O_3 \rightarrow ClO_3+O_2$	3.0 (−19)	2.1 (−12)	0	4,700
$CH_2ClO+O_2 \rightarrow CHClO+HO_2$	6.0 (−14)			
$CH_2ClO_2+HO_2 \rightarrow CH_2ClOOH+O_2$	5.2 (−12)	3.3 (−13)	0	−820
$CH_2ClO_2+NO \rightarrow CH_2ClO+NO_2$	1.9 (−11)	7.0 (−12)	0	−300
$CCl_3O_2+NO \rightarrow CCl_2O+NO_2+Cl$	1.8 (−11)	7.3 (−12)	0	−270
$CCl_2FO_2+NO \rightarrow CClFO+NO_2+Cl$	1.5 (−11)	4.5 (−12)	0	−350
$CClF_2O_2+NO \rightarrow CF_2O+NO_2+Cl$	1.5 (−11)	3.8 (−12)	0	−400
Bromine species				
$O+BrO \rightarrow Br+O_2$	4.1 (−11)	1.9 (−11)	0	−230
$O+HBr \rightarrow OH+Br$	3.8 (−14)	5.8 (−12)	0	1,500
$O+HOBr \rightarrow OH+BrO$	2.8 (−11)	1.2 (−10)	0	430
$O+BrONO_2 \rightarrow NO_3+BrO$	3.9 (−11)	1.9 (−11)	0	−215
$OH+Br_2 \rightarrow HOBr+Br$	4.6 (−11)	2.1 (−11)	0	−240
$OH+BrO \rightarrow products$	4.1 (−11)	1.8 (−11)	0	−250
$OH+HBr \rightarrow H_2O+Br$	1.1 (−11)	5.5 (−12)	0	−205
$Br+O_3 \rightarrow BrO+O_2$	1.2 (−12)	1.7 (−11)	0	800
$Br+HO_2 \rightarrow HBr+O2$	1.7 (−12)	7.7 (−12)	0	450
$Br+HCHO \rightarrow HCO+HBr$	1.1 (−12)	1.7 (−11)	0	800
$Br+NO_2+N_2 \rightarrow BrNO_2+N_2$	4.2 (−31) (k_0)	4.2 (−31)	−2.4	0
	2.7 (−11) (k_∞)	2.7 (−11)	$F_c=0.55$	0
$Br+NO_3 \rightarrow BrO+NO_2$	1.6 (−11)			
$Br+OClO \rightarrow BrO+ClO$	3.5 (−13)	2.7 (−11)	0	1,300

(continued)

Table 7.25 (continued)

Reaction	k_{298}	$A*$	α	B
$Br+Cl_2O \rightarrow BrCl+ClO$	4.3 (−12)	2.1 (−11)	0	470
$Br+Cl_2O_2 \rightarrow BrCl+ClOO$	3.0 (−12)			
$BrO+HO_2 \rightarrow products^c$	2.4 (−11)	4.5 (−12)	0	−500
$BrO+NO \rightarrow Br+NO_2$	2.1 (−11)	8.7 (−12)	0	−260
$BrO+NO_2+N_2 \rightarrow BrONO_2+N_2$	4.7 (−31) (k_0)	4.7 (−31)	−3.1	0
	1.8 (−11) (k_∞)	1.8 (−11)	$F_c=0.4$	0
$BrO+NO_3 \rightarrow BrOO+NO_2$	1.0 (−12)			
$BrO+BrO \rightarrow Br_2+O_2$	4.8 (−13)	2.9 (−14)	0	−840
$\rightarrow 2\,Br+O_2$	2.7 (−12)	2.7 (−12)	0	0
$BrO+ClO \rightarrow OClO+Br$	6.8 (−12)	1.6 (−12)	0	−430
$\rightarrow ClOO+Br$	6.1 (−12)	2.9 (−12)	0	−220
$\rightarrow BrCl+O_2$	1.0 (−12)	5.8 (−13)	0	−170
$CH_2BrO_2+NO \rightarrow HCHO+NO_2+Br$	1.1 (−11)	4.0 (−12)	0	−300
Iodine species				
$O+I_2 \rightarrow IO+I$	1.25 (−10)			
$O+IO \rightarrow I+O_2$	1.4 (−10)			
$OH+I_2 \rightarrow HOI+I$	2.1 (−10)	2.1 (−10)	0	0
$OH+HI \rightarrow H_2O+I$	7.0 (−11)	1.6 (−11)	0	−440
$I+O_3 \rightarrow IO+O_2$	1.3 (−12)	2.1 (−11)	0	830
$I+HO_2 \rightarrow HI+O_2$	3.8 (−13)	1.5 (−11)	0	1,090
$I+NO+N_2 \rightarrow INO+N_2$	1.8 (−32) (k_0)	1.8 (−32)	−1.0	0
	1.7 (−11) (k_∞)	1.7 (−11)	$F_c=0.6$	0
$I+NO_2+N_2 \rightarrow INO_2+N_2$	3.0 (−31) (k_0)	3.0 (−31)	−1.0	0
	6.6 (−11) (k_∞)	6.6 (−11)	$F_c=0.63$	0
$I+BrO \rightarrow IO+Br$	1.2 (−11)			
$IO+HO_2 \rightarrow HOI+O_2$	8.4 (−11)	1.4 (−11)	0	−540
$IO+ClO \rightarrow I+OClO$	6.6 (−12)			
$\rightarrow ICl+O_2$	2.4 (−12)			
$\rightarrow Cl+I+O_2$	3.0 (−12)			
	1.2 (−11) overall	4.7 (−12)	0	−280
$IO+BrO \rightarrow products^d$	8.5 (−11)	1.5 (−11)	0	−510
$IO+IO \rightarrow products^e$	9.9 (−11)	5.4 (−11)	0	−180
$IO+NO \rightarrow I+NO_2$	1.95 (−11)	7.15 (−12)	0	−300
$IO+NO_2+N_2 \rightarrow IONO_2+N_2$	7.7 (−31)	7.7 (−31)	−5.0	0
	1.6 (−11)	1.6 (−11)	$F_c=0.4$	0
$IONO_2+M \rightarrow products$	2.9 (−3) (s^{-1}) (10^5 Pa)	1.1 (−15)	0	12,060
$INO+INO \rightarrow I_2+2\,NO$	1.3 (−14)	8.4 (−11)	0	2,620
$INO_2+INO_2 \rightarrow I_2+2\,NO_2$	6.7 (−12)	1.1 (−12)	0	−542

[a]Powers of ten in parentheses, $k=A*(T/T)\exp(−B/T)$ (cm^3 molecule^{-1} s^{-1}); From Atkinson et al. (2007), Sander et al. (2006). For reactions of OH radicals with halogenated hydrocarbons see Table 7.26; a more complete list of reactions of chlorine atoms is given in Table 7.27

[b]The dominant products are $HOCl+O_2$, the product channel $HCl+O_3$ contributes less than a few percent, at most

[c]The dominant products are $HOBr+O_2$, the product channel $HBr+O_3$, makes an insignificant contribution

[d]The major products are $Br+OIO$, they contribute 80% to the overall reaction

[e]The product channel leading to $I+OIO$, which is well established, contributes about 38% to the total reaction. The formation of I_2+O_2 is insignificant

Table 7.26 Rate parameters for reactions of OH radicals with halogenated hydrocarbons[a]

Reaction	k_{298}	A	B
$OH + CH_3Cl \rightarrow CH_2Cl + H_2O$	3.6 (−14)	2.4 (−12)	1,250
$OH + CH_2Cl_2 \rightarrow CHCl_2 + H_2O$	1.0 (−13)	1.9 (−12)	870
$OH + CHCl_3 \rightarrow CCl_3 + H_2O$	1.0 (−13)	2.2 (−12)	920
$OH + CH_2FCl \rightarrow CHClF + H_2O$	4.1 (−14)	2.4 (−12)	1,210
$OH + CHFCl_2 \rightarrow CFCl_2 + H_2O$	3.0 (−14)	1.2 (−12)	1,100
$OH + CHF_2Cl \rightarrow CF_2Cl + H_2O$	4.8 (−15)	1.05 (−12)	1,600
$OH + CH_3CH_2Cl \rightarrow$ products	3.7 (−13)	5.4 (−12)	800
$OH + CH_3CCl_3 \rightarrow CH_2CCl_3 + H_2O$	1.0 (−14)	1.64 (−12)	1,520
$OH + CH_3CFCl_2 \rightarrow CH_2CFCl_2 + H_2O$	5.8 (−15)	1.25 (−12)	1,600
$OH + CH_3CF_2Cl \rightarrow CH_2CF_2Cl + H_2O$	3.4 (−15)	1.3 (−12)	1,770
$OH + CH_2ClCF_2Cl \rightarrow CHClCF_2Cl + H_2O$	1.7 (−14)	3.6 (−12)	1,600
$OH + CH_2ClCF_3 \rightarrow CHClCF_3 + H_2O$	1.4 (−14)	5.6 (−13)	1,100
$OH + CHCl_2CF_2Cl \rightarrow CCl_2CF_2Cl + H_2O$	5.1 (−14)	7.7 (−13)	810
$OH + CHFClCFCl_2 \rightarrow CFClCFCl_2 + H_2O$	1.6 (−14)	7.1 (−13)	1,140
$OH + CHFClCF_2Cl \rightarrow CFClCF_2Cl + H_2O$	1.3 (−14)	8.6 (−13)	1,250
$OH + CHFClCF_3 \rightarrow CFClCF_3 + H_2O$	9.0 (−15)	7.1 (−13)	1,300
$OH + CH_3F_2CFCl_2 \rightarrow$ products	2.4 (−15)	7.7 (−13)	1,720
$OH + CHCl_2CF_2CF_3 \rightarrow$ products	2.5 (−14)	6.3 (−13)	960
$OH + CHFClCF_2CF_2Cl \rightarrow$ products	8.9 (−15)	5.5 (−13)	1,230
$OH + CH_2=CHCl \rightarrow$ products	6.9 (−12)	1.3 (−12)	−500
$OH + CH_2=CCl_2 \rightarrow$ products	1.1 (−11)	1.9 (−12)	−530
$OH + cis\text{-}CHCl=CHCl \rightarrow$ products	2.6 (−12)	1.9 (−12)	−90
$OH + trans\text{-}CHCl=CHCl \rightarrow$ products	2.3 (−12)	1.0 (−12)	−250
$OH + CHCl=CCl_2 \rightarrow$ products	2.2 (−12)	8.0 (−13)	−300
$OH + CCl_2=CCl_2 \rightarrow$ products	1.7 (−13)	4.7 (−12)	990
$OH + CCl_3CHO \rightarrow CCl_3O + H_2O$	1.3 (−12)	9.1 (−12)	580
$OH + CH_3Br \rightarrow CH_2Br + H_2O$	1.8 (−13)	2.35 (−12)	1,300
$OH + CH_2Br_2 \rightarrow CHBr_2 + H_2O$	1.2 (−13)	2.0 (−12)	840
$OH + CHBr_3 \rightarrow CBr_3 + H_2O$	1.8 (−13)	1.35 (−12)	600
$OH + CHF_2Br \rightarrow CF_2Br + H_2O$	1.0 (−14)	1.0 (−12)	1,380
$OH + CH_2ClBr \rightarrow CHClBr + H_2O$	1.1 (−13)	2.4 (−12)	920
$OH + CH_2BrCH_3 \rightarrow$ products	3.4 (−13)	2.9 (−12)	640
$OH + CH_2BrCF_3 \rightarrow CHBrCF_3 + H_2O$	1.6 (−14)	1.4 (−12)	1,340
$OH + CHFBrCF_3 \rightarrow CFBrCF_3 + H_2O$	1.7 (−14)	7.3 (−13)	1,120
$OH + CHClBrCF_3 \rightarrow CClBrCF_3 + H_2O$	4.7 (−14)	1.1 (−12)	940
$OH + CHFClCF_2Br \rightarrow CFClCF_2Br + H_2O$	1.4 (−14)	8.4 (−13)	1,220
$OH + CH_2BrCH_2CH_3 \rightarrow$ products	1.0 (−12)	3.0 (−12)	330
$OH + CH_3CHBrCH_3 \rightarrow$ products	7.5 (−13)	1.85 (−12)	270
$OH + I_2 \rightarrow HOI + I$	1.8 (−10)		
$OH + HI \rightarrow H_2O + I$	3.0 (−11)		
$OH + CH_3I \rightarrow H_2O + CH_2I$	7.2 (−14)	2.9 (−12)	1,100
$OH \rightarrow CF_3I \rightarrow HOI + CF_3$	2.4 (−14)	2.5 (−11)	2,070

[a]Powers of ten in parentheses, $k = A \exp(-B/T)$; Source: Sander et al. (2006), data for 1,2-dichloroethene from Atkinson (1994)

Table 7.27 Rate parameters for reactions of chlorine atoms with hydrocarbons, some oxygen-containing organic compounds and halogenated hydrocarbons[a]

Reaction	k_{298}	A	B
$Cl + CH_4 \rightarrow CH_3 + HCl$	1.0 (−13)	6.6 (−12)	1,240
$Cl + C_2H_6 \rightarrow C_2H_5 + HCl$	5.9 (−11)	8.3 (−11)	100
$Cl + C_3H_8 \rightarrow CH_3CHCH_3 + HCl$	8.0 (−11)	6.54 (−11)	−60
$\rightarrow CH_2CH_2CH_3 + HCl$	6.0 (−11)	7.85 (−11)	80
$Cl + CH_3(CH_2)_2CH_3 \rightarrow products$	2.2 (−10)	2.2 (−10)	0
$Cl + HCHO \rightarrow HCl + HCO$	7.3 (−11)	8.2 (−11)	34
$Cl + CH_3CHO \rightarrow HCl + CH_3CO$	7.9 (−11)	7.9 (−11)	0
$Cl + + CH_3COCH_3 \rightarrow HCl + CH_3COCH_2$	2.5 (−12)	3.9 (−11)	815
$Cl + CH_3OH \rightarrow HCl + CH_2OH$	5.5 (−11)	5.5 (−11)	0
$Cl + C_2H_5OH \rightarrow products$	9.6 (−11)	8.2 (−11)	−45
$Cl + CH_3Cl \rightarrow CH_2Cl + HCl$	4.9 (−13)	2.17 (−11)	1,130
$Cl + CH_2Cl_2 \rightarrow CHCl_2 + HCl$	3.5 (−13)	7.4 (−12)	910
$Cl + CHCl_3 \rightarrow CCl_3 + HCl$	1.2 (−13)	3.3 (−12)	990
$Cl + CH_3F \rightarrow CH_2F + HCl$	3.5 (−13)	1.96 (−11)	1,200
$Cl + CH_2F_2 \rightarrow CHF_2 + HCl$	3.2 (−14)	4.9 (−12)	1,500
$Cl + CHF_3 \rightarrow CF_3 + HCl$	<5.0 (−16)		
$Cl + CH_2FCl \rightarrow CHFCl + HCl$	1.05 (−13)	5.9 (−12)	1,200
$Cl + CHFCl_2 \rightarrow CFCl_2 + HCl$	2.0 (−14)	6.0 (−12)	1,700
$Cl + CHF_2Cl \rightarrow CF_2Cl + HCl$	1.6 (−15)	5.6 (12)	2,430
$Cl + CH_3CCl_3 \rightarrow CH_2CCl_3$	8.5 (−15)	3.23 (−12)	1,770
$Cl + CH_3CH_2F \rightarrow CH_3CHF + HCl$	6.0 (−12)	1.82 (−11)	330
$\rightarrow CH_2CH_2F + HCl$	6.0 (−13)	1.4 (−11)	940
$Cl + CH_2FCH_2F \rightarrow CHFCH_2F + HCl$	6.7 (−13)	2.27 (−11)	1,050
$Cl + CH_3CHF_2 \rightarrow CH_3CF_2 + HCl$	2.4 (−13)	5.8 (−12)	950
$\rightarrow CH_2CHF_2 + HCl$	2.6 (−15)	6.25 (−12)	2,320
$Cl + CH_3CFCl_2 \rightarrow CH_2CFCl_2 + HCl$	2.1 (−15)	3.4 (−12)	2,200
$Cl + CH_3CF_2Cl \rightarrow CH_2CF_2Cl + HCl$	4.3 (−16)	1.35 (−12)	2,400
$Cl + CH_3CF_3 \rightarrow CH_2CF_3 + HCl$	2.6 (−17)	1.44 (−11)	3,940
$Cl + CH_2FCHF_2 \rightarrow CH_2FCF_2 + HCl$	2.5 (−14)	6.8 (−12)	1,670
$\rightarrow CHFCHF_2 + HCl$	2.4 (−14)	9.1 (−12)	1,770
$Cl + CH_2ClCF_3 \rightarrow CHClCF_3$	6.5 (−15)	1.83 (−12)	1,680
$Cl + CH_2FCF_3 \rightarrow CHFCF_3 + HCl$	1.5 (−15)	2.4 (−12)	2,200
$Cl + CHF_2CHF_2 \rightarrow CF_2CHF_2 + HCl$	2.0 (−15)	7.0 (−12)	2,430
$Cl + CHCl_2CF_3 \rightarrow CCl_2CF_3 + HCl$	1.2 (−14)	5.0 (−12)	1,800
$Cl + CHFClCF_3 \rightarrow CFClCF_3 + HCl$	2.7 (−15)	1.13 (−12)	1,800
$Cl + CHF_2CF_3 \rightarrow CF_2CF_3 + HCl$	3.0 (−16)	1.8 (−12)	2,600
$Cl + CH_2ClBr \rightarrow CHClBr + HCl$	3.7 (−13)	6.8 (−12)	870
$Cl + CH_3Br \rightarrow CH_2Br + HCl$	4.4 (−13)	1.4 (−11)	1,030
$Cl + CH_2Br_2 \rightarrow CHBr_2 + HCl$	4.3 (−13)	6.3 (−12)	800
$Cl + CHBr_3 \rightarrow CBr_3 + HCl$	2.8 (−13)	4.85 (−12)	850
$Cl + CH_3I \rightarrow CH_2I + HCl$	1.0 (−12)	2.9 (−11)	1,000

[a]Powers of ten in parentheses, $k = A \exp(−B/T)$; Source of data: Atkinson et al. (2007), Sander et al. (2006)

Table 7.28 Rate coefficients and Arrhenius parameters for gas-phase reactions of NO_3 radicals with halogen-containing hydrocarbons[a]

	k_{298}	A	B	T-range
Chloromethane	3.0 (−17)			
Dichloromethane	3.2 (−17)			
Trichloromethane (chloroform)	2.6 (−17)	8.5 (−13)	2,815	370–520
Chloroethene	2.3 (−16)			
1,1-Dichloroethene	6.6 (−16)			
cis-1,2-Dichloroethene	8.0 (−17)			
trans-1,2-Dichloroethene	6.0 (−17)			
Trichloroethene	1.5 (−16)			
3-Chloro-1-propene	2.9 (−16)			
1-Chloromethylpropene	9.0 (−14)			
3-Chloromethylpropene	2.5 (−14)	1.7 (−12)	1,277	295–470
2-Chloro-1-butene	1.7 (−14)			
3-Chloro-1-butene	2.8 (−14)	2.4 (−13)	1,992	295–470
1-Chloro-2-butene	2.1 (−14)	6.0 (−13)	981	295–470
2-Chloro-2-butene	1.1 (−13)			
Bromomethane	3.0 (−17)			
Dibromomethane	3.2 (−17)			
Iodomethane	3.7 (−17)			
Diiodomethane	3.4 (−17)			
Tribromomethane	2.5 (−17)			
Triiodomethane	2.4 (−17)			

[a]Powers of ten in parentheses, $k = A \exp(-B/T)$; Source: Wayne et al. (1991), Atkinson (1994)

Table 7.29 Rate coefficients and Arrhenius parameters for self-reactions of halogenated peroxy radicals in the gas-phase[a]

	k_{298}	$k_a/(k_a+k_b)^b$	A	B	T-range
CH_2FOO	4.0 (−12)	>077	3.8 (−13)	−700	230–380
CH_2ClOO	3.7 (−12)	>0.8	2.0 (−13)	−875	250–590
CH_2BrOO	1.0 (−12)	~1			
CH_2IOO	9 (−11)				
$CHCl_2OO$	3.9 (−12)	>0.95	2.6 (−13)	−800	285–440
CCl_3OO	4.0 (−12)	>0.98	3.3 (−13)	−745	275–460
CHF_2OO	5.0 (−12)	~1			
CF_3OO	1.8 (−12)				
CH_2ClCH_2OO	2.7 (−12)	0.69	3.5 (−14)	−1,300	230–380
CH_2BrCH_2OO	5.8 (−12)	0.57	5.5 (−14)	−1,390	275–370
CHF_2CF_2OO	2.7 (−12)				
CCl_3CH_2OO	4.7 (−12)				
CF_3CH_2OO	8.4 (−12)				
$CF_3CClHOO$	4.4 (−12)				
CF_3CFHOO	5.9 (−12)	0.4	7.8 (−13)	−605	210–370
CF_3CCl_2OO	3.3 (−12)				
$CF_3CFClOO$	2.6 (−12)				
CF_3CF_2OO	2.1 (−12)				
$CH_3CF_2CH_2OO$	8.6 (−12)				
$CF_3CH(OO)CF_3$	5.6 (−12)				

[a]Powers of ten in parentheses, $k = A \exp(-B/T)$ (cm³ molecule⁻¹); Source Lesclaux (1997)
[b]Ratio of the rate coefficient k_a to the sum of both rate coefficients for the two channels of the reaction (a) $RO_2 + RO_2 \rightarrow RO + RO + O_2$ and (b) $RO_2 + RO_2 \rightarrow ROH + R_{-H}O + O_2$

Table 7.30 Rate coefficients and Arrhenius parameters for gas-phase reactions of halogenated alkyl peroxy radicals with nitric oxide[a]

Peroxy radical	k_{298}	$A*$	α	B	T-range
$CH_2FOO + NO \rightarrow$ products	1.25 (−11)				
$CHF_2OO + NO \rightarrow$ products	1.26 (−11)				
$CF_3OO + NO \rightarrow$ products	1.6 (−11)	1.45 (−11)	−1.2	0	230–430
$CF_2ClOO + NO \rightarrow$ products	1.6 (−11)	1.6 (−11)	−1.5	0	230–460
$CFCl_2OO + NO \rightarrow$ products	1.5 (−11)	1.45 (−11)	−1.3	0	230–430
$CCl_3OO + NO \rightarrow$ products	1.8 (−11)	1.7 (−11)	−1.0	0	230–430
$CH_2ClOO + NO \rightarrow$ products	1.9 (−11)				
$CH_2BrOO + NO \rightarrow$ products	1.1 (−11)				
$FC(O)OO + NO \rightarrow$ products	2.5 (−11)				
$CF_3CH_2OO + NO \rightarrow$ products	1.2 (−11)				
$CF_2ClCH_2OO + NO \rightarrow$ products	1.3 (−11)				
$CFCl_2CH_2OO + NO \rightarrow$ products	1.2 (−11)				
$CF_3CFHOO + NO \rightarrow$ products	1.2 (−11)				
$CF_3CClHOO + NO \rightarrow$ products	1.0 (−11)				
$CF_3C(O)OO + NO \rightarrow$ products	2.8 (−11)	4.0 (−12)	0	−563	220–325
$CH_3CF_2CH_2OO + NO \rightarrow$ products	8.5 (−12)				
$CF_3CH(OO)CF_3 + NO \rightarrow$ products	1.1 (−11)				

[a]Powers of ten in parentheses, $k = A*(T/T^{\ominus})^{\alpha} \exp(-B/T)$ (cm³ molecule⁻¹ s⁻¹); Source of data: Wallington et al. (1997)

References

Atkinson, R., D.L. Baulch, R.A. Cox, J.N. Crowley, R.F. Hampson, R.G. Hynes, M.E. Jenkin, M.J. Rossi, J. Troe. Atmos. Chem. Phys. **7**, 981–1191 (2007)

Lesclaux, R., in *Peroxy Radicals*, ed. by Z.B. Alfassi (Wiley, Chichester, 1997), pp. 81–112

Sander, S.P., R.R. Friedl, D.M. Golden, M.J. Kurylo, G.K. Moortgat, H. Keller-Rudek, P.H. Wine, A.R. Ravishankara, C.E. Kolb, M.J. Molina, B.J. Finlayson-Pitts, R.E. Huie, V.L. Orkin, NASA Chemical Kinetics and Photochemical Data for Use in Atmospheric Studies, Evaluation Number 15, JPL Publication 06–2, Jet Propulsion Laboratory, California Institute of Technology, Pasadena (2006)

Wallington, T.J., O.J. Nielsen, J. Sehested, in *Peroxy Radicals*, ed. by Z.B. Alfassi (Wiley, Chichester, 1997), pp. 113–172

Wayne, R.P., I. Barnes, P. Biggs, J.P. Burrows, C.E. Canosa-Mas, J. Hjorth, G. Le Bras, G.K. Moortgat, D. Perner, G. Poulet, G. Restrelli, H. Sidebotttom, Atmos. Environ. **25A**, 1–203 (1991)

7.6   Additional Reactions

Table 7.31 Gas-phase equilibrium constants for species that are unstable under atmospheric conditions, with temperature dependence: $K_{eq} = A \exp(B/T)$ (cm^3 molecule^{-1})^a

Reaction	K_{eq} (298)	A	B (K)
$NO_2 + NO \rightleftharpoons N_2O_3$	2.07 (−20)	3.3 (−27)	4,667
$NO_2 + NO_2 \rightleftharpoons N_2O_4$	2.80 (−19)	5.9 (−29)	6,643
$NO_2 + NO_3 \rightleftharpoons N_2O_5$	2.85 (−11)	2.7 (−27)	11,000
$NO_2 + OH \rightleftharpoons HOONO$	2.19 (−12)	3.9 (−27)	10,125
$NO_2 + HO_2 \rightleftharpoons NO_2NO_2$	1.58 (−11)	2.1 (−27)	10,900
$NO_2 + CH_3O_2 \rightleftharpoons CH_3O_2NO_2$	2.20 (−12)	9.5 (−29)	11,234
$NO_2 + CH_3C(O)O_2 \rightleftharpoons CH_3C(O)O_2NO_2$	2.22 (−8)	9.0 (−29)	14,000
$F + O_2 \rightleftharpoons FOO$	3.7 (−16)	4.5 (−25)	6,118
$Cl + O_2 \rightleftharpoons ClOO$	2.9 (−21)	6.6 (−25)	2,502
$Cl + CO \rightleftharpoons ClCO$	9.6 (−20)	3.5 (−25)	3,730
$ClO + ClO \rightleftharpoons Cl_2O_2$	7.0 (−15)	9.3 (−28)	8,835
$ClO + OClO \rightleftharpoons Cl_2O_3$	4.3 (−17)	1.6 (−27)	7,155
$OClO + NO_3 \rightleftharpoons O_2ClONO_2$	4.0 (−23)	6.6 (−29)	3,971
$OH + CS_2 \rightleftharpoons CS_2OH$	1.4 (−17)	4.5 (−25)	5,140
$CH_3S + O_2 \rightleftharpoons CH_3SO_2$	2.2 (−19)	1.8 (−27)	5,545

^{a}Powers of ten are shown in parentheses; Source: Sander et al. (2006). To convert from units of (cm^3 molecule^{-1}) to (m^3 mol^{-1}) multiply by $10^{-6} N_A = 6.022 \times 10^{17}$; to convert to units of (Pa^{-1}) multiply by $10^{-6}/k_B T = (7.243 \times 10^{16})/T$, where N_A is Avogadro's constant, k_B is the Boltzmann constant, and T is absolute temperature

Table 7.32 Rate coefficients and temperature parameters for gas-phase association reactions in air or nitrogen as carrier gas[a]

Reaction	k_0^{300}	n	k_∞^{300}	m	k_{atm}^{300}
$O+O_2+M\to O_3+M$	6.0 (−34)	2.4	−	−	1.5 (−14)
$O+NO+M\to NO_2+M$	9.0 (−32)	1.5	3.0 (−11)	0	1.6 (−12)
$O+NO_2+M\to NO_3+M$	2.5 (−31)	1.8	2.2 (−11)	0.7	3.2 (−12)
$O+SO_2+M\to SO_3+M$	1.8 (−33)	−2.0	4.2 (−14)	−1.8	1.3 (−14)
$O+CO+M\to CO_2+M$	3.3 (−36)	−6.3	−	−	8.1 (−17)
$O+OClO+M\to ClO_3+M$	2.9 (−31)	3.1	8.3 (−12)	0	3.0 (−12)
$H+O_2+M\to HO_2+M$	4.4 (−32)	1.3	4.7 (−11)	0.2	9.9 (−13)
$OH+OH+M\to H_2O_2+M$	6.9 (−31)	1.0	2.6 (−11)	0	6.3 (−12)
$OH+NO+M\to HONO+M$	7.0 (−31)	2.6	3.6 (−11)	0.1	7.3 (−12)
$OH+NO_2+M\to HNO_3+M$	1.8 (−30)	3.0	2.8 (−11)	0	1.0 (−11)
$OH+NO_2+M\to HOONO+M$	9.1 (−32)	3.9	4.2 (−11)	0.5	1.7 (−12)
$OH+SO_2+M\to HOSO_2+M$	3.3 (−31)	4.3	1.6 (−12)	0	9.5 (−13)
$OH+CS_2+M\to HOCS_2+M$	4.9 (−31)	3.5	1.4 (−11)	1.0	3.9 (−12)
$HO_2+NO_2+M\to HO_2NO_2+M$	2.0 (−31)	3.4	2.9 (−12)	1.1	1.1 (−12)
$NO_3+NO_2+M\to N_2O_5+M$	2.0 (−30)	4.4	1.4 (−12)	0.7	1.2 (−12)
$HS+NO+M\to HSNO+M$	2.4 (−31)	2.5	2.7 (−11)	0	3.4 (−12)
$CH_3S+NO+M\to CH_3SNO+M$	3.2 (−29)	4.0	3.5 (−11)	1.8	2.7 (−11)
$SO_3+NH_3+M\to NH_3SO_3+M$	3.6 (−30)	6.1	4.3 (−11)	0	1.8 (−11)
$CH_3+O_2+M\to CH_3O_2+M$	4.0 (−31)	3.6	1.2 (−12)	−1.1	8.1 (−13)
$CH_3O_2+NO_2+M\to CH_3O_2NO_2+M$	1.0 (−30)	4.8	7.2 (−12)	2.1	3.7 (−12)
$CH_3O+NO+M\to CH_3ONO+M$	2.3 (−29)	2.8	3.8 (−11)	0.6	2.9 (−11)
$CH_3O+NO_2+M\to CH_3ONO_2+M$	5.3 (−29)	4.4	1.9 (−11)	1.8	1.7 (−11)
$C_2H_5+O_2+M\to C_2H_5O_2+M$	1.5 (−28)	3.0	8.0 (−12)	0	7.5 (−12)
$C_2H_5O_2+NO_2+M\to C_2H_5O_2NO_2+M$	1.2 (−29)	4.0	9.0 (−12)	0	7.5 (−12)
$C_2H_5O+NO+M\to C_2H_5ONO+M$	2.8 (−27)	4.0	5.0 (−11)	0.2	4.8 (−11)
$C_2H_5O+NO_2+M\to C_2H_5ONO_2+M$	2.0 (−27)	4.0	2.8 (−11)	1.0	2.7 (−11)
$CH_3(O)O_2+NO_2+M\to CH_3(O)O_2NO_2+M$	9.7 (−29)	5.6	9.3 (−12)	1.5	8.6 (−12)
$C_2H_5(O)O_2+NO_2+M\to C_2H_5(O)O_2NO_2+M$	9.0 (−28)	8.9	7.7 (−12)	0.2	7.4 (−12)
$OH+C_2H_2+M\to CHCHOH+M$	5.5 (−30)	0	8.3 (−13)	−2.0	7.6 (−13)
$OH+C_2H_4+M\to CH_2CH_2OH+M$	1.0 (−28)	4.5	8.8 (−12)	0.85	8.1 (−12)
$F+O_2+M\to FO_2+M$	5.8 (−33)	1.7	1.0 (−10)	0	1.3 (−13)
$F+NO+M\to FNO+M$	1.2 (−31)	0.5	2.8 (−10)	0	2.6 (−12)
$F+NO_2+M\to FNO_2+M$	1.5 (−30)	2.0	1.0 (−11)	0	5.3 (−12)
$FO+NO_2+M\to FONO_2+M$	2.6 (−31)	1.3	2.0 (−11)	1.5	3.2 (−12)
$CF_3+O_2+M\to CF_3O_2+M$	3.0 (−29)	4.0	3.0 (−12)	1.0	2.8 (−12)
$CF_3O_2+NO_2+M\to CF_3O_2NO_2+M$	1.5 (−29)	2.2	9.6 (−12)	1.0	8.1 (−12)
$CF_3O+NO_2+M\to CF_3ONO_2+M$	1.7 (−28)	6.9	1.1 (−11)	1.0	1.0 (−11)
$CF_3O+CO+M\to CF_3OCO+M$	2.5 (−31)	2.0	6.8 (−14)	−1.2	6.0 (−14)
$Cl+O_2+M\to ClOO+M$	2.2 (−33)	3.1	1.8 (−10)	0	5.2 (−14)
$Cl+NO+M\to ClNO+M$	7.6 (−32)	1.8	−	−	1.9 (−12)
$Cl+NO_2+M\to ClONO+M$	1.3 (−30)	2.0	1.0 (−10)	1.0	1.6 (−11)
$\to ClNO_2+M$	7.6 (−32)	1.8	1.0 (−10)	1.0	2.9 (−13)
$Cl+CO+M\to ClCO+M$	1.3 (−33)	3.8	−	−	3.2 (−14)
$Cl+C_2H_2+M\to C_2H_2Cl+M$	5.2 (−30)	2.4	2.2 (−10)	0.7	5.0 (−11)
$Cl+C_2H_4+M\to C_2H_4Cl+M$	1.6 (−29)	3.3	3.1 (−10)	1.0	1.0 (−10)

(continued)

Table 7.32 (continued)

Reaction	k_0^{300}	n	k_∞^{300}	m	k_{atm}^{300}
$Cl+CS_2+M \rightarrow Cl-CS_2$	5.9 (−31)	3.6	4.6 (−10)	0	1.2 (−11)
$Cl+CH_3SCH_3+M \rightarrow (CH_3)_2SCl+M$	4.0 (−28)	7.0	2.0 (−10)	1.0	1.7 (−10)
$ClO+NO_2+M \rightarrow ClONO_2+M$	1.8 (−31)	3.4	1.5 (−11)	1.9	2.3 (−12)
$ClO+ClO+M \rightarrow Cl_2O_2+M$	1.6 (−32)	4.5	2.0 (−12)	2.4	2.3 (−13)
$ClO+OClO+M \rightarrow Cl_2O_3+M$	6.2 (−32)	4.7	2.4 (−11)	0	1.2 (−12)
$CH_2Cl+O_2+M \rightarrow CH_2ClO_2+M$	1.9 (−30)	3.2	2.9 (−12)	1.2	1.6 (−12)
$CHCl_2+O_2+M \rightarrow CHCl_2O_2+M$	1.3 (−30)	4.0	2.8 (−12)	1.4	2.0 (−12)
$CCl_3+O_2+M \rightarrow CCl_3O_2+M$	8.0 (−31)	6.0	3.5 (−12)	1.0	2.1 (−12)
$CFCl_2+O_2+M \rightarrow CFCl_2O_2+M$	5.0 (−30)	4.0	6.0 (−12)	1.0	4.7 (−12)
$CF_2Cl+O_2+M \rightarrow CF_2ClO_2+M$	1.0 (−29)	4.0	6.0 (−12)	1.0	5.1 (−12)
$CCl_3O_2+NO_2+M \rightarrow CCl_3O_2NO_2+M$	2.9 (−29)	6.8	1.3 (−11)	1.0	1.1 (−11)
$CFCl_2O_2+NO_2+M \rightarrow CFCl_2O_2NO_2+M$	2.2 (−29)	5.8	1.0 (−11)	1.0	8.6 (−12)
$CF_2ClO_2+NO_2+M \rightarrow CF_2ClO_2NO_2+M$	1.1 (−29)	4.6	1.7 (−11)	1.2	1.3 (−11)
$BrO+NO_2+M \rightarrow BrONO_2+M$	5.2 (−31)	3.2	6.9 (−12)	2.9	2.8 (−12)
$Br+NO_2+M \rightarrow BrNO_2+M$	4.2 (−31)	2.4	2.7 (−11)	0	4.8 (−12)
$Br+CH_3SCH_3+M \rightarrow (CH_3)_2SBr+M$	3.7 (−29)	5.3	1.5 (−10)	2	9.4 (−11)
$I+NO+M \rightarrow INO+M$	1.8 (−32)	1.0	1.7 (−11)	0	3.7 (−13)
$I+NO_2+M \rightarrow INO_2+M$	3.0 (−31)	1.0	6.6 (−11)	0	5.1 (−12)
$IO+NO_2+M \rightarrow IONO_2+M$	6.5 (−31)	3.5	7.6 (−12)	1.5	3.2 (−12)
$Na+O_2+M \rightarrow NaO_2+M$	3.2 (−30)	1.4	6.0 (−10)	0	b
$NaO+O_2+M \rightarrow NaO_3+M$	3.5 (−30)	2.0	5.7 (−10)	0	b
$NaO+CO_2+M \rightarrow NaCO_3+M$	8.7 (−28)	2.0	6.5 (−10)	0	b
$NaOH+CO_2+M \rightarrow NaHCO_3$	1.3 (−28)	2.0	6.8 (−10)	0	b

[a]Powers of ten are shown in parentheses; Source: Sander et al. (2006). The equation $k = 0.6^y k_0(T)n(M)/[1 + k_0(T)n(M)/k_\infty(T)]$ with $y = \{1 + [\log_{10} k_0(T)n(M)/k_\infty(T)]^2\}^{-1}$ can be used to calculate the second-order compound rate coefficient at temperature $T(K)$ and number density $n(M)$. The table lists parameters for the low pressure and high pressure rate coefficients $k_0(T) = k_0^{300}(T/300)^{-n}$ (cm^6 molecule^{-2} s^{-1}) and $k_\infty(T) = k_\infty^{300}(T/300)^{-m}$ (cm^3 molecule^{-1} s^{-1}), respectively, and the calculated rate coefficients k_{atm}^{300} at 300 K and atmospheric pressure (1 atm = 1.01325×10^5 Pa)
[b]These reactions occur in the upper atmosphere

Reference

Sander, S.P., R.R. Friedl, D.M. Golden, M.J. Kurylo, G.K. Moortgat, H. Keller-Rudek, P.H. Wine, A.R. Ravishankara, C.E. Kolb, M.J. Molina, B.J., Finlayson-Pitts, R.E. Huie, V.L. Orkin, NASA Chemical Kinetics and Photochemical Data for Use in Atmospheric Studies, Evaluation Number 15, JPL Publication 06–2, Jet Propulsion Laboratory, California Institute of Technology, Pasadena (2006)

Chapter 8
Aqueous Phase Chemistry

8.1 Physicochemical Properties of Water

Table 8.1 Physicochemical quantities for pure water[a]

Molar mass (kg mol⁻¹)	M_w	1.80153×10^{-2}
Density of water (kg m⁻³)	ρ_w	999.84
Maximum density (kg m⁻³)	$\rho_{w\,max}$	999.97 (4.0°C)
Triple point density (kg m⁻³)	$\rho_{w\,trp}$	999.78 (0.01°C)
Density of ice (kg m⁻³)	ρ_i	916.4
Melting temperature (°C)	T_m	0.00 (1013.25 hPa)
Specific heat, water vapor (J kg⁻¹ K⁻¹)	c_p (constant pressure)	1,850
	c_v (constant volume)	1,319
Specific heat, liquid water (J kg⁻¹ K⁻¹)	c_w	4217.6
Specific heat of ice (J kg⁻¹ K⁻¹)	c_w	2,110
Latent heat of evaporation (J kg⁻¹)	L_e	2.501×10^6
Latent heat of sublimation (J kg⁻¹)	L_s	2.834×10^6
Saturation vapor pressure (hPa)	e_{sat}	6.113
Triple point vapor pressure (hPa)	e_{trp}	6.1173
Thermal conductivity (mW K⁻¹ m⁻¹)	κ_T	561.0
Viscosity (mPa s)	μ	1.793
Dielectric constant, water[b]	ε_r	87.90
Dielectric constant, ice[b]	ε_r	91.6
Electrical conductance (μS cm⁻¹)	λ	0.0115
Surface tension (mN m⁻¹)	σ	75.64

[a]All quantities except mass are temperature-dependent (see the following tables). Numerical values refer to $T = 273.15$ K and $p = 1.0 \times 10^5$ Pa (1 bar) except where indicated
[b]The dielectric constant is defined as the relative permittivity $\varepsilon_r = \varepsilon/\varepsilon_0$, where ε is the permittivity of water or ice and ε_0 is the permittivity of vacuum

P. Warneck and J. Williams, *The Atmospheric Chemist's Companion: Numerical Data for Use in the Atmospheric Sciences*, DOI 10.1007/978-94-007-2275-0_8, © Springer Science+Business Media B.V. 2012

Table 8.2 Density of liquid water and of ice (kg m^{-3}) as a function of temperature (°C)[a]

Water				Ice	
−34	977.5	4	999.97	0	916.4
−32	980.9	6	999.94	−10	917.5
−30	983.9	8	999.85	−20	918.6
−28	986.4	10	999.70	−30	919.7
−26	988.6	12	999.50	−40	920.8
−24	990.4	14	999.25	−50	921.9
−22	992.1	16	998.95	−60	923.0
−20	993.5	18	998.60	−70	924.1
−18	994.7	20	998.21	−80	925.2
−16	995.8	22	997.77	−90	926.3
−14	996.7	24	997.30	−100	917.5
−12	997.5	26	996.79	−110	928.5
−10	998.2	28	996.24	−120	929.6
−8	998.7	30	995.65	−130	930.7
−6	999.1	32	995.03	−140	931.8
−4	999.4	34	994.38		
−2	999.7	36	993.69	−200 ⎫	
0	999.84	38	992.97	⎬ Average 933.1	
2	999.94	40	992.22	−260 ⎭	

[a]Sources: Super-cooled water, Hare and Sorensen (1987); water above 0°C, Laube and Höller (1988), see also Linstrom and Mallard (2005); ice, Hobbs (1974); see comments

Comments: Densities for super-cooled water in the temperature range 0>T>−34°C are smoothed data from Hare and Sorensen (1987) who fitted their measurements by a sixth order polynomial:

$$\rho_{ice}\left[\text{g cm}^{-3}\right] = 0.99986 + 6.69 \times 10^{-5}T + 8.486 \times 10^{-6}T^2 + 1.518 \times 10^{-7}T^3$$
$$-6.9484 \times 10^{-9}T^4 - 3.6449 \times 10^{-10}T^5 - 7.497 \times 10^{-12}T^6 \ (T \text{ in °C})$$

Densities for ice in the range 0> T >−140°C are averages of three data sets summarized by Hobbs (1974), which were derived from X-ray diffraction measurements of Lonsdale (1958), La Placa and Post (1960), evaluated by Eisenberg and Kauzmann (1969), and Brill and Tippe (1967) evaluated by Hobbs (1974). These data fall on a straight line, and linear regression analysis gives ρ_{ice} [kg m^{-3}] = (916.4±0.4) − (0.110±0.005)T [°C]. At temperatures below 140°C the density approaches a limiting value ρ_{ice} = 933.1±0.2 kg m^{-3} at $T \leq −200$°C.

Table 8.3 Structure of ice[a]

Crystal system	Occurrence	Temperature range
Hexagonal	Water frozen in bulk or deposited on a surface	0>T>−80 (°C)
Cubic	Deposition of water vapor onto a surface	−80>T>−130 (°C)
Vitreous/amorphous	Deposition of water vapor onto a surface	T<140 (°C)

[a]As described by Hobbs (1974)

Table 8.4 Thermodynamic properties of condensed water as a function of temperature[a]

T (°C)	c_w (J kg^{-1} K^{-1})	L_e^b (10^6 J kg^{-1})	T (°C)	c_i (J kg^{-1} K^{-1})	L_s^b (10^6 J kg^{-1})	L_m^b (10^6 J kg^{-1})
−50	5,400	2.6348	−100	1,382	2.8236	
−40	4,770	2.6030	−90	1,449	2.8278	
−30	4,520	2.5749	−80	1,520	2.8316	
−20	4,350	2.5494	−70	1,591	2.8345	
−10	4,270	2.5494	−60	1,662	2.8366	
0	4217.8	2.50084	−50	1,738	2.8383	0.2035
10	4192.3	2.7474	−40	1,813	2.8387	0.2357
20	4181.8	2.4535	−30	1,884	2.8387	0.2638
30	4178.5	2.4300	−20	1,959	2.8383	0.2889
40	4178.5	2.3823	−10	2,031	2.8366	0.3119
50	4180.6	2.3823	0	2,106	2.8345	0.3337

[a]Symbols: c_w, specific heat of water; L_e, latent heat of evaporation; c_i, specific heat of ice; L_s, latent heat of sublimation; L_m, latent heat of melting; Source: Laube and Höller (1988)
[b]To convert (J kg^{-1}) to (J mol^{-1}) multiply with $M_w = 1.80153 \times 10^{-2}$ (kg mol^{-1})

Table 8.5 Saturation vapor pressures (Pa) of liquid water and of ice as a function of temperature (°C)[a]

t	e_{sw}	t	e_{sw}	e_{si}	t	e_{si}
0	611.2	0	611.2	611.2	−42	10.28
2	706.0	−2	527.9	517.6	−44	8.07
4	813.5	−4	454.9	437.3	−46	6.37
6	935.2	−6	391.0	368.5	−48	5.01
8	1072.8	−8	335.2	309.7	−50	3.92
10	1227.9	−10	286.6	259.7	−52	3.05
12	1402.5	−12	244.4	217.1	−54	2.37
14	1598.6	−14	207.8	181.0	−56	1.83
16	1818.3	−16	176.2	150.4	−58	1.41
18	2064.1	−18	149.0	124.7	−60	1.08
20	2338.6	−20	125.6	103.1	−62	0.82
22	2644.4	−22	105.6	84.9	−64	0.62
24	2984.7	−24	88.5	69.7	−66	0.47
26	3362.6	−26	73.9	57.1	−68	0.35
28	3781.6	−28	61.5	46.6	−70	0.26
30	4245.2	−30	51.1	37.9	−75	0.12
32	4757.8	−32	42.2	30.7	−80	0.055
34	5322.9	−34	34.8	24.8	−85	0.023
36	5945.3	−36	28.6	20.0	−90	0.0096
38	6629.8	−38	23.4	16.0	−95	0.0038
40	7381.4	−40	19.0	12.8	−100	0.0014

[a]Data for temperatures $t > -40$°C from Laube and Höller (1988). Vapor pressures for ice at temperatures $t < -40$°C were calculated from the formula given below. These values agree with those recommended by Wagner et al. (1994). Vapor pressure refers to pure water and a plane surface. The data are based on accurate values obtained by integration of the Clausius-Clapeyron Equation. The vapor pressures over water and ice, respectively, are fitted by

$$\ln e_w \, (\text{Pa}) = 21.1249952 - 6094.4642/T - 0.027245552\, T + 1.6853396 \times 10^{-5} T^2$$
$$+ 2.4575506 \ln(T) \qquad 273\ \text{K} < T < 373\ \text{K}$$

$$\ln e_i \, (\text{Pa}) = -3.5704628 - 5504.4088/T - 0.017337458\, T + 6.5204209 \times 10^{-6} T^2$$
$$+ 6.1295027 \ln(T) \qquad 173\ \text{K} < T < 273\ \text{K}$$

Table 8.6 Miscellaneous properties of pure water and ice as a function of temperature[a]: Thermal conductivity, κ, shear viscosity, η, coefficient of self-diffusion in liquid water, D_{H_2O}, dielectric constant, ε, surface tension of water against air, σ_s

t (°C)	κ_{liquid} (W m⁻¹ K⁻¹)	κ_{ice}	η (mPa s)	D_{H_2O} (10^5 cm² s⁻¹)	ε_{liquid}	ε_{ice}	σ_s (N m⁻¹)
−35		2.55	18.70	0.128	106.8	101.6	
−30		2.50	10.20	0.204	103.7	100.2	
−25		2.45	6.45	0.310	100.8	98.7	
−20		2.40	4.34	0.445	97.9	97.3	
−15		2.35	3.30	0.555	95.2	95.9	
−10		2.30	2.63	0.704	92.7	94.4	0.07710
−5		2.21	2.15	0.899	90.2	93.0	0.07640
±0	0.561	2.14	1.793	1.098	87.9	91.6	0.07562
5	0.571		1.535	1.313	85.9		0.07490
10	0.580		1.307	1.532	84.0		0.07420
15	0.589		1.142	1.777	82.1		0.07348
20	0.598		1.002	2.017	80.2		0.07275
25	0.607		0.893	2.299	78.4		0.07275
30	0.615		0.798	2.596	76.6		0.07115
35	0.623		0.723	2.919	74.9		0.07035
40	0.631		0.653	3.236	73.2		0.06955

[a]Sources of data: Angell (1982), Sengers and Watson (1986), Archer and Wang (1990), Laube and Höller (1988) Slack (1980), Auty and Cole (1952), Humbel et al. (1953)

Table 8.7 Different expressions for the concentration of water vapor in the atmosphere[a]

Absolute humidity	$\rho_v = eM_w/R_g T$	Specific humidity	$q = \rho_v/(\rho_v + \rho_{dry\,air})$
Mass mixing ratio	$w = \rho_v/\rho_{dry\,air}$	Relative humidity	$f = e/e_s$

[a]Parameters: ρ_v (kg m⁻³) = density of water vapor; M_w = 0.0180153 (kg mol⁻¹) molar mass of water; R_g = 8.3145 (J K⁻¹ mol⁻¹) gas constant; T (K) absolute temperature; e, e_s (Pa), partial pressure and saturation vapor pressure of water, respectively

Diffusion coefficient of water in air derived from measurements shown below:

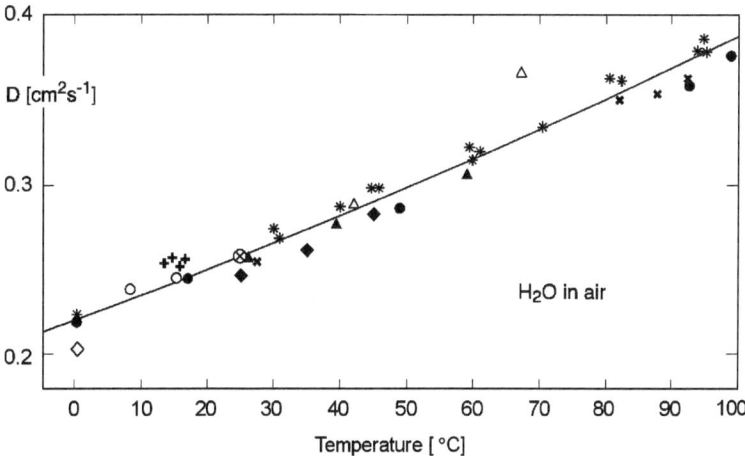

Fig. 8.1 Diffusion coefficient for H_2O in air at 1 atm pressure (1013.25 hPa) as a function of temperature

Key: • Winkelmann, A., *Wiedemanns Annal. Phys.* **22**, 1–31; 152–161 (1884); **33**, 445–453 (1888); **36**, 93–114 (1889). Corrections: Trautz, M., W. Müller, *Annal. Phys.* **22**, 333–352 (1935)

    o Gugliemo, G., *Atti Acad. di Torino* **17**, 54–72 (1881); **18**, 93–107 (1882)

    ◊ Houdaille, F., *Fortschr. Physik* **52**, 442–443 (1897)

    + Brown, H.T., F. Escombe, *Philos. Trans. R. Soc. Lond.* **B 190**, 223–291 (1900)

    × Mache, H., *Ber. Kaiserl. Acad. Wien* **119**, 1399–1423 (1910)

    Δ Le Blanc, M., G. Wuppermann, *Z. Physik. Chem.* **91**, 143–154 (1916)

    ▲ Gilliland, E.R. *Ind. Eng. Chem.* **26**, 681–685 (1934)

    * Schirmer, R., *Z. Verein Deutscher Ing. Beihefte Verfahrenstechn.* **6**, 170–177 (1938)

    ♦ Brookfield, K.J., H.D.N. Fitzpatrick, J.F. Jackson, J.B. Matthews, E.A. Moelwyn-Hughes, *Proc. R. Soc. Lond.* **A 190**, 59–67 (1947)

    ⊗ Average of data at 25°C from 5 investigators: Gugliemo ($D_{25} = 0.257$), Le Blanc and Wuppermann ($D_{25} = 0.258$), both cited above; Kimpton, D.D., F.T. Wall, *J. Phys. Chem.* **56**, 715–717 (1952) [$D_{25} = 0.257$]; Lee, C.Y., C.R. Wilke, *Ind. Eng. Chem.* **46**, 2381–2387 (1954) [$D_{25} = 0.260$]; Nelson, E.T., *J. Appl. Chem.* **6**, 286–292 (1956) [$D_{25} = 0.257$]. The average value is: $D_{25av} = 0.2578 \pm 0.0013$ (cm^2 s^{-1})

Comments: Only statistically meaningful data identified by Massman (1998) are used in Fig. 8.1. The dependence of diffusivity on pressure and temperature can be expressed by

$$D = D_o \left(p_o/p \right) \left(T/T_o \right)^n$$

where D_o is the diffusion coefficient at standard pressure $p_o = 1$ atm (1013.25 hPa) and a standard absolute temperature T_o. In view of the excellent agreement of 5 independent values obtained at 298.15 K it is recommended to use of $T_o = 298.15$ K and $D_o = 0.2578$ [cm^2 s^{-1}]. The exponent n is obtained from a two parameter logarithmic regression analysis, which after omitting obvious outliers yields n = 1.830 ± 0.054 ($R^2 = 0.975$) with $D_0(298.15) = 0.258$ (shown by the solid line). No data exist for $T_0 \leq 273.15$ K so that extrapolation is required.

References

Angell, C.A., Super-cooled water, in *Water, A Comprehensive Treatise*, ed. by F. Franks. Water and Aqueous Solutions at Subzero Temperatures, vol. 7 (Plenum Press, New York, 1982), pp. 1–81

Archer, D.G., P. Wang, J. Phys. Chem. Ref. Data **19**, 371– (1990)

Auty, R.P., R.H. Cole, J. Chem. Phys. **20**, 1309–1314 (1952)

Brill, R., A. Tippe, Acta Crystallogr. **23**, 343–345 (1967)

Eisenberg, D., W. Kauzmann, *The Structure and Properties of Water* (Oxford University Press, London, 1969)

Hare, D.E., C.M. Sorensen, J. Chem. Phys. **87**, 4840–4845 (1987)

Hobbs, P.V., *Ice Physics* (Clarendon Press, Oxford, 1974)

Humbel, F., F. Jona, P. Scherrer, Helv. Phys. Acta **26**, 17–32 (1953)

La Placa, S.J., B. Post, Acta Crystallogr. **13**, 503–505 (1960)

Laube, M., H. Höller, *Cloud Physics*, in *Meteorology*, ed. by G. Fischer. Landolt-Börnstein, New Series, vol. V/4b (Springer, Berlin, 1988), pp. 1–100

Linstrom, P.J., W.G. Mallard, *NIST Standard Reference Database* No. 69, (National Institute of Standards and Technology, Gaithersburg, 2005), http://webbook.nist.gov

Lonsdale, K., Proc. Roy. Soc. A **247**, 424–434 (1958)

Massman, W.J., Atmos. Environ. **32**, 1111–1127 (1998)

Sengers, J.V., J.T.R. Watson, J. Phys. Chem. Ref. Data **15**, 1291–1314 (1986)

Slack, G.A., Phys. Rev. B **22**, 3065–3071 (1980)

Wagner, W., A. Saul, A. Pruss, J. Phys. Chem. Ref. Data **23**, 515– (1994)

8.2   Global Distribution of Clouds, Precipitation and Chemical Constituents

Table 8.8 Zonal mean distribution of atmospheric water vapor and annual precipitation rate[a]

Latitude belt	W (kg m^{-2})		P (kg m^{-2} a^{-1})		
	A	B	A	C	D
80–90°N	4.90	4.21	120	46	146
70–80	6.48	5.68	185	200	310
60–70	8.52	8.02	415	507	639
50–60	11.64	10.33	789	843	865
40–50	15.21	14.55	907	874	938
30–40	18.95	22.13	872	761	876
20–30	26.37	29.04	790	675	646
10–20	36.73	38.03	1,151	1,117	989

(continued)

Table 8.8 (continued)

Latitude belt	W (kg m⁻²)		P (kg m⁻² a⁻¹)		
	A	B	A	C	D
0–10	41.07	46.66	1,934	1,885	1,708
0–10°S	40.90	44.15	1,445	1,435	1,343
10–20	36.66	36.29	1,132	1,109	931
20–30	29.86	27.78	857	777	774
30–40	23.81	19.67	932	875	956
40–50	18.10	12.76	1,226	1,128	1,040
50–60	12.61	8.45	1,046	1,003	1,124
60–70	6.84	4.84	418	549	394
70–80	2.87	2.79	82	230	274
80–90	1.56	2.05	30	73	164
Global average	24.67	28	1,004	975	953

[a]A: from meteorological stations, data presented by Sellers (1965); B: satellite-derived data 1990–1995 summarized by Raschke and Stubenrauch (2005); C: Baumgartner and Reichel (1975); D: satellite-derived data 1979–2003, Global Precipitation Climatology Project (Adler et al. 2003)

Table 8.9 Zonal means of cloud type and clear sky frequencies (%)[a]

Latitude belt		Cb	Ci	Thin Ci	As	Ac	St	Cu	Clear sky
Over the oceans									
NH	60–90°	1	6	3	16.5	3.5	37	13.5	31
	30–60°	2	15.5	10	10	3	22.5	11.5	20.5
	30–15°	1	11	17.5	2	2.5	12	13.5	36.5
	Tropics	2	19.5	23	2	1.5	11.5	11	30
SH	15–30°	1.5	10	12	2.5	3.5	17.5	18.5	34
	30–60°	2	14.5	7.5	12	3.5	25	15	18.5
	60–90°	1	7.5	2.5	20	2.5	42	8	13.5
Over land									
NH	60–90°	2	6	6	21	10	21.5	10	20.5
	30–60°	3	12	10	12.5	5.5	12	8	33
	30–15°	2	12	17.5	3.5	1.5	11.5	11.5	40
	Tropics	3.5	28	23	3	2	9	4	27
SH	15–30°	2	13.5	13.5	4.5	2.5	6	7	48.5
	30–60°	4	13.5	10	6	3.5	8.5	9.5	44.5
	60–90°	12.5	10	9	30	11.5	15	7.5	10

[a]Frequencies of occurrence: cumulonimbus, Cb, cirrus, Ci, altostratus, As, altocumulus, Ac, stratus, St, cumulus, Cu, and clear sky; 5 year averages derived from TIROS-N Operational Vertical Sounder (TOVS) instruments on NOAA Polar Orbiting Environmental Satellites. Source: Stubenrauch (2005)

Table 8.10 Frequency of occurrence and areal coverage of common cloud types[a]

Type of cloud	Oceans		Continents	
	Frequency of occurrence (%)	Area covered (%)	Frequency of occurrence (%)	Area covered (%)
Stratus and Stratocumulus	45	34	27	18
Cumulus	33	12	14	5
Cumulonimbus	10	6	7	4
Altostratus and altocumulus	6	6	6	5
Nimbostratus	46	22	35	21
Cirrus, cirrostratus, cirrocumulus	37	13	47	23
Global average		64.8		52.4

[a]Source of data: Heymsfield (1993)

Table 8.11 Regional wet deposition rates of sulfate, nitrate and ammonium (g m^{-2} a^{-1})[a]

	Sulfate	Nitrate	Ammonium
Europe	3.0 (1.4–4.3)	1.9 (0.6–2.5)	0.6 (0.3–1.1)
Eastern North America	1.4 (0.5–3.0)	0.8 (0.5–1.9)	0.2 (0.1–0.4)
Northern South America	0.7 (0.1–1.9)	0.2 (0.06–0.6)	0.07 (0.02–0.3)
Eastern Asia	6.1 (1.1–8.1)	2.1 (0.8–1.9)	0.9 (0.6–1.7)
Africa	0.8 (0.2–1.6)	1.5 (0.3–2.4)	0.3 (0.1–0.7)
Australia	0.4 (0.19–1.2)	0.25 (0.1–0.5)	0.06 (0.04–0.2)

[a]Median values, range (10–90 percentile) in parentheses. Source of data: Whelpdale and Kaiser (1996)

Comments: The origins of inorganic ions in rainwater are aerosol particles incorporated during the process of cloud formation and, in the case of sulfate and nitrate, chemical oxidation reactions occurring in the aqueous phase as well as in the gas phase.

Table 8.12 Concentrations (μmol dm^{-3}) of major inorganic ions in cloud and fog waters[a]

	East Europe		Southern England non-precipitating	Whiteface Mt., New York intercepted clouds	Kleiner Feldberg Germany intercepted clouds	Brocken., Germany mountain fog	Pasadena California ground fog	Laegeren Mt. Switzerland ground fog
	Precipitating	Non-precipitating						
SO$_4^{2-}$	29	117	40	26–70	61.7	387	240–472	40–1,719
Cl$^-$	22.2	76.8	94	1.7–3.1	84.7	205	480–730	68–4,316
NO$_3^-$	3.2	16.1	18.6	140–215	183.7	225	1,220–3,250	44–4,420
HCO$_3^-$	11.5	16.4	0.14–91	0.023–0.045	0.03	0.72	0.005–0.4	0.004–57
NH$_4^+$	28.8	11.1	22.1	32–89	185.6	710	1,290–2,380	102–9,215
Na$^+$	17	29.8	95.2	2.3–11	46.3	295	320–500	14–789
K$^+$	5.1	20.4	12.5	13–20	8.7	85	33–53	9–424
Ca^{2+}	10	29.3	33.2	5–10	6.0	110	70–265	7–11,230
Mg^{2+}	12.3	40.0	12.3	1.1–3.1	8.7	–	45–160	0–135
H$^+$	5.0	94.4	0.06–40	126–251	168	7.9	14–1,200	0.1–1,380
pH	5.3	4.02	4.4–7.2	3.6–3.9	3.77	5.1	2.92–4.85	2.9–7.1
Sum of anions[b]	95	327	193	194–358	407	1,429	2,180–4,924	192–12,074
Sum of cations[b]	100	294	220	185–397	438	1,323	1,887–4,983	139–34,538
Sum of charges								

[a]From the summary presented by Warneck (2000), slightly abbreviated, with permission of Elsevier
[b]Sum of charges

Table 8.13 Inorganic ion composition ($\mu mol\ dm^{-3}$) of rainwater at various Locations[a]

	North Sweden 1969	Belgium 1969	East Europe 1961–1964	New Hampshire 1963–1974	California 1978–1979	San Carlos, Venezuela 1979–1980	Katherine Australia 1980–1984	Cape Grim Tasmania 1977–1981	Amsterdam Island 1980–1987	Bermuda 1979–1980
SO_4^{2-}	21	63	82	29.8±1.2	19.5	1.6±1.2	2	79.1	37.2	24.4±23.8
Cl^-	11	55	60	14.4±2.5	28	4.3±3.4	8	1,349	317.6	264±337
NO_3^-	5	36	17.7	23.1±1.7	31	3.5±3.6	4.1	5	1.6	7.9±9.1
HCO_3^-	21	0.15	71.7	0.077	0.15	0.33	0.32	5.6	0.67	0.27
NH_4^+	6	25	52.8	12.1±0.7	21	17.0±10	2.9	2	2.4	4.8±7.1
Na^+	13	42	65.2	5.4±0.6	24	2.7±2.0	4.5	1,297	268.8	221±282
K^+	5	6	17.3	1.9±0.5	2	1.1±1.7	1.1	32.3	5.9	6.5±8.0
Ca^{2+}	16	33	38.1	4.3±0.6	3.5	0.25±0.35	0.95	42.9	12.1	7.2±7.1
Mg^{2+}	5	15	52.5	1.8±10.4	3.5	0.4±0.45	0.7	122	60.3	24.6±30.4
H^+	–	38	3.3	73.9±3.0	39	17±10	16.9	1	8.4	21.1±26.9
pH (av.)	–	4.42	5.48	4.13	4.41	4.77	4.77	5.99	5.08	4.68
Sum of anions[b]	79	217	313	97.1	98	11	16.1	1,512	397	321
Sum of cations[b]	66	207	320	105	93	39	28.7	1,662	430	273
Annual precip. (mm)	360	648	–	1,310	697	4,000	104	1,120	1,120	1,131

[a]From the summary presented by Warneck (2000), slightly abbreviated, with permission of Elsevier
[b]Sum of charges (2000)

Table 8.14 Volume-averaged formic and acetic acid concentrations in precipitation at various locations of the world and the corresponding gas-phase mixing ratios[a]

	HCOOH			CH_3COOH		
	Aqueous[b] (μmol dm^{-3})	Gaseous[c] (nmol mol^{-1})	x_0^c	Aqueous[b] (μmol dm^{-3})	Gaseous[c] (nmol mol^{-1})	x_0^c
Continental						
Poker Flat, Alaska	4.3	0.07	0.29	1.2	0.07	0.13
Charlottesville, Virginia						
April-September	11.6	0.55	1.15	4.4	0.39	0.62
October-March	2.7	0.08	0.22	1.6	0.12	0.20
Wilmington, North Carolina	7.4	0.13	0.31	3.6	0.57	0.75
Calabozo, Venezuela[d]	6.5	0.03	0.34	3.5	0.22	0.39
Altos de Pipe, Venezuela[d]	6.4	0.01	0.32	5.6	0.17	0.44
Lago Colado, Brazil	19.0	0.39	1.37	8.8	0.58	1.04
Torres del Paine, Chile	0.5	<0.02	0.05	0.4	0.01	0.03
Katherine, Australia	10.5	0.16	0.70	4.2	0.24	0.46
Dimonika, Congo	6.6	0.07	0.39	3.0	0.37	0.51
Agra, India	12.7	<0.01	0.62	10.6	0.02	0.51
Marine						
Mauna Loa, Hawaii	2.0	0.03	0.13	0.6	0.03	0.06
North Pacific	3.4	0.05	0.23	1.9	0.10	0.20
90 Mile Beach, New Zealand	1.0	<0.02	0.08	0.2	0.01	0.02
Amsterdam Island	2.2	0.01	0.12	0.6	0.02	0.05
High Point, Bermuda	2.2	0.03	0.14	1.3	0.06	0.13
North Atlantic	4.2	0.05	0.27	1.2	0.14	0.28
Cape Point, South Africa	1.8	0.02	0.10	0.6	0.01	0.04

[a]Data sources: Warneck (2000) with permission of Elsevier, Keene and Galloway (1986), Avery et al. (1991), Lacaux et al. (1992), Kumar et al. (1993), Sanhueza et al. (1996)
[b]Volume-weighted aqueous concentration of anions and undissociated acid combined
[c]Gas-phase mixing ratio in equilibrium with the aqueous phase as calculated from the observed anion and hydrogen ion concentrations and the Henry's law coefficients normalized to 25 K and 101,325 Pa pressure, x_0 (nmol mol^{-1}) in the absence of liquid water assuming a liquid water volume fraction of 2×10^{-6}
[d]Calabozo is located in the savannah region, Altos de Pipe is a high altitude forested site

Comments: The main origin of formaldehyde and acetaldehyde in rain water is the absorption of these compounds into cloud water from the gas phase. The concentration of these compounds bound to aerosol particles is low compared to that in the gas phase.

Table 8.15 Percentage contributions of anion equivalents to the acidity of rainwater and cloud water at various locations[a]

	Cl⁻	SO₄²⁻	NO₃⁻	HCOOH	CH₃COOH
Rain water					
Joaquin del Tigre, Venezuela (1)	13	8	15	33	31
Chaguaramas, Venezuela (1)	8	18	16	30	29
La Paragua, Venezuela (1)	11	15	11	34	29
Amazonia, Brazil (2)	30	15	10	26	19
Torres del Paine, Chile (3)	18	14	6	55	7
Dimonika, Congo (4)	–	23	41	29	7
Katherine, Australia (5)	25	12	15	39	9
Barrington Tops, Australia (6)	7	23	26	24	20
Dayalbagh (Agra) India (7)	29	31	23	9	8
Bermuda (8)	–	66	21	9	3
Cloud water					
Redwood Natl. Park, Calif. (9)	–	71	13	11	5
Loft Mountain, Virginia (9)	4.3	67	27	1	1
Mohonk Mountain, New York (9)	5	53	34	6	2
Whiteface Mountain, New York (9)	–	81	16	2	1

[a]Data sources: (1) Sanhueza et al. (1992), savannah regions; (2) Andreae et al. (1990), (3) Galloway et al. (1996); (4) Lacaux et al. (1992); (5) Likens et al. (1987); (6) Post et al. (1991); (7) Kumar et al. (1993); (8) Galloway et al. (1989); (9) Weathers et al. (1988)

References

Adler, R.F., G.J. Huffman, A. Chang, R. Ferraro, P.-P. Xie, J. Janowiak, B. Rudolf, U. Schneider, S. Curtis, D. Bolvin, A. Gruber, J. Susskind, P. Arkin, E. Nelkin, J. Hydrometeorol. **4**, 1147–1167 (2003)
Andreae, M.O., R.W. Talbot, H. Berresheim, K.M. Beecher, J. Geophys. Res. **95**, 16987–16999 (1990)
Avery, G.B., J.D. Willey, C.A. Wilson, Environ. Sci. Technol. **25**, 1875–1880 (1991)
Baumgartner, A., E. Reichel, *The world Water Balance: Mean Annual Global Continental and Maritime precipitation, Evaporation and Runoff* (Elsevier, Amsterdam, 1975)
Galloway, J.N., W.C. Keene, R.S. Artz, J.M. Miller, T.M. Church, A.H. Knap, Tellus **41B**, 427–443 (1989)
Galloway, J.N., W.C. Keene, G.E. Likens, J. Geophys. Res. **101**, 6883–6897 (1996)
Heymsfield, A.J., Microphysical structures of stratiform and cirrus clouds, in *Aerosol-Cloud-Climate Interactions*, ed. by P.V. Hobbs (Academic Press, San Diego, 1993), pp. 97–121
Keene, W.C., J.N. Galloway, J. Geophys. Res. **91**, 14466–14474 (1986)
Kumar, N., U.C. Kulshrestha, A. Saxena, K.M. Kumari, S.S. Srivastava, J. Geophys. Res. **98**, 5135–5137 (1993)
Lacaux, J.P., R. Delmas, G. Kouado, B. Cros, M.O. Andreae, J. Geophys. Res. **97**, 6195–6206 (1992)
Likens, G.E., W.C. Keene, J.M. Miller, J.N. Galloway, J. Geophys. Res. **92**, 13299–13314 (1987)
Post, D., H.A. Bridgman, G.P. Ayers, J. Atmos. Chem. **13**, 83–95 (1991)
Raschke, E., C. Stubenrauch, Water Vapor in the Atmosphere, in *Observed Global Climate*, ed. by M. Hantel. Landolt-Börnstein, New Series, Group V, Geophysics, vol. 6 (Springer, Berlin, 2005), pp. 5–1 – 5–18
Sanhueza, E., M.C. Arias, L. Donoso, N. Graterol, M. Hermosos, I. Marti, J. Romero, A. Rondón, M. Santana, Tellus **44B**, 54–62 (1992)
Sanhueza, E., L. Figueroa, M. Santana, Atmos. Environ. **30**, 1861–1873 (1996)

Sellers, W.D., *Physical Climatology* (University of Chicago Press, Chicago, 1965)

Stubenrauch, C., Clouds, in *Observed Global Climate*, ed. by M. Hantel. Landolt-Börnstein, New Series, Group V, Geophysics, vol. 6 (Springer, Berlin, 2005), pp. 6–1 – 6–24

Warneck, P., *Chemistry of the Natural Atmosphere*, 2nd edn. (Academic Press, San Diego, 2000), Copyright Elsevier

Weathers, K.C., G.E. Likens, F.H. Bormann, S.H. Bicknell, B.T. Bormann, B.C. Daube Jr., J.S. Eaton, J.N. Galloway, W.C. Keene, K.D. Kimball, W.H. McDowell, T.G. Siccama, D. Smiley, R.A. Tarrant, Environ. Sci. Technol. **22**, 1018–1026 (1988)

Whelpdale, D.M., M.S. Kaiser (eds.), *Global Acid Deposition Assessment*. WMO *Global Atmosphere Watch Report* 106, World Meteorological Organization, Geneva, 1996

8.3 Appearance and Microstructure of Clouds

Characterization of clouds:

Clouds are differentiated by appearance and internal structure. The following genera of clouds are commonly distinguished (Laube and Höller 1988):

Cirrus

Non-precipitating, detached clouds in the form of white, delicate filaments, or white patches or narrow bands occurring in the 5–13 km altitude region. Cirrus is composed exclusively of ice crystals that are large enough to acquire an appreciable terminal velocity so that vertical trails may form.

Cirrocumulus

Thin, white patch, sheet or layer cloud in the form of more or less regular grains or ripples, occurring most frequently at altitudes above 5 km. Cirrocumulus contains ice crystals and super-cooled water drops (non-precipitating).

Cirrostratus

Whitish cloud veil of fibrous or smooth appearance occurring at altitudes above 5 km, totally or partly covering the sky, generally associated with halo phenomena; composed mainly of ice crystals (non-precipitating).

Altocumulus

White or gray (or both) patch, sheet or layer cloud, generally with shading, occurring most frequently at altitudes between 2 and 7 km. It consists of water drops; at very low temperature ice crystals may form (non-precipitating).

Altostratus

Grayish or bluish cloud sheet or layer of striated, fibrous or uniform appearance, totally or partly covering the sky and usually occurring at altitudes between 2 and 7 km. Altostratus nearly always exists as a layer of considerable horizontal extent (10–100 km) with appreciable vertical depth (several 100–1,000 m). The lower part is composed of water drops, the upper part of ice crystals, and the middle part of a mixture of both. Precipitation as rain is common, snow or ice pellets are possible.

Nimbostratus

Gray cloud layer, often dark, of diffuse appearance due to falling rain or snow, with the main part occurring at altitudes between 2 and 8 km. Nimbostratus consist of water drops and raindrops, snow crystals and flakes. Precipitation as rain is usual, snow and ice pellets are possible.

Stratocumulus

Gray or whitish (or both) patch, sheet or layer cloud, usually occurring at altitudes below 2 km, with thickness ranging from 500 to 1,000 m. It is composed of water drops sometimes accompanied by raindrops, snow pellets, crystals or flakes. Precipitation in the form of rain, snow and snow pellets is possible.

Stratus

A usually gray cloud layer with fairly uniform base, occurring at altitudes between the earth surface and 2 km and consisting of small water drops; at very low temperatures small ice particles may be present. Precipitation occurs as drizzle, snow grains or ice prisms may be possible.

Cumulus

Detached clouds, usually dense with sharp boundaries and developing in the form of mounds, domes or towers. The base is fairly dark and often well-defined. Cumulus occurs in various sizes depending on the degree of development with vertical extent ranging from about 100 m up to >5 km altitude. Cumulus is composed of water drops. Precipitation as rain is possible.

Cumulonimbus

Heavy and dense cloud with large vertical extent in the form of a mountain or huge tower reaching beyond 6 km (often up to 15 km) altitude. The appearance of the upper part is smooth, fibrous or striated and nearly always flattened. This part generally spreads out in the form of an anvil or vast plume. Cumulonimbus is composed of water drops, the upper part of ice crystals. Precipitation as rain is common, snow, snow pellets or hail are possible.

Species:

Further subdivision is made to identify observed peculiarities in shape and internal structure.

Examples:

Congestus: cumulus clouds of great vertical extent,
Fibratus: detached clouds or a thin veil;
Lenticularis: clouds having the shape of lenses or almonds; commonly seen in mountainous regions;
Nebulosus: a cloud like a nebulous veil or layer without distinct details;
Stratiformis: clouds spread out in an extensive horizontal sheet or layer.

Formation of clouds:

In the atmosphere, water drops are formed by the activation of aerosol particles when, due to the cooling of air, the relative humidity rises beyond the saturation level (r.h. > 100%). The deliquescence of water-soluble salts associated with aerosol particles leads to the addition of water in accordance with Raoult's law. At r.h. > 100% a critical size exists, depending on surface curvature (Kelvin effect) and degree of supersaturation. Particles that overcome the critical size enter a region of instability and grow further to form a cloud drop.

Table 8.16 Equations governing the condensation (and evaporation) of water vapor[a]

$$\frac{dm}{dt} = 4\pi r \xi_D D \left(\rho_{va} - \rho_{vr} \right) \qquad\qquad \xi_D = \left(\frac{r}{r + \Delta r} + \frac{4D}{\alpha r \bar{v}} \right)^{-1}$$

$$L\frac{dm}{dt} = 4\pi r \xi_T K \left(T_r - T_a \right) \qquad\qquad \xi_T = \left(\frac{r}{r + \Delta r} + \frac{4K}{\alpha_T r \rho_a c_{pa} \bar{v}} \right)^{-1}$$

[a]*Assumptions*: spherical symmetry, constant ambient density of air, ρ_a, of water vapor, ρ_{va}, and of temperature, T_a
Quantities: $r(t)$=radius, ρ_w=density of liquid water, $m(t)=(4/3)\pi\rho_w r^3$=mass of the water drop, $\rho_{vr}(t, T_r)$=density of water vapor at the drop's surface, $T_r(t)$=temperature at the drop's surface, $D(T)$=diffusion coefficient for water in air, $L(T)$=latent heat of vaporization, $K(T)$=coefficient of thermal conductivity, $\rho_{va}=e_{sa}(M_w/R_g T_a)$, e_{sa}=water vapor pressure (determined by the Clausius-Clapeyron equation), M_w=molar mass of water; R_g=gas constant; ξ_D and ξ_T are correction factors required on the one hand, for the initial growth phase when the size of drops is comparable to the mean free path λ between collisions of molecules in the gas phase, and on the other hand to make allowance for mass and thermal accommodation; $\Delta r \approx \lambda$, α=mass accommodation coefficient, α_T=thermal accommodation coefficient, $\bar{v} = (8 R_g T_a/\pi M_w)^{\frac{1}{2}}$=mean thermal velocity of water molecules in air, $c_{pa}=1005.7$ (J kg^{-1} K^{-1})=(specific) heat capacity of air at constant pressure. Source: Pruppacher and Klett (1997)

Comments: Both equations are coupled and should be solved together. No measurements exist for the thermal accommodation coefficient, which usually is taken to be close to unity. Table 8.17 summarizes the other parameters as a function of temperature.

Table 8.17 Saturation vapor pressure over water, e_s (Pa), the corresponding water vapor density ρ_{ws} (kg m^{-3}), latent heat of condensation, L (kJ kg^{-1}), coefficient of thermal conductivity of dry air K (J m^{-1} s^{-1} K^{-1}), diffusion coefficient D (m^2 s^{-1}) of water vapor in air ($p=101.325$ kPa), mass accommodation coefficient α, as a function of temperature

$T(°C)$	e_s[a]	$10^2\rho_{ws}$	L[b]	$10^2 K$[c]	$10^5 D$[d]	α^e
−30	51.06	0.04550	2,575	2.16	1.776	0.430
−20	125.63	0.10753	2,549	2.23	1.912	0.337
−10	286.57	0.23596	2,525	2.32	2.053	0.262
0	611.21	0.48484	2,501	2.40	2.198	0.202
5	872.47	0.67964	2,489	2.44	2.272	0.178
10	1227.94	0.93965	2,477	2.48	2.347	0.156
15	1705.32	1.28231	2,466	2.52	2.424	0.138

(continued)

Table 8.17 (continued)

$T(°C)$	e_s^a	$10^2\rho_{ws}$	L^b	10^2K^c	10^5D^d	α^e
20	2338.54	1.728486	2,454	2.55	2.501	0.122
25	3168.74	2.30280	2,442	2.60	2.580	0.108
30	4245.20	3.03421	2,430	2.63	2.660	0.096
35	5626.45	3.95619	2,418	2.68	2.741	0.085

[a]Values for $t \geq 0°C$ are from Wexler (1976), values of e_s below 0°C were obtained by extrapolation. Compare this with Table 8.5

[b]To convert J g^{-1} to J mol^{-1} multiply with $M_w = 18.0153$ g mol^{-1}

[c]For typical atmospheric conditions, the conductivity of moist air is essentially the same as for dry air; K (mJ m^{-1} s^{-1}°C^{-1}) = 23.822 + 0.070338 (t/°C) (Beard and Pruppacher 1971); the formula is based on theoretical data of Mason and Monchick (1962) and experimental data of Taylor and Johnston (1949) and Franck (1951)

[d]$D = D_0 \times (T/T_0)^{1.83}$, $T_0 = 298.15$ K, $D_0 = 2.58 \times 10^{-5}$ (m^2 s^{-1})

[e]Calculated from: $\alpha/(1+\alpha) = \exp(-\Delta G_s/R_g T)$ with $\Delta G_s = \Delta H_s - T\Delta S_s$; where $\Delta H_s = (20.083 \pm 2.092)$ kJ mol^{-1}, $\Delta S_s = (84.94 \pm 7.53)$ J mol^{-1} K^{-1}, $R_g = 8.3145$ J mol^{-1} K^{-1} (Li et al. 2001)

Table 8.18 Microstructure of clouds: characteristic values:[a] number concentration of particles, N, size range, Δr, liquid water content/ice content, LWC/IC, altitude region of occurrence, cloud depth Δz, updraft velocity v_z

Cloud type		N (cm^{-3})	LWC/IC (g m^{-3})	Altitude (km)	Δz (km)	v_z (m s^{-1})
Cumulus	Continental	200–1,000[b]	0.1–0.4	1–5	0.3–4	0.5–2
	Marine	80–250	0.4–0.5	1–3	1	
Stratus	Continental	100–400	0.1–0.9	0–2	0.4	0.005–0.5
	Marine	50–350	0.3–0.8	0–2	0.4–0.9	
Altostratus/altocumulus		20–250	0.05–0.3	2–7	1–2	0.005–0.5
Cumulonimbus		~100	1–5	1–15	1–15	5–10
Cirrus		0.01–10	0.001–0.1	7–15	Thin	0.02–0.4

[a]Source of data: Heymsfield (1993), Pruppacher and Klett (1997), Seidl (1994), Lelieveld et al. (1989)

[b]High numbers occur in polluted regions with high aerosol content

Comments: The number concentration of water drops depends on the concentration of aerosol particles and on the level of supersaturation reached. High aerosol concentrations as found over the continents lead to larger concentrations of cloud drops than low concentrations of aerosol particles over the ocean. The supersaturation levels resulting from adiabatic cooling are higher in cumulus clouds than in stratus, owing to larger updraft velocities occurring in cumuli (~1 m s^{-1}) versus stratus (~0.1 m s^{-1}). The growth of cloud drops is initially rapid but it slows down with increasing drop size (the surface to volume ratio decreases with radius^{-1}), so that size distributions at the cloud base are quite narrow. Drop sizes in continental cumuli are centered near 5 μm radius, maritime cumuli feature broader size distributions centered near 8 μm radius. In all clouds, the size distribution broadens with height above cloud base, while the number concentration of cloud drops is determined by the peak supersaturation. Bimodal size distributions are frequently observed, usually near the top of the cloud. This feature is thought to be caused by turbulence and/or entrainment of ambient air. The further growth of cloud drops toward sizes in the precipitation range requires the coalescence of cloud drops (in warm clouds) or the formation of ice particles that subsequently scavenge liquid water drops.

Table 8.19 Observed size spectra of cloud drops, frequency distribution (% μm^{-1})[a]

Radius (μm)	Continental					Maritime			
	A	B	C	D	E	F	G	H	I
2	5.2	0.3	6.0	4.9	0.2		4.2	5.0	
3	10.9	1.5	8.9	15.4	0.8	0.2	2.9	11.0	3.9
4	15.9	2.4	12.7	22.2	2.7	0.7	2.9	13.9	4.9
5	15.9	4.2	13.4	20.4	6.8	1.1	3.3	14.3	5.7
6	13.7	5.1	12.1	15.4	11.8	2.8	3.9	16.7	6.2
7	12.0	6.3	7.9	10.2	15.0	4.8	4.5	16.4	6.9
8	9.3	8.4	4.6	6.2	16.8	6.9	5.4	9.5	8.3
9	7.1	10.8	3.5	3.2	16.7	10.2	6.6	4.8	10.3
10	4.4	12.0	2.8	1.5	12.5	12.4	7.8	2.7	14.2
11	2.7	11.4	2.3	0.6	8.6	12.4	9.6	1.9	15.2
12	1.6	9.6	1.8		5.1	11.7	11.0	1.6	12.0
13	0.7	7.8	1.9		2.0	10.0	12.4	1.2	5.9
14	0.5	6.0	2.3		0.7	8.7	11.2	0.6	2.9
15	0.4	4.2	4.2		0.2	7.2	7.6	0.3	1.4
16		3.3	5.2			5.2	4.7		1.0
17		2.4	7.3			3.2	2.0		0.7
18		1.8	3.3			1.7	0.5		0.5
19		1.2	0.3			0.7	0.1		0.1

[a]Derived by interpolation and smoothing of published data

A, B: cumuli over England up to 2,100 m in depth; averages of 3 samples each taken near 1 and 2 km altitude, respectively; data selected from Durbin (1959)

C: sample taken near the top of a 1,400 m deep cumulus cloud over Australia demonstrating bimodal distribution (Warner 1969)

D, E: wintertime stratus over southern Germany, cloud base ~700 m, cloud top ~1,200 m, sampled at 700 and 970 m, respectively (Hoffmann and Roth 1989)

F: Trade wind cumuli near Hawaii sampled near cloud base at 1 km altitude (Squires 1958; Jiusto 1967)

G: wintertime cumuli of ~2 km depth near Tasmania, average of 2 samples at 2.4 and 2.9 km altitude, $-5/-6°C$; concentrations of water and ice particles ≥ 300 μm were in the range $(0.2–2.9) \times 10^{-5}$ of total droplet concentrations (Mossop 1985)

H, I: stratocumulus over the North Sea, cloud base ~380 m, cloud top ~830 m; samples taken at 480 and 730 m, respectively (Nicholls 1984)

Table 8.20 Parameters for modeling cloud drop size distributions[a]

Cloud type	b	r_c	α	γ
Cumulus (1)	1.1046×10^{-3}	4.90	5	2.16
Cumulus (2)	4.5880×10^{-3}	7.72	2	2.39
Orographic[b]				
1st mode	8.1124×10^{-5}	12.35	4	1.22
2nd mode	1.9576×10^{-5}	21.87	3	2.01
Stratus, base	9.7923×10^{-3}	4.70	5	1.05
Stratus, top	3.8180×10^{-3}	6.75	3	1.30
Stratocumulus, base	2.8230×10^{-3}	5.33	5	1.19
Stratocumulus, top	1.9779×10^{-3}	10.19	2	2.46

(continued)

Table 8.20 (continued)

Cloud type	b	r_c	α	γ
Nimbostratus, base	8.0606×10^{-4}	6.41	5	1.24
Nimbostratus, top	1.0969×10^{-2}	9.67	1	2.41
Radiation fog	3.0410×10^{-4}	8.06	4	1.77
Advection fog	4.2028×10^{-3}	5.04	5	1.17
Arctic marine fog	2.6209×10^{-3}	7.85	3	1.15

[a]Model size distributions based on the modified Gamma function, parameters selected from Tampieri and Tomasi (1976)
[b]Bimodal distributions resulting from droplet coalescence are frequently observed in longer-lived cumuli, and especially in rain clouds

Comments: The modified Gamma distribution function, originally introduced by Deirmendjian (1964), provides the most general form of model to describe the (monomodal) volume concentration of droplets as a function of radius (r)

$$n(r) = N\, b\, r^{\alpha} \exp\left\{ -(\alpha/\gamma)(r/r_c)^{\gamma} \right\}, \qquad 0 \le r < \infty.$$

Here, N is the total number concentration of cloud droplets, b is a normalization factor, r_c is the radius where the distribution has its maximum, and α (an integer) as well as γ are positive constants that determine the shape of the distribution function. Bimodal distributions may be expressed by the sum of two suitably chosen modified Gamma distributions.

Table 8.21 Median values of effective diameter, D_{eff}, total water content, W, extinction coefficient β, and effective cloud drop concentration, N_{eff}, as a function of temperature[a]

Temperature range (°C)	D_{eff} (µm)	W (g m^{-3})	β (km)	N_{eff} (cm^{-3})
$0 < T < 10$	16	0.16	31	95
$0 < T < -10$	21	0.10	11	22
$-10 < T < -20$	23	0.048	5	6
$-20 < T < -30$	20	0.021	2.5	4
$-30 < T < -40$	15	0.011	2	7
$-40 < T < -50$	18	0.007	1.5	4

[a]Definitions: $D_{eff} = \bar{D}^3 / \bar{D}^2$, $W = (\pi \rho_w / 6) \int f(D) D^3 dD$, $\beta = (3/\rho_w) W/D_{eff}$, N_{eff} is defined as the concentration of the monodisperse size distribution having the same W, β, and D_{eff} as the drop size spectrum $f(D)$; Source of data: Korolev et al. (2001)

Table 8.22 Median values of effective diameter, D_{eff}, total water content, W, extinction coefficient, β, and effective concentration of cloud drops, N_{eff}, for different cloud types[a]

Cloud type	D_{eff} (µm)	W (g m^{-3})	β (km)	N_{eff} (cm^{-3})
Cumulus	17	0.15	25	62
Stratus, stratocumulus	12	0.11	30	145
Nimbostratus	26	0.1	9	7
Altostratus, altocumulus	22	0.05	4.5	6
Cirrus	16	0.01	2.2	5

[a]For definitions see footnote to previous table; source of data: Korolev et al. (2001)

References

Beard, K.V., H. Pruppacher, J. Atmos. Sci. **28**, 1455–1464 (1971)

Deirmendjian, D., Appl. Opt. **3**, 187–196 (1964)

Durbin, W.G., Tellus **11**, 203–215 (1959)

Franck, E.U., Z. *Elektrochemie* **55**, 636–647 (1951)

Heymsfield, A.J., in *Aerosol-Cloud-Climate Interactions*, ed. by P.V. Hobbs (Academic Press, San Diego, 1993), pp. 97–121

Hoffmann, H.E., R. Roth, Meteorol. Atmos. Phys. **41**, 24–254 (1989)

Jiusto, J.E., Tellus **19**, 359–368 (1967)

Korolev, A.V., G.A. Isaak, I.P. Mazin, H.W. Barker, Quart. J. R. Met. Soc. **127**, 2117–2151 (2001)

Laube, M., H. Höller, *Cloud Physics*. In *Meteorology*, ed. by G. Fischer, Landolt-Börnstein, New Series, vol. V/4b (Springer, Berlin, 1988), pp. 1–100

Lelieveld, J., P.J. Crutzen, H. Rodhe, *Zonal average Cloud Characteristics for Global Atmospheric Chemistry Modelling*. Report CM-76 (Department of Meteorology, University of Stockholm, Sweden, 1989)

Li, Y.Q., P. Davidovits, Q. Shi, J.T. Jayne, C.E. Kolb, D.R. Worsnop, J. Phys. Chem. **A105**, 10627–10634 (2001)

Mason, E.A., L. Monchick, J. Chem. Phys. **36**, 1622–1639 (1962)

Mossop, S.C., Quart. J. Roy. Meteor. Soc. **111**, 183–198 (1985)

Nicholls, S., Quart. J. Roy. Meteor. Soc. **110**, 783–820 (1984)

Pruppacher, H., J.D. Klett, *Microphysics of Clouds and Precipitation*, 2nd edn. (Kluwer, Dordrecht, 1997)

Seidl, W., Atmos. Res. **31**, 157–185 (1994)

Squires, P., Tellus **10**, 256–261 (1958)

Tampieri, F., C. Tomasi, Tellus **28**, 333–347 (1976)

Taylor, W.J., H.L. Johnston, J. Chem. Phys. **14**, 219–233 (1949)

Warner, J., J. Atmos. Sci. **26**, 1049–1059 (1969)

Wexler, A., J. Res. Natl. Bur. Stand. **80A**, 775–785 (1976)

8.4   Solubilities of Gases and Vapors in Water (Henry's Law Coefficients)

The dissolution of atmospheric gaseous species in liquid water occurring in clouds, fog and rain is governed by Henry's law, which states that under conditions of equilibrium (and in the ideal case) the concentration of a dissolved substance is proportional to its vapor pressure in the gas phase, and that each component acts independently of other substances simultaneously present in the system. The solubility is expressed by a coefficient that depends on the units used to describe concentration and pressure. Units frequently used in atmospheric chemistry are atmosphere for pressure (1 atm = 1.01325×10^5 Pa) and mol dm^{-3} for concentration in the aqueous phase. If c_{aq} denotes the concentration and p the pressure, Henry's law coefficient takes the form

$$K_H = c_{aq} / p \ (\text{mol dm}^{-3} \text{ atm}^{-1}) \tag{8.1}$$

The temperature dependence of K_H – over not too wide a range – can be expressed by an exponential function of T (K) such that

$$\ln(K_H) = -A + B / T \tag{8.2}$$

where A and B are constants. This empirical form is used in Table 8.24 presented below. Also shown are the temperature range for the experimental data that were used to calculate A and B, and the value of K_H at 25°C (298.15 K). Because the measurements are confined to a temperature range above the freezing point of water, extrapolation is required if one deals with super-cooled solutions. Very precise measurements have been made in a number of cases, including the main atmospheric constituents and the rare gases. These data are better represented by an equation derived from thermodynamic considerations

$$\ln(x) = A* + B*/T + C* \ln T \qquad (8.3)$$

where x is the chemical amount fraction of the solute in the aqueous phase, and the coefficient A*, B* and C* are related, respectively, to entropy and heat of solution and the difference in heat capacity of the substances between liquid and gaseous phases. Selected data sets have been evaluated on the basis of Eq. 8.3 and are discussed in the IUPAC Solubility Data Series. The equations were reformulated in terms of $\ln(K_H)$ and are presented in Table 8.23. Because the conversion involves the density of water, which is a moderate function of T, an average density for the temperature range 0–35°C is used. Gases that react with water to form ions, such as carbon dioxide, sulfur dioxide or ammonia, which involve dissociation equilibria and depend on the pH of the solution must be treated separately (see Table 8.25). For a more extensive list of Henry's law coefficients see www.henrys-law.org.

Table 8.23 Interpolation formulae for the solubility of gases in water, K_H (mol dm^{-3} atm^{-1}), based on Eq. 8.3, range 0–35°C; values for $K_H(298.15)$ are averages from experimental data

Constituent	Interpolation formula	$K_H(298.15)$	Ref.
Helium He	$\ln(K_H) = -101.96 + 4259.6/T + 14.009 \times \ln(T)$	$(3.88 \pm 0.04) \times 10^{-4}$ $n = 5$	(1)
Neon Ne	$\ln(K_H) = -135.95 + 6104.9/T + 18.916 \times \ln(T)$	$(4.54 \pm 0.05) \times 10^{-4}$ $n = 5$	(1)
Argon Ar	$\ln(K_H) = -146.40 + 7476.3/T + 20.140 \times \ln(T)$	$(1.39 \pm 0.02) \times 10^{-3}$ $n = 5$	(2)
Krypton Kr	$\ln(K_H) = -174.52 + 9101.7/T + 24.221 \times \ln(T)$	–	(3)
Xenon Xe	$\ln(K_H) = -197.21 + 10521.0/T + 27.466 \times \ln(T)$	–	(3)
Radon-222 ^{222}Rn	$\ln(K_H) = -247.74 + 13003.0/T + 35.005 \times \ln(T)$	$(9.09 \pm 0.41) \times 10^{-3}$ $n = 4$	(3)
Oxygen O_2	$\ln(K_H) = -175.33 + 8747.5/T + 24.453 \times \ln(T)$	$(1.26 \pm 0.02) \times 10^{-3}$ $n = 5$	(4)
Nitrogen N_2	$\ln(K_H) = -177.57 + 8632.1/T + 24.798 \times \ln(T)$	$(6.49 \pm 0.06) \times 10^{-4}$ $n = 4$	(5)
Nitrous oxide N_2O	$\ln(K_H) = -154.61 + 8882.8/T + 21.253 \times \ln(T)$	$(2.42 \pm 0.03) \times 10^{-2}$ $n = 8$	(6)
Nitric oxide NO	$\ln(K_H) = -163.86 + 8234.2/T + 22.815 \times \ln(T)$	$(1.95 \pm 0.06) \times 10^{-3}$ $n = 2$	(6)
Hydrogen H_2	$\ln(K_H) = -121.92 + 5528.5/T + 16.889 \times \ln(T)$	$(7.90 \pm 0.04) \times 10^{-4}$ $n = 6$	(7)

(continued)

Table 8.23 (continued)

Constituent	Interpolation formula	$K_H(298.15)$	Ref.
Carbon monoxide CO	$\ln(K_H) = -178.00 + 8750.0/T + 24.875 \times \ln(T)$	$(9.84 \pm 0.05) \times 10^{-4}$ $n = 4$	(8)
Methane CH_4	$\ln(K_H) = -211.28 + 10447.9/T + 29.780 \times \ln(T)$	$(1.407 \pm 0.006) \times 10^{-3}$ $n = 5$	(9)
Ethane C_2H_6	$\ln(K_H) = -246.80 + 12695.6/T + 34.474 \times \ln(T)$	$(1.872 \pm 0.014) \times 10^{-3}$ $n = 5$	(10)
Propane C_3H_8	$\ln(K_H) = -279.81 + 14435.5/T + 39.760 \times \ln(T)$	$(1.505 \pm 0.06) \times 10^{-3}$ $n = 7$	(11)
Butane C_4H_{10}	$\ln(K_H) = -276.53 + 14604.4/T + 38.760 \times \ln(T)$	$(1.206 \pm 0.032) \times 10^{-3}$ $n = 8$	(11)
Isobutane $(CH_3)_2CHCH_3$	$\ln(K_H) = -371.92 + 18304.4/T + 53.465 \times \ln(T)$	$(8.94 \pm 0.50) \times 10^{-4}$	(11)

References

(1) Clever, H.L. (ed.), *IUPAC Solubility Data Series*. Helium and Neon, vol. 1 (Pergamon Press, Oxford, 1979)

(2) Clever, H.L. (ed.), *IUPAC Solubility Data Series*. Argon, vol. 4 (Pergamon Press, Oxford, 1980)

(3) Clever, H.L. (ed.), *IUPAC Solubility Data Series*. Krypton, Xenon and Radon, vol. 2 (Pergamon Press, Oxford, 1979)

(4) Battino, R. (ed.), *IUPAC Solubility Data Series*. Oxygen and Ozone, vol. 7 (Pergamon Press, Oxford, 1981)

(5) Battino, R. (ed.), *IUPAC Solubility Data Series*. Nitrogen and Air, vol. 10 (Pergamon Press, Oxford, 1982)

(6) Young, C.E. (ed.), *IUPAC Solubility Data Series*. Oxides of Nitrogen, vol. 8 (Pergamon Press, Oxford, 1981)

(7) Young, C.E. (ed.), *IUPAC Solubility Data Series*. Hydrogen and Deuterium, vol. 5/6 (Pergamon Press, Oxford, 1981)

(8) Rettich, T.R., R. Battino, E. Wilhelm, Ber. Bunsenges. Phys. Chem. **86**, 1128–1132 (1982)

(9) Clever, H.L., C.L. Young (eds.), *IUPAC Solubility Data Series*. Methane, vol. 27/28 (Pergamon Press, Oxford, 1987)

(10) Hayduk, W. (ed.), *IUPAC Solubility Data Series*. Ethane, vol. 9, (Pergamon Press, Oxford, 1982)

(11) Hayduk, W. (ed.), *IUPAC Solubility Data Series*. Propane, Butane, and 2-Methylpropane, vol. 24 (Pergamon Press, Oxford, 1986)

Table 8.24 Henry's law coefficients K_H (mol dm^{-3} atm^{-1}) for atmospheric gases: parameters A and B in Eq. 8.2, temperature range, values at 25°C, references used for evaluation

Substance	A	B	$\Delta T(°C)$	$K_H(298.15)$	Ref[a]
Main atmospheric constituents					
Nitrogen N$_2$	12.59±0.06	1,563±16	0–30	$(6.482±0.054)×10^{-4}$	(1–3, 8)
Oxygen O$_2$	12.51±0.02	1,738±18	0–30	$(1.246±0.018)×10^{-3}$	(1, 2) (4–8)
Argon Ar	12.22±0.08	1,683±22	0–30	$(1.397±0.003)×10^{-3}$	(1, 2) (8, 9)
Several trace gases					
Methane CH$_4$	12.83±0.24	1,864±70	0–25	$(1.407±0.006)×10^{-3}$	(10–13)
Carbon monoxide CO	11.44±0.34	1,348±101	0–40	$(9.857±0.006)×10^{-4}$	(14, 15)
Nitrous oxide N$_2$O	12.64±0.37	2,665±48	0–40	$(2.42±0.03)×10^{-2}$	(16–23)
Oxidants					
Ozone O$_3$	12.45±0.04	2,367±68	0–40	$(1.10±0.25)×10^{-2}$	(24)
Hydrogen peroxide H$_2$O$_2$	10.72±0.54	6,605±154	3–24	$(9.17±0.39)×10^{4}$	(25–29)
Methyl hydroperoxide, CH$_3$OOH	11.79±0.59	5,219±145	4–25	$(2.94±0.36)×10^{2}$	(27, 29)
Ethyl hydroperoxide, C$_2$H$_5$OOH	14.28±0.70	5,995±200	5–25	$(3.36±0.20)×10^{2}$	(29)
Hydroxymethyl hydroperoxide, HOCH$_2$OOH	18.04±0.18	9,652±53	5–25	$(1.67±0.35)×10^{6}$	(29)
Acetic peroxoic acid, CH$_3$(CO)OOH	11.07±2.34	5,308±672	5–25	$(8.37±1.75)×10^{2}$	(29)
Nitrogen compounds					
Nitric oxide NO	11.60±0.37	1,598±106	0–30	$(1.95±0.06)×10^{-3}$	(30, 31)
Nitrogen dioxide NO$_2$		no data^b		$(1.0±0.3)×10^{-2}$	(32, 33)
Acetonitrile CH$_3$CN	9.82±0.30	4,106±101	0–30	51.2±3.0	(34)
Peroxyacetyl nitrate PAN, CH$_3$C(O)OONO$_2$	18.04±0.30	5,701±96	1–24	2.95±0.11	(35, 36)
Peroxypropionyl nitrate C$_2$H$_5$C(O)OONO$_2$	19.24±0.11	5,941±108	1–24	1.98±0.04	(36)
Sulfur compounds					
Carbonyl sulfide COS	14.78±0.36	3,262±104	0–30	$(2.1±0.8)×10^{-2}$	(37, 39)
Carbon disulfide CS$_2$	15.88±0.28	3,902±82	0–32	$(6.2±1.8)×10^{-2}$	(38, 39)
Dimethyl sulfide CH$_3$SCH$_3$	12.34±0.20	3,502±57	0–35	0.565±0.017	(40–42)
Sulfur hexafluoride SF$_6$	18.60±0.24	3,060±71	3–30	$(2.44±0.02)×10^{-4}$	(8, 43)

Carbonyl compounds					
Formaldehyde HCHO	14.68 ± 0.60	$6,775 \pm 181$	$10\text{--}45$	$(3.2 \pm 0.3) \times 10^3$	(44, 45)
Acetaldehyde CH_3CHO	16.4 ± 0.1	$5,671 \pm 22$	$0\text{--}40$	13.5 ± 1.5	(34)
Acetone CH_3COCH_3	14.4 ± 0.3	$5,286 \pm 100$	$0\text{--}45$	27.8 ± 0.3	(34)
Glyoxal OHCCHO	13.69 ± 0.3	$7,808 \pm 556$	$2\text{--}55$	$(2.7 \pm 0.6) \times 10^5$ $(n=10)$	(55)
Alcohols					
Methanol	12.46 ± 0.25	$5,312 \pm 76$	$0\text{--}80$	$(2.16 \pm 0.14) \times 10^2$ $(n=8)$	(46)
Ethanol C_2H_5OH	15.87 ± 0.82	$6,274 \pm 242$	$0\text{--}60$	$(1.94 \pm 0.13) \times 10^2$ $(n=8)$	(46)
Halogenated hydrocarbons					
Chloromethane CH_3Cl	13.16 ± 0.47	$3,262 \pm 139$	$0\text{--}50$	$(1.10 \pm 0.05) \times 10^{-1}$ $(n=2)$	(47)
Dichloromethane CH_2Cl_2	13.23 ± 0.78	$3,665 \pm 230$	$10\text{--}40$	$(3.80 \pm 0.55) \times 10^{-1}$ $(n=4)$	(47)
Trichloromethane $CHCl_3$	15.63 ± 0.38	$4,261 \pm 112$	$2\text{--}50$	$(2.55 \pm 0.17) \times 10^{-1}$ $(n=6)$	(47)
Tetrachloromethane CCl_4	17.69 ± 0.40	$4,284 \pm 119$	$10\text{--}47$	$(3.53 \pm 0.23) \times 10^{-2}$ $(n=5)$	(47)
Chloroethane C_2H_5Cl	11.88 ± 1.04	$2,803 \pm 305$	$10\text{--}35$	$(8.64 \pm 0.53) \times 10^{-2}$ $(n=2)$	(47)
1,1-Dichloroethane CH_3CHCl_2	15.56 ± 0.74	$4,121 \pm 219$	$2\text{--}50$	$(1.79 \pm 0.19) \times 10^{-1}$ $(n=3)$	(47)
1,2-Dichloroethane CH_3ClCH_2Cl	14.65 ± 0.56	$4,340 \pm 165$	$2\text{--}50$	$(8.5 \pm 2.0) \times 10^{-1}$ $(n=2)$	(47)
1,1,1,-Trichloroethane CH_3CCl_3	15.19 ± 0.32	$3,697 \pm 96$	$1\text{--}60$	$(5.80 \pm 0.54) \times 10^{-2}$ $(n=6)$	(47)
1,1,2-Trichloroethane $CH_2ClCHCl_2$	13.48 ± 0.40	$4,060 \pm 118$	$3\text{--}50$	1.10 ± 0.10	(47)

(continued)

Table 8.24 (continued)

Substance	A	B	$\Delta T(°C)$	$K_H(298.15)$	Ref[a]
1,1,1,2-Tetrachloroethane, CH_2ClCCl_3	16.36±1.30	4,626±394	20–40	$(5.74\pm0.20)\times10^{-1}$ (n=2)(at 20°C)	(47)
1,1,2,2-Tetrachloroethane, $CHCl_2CHCl_2$	15.18±0.94	4,784±281	11–50	3.01±0.37	(47)
Hexachloroethane C_2Cl_6	20.26	5,634	10–30	0.355	(48)
Chloroethene C_2H_3Cl	13.79±0.66	3,144±194	10–35	$(3.69\pm0.13)\times10^{-2}$ (n=2)	(47)
1,1-Dichloroethene CH2CCl2	14.82±0.65	3,440±191	10–40	$(3.85\pm0.02)\times10^{-2}$ (n=2)	(47)
cis-1,2-Dichloroethene CHClCHCl	15.10±0.59	4,099±176	10–40	$(2.33\pm0.18)\times10^{-1}$ (n=2)	(47)
trans-1,2-Dichloro-ethene, CHClCHCl	15.64±0.43	3,996±130	10–46	$(1.06\pm0.01)\times10^{-1}$ (n=2)	(47)
Trichloroethene $CHCl_3$	16.99±0.50	4,410±147	1–60	$(1.05\pm0.08)\times10^{-1}$ (n=7)	(47)
Tetrachloroethene C_2Cl_4	18.10±0.30	4,573±89	1–60	$(6.09\pm0.39)\times10^{-1}$ (n=7)	(47)
Fluorine compounds					
Trichlorofluoromethane, $CFCl_3$	16.11±0.42	3,449±125	1–40	$(1.11\pm0.11)\times10^{-2}$ (n=4)	(49–53)
Dichlorodifluoromethane, CF_2Cl_2	17.25±0.22	3,420±66	1–40	$(3.03\pm0.09)\times10^{-3}$ (n=3)	(48) (51–53)
Tetrafluoromethane CF_4	16.04±0.21	2,258±60	2.5–30	$(2.09\pm0.02)\times10^{-4}$ (n=3)	(8) (43, 54)
Hexafluoroethane C_2F_6	19.34±0.65	2,859±189	5–30	$(5.72\pm0.06)\times10^{-5}$	54

[a]See references below
[b]In aqueous solution nitrogen dioxide undergoes the reactions $2NO_2 = N_2O_4$ followed by $N_2O_4 + H_2O = HNO_2 + H^+ + NO_3^-$

References

(1) Klots, C.E., B.B. Benson, *J. Mar. Res.* **21**, 48–57 (1963)

(2) Douglas, E., *J. Phys. Chem.* **68**, 169–174 (1964)

(3) Murray, C.N., J.P. Riley, T.R.S. Wilson, *Deep-Sea Res.* **16**, 297–310 (1969)

(4) Montgomery, H.A.C., N.S. Thom, A. Cockburn, *J. Appl. Chem.* **14**, 280–296 (1964)

(5) Carpenter, J.H., *Limnol. Oceanogr.* **11**, 264–277 (1966)

(6) Murray, C.N., J.P. Riley, *Deep-Sea Res.* **16**, 311–320 (1969)

(7) Benson, B.B., D. Krause Jr., *J. Chem. Phys.* **64**, 689–709 (1976)

(8) Cosgrove, B.A., J. Walkley, *J. Chromatogr.* **216**, 161–167 (1981)

(9) Murray, C.N., J.P. Riley, *Deep-Sea Res.* **17**, 203–209 (1970)

(10) Ben Naim, A., J. Wilf, M. Yaacobi, *J. Phys. Chem.* **77**, 95–102; **78**, 170–175 (1973)

(11) Yamamoto, S., J.B. Alcanskas, T.E. Crozier, *J. Chem. Eng. Data* **21**, 78–80 (1976)

(12) Muccitelli, J.A., W.-Y. Wen, *J. Solution Chem.* **9**, 141–161 (1980)

(13) Rettich, T.R., Y.P. Handa, R. Battino, E. Wilhelm, *J. Phys. Chem.* **85**, 3230–3237 (1981)

(14) Winkler, L.W., Ber. Deutsch. *Chem. Ges.* **34**, 1408–1422 (1901)

(15) Rettich, T.R., R. Battino, E. Wilhelm, *Ber. Bunsenges. Phys. Chem.* **86**, 1128–1132 (1982)

(16) Winkler, L.W., *Ber. Deutsch. Chem. Ges.* **34**, 1408–1422 (1901)

(17) Geffcken, G., *Z. Phys. Chem.* **49**, 257–302 (1904)

(18) Knopp, W., *Z. Phys. Chem.* **48**, 97–108 (1904)

(19) Findlay, A., H.J.M. Creighton, *J. Chem. Soc.* **97**, 536–561 (1910)

(20) Kunerth, W., *Phys. Rev.* **19**, 512–524 (1922)

(21) Markham, A.E., K.A. Kobe, *J. Am. Chem. Soc.* **63**, 449–454 (1944)

(22) Joosten, G.E.H., P.V. Danckwerts, *J. Chem. Eng. Data* **17**, 452–454 (1972)

(23) Weiss, R.F., B.A. Price, *Mar. Chem.* **8**, 347–359 (1980)

(24) Warneck, P., in *Chemicals in the Atmosphere – Solubility, Sources and Reactivity*, ed. by P.G.T. Fogg, J.M. Sangster (Wiley, Chichester, 2003), pp. 225–228

(25) Yoshizumi, K., K. Aodi, I. Nouchi, T. Okita, T. Kobayashi, S. Kawamura, M. Tajima, *Atmos. Environ.* **18**, 395–401 (1984)

(26) Hwang, H., P.K. Dasgupta, *Environ. Sci. Technol.* **19**, 255–258 (1985)

(27) Lind, J.A., G.L. Kok, *J. Geophys. Res.* **91**, 7889–7895 (1986); Erratum **99**, 21119 (1994)

(28) Zhou, X., Y.-N. Lee, *J. Phys. Chem.* **96**, 265–272 (1992)

(29) O'Sullivan, D.W., M. Lee, B.C. Noone, B.G. Heikes, *J. Phys. Chem.* **100**, 3241–3247 (1996)

(30) Winkler, L.W., *Ber. Deutsch. Chem. Ges.* **34**, 1408–1422 (1901)

(31) Armor, J.N., *J. Chem. Eng. Data* **19**, 82–84 (1974)

(32) Schwartz, S.E., W.H. White, *Adv. Environ. Sci. Eng.* **4**, 1–45 (1981)

(33) Lee, Y.-N., S.E. Schwartz, *J. Phys. Chem.* **85**, 840–848 (1981)

(34) Benkelberg, H.-J., S. Hamm, P. Warneck, *J. Atm. Chem.* **20**, 17–34 (1995)

(35) Kames, J., U. Schurath, *J. Atm. Chem.* **21**, 151–164 (1995)

(36) Kames, J., S. Schweighoefer, U. Schurath, *J. Atm. Chem.* **12**, 169–180 (1991)

(37) Winkler, L.W., *Z. phys. Chem.* **55**, 344–354 (1906)

(38) Elliot, S., *Atmos. Environ.* **23**, 1977–1980 (1989)

(39) De Bruyn, W.J., E. Swartz, J.H. Hu, J.A. Shorter, P. Davidovits, *J. Geophys. Res.* **100**, 7245–7251 (1995) (only 25°C value)

(40) Przyjazny, A., W. Janicki, W. Chrzanowski, R. Staszewski, *J. Chromatogr.* **280**, 249–260 (1983)

(41) Dacey, J.W.H., S.G. Wakeham, B.L. Howes, *Geophys. Res. Lett.* **11**, 991–994 (1984)

(42) Wong, P.K., Y.H. Wang, *Chemosphere* **35**, 535–544 (1997)

(43) Ashton, J.T., R.A. Dawe, K.W. Miller, E.B. Smith, B.J. Stickings, *J. Chem. Soc.* **A 1968**, 1793–1796 (1968)

(44) Betterton, E.A., M.R. Hoffmann, *Environ. Sci. Technol.* **22**, 1415–1418 (1988)

(45) Zhou, X., K. Mopper, *Environ. Sci. Technol.* **24**, 1864–1869 (1990)

(46) Warneck, P., *Atmos. Environ.* **40**, 7146–7151 (2006)

(47) Warneck, P., *Chemosphere* **69**, 347–361 (2007)

(48) Munz, C., P.V. Roberts, *J. Am. Water Works Assoc.* **79**, 62–69 (1987)

(49) Hunter-Smith, R.J., P.W. Balls, P.S. Liss, *Tellus* **35B**, 170–176 (1983)

(50) Ashworth, R.A., G.B. Howe, M.E. Mullins, T.N. Rogers, *J. Hazard. Mater.* **18**, 25–36 (1988)

(51) Warner, M.J., R.F. Weiss, *Deep-Sea Res.* **32**, 1485–1497 (1985)

(52) Wisegarver, D.P., J.D. Cline, *Deep-Sea Res.* **32**, 97–106 (1985)

(53) Park, T., T.R. Rettich, R. Battino, D. Peterson, E. Wilhelm, *J. Chem. Eng. Data* **27**, 324–326 (1982)

(54) Wen, W.-Y., J.A. Muccitelli, *J. Solut. Chem.* **8**, 225–245 (1979)

(55) Ip, H.S.S., X.H.H. Huang, J.Z. Yu, *Geophys. Res. Lett.* **36**, L01802 (2009). doi:10.1024/2008GL036212

Table 8.25 Henry's law coefficients[a] for atmospheric constituents that react with water, parameters A and B in Eq. 8.2, temperature range, values at 25°C

Substance	A	B	$\Delta T(°C)$	$K_H(298.15)$	Ref.[d]
Carbon dioxide CO_2	11.55	2,440	0–40	3.45×10^{-2}	(1)
Sulfur dioxide SO_2	10.66 ± 0.16	$3,243 \pm 36$	0–50	1.24 ± 0.03	(2, 3)
Ammonia NH_3	9,935	4,186	0–40	60.7	(4, 5)
Hydrochloric acid[b] HCl	15.29 ± 0.15	$8,886 \pm 43$	0–40	2.04×10^6	(1, 6)
Nitric acid[b] HNO_3	14.46 ± 0.13	$8,694 \pm 38$	0–40	2.45×10^6	(1, 6)
Nitrous acid HNO_2	-12.48 ± 0.13	$4,912 \pm 37$	0–30	54.3	(7)
Formic acid HCOOH	11.4 ± 0.7	$6,100 \pm 200$	2–35	$(8.9 \pm 1.3) \times 10^3$	(8)
Acetic acid CH_3COOH	12.5 ± 0.4	$6,200 \pm 100$	2–35	$(4.1 \pm 0.4) \times 10^3$	(8)
Glycolic acid $CH_2OHCOOH$	3.03 ± 0.14	$3,946 \pm 548$	5–35	$(2.8 \pm 0.4) \times 10^4$	(9)
Glyoxylic acid[c] OHCCOOH	7.56 ± 0.22	$4,933 \pm 838$	5–35	$(8.0 \pm 0.2) \times 10^3$	(9)

[a] See the additional comments below
[b] Values for hydrochloric acid and nitric acid, which are fully dissociated at pH ≥ 1, are not Henry's law coefficients but equilibrium constants K_{HX} (mol^2 dm^{-6} atm^{-1}) for the reaction $p_{HX} = X^- + H^+$
[c] Glyoxylic acid is essentially fully hydrated
[d] References: (1) Wagman et al. (1982a), (2) Maahs (1982), (3) Goldberg and Parker (1985), (4) Clegg and Brimblecombe (1989), (5) Dasgupta and Dong (1986), (6) Brimblecombe and Clegg (1988), (7) Park and Lee (1988), (8) Johnson et al. (1996), (9) Ip et al. (2009)

Comments: Henry's law coefficients in Table 8.25 refer to the equilibrium between the compound in the gas phase and the undissociated compound in solution. Dissociation adds one or two more coupled equilibria that must be taken into account. In the case of SO_2, for example, the equilibria are

$$SO_{2\,gas} \rightleftharpoons SO_{2\,aq} \quad K_H$$

$$SO_{2\,aq} \rightleftharpoons H^+ + HSO_3^- \quad K_1$$

$$HSO_3^- \rightleftharpoons H^+ + SO_3^{2-} \quad K_2$$

The total concentration of sulfite species in solution in equilibrium with SO_2 in the gas phase depends on the hydrogen ion concentration [H^+], which is generally determined by other factors

$$\left[SO_{2aq}\right]+\left[HSO_3^-\right]+\left[SO_3^{2-}\right] = K_H p \left\{1+K_1 / \left[H^+\right]+K_1 K_2 / \left[H^+\right]^2\right\}$$

Here, brackets are used to indicate concentrations in units of [mol dm^{-3}]. Dissociation constants are listed in Table 8.26. For strong acids such as HCl and HNO$_3$ that are fully dissociated in aqueous solution values for K_H are not known. In these cases, the equilibrium constants are known only for the product of Henry's law coefficient and dissociation constant

$$K_{HX} = K_{diss} K_H = \left[H^+\right]\left[X^-\right] / p_{HX}$$

where X is either Cl or NO$_3$ and K_{diss} denotes the dissociation constant. Values for K_{HX} are entered in the above Table 8.25.

References

(1) Wagman, D.D., W.H. Evans, V.B. Parker, I. Halow, S.M. Bailey, R.H. Schumm, K.L. Churney, R.L. Nuttall, The NBS tables of thermodynamic properties. J. Phys. Chem. Ref. Data **11**(Suppl 2), 1–392 (1982)
(2) Maahs, H.G., in *Heterogeneous Atmospheric Chemistry*, ed. by D.R. Schryer. Geophysical Monograph Series 26 (American Geophysical Union, Washington, DC, 1982), pp. 187–195
(3) Goldberg, R.N., V.B. Parker, J. Res. Natl. Bur. Stand. **90**, 341–358 (1985)
(4) Clegg, S.L., P. Brimblecombe, J. Phys. Chem. **93**, 7237–7248 (1989)
(5) Dasgupta, P.K., S. Dong, Atmos. Environ. **20**, 565–570 (1986)
(6) Brimblecombe, P., S.L. Clegg, J. Atmos. Chem. **7**, 1–18 (1988); Erratum: **8**, 95
(7) Park, J.-Y., Y.-N. Lee, J. Phys. Chem. **92**, 6294–6302 (1988)
(8) Johnson, B.J., E.A. Betterton, D. Craig, J. Atmos. Chem. **24**, 113–119 (1996)
(9) Ip, H.S.S., X.H.H. Huang, J.Z. Yu, Geophys. Res. Lett. **36**, L01802 (2009). doi:10.1024/2008GL036212

8.5   Chemical Equilibria in Aqueous Solution, Dissociation Constants

Aqueous dissociation reactions important to atmospheric chemistry include acid and base dissociation processes leading to the formation of ions, as well as hydration and other addition reactions. Ion reactions generally are rapid so that equilibria are always established, whereas hydration reactions approach equilibrium more slowly. Time constants for reaching equilibrium can be estimated from the rates of forward and reverse reactions. Unless indicated otherwise the following tables list equilibrium constants and reaction rates (as far as known) for *dilute solutions*. Ion dissociation equilibria depend on the presence of other ions in the system due to the action of Coulomb forces, which generally begin to influence the value of the equilibrium constant when total ion concentrations exceed 10^{-3} mol dm^{-3}. The effect will be most pronounced in the concentrated solutions associated with aqueous aerosols so that appropriate caution must be exercised and suitable corrections applied when using ion dissociation constants in connection with aerosols. A commonly applied correction is the approximation suggested by Davies (1961): $\log K_{corr} = \log K_0 + 2Az_a z_b \{(\mu^{1/2}/(1+\mu^{1/2})) - C\mu\}$, where z_a and z_b are the charges of the ions involved, μ is the ionic strength defined by $\mu = \frac{1}{2}\Sigma\, z_i^2 c_i$, $A \approx 0.5$, and C is an adjustable constant, with $C = 0.2$ being frequently applicable. This equation is said to be valid up to ionic strengths $\mu \approx 0.1$. (Davies 1961)

Table 8.26 Dissociation reactions in dilute aqueous solutions: equilibrium constants at 298.15 K; forward and reverse rate coefficients (units: mol, dm³, s⁻¹); temperature dependence parameter B defined by $K(T) = K_{298}\exp\{-B[(1/T)-(1/298.15)]\}$

Reaction	K_{298}	B^a	k_f	k_r	Ref.[m]
Inorganic species					
$H_2O \rightleftharpoons H^+ + OH^-$	1.0×10^{-14b}	6,955	2.5×10^{-5c}	1.4×10^{11}	(1–3)
$H_2O_2 \rightleftharpoons H^+ + HO_2^-$	2.0×10^{-12}	3,710	d		(1, 2)
$HO_2 \rightleftharpoons H^+ + O_2^-$	1.6×10^{-5}	1,350	d		(4)
$OH \rightleftharpoons H^+ + O^-$	1.3×10^{-12}	2,045	d		(5)
$CO_{2aq} \rightleftharpoons H^+ + HCO_3^{-e}$	4.3×10^{-7}	995	2.7×10^{-2c}	6.4×10^4	(2, 6)
$CO_{2aq} + OH^- \rightleftharpoons HCO_3^{-e}$	4.25×10^7	−5,940	8.0×10^3	1.9×10^{-4c}	(2, 6)
$HCO_3^- \rightleftharpoons H^+ + CO_3^{2-}$	4.68×10^{-11}	1,785	d		(1, 2)
$HNO_2 \rightleftharpoons H^+ + NO_2^-$	7.0×10^{-4}	1,755	d		(1, 2)
$HNO_3 \rightleftharpoons H^+ + NO_3^-$	1.7×10^1	NA	d		(7)
$HOONO \rightleftharpoons H^+ + ONO_2^{-f}$	1.6×10^{-7}	NA	d		(8)
$HOONO_2 \rightleftharpoons H^+ + O_2NO_2^{-f}$	1.3×10^{-6}	NA	d		(9)
$NO_2 + NO_2 \rightleftharpoons N_2O_4^f$	6.4×10^4	NA	4.5×10^8	7.0×10^3	(10)
$NH_{3aq} \rightleftharpoons OH^- + NH_4^{+e}$	1.77×10^{-5}	710	6.0×10^5	3.4×10^{10}	(3, 11)
$HCN \rightleftharpoons H^+ + CN^-$	6.2×10^{-10}	5,230	d		(1, 2)
$SO_{2aq} \rightleftharpoons H^+ + HSO_3^{-e}$	1.39×10^{-2}	−1,870	2.8×10^{6c}	2.0×10^8	(3, 12)
$HSO_3^- \rightleftharpoons H^+ + SO_3^{2-}$	6.72×10^{-8}	−355	d		(12)
$HSO_4^- \rightleftharpoons H^+ + SO_4^{2-}$	1.02×10^{-2}	−2,445	1.5×10^9	1.0×10^{11}	(1–3)
$HOOSO_3^- \rightleftharpoons H^+ +^- OOSO_3^-$	7.4×10^{-10}	2,525	d		(13, 14)
$H_2S \rightleftharpoons H^+ + HS^-$	1.0×10^{-7}	2,660	4.3×10^3	7.5×10^{10}	(1–3)
$HOCl \rightleftharpoons H^+ + ClO^-$	2.8×10^{-8}	1,660	d		(2)
$HOBr \rightleftharpoons H^+ + BrO^-$	2.6×10^{-9}	2,275	d		(2)
$HOI \rightleftharpoons H^+ + IO^-$	2.4×10^{-11}	3,680	d		(2)
$H_3PO_4 \rightleftharpoons H^+ + H_2PO_4^-$	7.1×10^{-3}	960	d		(1)
$H_2PO_4^- \rightleftharpoons H^+ + HPO_4^{2-}$	4.5×10^{-13}	−1,760	d		(1)
$HPO_4^{2-} \rightleftharpoons H^+ + PO_4^{3-}$	6.3×10^{-8}	−400	d		(1)
$H_4SiO_4 \rightleftharpoons H^+ + H_3SiO_4^-$	1.4×10^{-10}		d		(1)
$Fe^{3+} + H_2O \rightleftharpoons Fe(OH)^{2+} + H^+$	6.73×10^{-3}	−5,230	d		(1)
$Fe^{3+} + SO_4^{2-} \rightleftharpoons Fe(SO_4)^+$	1.4×10^4	−3,120	2.3×10^{5g}	1.6×10^{1c}	(1, 15)
$Fe^{3+} + C_2O_4^{2-} \rightleftharpoons Fe(C_2O_4)^+$	4.0×10^{7h}	NA	$2.0 \times 10^{4g,h}$	5.0×10^{-4c}	(1, 15)
$Fe(C_2O_4)^+ + C_2O_4^{2-} \rightleftharpoons Fe(C_2O_4)_2^-$	$1.1 \times 10^{6h,i}$	NA			(1)

(continued)

Table 8.26 (continued)

Reaction	K_{298}	B^a	k_f	k_r	Ref.[m]
$Fe(C_2O_4)_2^- + C_2O_4^{2-} \rightleftharpoons Fe(C_2O_4)_3^{3-}$	$7.1 \times 10^{4h,i}$	NA			(1)
Organic compounds					
$HCOOH \rightleftharpoons H^+ + HCOO^-$	1.8×10^{-4}	−20	$9.0 \times 10^{6\,c}$	5.0×10^{10}	(1, 3)
$CH_3COOH \rightleftharpoons H^+ + CH_3COO^-$	1.75×10^{-5}	−50	$7.9 \times 10^{5\,c}$	4.5×10^{10}	(1, 3)
$CH_3C(O)OOH \rightleftharpoons H^+ + CH_3CO_3^-$	$1.1 \times 10^{-8\,j}$	NA		d	(16)
$C_2H_5COOH \rightleftharpoons H^+ + C_2H_5COO^-$	1.34×10^{-5}	−100		d	(1, 3)
$HOCH_2COOH \rightleftharpoons H^+ + HOCH_2COO^-$	1.48×10^{-4}	80		d	(1)
$(HO)_2CHCOOH \rightleftharpoons H^+ + (HO)_2CHCOO^-$	3.47×10^{-4}	265		d	(1)
$(HO)_2CHCOOH \rightleftharpoons OCHCOOH$	3.33×10^{-3}	NA	2.5×10^{-2}	7.5	(17)
$(HO)_2CHCOO^- \rightleftharpoons OCHCOO^-$	6.6×10^{-2}	NA	5.5×10^{-3}	8.3×10^{-2}	(17)
$CH_3CHOHCOOH \rightleftharpoons H^+ + CH_3CHOHCOO^-$	1.38×10^{-4}	40		d	(1)
$CH_3COCOOH \rightleftharpoons H^+ + CH_3COCOO^-$	3.37×10^{-3}	1,540		d	(18)
$H_2C(OH)_2 \rightleftharpoons HCHO \ (+H_2O)$	4.5×10^{-4}	4,030	5.1×10^{-3}	12.8^k	(19, 20)
$CH_3CH(OH)_2 \rightleftharpoons CH_3CHO$	9.5×10^{-1}	1,220	4.2×10^{-3}	4.3×10^{-3k}	(21)
$HOCH_2CH(OH)_2 \rightleftharpoons HOCH_2CHO$	1.1×10^{-1}	NA	9.6×10^{-3}	8.6×10^{-2}	(17)
$HSO_3^- + H_2C(OH)_2 \rightleftharpoons H_2C(OH)SO_3^-$	3.6×10^6	6,570	2.0×10^{-11}	$5.6 \times 10^{-8\,c}$	(22–24)
$SO_3^{2-} + H_2C(OH)_2 \rightleftharpoons H_2C(OH)SO_3^- + OH^-$	5.4×10^{-1}		2.4×10^{31}	4.5×10^3	(23–25)

[a]Temperature range 0–35°C; Values are accurate to ±50 K. They were calculated from the heat of reaction ΔH (298.15) and the change with heat capacity at other temperatures if such data were available; *NA* not available

[b]This value of the equilibrium constant corresponds to the product $K_w = [H^+][OH^-]$. The true equilibrium constant is $K = K_w/[H_2O] = 1.8 \times 10^{-16}$, and k_f is calculated accordingly

[c]The rate coefficient was calculated from the equilibrium constant and the counter reaction

[d]Rates of ion recombination reactions are at the diffusion-controlled limit. In analogy to other reactions listed, one may assume a value of $\sim 5 \times 10^{10}$ dm³ mol⁻¹ s⁻¹ if direct measurements are lacking

[e]In aqueous solution CO_2, SO_2 and NH_3 exist as physically dissolved gases as well as carbonic acid, sulfurous acid, and ammonium hydroxide, respectively. Thus, $[CO_{2aq}] = [CO_2 \cdot H_2O] + [H_2CO_3]$; $[SO_{2aq}] = [SO_2 \cdot H_2O] + [H_2SO_3]$; $[NH_{3aq}] = [NH_3 \cdot H_2O] + [NH_4OH]$

[f]HOONO is unstable and decomposes in acidic solution to form nitrate. The product $O_2NO_2^-$ decomposes toward $NO_2^- + O_2$. The dimer N_2O_4 reacts with water to form HNO_2 and HNO_3

[g]One cannot distinguish between the reaction path shown and the reactions of HSO_4^- or $HC_2O_4^-$ with $Fe(OH)^{2+}$

[h]Data refer to ionic strength $\mu = 0.5$

[i]Recalculated from the original data, which refer to $Fe^{3+} + 2\ C_2O_4^{2-}$ and $Fe^{3+} + 3\ C_2O_4^{2-}$, respectively

[j]Data refer to $T = 285$ K and ionic strength $\mu = 0.5$

[k]First order rate constant, pH range 4–6. Forward and reverse reactions are catalyzed by protons and hydroxyl ions, so that they increase somewhat at higher and lower pH

[l]The formation of hydroxyl-methane-sulfonate proceeds via non-hydrated formaldehyde, HCHO. Reaction rates are defined as $k_{HCHO} \times [HCHO]/\{[H_2C(OH)_2] + [HCHO]\}$

References

(1) Smith, R.M., A.E. Martell, *Critical Stability Constants*, vols. 3 and 4 (Plenum Press, New York, 1976)

(2) Wagman, D.D., W.H. Evans, V.B. Parker, R.H. Schumm, I. Halow, S.M. Bailey, K.L. Churney, R.L. Nuttal, J. Phys. Chem. Ref. Data **11**(Suppl 2), 1–392 (1982)

(3) Eigen, M., W. Kruse, G. Maass, L. De Maeyer, Progr. React. Kinet. **2**, 285–318 (1964)

(4) Bielski, B.H.J., D.E. Cabelli, R.L. Arudi, A.B. Ross, J. Phys. Chem. Ref. Data **14**, 1041–1100 (1985)

(5) Buxton, G.V., C.L. Greenstock, W.P. Helman, A.B. Ross, J. Phys. Chem. Ref. Data **17**, 513–886 (1988)

(6) Johnson, K.S., Limn. Oceanogr. **27**, 849–855 (1982)

(7) Davis Jr., W., H.J. De Bruin, J. Inorg. Nucl. Chem. **26**, 1069–1083 (1964)

(8) Drexler, C., H. Elias, B. Fechner, K.J. Wannowius, Fresenius J. Anal. Chem. **340**, 605–615 (1991)

(9) Goldstein, S., G. Czapski, Inorg. Chem. **36**, 4156–4162 (1997); Lρgager, T., K. Sehested, J. Phys. Chem. **97**, 10047–10052 (1993)

(10) Grätzel, M., A. Henglein, J. Lilie, G. Beck, Ber. Bunsenges. Physik. Chem. **73**, 646–653 (1969)

(11) Bates, R.G., G.D. Pinching, J. Am. Chem. Soc. **72**, 1393–1396 (1950)

(12) Goldberg, R.N., V.B. Parker, J. Res. Natl. Bur. Stand. **90**, 341–358 (1985)

(13) Elias, H., U. Götz, K.J. Wannowius, Atmos. Environ. **28**, 439–448 (1994)

(14) Ball, D.L., J.O. Edwards, J. Am. Chem. Soc. **78**, 1125–1129 (1956)

(15) Biruš, M., N. Kujundžić, M. Pribanić, Progr. React. Kinet. **18**, 171–271 (1993)

(16) Fecher, B., *Kinetik und Mechanismus der S(IV) Oxidation mit Wasserstoffperoxid und organischen Peroxiden in wässriger Lösung*. Dissertation, Technische Hochschule Darmstadt, Germany, 1995

(17) Sørensen, P.E., K. Bruhn, F. Lindeløv, Acta. Chem. Scand. **A28**, 162–168 (1974)

(18) Fischer, M., P. Warneck, Ber. Bunsenges. Physik. Chem. **95**, 523–527 (1991)

(19) Bell, R.P., P.G. Evans, Proc. Roy. Soc. Lond. **A291**, 297–323 (1966)

(20) Schecker, H.G., G. Schulz, Z. Phys. Chem. N. F. **65**, 221–224 (1969)

(21) Kurz, J.L., J.I. Coburn, J. Am. Chem. Soc. **89**, 3524–3528; 3528–3537 (1967)

(22) Deister, U., R. Neeb, G. Helas, P. Warneck, J. Phys. Chem. **90**, 3213–3217 (1986)

(23) Boyce, S.D., M.R. Hoffmann, J. Phys. Chem. **88**, 4740–4746 (1984)

(24) Kok, G.L., S.N. Gitlin, A.L. Lazrus, J. Geophys. Res. **91**, 2801–2804 (1986)

(25) Sørensen, P.E., V.S. Andersen, Acta Chem. Scand. **24**, 1301–1306 (1970)

Table 8.27 Dissociation constants for dicarboxylic acids in dilute aqueous solution: equilibrium constants at 298.15 K; temperature parameter B defined by $K(T)=K_{298}\exp{-B[(1/T)-(1/298.15)]}$[a]

Reaction	K_{298} (1st)	B	K_{298} (2nd)	B
Ethanedioic (oxalic) acid, HOOCCOOH	5.59×10^{-2}	-455	5.42×10^{-5}	-805
Propanedioic (malonic) acid, HOOCCH$_2$COOH	1.42×10^{-3}	20	2.01×10^{-6}	-580
Methylpropanedioic acid, HOOCCH(CH$_3$)COOH	9.77×10^{-4}		1.74×10^{-6}	
Butanedioic (succinic) acid, HOOC(CH$_2$)$_2$COOH	6.21×10^{-5}	350	2.31×10^{-6}	-55
Methylbutanedioic acid, HOOCCH$_2$CH(CH$_3$)COOH	7.94×10^{-5}		1.62×10^{-6}	
Hydroxybutanedioic (malic) acid, HOOCCH$_2$CHOHCOOH	3.48×10^{-4}	360	8.00×10^{-6}	-140
Pentanedioic (glutaric) acid, HOOC(CH$_2$)$_3$COOH	4.57×10^{-5}	-50	3.71×10^{-6}	-300
Hexanedioic (adipic) acid HOOC(CH$_2$)$_4$COOH	3.80×10^{-5}	-150	3.80×10^{-6}	-300
Heptanedioic (pimelic) acid, HOOC(CH$_2$)$_5$COOH	3.24×10^{-5}	-150	3.72×10^{-6}	-450
Nonanedioic (azelaic) acid HOOC(CH$_2$)$_7$COOH	2.82×10^{-5}		3.80×10^{-6}	
cis-Butenedioic (maleic) acid, HOOCCH=CHCOOH	1.23×10^{-2}	50	4.65×10^{-7}	-400
trans-Butenedioic (fumaric) acid, HOOCCH=CHCOOH	8.85×10^{-4}	50	3.21×10^{-6}	-350
1,2-Benzene dicarboxylic (phtalic) acid, C$_6$H$_4$(COOH)$_2$	1.12×10^{-3}	-320	3.91×10^{-6}	-250

[a]Data assembled from Smith and Martell (1976)

8.6 Aqueous Phase Photochemical Processes

The following tables provide absorption coefficients and quantum yields for several photochemical processes in the aqueous phase that have been recognized to participate in the chemistry of atmospheric aqueous systems. The first table gives an overview on important processes. The other tables present the decadic absorption coefficient, ε (dm^3 mol^{-1} cm^{-1}), the corresponding absorption cross section σ (cm^2 molecule^{-1}) and the quantum yield φ for radicals resulting from the photochemical process if wavelength-dependent. Absorption coefficients are defined by $\varepsilon=(1/cd)\times\log(I_0/I)$ and $\sigma=(1/nd)\ln(I_0/I)$, where c is the concentration (mol dm^{-3}), n is the number density (molecule cm^{-3}), I_0 and I are the incident and transmitted monochromatic light intensities, respectively, for an optical path length d and uniform concentration of the absorbing compound.

Table 8.28 Summary of aqueous photochemical processes: photolysis frequency j (s^{-1}), radical quantum yield, wavelength or wavelength range of measurement.

Reaction/rate[a]	φ	λ, $\Delta\lambda$/nm	Ref.
$H_2O_2 + h\nu \rightarrow OH + OH$[b]	1.0	254	(2)
$j = 1.0 \times 10^{-5}$	0.98, 0.96[c]	308, 351	(3)
$NO_3^- + h\nu\,(+H^+) \rightarrow NO_2 + OH$	0.0092	305	(4)
$j = 5.6 \times 10^{-7}$	0.017	308	(3)
	0.013–0.017	313	(5)
$NO_2^- + h\nu\,(+H^+) \rightarrow NO + OH$	0.023–0.068[d]	280–390	(6)
$j = 3.8 \times 10^{-5}$			
$HNO_2 + h\nu \rightarrow NO_2 + OH$	0.46, 0.45	254, 365	(7)
$j = 6.7 \times 10^{-4}$	0.35[c]	280–390	(6)
$Fe(OH)^{2+} + h\nu \rightarrow Fe^{2+} + OH$	0.105	340	(8)
$j = 5.2 \times 10^{-3}$	0.14, 0.017	313, 360	(9)
	0.074–0.320[d]	280–370	(10)
$Fe(SO_4)^+ + h\nu \rightarrow Fe^{2+} + SO_4^-$	$(1.75–7.5) \times 10^{-3d}$	280–350	(10)
$j = 1.0 \times 10^{-4}$			
$Fe(C_2O_4)_x^{\,n} + h\nu \rightarrow$	0.62[c], $x = 3$	250–400	(11)
$Fe^{2+} + C_2O_4^- + (x-1)C_2O_4^{2-e}$	~0.6, $x = 2$	313	(12)
$j = 6.2 \times 10^{-2}$			

[a]Photolysis frequencies from Warneck et al. (1996); (1) Buxton and Wilmarth (1963), (2) Zellner et al. (1990), (3) Warneck and Wurtzinger (1988), (4) Zepp et al. (1987), (5) Fischer and Warneck (1996), (6) Alif and Boule (1991), (7) David and David (1976), (8) Faust and Hoigné (1990), (9) Benkelberg and Warneck (1995), (10) Hatchard and Parker (1956), (11) Zuo and Hoigné (1992)
[b]The OH quantum yield is twice the quantum yield for dissociation
[c]Essentially independent of wavelength
[d]For details see the following tables
[e]Three Fe(III)-oxalato complexes exist: $x = 1$, 2, 3, with charge numbers $n = (3 - 2x)^+$. Photodecomposition of $Fe(C_2O_4)_3^{3-}$ in sulfuric acid solution is widely used as an actinometer with a quantum yield of Fe^{2+} formation $\phi_{prod} = 1.24$, essentially independent of wavelength between 250 and 400 nm. The radical quantum yields shown take into account the subsequent reaction $C_2O_4^- + Fe(C_2O_4)_x^{\,n} \rightarrow Fe^{2+} + xC_2O_4^{2-} + 2CO_2$, which raises the Fe^{2+} quantum yield by a factor of two. No data exist for the $Fe(C_2O_4)^+$ complex

Table 8.29 Absorption coefficients (ε ($M^{-1}cm^{-1}$), σ (cm^2 molecule^{-1})) for aqueous hydrogen peroxide and nitrate anion as function of wavelength (nm)[a]

	H_2O_2		NO_3^-			H_2O_2		NO_3^-	
λ	ε	$10^{20}\sigma$	ε	$10^{20}\sigma$	λ	ε	$10^{20}\sigma$	ε	$10^{20}\sigma$
280	5.23	2.00	3.73	1.43	320	0.52	0.20	3.17	1.21
285	3.92	1.50	4.75	1.81	325	0.42	0.16	1.80	0.69
290	2.96	1.13	5.82	2.23	330	0.31	0.12	0.87	0.33
295	2.28	0.87	6.75	2.58	335	0.24	0.09	0.37	0.14
300	1.73	0.66	7.24	2.77	340	0.18	0.07	0.153	0.058
305	1.28	0.49	7.06	2.70	345	0.13	0.05	0.069	0.026
310	0.97	0.37	6.20	2.37	350	0.08	0.03	0.028	0.011
315	0.73	0.28	4.77	1.82					

[a]For quantum yields see Table 8.28

Table 8.30 Absorption coefficients for nitrite and nitrous acid, quantum yields for the process $NO_2^- + h\nu\,(+H^+) \rightarrow NO + OH$ as a function of wavelength[a]

λ (nm)	Nitrite anion, NO_2^-			Nitrous acid, HNO_2[a]	
	$\varepsilon\,(M^{-1}cm^{-1})$	$10^{20}\sigma$ (cm^2 molecule^{-1})	φ	$\varepsilon\,(M^{-1}cm^{-1})$	$10^{20}\sigma$ (cm^2 molecule^{-1})
280	7.97	3.05	0.0678	0.96	0.37
285	8.31	3.17	0.0677	0.97	0.37
290	8.50	3.25	0.0675	1.08	0.41
295	8.65	3.31	0.0672	1.41	0.54
300	8.80	3.36	0.0667	2.11	0.81
305	9.05	3.46	0.0659	2.77	1.06
310	9.45	3.61	0.0646	3.79	1.45
315	10.09	3.86	0.0626	5.38	2.06
320	11.14	4.26	0.0595	7.21	2.76
325	12.59	4.81	0.0554	10.39	3.98
330	14.61	5.59	0.0501	12.95	4.97
335	16.66	6.37	0.0443	18.46	7.08
340	18.91	7.23	0.0387	20.38	7.81
345	20.88	7.98	0.0338	27.94	10.71
350	22.29	8.52	0.0301	26.35	10.10
355	22.71	8.68	0.0275	36.24	13.90
360	21.79	8.33	0.0258	37.76	14.48
365	19.66	7.52	0.0247	31.03	11.90
370	16.48	6.30	0.0240	41.78	16.02
375	12.64	4.83	0.0236	29.50	11.31
380	8.91	3.41	0.0234	20.39	7.82
385	5.67	2.17	0.0232	24.72	9.48
390	3.13	1.20	0.0231	15.63	5.98
395	1.39	0.53	0.0231	3.71	1.42
400	0.45	0.17	0.0230	0.69	0.26

[a]The quantum yield for decomposition of nitrous acid is constant over the entire wavelength range with $\varphi = 0.35 \pm 0.02$ (see Table 8.28). Source of data: Fischer and Warneck (1996a)

Table 8.31 Absorption coefficients ($\varepsilon\,(M^{-1}cm^{-1})$, σ (cm^2 molecule^{-1})) and quantum yields for the processes $Fe(OH)^{2+} + h\nu \rightarrow Fe^{2+} + OH$ and $Fe(SO_4)^+ + h\nu \rightarrow Fe^{2+} + SO_4^-$ as a function of wavelength (nm)[a]

λ	Fe(OH)$^{2+}$			Fe(SO$_4$)$^+$		
	ε	$10^{18}\sigma$	φ	ε	$10^{18}\sigma$	φ
280	1,717	6.564	0.312	1,552	5.933	0.0073
290	1,958	7.485	0.288	1,722	6.583	0.0060
300	1,985	7.588	0.218	1,852	7.080	0.0042
310	1,826	6.981	0.195	1,800	6.881	0.0034
320	1,565	5.983	0.160	1,552	5.933	0.0031
330	1,244	4.756	0.134	1,187	4.538	0.0022
340	920	3.517	0.112	822	3.142	0.0018
350	631	2.412	0.085	522	1.995	0.0015
360	431	1.648	0.073			
370	257	0.983	0.074			

[a]Source of data: Benkelberg and Warneck (1995)

References

Alif, A., P. Boule, J. Photochem. Photobiol. **A59**, 357–367 (1991)

Benkelberg, H.-J., P. Warneck, J. Phys. Chem. **99**, 5214–5221 (1995)

Buxton, G.V., W.K. Wilmarth, J. Phys. Chem. **67**, 2835–2841 (1963) (and earlier papers cited therein)

David, F., P.G. David, J. Phys. Chem. **80**, 579–583 (1976)

Faust, B.C., J. Hoigné, Atmos. Environ. **24A**, 79–89 (1990) (and earlier papers cited therein)

Fischer, M., P. Warneck, J. Phys. Chem. **100**, 18749–18756 (1996) (and earlier papers cited therein)

Hatchard, C.G., C.A. Parker, Proc. Roy. Soc. (Lond.) **A235**, 518–536 (1956)

Warneck, P., C. Wurzinger, J. Phys. Chem. **92**, 6278–6283 (1988)

Warneck, P., P. Mirabel, G.A. Salmon, R. van Eldik, C. Vinckier, K.J. Wannowius, C. Zetzsch, in *Heterogeneous and Liquid-Phase Processes*, Transport and Chemical Transformation of Pollutants in the Troposphere, vol. 2. (Springer, Berlin, 1996), pp. 7–71

Zellner, R., M. Exner, H. Herrmann, J. Atm. Chem. **10**, 411–425 (1990)

Zepp, R.G., J. Hoigné, H. Bader, Environ. Sci. Technol. **21**, 443–450 (1987)

Zuo, Y., J. Hoigné, Environ. Sci. Technol. **26**, 1014–1022 (1992)

8.7 Rate Coefficients for Elementary Reactions in the Aqueous Phase

Rate coefficients for aqueous phase reactions are given in units ($dm^3 \, mol^{-1} \, s^{-1}$). The temperature dependence, if known, is expressed by $k(T)=k_{298}\exp\{-B[(1/T)-(1/298.15)]\}$. The parameter B is related to the activation energy E_a by $B=E_a/R_g$, where R_g is the gas constant.

Table 8.32 Basic chemistry involving HO_x and CO_x reactions

Reaction	k_{298}	B	Ref.
$e_{aq}^- + O_2 \rightarrow O_2^-$	2.0×10^{10}		(1)
$e_{aq}^- + H^+ \rightarrow H$	2.2×10^{10}		(1)
$H + O_2 \rightarrow HO_2$	2.0×10^{10}		(1)
$HO_2 + HO_2 \rightarrow H_2O_2 + O_2$	8.3×10^5	2,480	(1, 2)
$HO_2 + O_2^- \, (+ \, H^+) \rightarrow H_2O_2 + O_2$	9.7×10^7	980	(1, 2)
$OH + HO_2 \rightarrow O_2 + H_2O$	1.0×10^{10}		(3)
$OH + O_2^- \rightarrow OH^- + O_2$	1.1×10^{10}	2,120	(3)
$OH + OH \rightarrow H_2O_2$	5.5×10^9		(4)
$OH + H_2 \rightarrow H + H_2O$	4.2×10^7	2,290	(4)
$OH + H_2O_2 \rightarrow HO_2 + H_2O$	2.7×10^7	1,680	(4)
$O_3 + O_2^- \rightarrow O_3^- + O_2$	1.5×10^9		(5)
$O_3^- + H^+ \rightleftharpoons HO_3$	$5.2 \times 10^{10}/3.3 \times 10^2$		(5)
$HO_3 \rightarrow OH + O_2$	1.1×10^5		(5)
$O_3 + OH \rightarrow HO_2 + O_2$	1.1×10^8		(4)

(continued)

Table 8.32 (continued)

Reaction	k_{298}	B	Ref.
$OH + HCOOH \rightarrow CO_2 + H + H_2O$	1.2×10^8	990	(4, 6)
$OH + HCOO^- \rightarrow CO_2^- + H_2O$	3.1×10^9	1,240	(4, 6)
$OH + HCO_3^- \rightarrow CO_3^- + H_2O$	1.0×10^7	1,900	(4, 7)
$OH + CO_3^{2-} \rightarrow CO_3^- + OH^-$	3.9×10^8	2,840	(4, 7)
$CO_3^- + O_2^- (+ H^+) \rightarrow HCO_3^- + O_2$	6.5×10^8		(8)
$CO_3^- + H_2O_2 \rightarrow HCO_3^- + HO_2$	4.3×10^5		(8)
$CO_3^- + HCOO^- \rightarrow HCO_3^- + CO_2^-$	1.5×10^5		(8)
$CO_2^- + O_2 \rightarrow O_2^- + CO_2$	2.4×10^9		(1)

References

(1) Bielski, B.H.J., D.E. Cabelli, R.L. Arudi, A.B. Ross, J. Phys. Chem. Ref. Data **14**, 1041–1100 (1985)

(2) Christensen, H., K. Sehested, J. Phys. Chem. **92**, 3007–3011 (1988)

(3) Christensen, H., K. Sehested, E. Bjergbakke, Water Chem. Nucl. React. Syst. **5**, 141–144 (1989); Elliot, A.J., G.V. Buxton, J. Chem. Soc. Faraday Trans. **88**, 2465–2470 (1992)

(4) Buxton G.V., C.L. Greenstock, W.P. Helman, A.B. Ross, J. Phys. Chem. Ref. Data **17**, 513–886 (1988)

(5) Sehested, K., J. Holcman, E.J. Hart, J. Phys. Chem. **87**, 1951–1954 (1983); Staehelin, J., R.E. Bühler, J. Staehelin, J. Hoigné, J. Phys. Chem. **88**, 2560–2564; 5450; 5999–6004 (1984)

(6) Chin, M., P.H. Wine, in *Aquatic and Surface Photochemistry*, ed. by G.R. Helz, R.G. Zepp, D.G. Crosby (Lewis Publication, Boca Raton, 1994), pp. 85–96

(7) Herrmann, H., B. Ervens, H.-W. Jacobi, R. Wolke, P. Nowacki, R. Zellner, J. Atmos. Chem. **36**, 231–284 (2000)

(8) Neta, P., R.E. Huie, A.B. Ross, J. Phys. Chem. Ref. Data **17**, 1027–1284 (1988)

Table 8.33 Reactions involving sulfur species

Reaction	k_{298}	B	Ref.
$O_3 + SO_2 + H_2O \rightarrow HSO_4^- + O_2 + H^+$	2.4×10^4		(1)
$O_3 + HSO_3^- \rightarrow HSO_4^- + O_2$	3.7×10^5	5,530	(1)
$O_3 + SO_3^{2-} \rightarrow SO_4^{2-} + O_2$	1.5×10^9	5,280	(1)
$HSO_3^- + H_2O_2 \rightarrow SO_4^{2-} + H^+ + H_2O$	$9.1 \times 10^7 [H^+]^a$	3,570	(2)
$HSO_3^- + CH_3OOH \rightarrow HSO_4^- + CH_3OH$	$1.9 \times 10^7 [H^+]^a$	3,800	(3)
$HSO_3^- + CH_3C(O)OOH \rightarrow HSO_4^- + CH_3COOH$	$4.5 \times 10^7 [H^+]^a$	3,990	(3)
$HSO_3^- + HOONO_2 \rightarrow HSO_4^- + NO_3^- + H^+$	$1.1 \times 10^7 [H^+] + 3.3 \times 10^{5a,b}$		(4)
$HSO_3^- + HOONO \rightarrow HSO_4^- + HNO_2$	$2.0 \times 10^7 [H^+] + 4.0 \times 10^{4a}$		(5)
$HSO_3^- + HSO_5^- \rightarrow 2\,HSO_4^-$	$6.2 \times 10^6 [H^+] + 189^a$		(6)
$SO_3^{2-} + HSO_5^- \rightarrow 2\,SO_4^{2-} + H^+$	2.3×10^3		(6)
$OH + HSO_3^- \rightarrow SO_3^- + H_2O$	2.7×10^9		(7)
$OH + SO_3^{2-} \rightarrow SO_3^- + OH^-$	4.6×10^9		(7)
$OH + HSO_5^- \rightarrow SO_5^- + H_2O$	1.7×10^7		(8)
$OH + HSO_4^- \rightleftharpoons SO_4^- + H_2O$	$7.0 \times 10^5/6.6 \times 10^2 \, (s^{-1})$		(8, 9)

(continued)

Table 8.33 (continued)

Reaction	k_{298}	B	Ref.
$SO_3^- + O_2 \rightarrow SO_5^-$	2.5×10^9		(7)
$SO_5^- + HSO_3^- \rightarrow HSO_5^- + SO_3^-$	8.6×10^3		(7)
$SO_5^- + HSO_3^- \rightarrow HSO_4^- + SO_4^-$	3.6×10^2		(7)
$SO_5^- + SO_3^{2-} (+ H^+) \rightarrow HSO_5^- + SO_3^-$	2.1×10^5		(7)
$SO_5^- + SO_3^{2-} \rightarrow SO_4^{2-} + SO_4^-$	5.5×10^5		(7)
$SO_5^- + SO_5^- \rightarrow S_2O_8^{2-} + O_2$	4.8×10^7		(7)
$SO_5^- + HO_2 \rightarrow HSO_5^- + O_2$	1.8×10^9		(10)
$SO_5^- + O_2^- (+ H^+) \rightarrow HSO_5^- + O_2$	2.3×10^8		(7)
$SO_4^- + HSO_3^- \rightarrow HSO_4^- + SO_3^-$	6.8×10^8		(7)
$SO_4^- + SO_3^{2-} \rightarrow SO_4^{2-} + SO_3^-$	3.1×10^8		(7)
$SO_4^- + H_2O_2 \rightarrow HSO_4^- + HO_2$	1.2×10^7		(11)
$SO_4^- + S_2O_8^{2-} \rightarrow S_2O_8^- + SO_4^{2-}$	$< 1.5 \times 10^3$		(9)
$SO_4^- + OH^- \rightarrow SO_4^{2-} + OH$	1.4×10^7		(12)
$SO_4^- + HO_2 \rightarrow HSO_4^- + O_2$	3.0×10^9		(13)
$SO_4^- + NO_3^- \rightleftharpoons SO_4^{2-} + NO_3$	$5.0 \times 10^4/1.8 \times 10^5$		(13, 14)
$SO_4^- + NO_2^- \rightarrow SO_4^{2-} + NO_2$	9.8×10^8		(11)
$SO_4^- + HCO_3^- \rightarrow SO_4^{2-} + H^+ + CO_3^-$	2.8×10^6		(15)
$NO_3 + SO_2 (+ H_2O) \rightarrow SO_3^- + NO_3^- + 2\,H^+$	2.3×10^8	3,700	(14)
$NO_3 + HSO_3^- \rightarrow SO_3^- + NO_3^- + H^+$	1.3×10^9	2,000	(14)
$NO_3 + SO_3^{2-} \rightarrow SO_3^- + NO_3^-$	3.0×10^8		(14)
$HSO_3^- + HCHO \rightarrow HOCH_2SO_3^-$	4.5×10^2	2,660	(16)
$SO_3^{2-} + HCHO (+ H^+) \rightarrow HOCH_2SO_3^-$	5.4×10^6	2,530	(16)
$HOCH_2SO_3^- + OH^- \rightarrow CH_2(OH)_2 + SO_3^{2-}$	4.6×10^3	4,880	(17)
$OH + HOCH_2SO_3^- \rightarrow HOCHSO_3^- + H_2O$	3.0×10^8		(18)
$HOCHSO_3^- \rightleftharpoons {}^-OCHSO_3^- + H^+$	$5.9 \times 10^4/4.4 \times 10^{10}$		(18)
$HOCHSO_3^- + O_2 \rightarrow O_2CH(OH)SO_3^-$	1.6×10^9		(18)
$^-OCHSO_3^- + O_2 \rightarrow O_2^- + OCHSO_3^-$	2.6×10^9		(18)
$O_2CH(OH)SO_3^- \rightarrow O_2CHO + HSO_3^-$	$7.0 \times 10^3 (s^{-1})$		(19)
$O_2CH(OH)SO_3^- \rightarrow HO_2 + OCHSO_3^-$	$1.7 \times 10^4 (s^{-1})$		(18)
$O_2CHO (+ H_2O) \rightarrow HO_2 + HCOOH$	$2.5 \times 10^3 (s^{-1})$		(18)
$OCHSO_3^- (+ H_2O) \rightarrow HCOOH + HSO_3^-$	$3.4 (s^{-1})$		(18)
$SO_4^- + HOCH_2SO_3^- \rightarrow products$	1.3×10^6		(18)
$NO_3 + HOCH_2SO_3^- \rightarrow products$	4.2×10^6		(19)

[a]The reaction is proton-catalyzed as well as general-acid-catalyzed. In dilute solutions H_2O assumes the role of an acid, so that the rate coefficient: $k = k_p [H^+] + k_a [H_2O]$ consists of two terms. If the contribution of the second term is negligible, it is not shown
[b]At $T = 285$ K, $\mu = 1.0$ M $NaClO_4$

References

(1) Hoffmann, M.R., Atmos. Environ. **20**, 1145–1154 (1986)
(2) Maaß, F., H. Elias, K.J. Wannowius, Atmos. Environ. **33**, 4413–4419 (1999)
(3) Lind, J.A., A.L. Lazrus, G.L. Kok, J. Geophys. Res. **92**, 4171–4177 (1987)
(4) Amels, P., H. Elias, U. Götz, U. Steingens, K.J. Wannowius, in *Heterogeneous and Liquid Phase Processes*, ed. by P. Warneck. Transport and Chemical Transformation of Pollutants in the Troposphere, vol. 2 (Springer, Berlin, 1996), pp. 77–88

(5) Drexler, C., H. Elias, B. Fecher, K.J. Wannowius, Fresenius. J. Anal. Chem. **340**, 605–615 (1991)

(6) Elias, H., U. Götz, K.J. Wannowius, Atmos. Environ. **28**, 439–448 (1994)

(7) Buxton, G.V. S. McGowan, G.A. Salmon, J.E. Williams, N.D. Wood, Atmos. Environ. **30**, 2483–2493 (1996)

(8) Buxton, G.V., C.L. Greenstock, W.P. Helman, A.B. Ross, J. Phys. Chem. Ref. Data **17**, 513–886 (1988)

(9) Buxton, G.V., M. Bydder, G.A. Salmon, Phys. Chem. Chem. Phys. **1**, 269–273 (1999)

(10) Buxton, G.V., T.N. Malone, G.A. Salmon, J. Chem. Soc. Faraday Trans. **92**, 1287–1289 (1996); Fischer, M., P. Warneck, J. Phys. Chem. **100**, 15111–15117 (1996)

(11) Wine, P.H., Y. Tang, R.P. Thorn, J.R. Wells, D.D. Davis, J. Geophys. Res. **94**, 1085–1094 (1989)

(12) Herrmann, H., A. Reese, R. Zellner, J. Molec. Struct. **348**, 183–186 (1995)

(13) Løgager, T., K. Sehested, J. Holcman, Radiat. Phys. Chem. **41**, 539–543 (1993)

(14) Exner, M., H. Herrmann, R. Zellner, Ber. Bunsenges. Phys. Chem. **96**, 470–477 (1992)

(15) Huie, R.E., C.L. Clifton, J. Phys. Chem. **94**, 8561–8567 (1990)

(16) Boyce, S.D., M.R. Hoffmann, J. Phys. Chem. **88**, 4740–4746 (1984); Kok, G.L., S.N. Gitlin, A.L. Lazrus, J. Geophys. Res. **91**, 2801–2804 (1986)

(17) Sørensen, P.E., V.S. Andersen, Acta Chem. Scand. **24**, 1301–1306 (1970)

(18) Barlow, S., G.V. Buxton, S.A. Murray, G.A. Salmon, J. Chem. Soc. Faraday Trans. **93**, 3637–3640; 3641–3645 (1997). See also: Barlow, S., G.V. Buxton, S.A. Murray, G.A. Salmon, in *Proc. EUROTRAC Symp. '96*, ed. by P.M. Borrell, P. Borrell, K. Kelly, T. Cvitaš, W. Seiler. Clouds, Aerosols, Modeling and Photo-Oxidants, vol. 1 (Computational Mechanics Publications, Southampton, 1997), pp. 360–365

(19) Herrmann, H., R. Zellner, in *N-Centered Radicals*, ed. by Z.B. Alfassi (Wiley, London, 1998), pp. 291–343

Table 8.34 Reactions involving nitrogen species

Reaction	k_{298}	B	Ref.[a]
$2\,NO_2 \rightleftharpoons N_2O_4(+H_2O) \to HNO_2 + NO_3^- + H^+$	8.4×10^7		(1)
$NO + NO_2 \rightleftharpoons N_2O_3(+H_2O) \rightleftharpoons 2\,HNO_2$	$1.6 \times 10^8/13.4$		(1)
$NO_3 + NO_2 \to N_2O_5\ (+\,H_2O \to 2NO_3^- + 2\,H^+)$	1.7×10^9		(2)
$2\,NO + O_2 \to 2\,NO_2$	2.1×10^6		(3)
$O_2^- + NO \to ONOO^-$	6.7×10^9		(4)
$OH + NO_2 \to HOONO$	4.5×10^9		(5)
$HOONO \to H^+ + NO_3^-$	$1.0\ (s^{-1})$		(5)
$OH + NO_2^- \to NO_2 + OH^-$	6.0×10^9		(5)
$NO_2^- + O_3 \to NO_3^- + O_2$	5.0×10^5	7,000	(6)
$HO_2 + NO_2 \to O_2NOOH$	1.8×10^9		(7)
$O_2^- + NO_2 \to O_2NOO^-$	4.5×10^9		(7)
$O_2NOOH \to HNO_2 + O_2$	7.0×10^{-4}		(7)
$O_2NOO^- \to NO_2^- + O_2$	$1.0\ (s^{-1})$		(7)
$O_2NOOH + HNO_2 \to 2\,H^+ + 2NO_3^-$	12.0		(7)
$NO_3 + OH^- \to NO_3^- + OH$	1.4×10^7	2,700	(8)
$NO_3 + HO_2 \to NO_3^- + H^+ + O_2$	3.0×10^9		(9)
$NO_3 + NO_2^- \to NO_3^- + NO_2$	1.5×10^9		(10)

References

(1) Park, J.-Y., Y.-N. Lee, J. Chem. Phys. **92**. 6294–6302 (1988)
(2) Katsumura, Y., P.Y. Jiang, R. Nagaishi, T. Oishi, K. Ishigure, Y. Yoshida, J. Phys. Chem. **95**, 4435–4439 (1991)
(3) Awad, H.H., D.M. Stanbury, Int. J. Chem. Kinet. **25**, 375–381 (1993); Pires, M., J. Rossi, D.S. Ross, Int. J. Chem. Kinet. **26**, 1207–1227 (1994)
(4) Huie, R.E., S. Padmaja, Free Rad. Res. Commun. **18**, 195–199 (1993)
(5) Løgager, T., K. Sehested, J. Phys. Chem. **97**, 6664–6669 (1993)
(6) Damschen, D.E., L.R. Martin, Atmos. Environ. **17**, 2005–2011 (1983)
(7) Løgager, T., K. Sehested, J. Phys. Chem. **97**, 10047–10052 (1993)
(8) Exner, M., H. Herrmann, R. Zellner, Ber. Bunsenges. Phys. Chem. **96**, 470–477 (1992)
(9) Sehested, K., T. Løgager, J. Holcman, O.J. Nielsen, in *Transport and Transformation of Pollutants in the Troposphere, Proc. EUROTRAC Symp. '94*, ed. by P.M. Borrell, P. Borrell, T. Cvitaš, W. Seiler (SPB Academic Publishing bv, Den Haag, 1994), pp. 999–1004
(10) Herrmann, H., R. Zellner, in *N-Centered Radicals*, ed. by Z.B. Alfassi (Wiley, London, 1998), pp. 291–343

Table 8.35 Reactions involving metal ion species

Reaction	k_{298}	B	Ref.
$OH + Fe^{2+} \rightarrow FeOH^{2+}$	4.3×10^8	1,100	(1)
$HO_2 + Fe^{2+} (+H_2O) \rightarrow FeOH^{2+} + H_2O_2$	1.2×10^6	5,050	(2, 3)
$O_2^- + Fe^{2+} (+ H_2O, H^+) \rightarrow FeOH^{2+} + H_2O_2$	1.0×10^7		(2, 3)
$HO_2 + FeOH^{2+} \rightarrow Fe^{2+} + O_2 + H_2O$	1.0×10^4		(2, 3)
$O_2^- + FeOH^{2+} \rightarrow Fe^{2+} + O_2 + OH^-$	1.5×10^8		(2, 3)
$Fe^{2+} + H_2O_2 \rightarrow FeOH^{2+} + OH$	7.6×10^1		(4)
$Fe^{2+} + O_3 \rightarrow FeO^{2+} + O_2$	8.5×10^5	4,690	(5)
$FeO^{2+} + H_2O \rightarrow FeOH^{2+} + OH$	1.3×10^{-2}	4,090	(5)
$FeO^{2+} + H_2O_2 \rightarrow FeOH^{2+} + HO_2$	9.5×10^3	2,770	(5)
$FeO^{2+} + OH (+H_2O) \rightarrow FeOH^{2+} + H_2O_2$	1.0×10^7		(5)
$FeO^{2+} + HO_2 \rightarrow FeOH^{2+} + O_2$	2.0×10^6		(5)
$FeO^{2+} + HNO_2 \rightarrow FeOH^{2+} + O_2$	1.1×10^4		(5)
$FeO^{2+} + Cl^- (+H_2O) \rightarrow FeOH^{2+} + HOCl^-$	1.0×10^2		(5)
$FeO^{2+} + SO_{2aq} (+H_2O) \rightarrow FeOH^{2+} + SO_3^- + H^+$	4.5×10^5		(5)
$FeO^{2+} + HSO_3^- \rightarrow FeOH^{2+} + SO_3^-$	2.5×10^5		(5)
$FeO^{2+} + Mn^{2+} (+ H_2O) \rightarrow FeOH^{2+} + MnOH^{2+}$	1.0×10^4	2,560	(5)
$FeOH^{2+} + HSO_3^- \rightarrow Fe^{2+} + SO_3^- + H_2O$	4.0×10^1	8,300	(6)
$Fe^{2+} + HSO_5^- \rightarrow FeOH^{2+} + SO_4^-$	3.1×10^4		(7)
$Fe^{2+} + S_2O_8^{2-} \rightarrow FeOH^{2+} + SO_4^{2-} + SO_4^-$	1.7×10^1		(8)
$Fe^{2+} + SO_4^- (+ H_2O) \rightarrow FeOH^{2+} + H^+ + SO_4^{2-}$	1.1×10^9	−2,165	(8)
$Fe^{2+} + SO_5^- (+ H_2O) \rightarrow FeOH^{2+} + HSO_5^-$	$8.0 \times 10^{5\,a}$		(9)
$Fe^{2+} + NO_3 (+ H_2O) \rightarrow FeOH^{2+} + NO_3^- + H^+$	8.0×10^6		(10)
$FeOH^{2+} + Cu^+ \rightarrow Cu^{2+} + Fe^{2+} + OH^-$	3.0×10^7		(11, 12)
$Cu^+ + OH \rightarrow Cu^{++} + OH^-$	3.0×10^9		(13)
$Cu^{++} + OH \rightarrow CuOH^{++}$	3.1×10^8		(1)
$Cu^+ + HO_2 (+ H^+) \rightarrow Cu^{2+} + H_2O_2$	2.3×10^9		(2)
$Cu^+ + O_2^- (+ 2 H^+) \rightarrow Cu^{2+} + H_2O_2$	9.4×10^9		(14)

(continued)

Table 8.35   (continued)

Reaction	k_{298}	B	Ref.
$Cu^{2+} + HO_2 \rightarrow Cu^+ + H^+ + O_2$	5.0×10^7		(2)
$Cu^{2+} + O_2^- \rightleftharpoons Cu^+ + O_2$	$8.0 \times 10^9 / 4.6 \times 10^5$		(12, 14)
$Cu^+ + H_2O_2 \rightarrow Cu^+ \cdot H_2O_2$	4.1×10^3		(15)
$Cu^+ \cdot H_2O_2 \rightarrow Cu^{2+} + OH + OH^-$	$1.0 \times 10^{2\,b}$		(15)
$Cu^+ + O_3 (+ H_2O) \rightarrow Cu^{2+} + OH + OH^-$	3.0×10^7		(16)
$MnOH^{2+} + Fe^{2+} \rightarrow Mn^{2+} + FeOH^{2+}$	1.5×10^4		(17)
$Mn^{2+} + OH \rightarrow MnOH^{2+}$	3.0×10^7		(1)
$Mn^{2+} + O_2^- \rightleftharpoons MnO_2^+$	$1.5 \times 10^8 / 6.5 \times 10^3$		(18)
$Mn^{2+} + HO_2 \rightleftharpoons MnO_2^+ + H^+$	$1.1 \times 10^6 / 6.5 \times 10^6$		(18)
$MnO_2^+ + HO_2 (+H^+) \rightarrow Mn^{2+} + H_2O_2 + O_2$	1.0×10^7		(18)
$MnOH^{2+} + H_2O_2 \rightarrow MnO_2^+ + H^+ + H_2O$	2.8×10^3		(18)
$Mn^{2+} + O_3 \rightarrow MnO^{2+} + O_2$	1.5×10^3	4,750	(19)
$MnO^{2+} + Mn^{2+} (+H_2O) \rightarrow 2MnOH^{2+}$	$\geq 1.0 \times 10^5$		(19)
$Mn^{2+} + SO_4^- (+ H_2O) \rightarrow MnOH^{2+} + H^+ + SO_4^{2-}$	2.6×10^7	4,090	(8)
$Mn^{2+} + SO_5^- (+ H_2O) \rightarrow products$	4.6×10^6		(20)
$Mn^{2+} + NO_3 (+ H_2O) \rightarrow products$	1.5×10^6		(10)

[a]The reaction proceeds via an intermediate complex: $Fe(H_2O) + SO_5^- \rightleftharpoons Fe(H_2O)(SO_5)^+ \rightarrow Fe(OH)^{2+} + HSO_5^-$. Herrmann et al. (1996) reported a rate coefficient 4.3×10^7 (20), which presumably refers to the first step of the overall process
[b]The extent of the reaction leading to OH radicals is not exactly known. The value given was derived by Moffet and Zika (1987)

References

(1) Buxton G.V., C.L. Greenstock, W.P. Helman, A.B. Ross, J. Phys. Chem. Ref. Data **17**, 513–886 (1988)
(2) Bielski, B.H.J., D.E. Cabelli, R.L. Arudi, A.B. Ross, J. Phys. Chem. Ref. Data **14**, 1041–1100 (1985)
(3) Rush, J.D., B.H.J. Bielski, J. Phys. Chem. **89**, 5062–5066 (1985)
(4) Walling, C., Acc. Chem. Res. **8**, 125–131 (1975)
(5) Løgager, T., J. Holcman, K. Sehested, T. Pedersen, Inorg. Chem. **31**, 3523–3529 (1992); Jacobsen, F., J. Holcman, K. Sehested, Int. J. Chem. Kinet. **29**, 17–24; **30**, 215–221 (1997/1998)
(6) Kraft, J., R. Van Eldik, Inorg. Chem. **28**, 2306–2312 (1989)
(7) Gilbert, B.C., J.K. Stell, J. Chem. Soc. Perkin Trans. **2**, 1281–1288 (1990)
(8) Buxton, G.V., T.N. Malone, G.A. Salmon, J. Chem. Soc. Faraday Trans. **93**, 2893–2897 (1997)
(9) Ziajka, J., F. Beer, P. Warneck, Atmos. Environ. **28**, 2549–2552 (1994); Warneck, P., J. Ziajka, Ber. Bunsenges. Phys. Chem. **99**, 59–65 (1995)
(10) Herrmann, H., R. Zellner, in *N-Centered Radicals*, ed. by Z.B. Alfassi (Wiley, London, 1998), pp. 291–343
(11) Sedlak, D.L., J. Hoigné, Atmos. Environ. **27A**, 2173–2185 (1993)
(12) Buxton, G.V., Q.G. Mulazzani, A.B. Ross, J. Phys. Chem. Ref. Data **24**, 199–1349 (1995)
(13) Goldstein, S., G. Czapski, H. Cohen, D. Meyerstein, Inorg. Chim. Acta **192**, 87–93 (1992)
(14) Von Piechowski, M., T. Nauser, J. Hoigné, R.E. Bühler, Ber. Bunsenges. Phys. Chem. **97**, 762–771 (1993)

(15) Marsawa, M., H. Cohen, D. Meyerstein, D.L. Hickman, A. Bacac, J.H. Espensen, J. Am. Chem. Soc. **110**, 4293–4297 (1988); Kozlov, Y.N., V.M. Berdnikov, Russ. J. Phys. Chem. **47**, 338–340 (1973); see also: Moffet, W., R.G. Zika, Environ. Sci. Technol. **21**, 804–810 (1987)

(16) Hoigné, J., R. Bühler, in *Heterogeneous and Liquid Phase Processes*, ed. by P. Warneck. Transport and Chemical Transformation of Pollutants in the Troposphere, vol. 2 (Springer, Berlin, 1996), pp. 110–115

(17) Diebler, H., N. Sutin, J. Phys. Chem. **68**, 174–180 (1964)

(18) Jacobsen, F., J. Holcman, K. Sehested, J. Phys. Chem. **A101**, 1324–1328 (1997)

(19) Jacobsen, F., J. Holcman, K. Sehested, Int. J. Chem. Kinet. **30**, 207–214 (1998)

(20) Herrmann, H., H.-W. Jacobi, G. Raabe, A. Reese, R. Zellner, Fresenius J. Anal. Chem. **355**, 343–344 (1996)

Table 8.36 Reactions involving halogen species

Reaction	k_{298}	B	Ref.
Chlorine species			
$SO_4^- + Cl^- \rightleftharpoons SO_4^{2-} + Cl$	$2.5 \times 10^8 / 2.1 \times 10^8$		(1)
$Cl + Cl^- \rightleftharpoons Cl_2^-$	$8.5 \times 10^9 / 6.0 \times 10^4$		(2)
$OH + Cl^- \rightleftharpoons HOCl^-$	$4.3 \times 10^9 / 6.1 \times 10^9$		(3)
$NO_3 + Cl^- \rightleftharpoons NO_3^- + Cl$	$3.4 \times 10^8 / 1.0 \times 10^8$		(4)
$HOCl^- + H^+ \rightleftharpoons Cl + H_2O$	$5.0 \times 10^{10} / 2.5 \times 10^5$		(5)
$Cl + NO_2^- \rightarrow Cl^- + NO_2$	5.0×10^9		(5)
$Cl + HCO_3^- \rightarrow Cl^- + H^+ + CO_3^-$	2.4×10^9		(5)
$Cl_2^- + HO_2 \rightarrow 2\,Cl^- + H^+ + O_2$	3.0×10^9		(6)
$Cl_2^- + O_2^- \rightarrow 2\,Cl^- + O_2$	6.0×10^9		(7)
$Cl_2^- + H_2O \rightarrow 2Cl^- + H^+ + OH$	$1.3 \times 10^3 \ (s^{-1})$		(2)
$Cl_2^- + HSO_3^- \rightarrow 2\,Cl^- + H^+ + SO_3^-$	2.0×10^8	1,080	(8, 9)
$Cl_2^- + SO_3^{2-} \rightarrow 2\,Cl^- + SO_3^-$	6.2×10^7		(9)
$Cl_2^- + NO_2^- \rightarrow 2\,Cl^- + NO_2$	2.5×10^8		(6)
$Cl_2^- + HCOOH \rightarrow 2\,Cl^- + 2\,H^+ + CO_2^-$	6.7×10^3		(6)
$Cl_2^- + HCOO^- \rightarrow 2\,Cl^- + H^+ + CO_2^-$	1.9×10^6		(6)
$Cl_2^- + H_2O_2 \rightarrow 2\,Cl^- + H^+ + HO_2$	1.4×10^5		(6)
$Cl_2^- + Cl_2^- \rightarrow 2\,Cl^- + Cl_2$	7.0×10^8		(10)
$Cl^- + HOCl + H^+ \rightarrow Cl_2 + H_2O$	2.84×10^4	3,250	(11)
$Cl_2 + H_2O \rightarrow Cl^- + HOCl + H^+$	$1.1 \times 10^1 \ (s^{-1})$	7,580	(11)
$Cl_2 + HO_2 \rightarrow Cl_2^- + H^+ + O_2$	1.0×10^9		(12)
$HOCl + O_2^- \rightarrow Cl^- + OH + O_2$	7.5×10^6		(12)
$HOCl + SO_3^{2-} \rightarrow ClSO_3^- + OH^-$	7.6×10^8		(13)
$ClSO_3^- + H_2O \rightarrow SO_4^{2-} + Cl^- + 2\,H^+$	$2.7 \times 10^2 \ (s^{-1})$		(13)
Bromine species			
$OH + Br^- \rightleftharpoons HOBr^-$	$1.1 \times 10^{10} / 3.3 \times 10^7$		(14)
$HOBr^- + H^+ \rightleftharpoons Br + H_2O$	$4.4 \times 10^{10} / 1.4 \ (s^{-1})$		(14)
$HOBr^- \rightarrow Br + OH^-$	4.2×10^6		(14)
$SO_4^- + Br^- \rightarrow Br + SO_4^{2-}$	3.5×10^9		(6)
$NO_3 + Br^- \rightarrow Br + NO_3^-$	4.0×10^9		(15)
$Br + Br^- \rightleftharpoons Br_2^-$	$1.1 \times 10^{10} / 1.9 \times 10^4 \ (s^{-1})$		(16)

(continued)

Table 8.36 (continued)

Reaction	k_{298}	B	Ref.
$Br_2^- + HO_2 \rightarrow 2Br^- + H^+ + O_2$	1.6×10^9		(6)
$Br_2^- + O_2^- \rightarrow 2Br^- + O_2$	1.7×10^8		(17)
$Br_2^- + H_2O_2 \rightarrow 2Br^- + H^+ + HO_2$	1.0×10^5		(7)
$Br_2^- + HSO_3^- \rightarrow 2Br^- + H^+ + SO_3^-$	6.3×10^7	780	(8)
$Br_2^- + SO_3^{2-} \rightarrow 2Br^- + SO_3^-$	2.2×10^8	650	(8)
$Br_2^- + NO_2^- \rightarrow 2Br^- + NO_2$	2.0×10^7		(6)
$Br_2^- + Br_2^- \rightarrow 2Br^- + Br_2$	2.0×10^9		(6)
$Br^- + HOBr + H^+ \rightleftharpoons Br_2 + H_2O$	$1.6 \times 10^{10}/9.7 \times 10^1 \ (s^{-1})$		(18)
$HO_2 + Br_2 \rightarrow Br + Br^- + H^+ + O_2$	1.1×10^8		(12)
$O_2^- + Br_2 \rightarrow Br_2^- + O_2$	5.6×10^9		(12)
$O_2^- + HOBr \rightarrow Br + + OH^- + O_2$	9.5×10^8		(12)
$Br^- + HOCl + H^+ \rightarrow BrCl + H_2O$	1.3×10^6		(19)
$Cl^- + HOBr + H^+ \rightleftharpoons BrCl + H_2O$	$3 \times 10^{10}/\sim 6 \times 10^5 \ (s^{-1})$		(20)
$Cl_2 + Br^- \rightarrow BrCl_2^-$	7.7×10^9		(20)
$BrCl + Br^- \rightleftharpoons Br_2Cl^-$	$> 1 \times 10^8 / > 6 \times 10^3 \ (s^{-1})^a$		(20)
$O_3 + Br^- \ (+ H^+) \rightarrow HOBr + O_2$	1.6×10^2		(6)

Iodine species

Reaction	k_{298}	B	Ref.
$OH + I^- \rightarrow HOI^-$	1.1×10^{10}		(21)
$HOI^- \rightarrow I + OH^-$	$1.2 \times 10^8 \ (s^{-1})$		(22)
$I + I \rightarrow I_2$	8.0×10^9		(6)
$I + I^- \rightleftharpoons I_2^-$	$1.0 \times 10^{10}/9.0 \times 10^5 \ (s^{-1})$		(23)
$I_2^- + I^- \rightleftharpoons I_3^-$	$6.2 \times 10^9/8.5 \times 10^6 \ (s^{-1})$		(24)
$I_2^- + I_2^- \rightarrow I^- + I_3^-$	3.2×10^9		(6)
$I_2^- + HOI \rightarrow 2I^- + H^+ + IO$	$\sim 1.0 \times 10^5$		(6)
$I_2^- + HSO_3^- \rightarrow 2I^- + H^+ + SO_3^-$	1.4×10^6	2,320	(8)
$I_2^- + SO_3^{2-} \rightarrow 2I^- + SO_3^-$	1.7×10^8	1,420	(8)
$O_3 + I^- \ (+ H^+) \rightarrow HOI + O_2$	2.4×10^9	8,790	(25)
$HOI + I^- + H^+ \rightleftharpoons I_2 + H_2O$	$4.4 \times 10^{12}/3 \ (s^{-1})$		(26)
$OH + I_2 \rightarrow HOI + I$	1.1×10^{10}		(21)
$OH + HOI \rightarrow IO + H_2O$	1.3×10^5		(6)
$IO + IO \ (+ H_2O) \rightarrow HOI + IO_2H$	1.5×10^9		(6)
$O_2^- + I_2 \rightarrow I_2^- + O_2$	5.5×10^9		(12)
$O_2^- + I_3^- \rightarrow I_2^- + I^- + O_2$	8.0×10^8		(12)
$HSO_3^- + I_2 \ (+ H_2O) \rightarrow 2I^- + HSO_4^- + 2\,H^+$	1.0×10^6		(27)
$HOI + Br^- + H^+ \rightleftharpoons IBr + H_2O$	$3.3 \times 10^{12}/8.0 \times 10^5 \ (s^{-1})$		(25)
$HOI + Cl^- + H^+ \rightleftharpoons ICl + H_2O$	$2.9 \times 10^{10}/2.4 \times 10^6 \ (s^{-1})$		(28)
$HOCl + I^- + H^+ \rightarrow ICl + H_2O$	3.5×10^{11}		(29)
$ICl + I^- \rightleftharpoons I_2Cl^{-\ b}$	$5.1 \times 10^8/0.7 \ (s^{-1})$		(24)
$HOBr + I^- \rightarrow IBr + OH^-$	5.0×10^9		(30)
$IBr + I^- \rightleftharpoons I_2Br^{-\ c}$	$2.0 \times 10^9/8.0 \times 10^2 \ (s^{-1})$		(24)
$HOI + I^- \rightleftharpoons I_2OH^-$	$6.7 \times 10^5/9.9 \times 10^2 \ (s^{-1})$		(30)
$I_2OH^- \rightleftharpoons I_2 + OH$	$3.0 \times 10^5 \ (s^{-1})/1.0 \times 10^{10}$		(30)
$HOBr + HOI \ (+ H_2O) \rightarrow IO_2^- + Br^- + 2\,H^+$	1.0×10^6		(31)
$HOBr + IO_2^- \rightarrow IO_3^- + Br^- + H^+$	1.0×10^6		(31)
$HOI + IO_2^- \rightarrow IO_3^- + I^- + H^+$	6.0×10^2		(31)

[a]The equilibrium constant is $K = [Br_2Cl^-]/[BrCl][Br^-] = 1.8 \times 10^4 \ [dm^3 \ mol^{-1}]$

[b]The equilibrium $I_2Cl^- \rightleftharpoons I_2 + Cl^-$ leads to the formation of iodine. Other equilibria involved are: $ICl + Cl^- \rightleftharpoons ICl_2^-$, $K = [ICl_2^-]/[ICl][Cl^-] = 77 \ (dm^3 \ mol^{-1})$, $HOI + 2Cl^- + H^+ \rightleftharpoons ICl_2^- + H_2O$, $K = [ICl_2^-]/[HOI][Cl^-]^2[H^+] = 9.4 \times 10^5 \ (dm^3 \ mol^{-1})$ Wang et al. (1989)

[c]The equilibrium constant for $I_2Br^- \rightleftharpoons I_2 + Br^-$ is $K = 0.08 \ (mol \ dm^{-3})$ Troy et al. (1991)

References

(1) Buxton, G.V., M. Bydder, G.A. Salmon, Phys. Chem. Chem. Phys. **1**, 269–273 (1999)

(2) Buxton, G.V., M. Bydder, G.A. Salmon, J. Chem. Soc. Faraday Trans. **94**, 653–657 (1998)

(3) Jayson, G.G., B.J. Parsons, A.J. Swallow, J. Chem. Soc. Faraday Trans. **69**, 1597–1607 (1973)

(4) Buxton, G.V., G.A. Salmon, J. Wang, Phys. Chem. Chem. Phys. **1**, 3589–3594 (1999)

(5) Buxton, G.V., M. Bydder, G.A. Salmon, J.E. Williams, Phys. Chem. Chem. Phys. **2**, 237–245 (2000)

(6) Neta, P., R.E. Huie, A.B. Ross, J. Phys. Chem. Ref. Data **17**, 1027–1284 (1988)

(7) Herrmann, H., B. Ervens, H.-W. Jacobi, R. Wolke, P. Nowacki, R. Zellner, J. Atmos. Chem. **36**, 231–284 (2000)

(8) Shoute, L.C.T., Z.B. Alfassi, P. Neta, R.E. Huie, J. Phys. Chem. **95**, 3238–3242 (1991)

(9) Herrmann, H., H.-W. Jacobi, G. Raabe, A. Reese, R. Zellner, Fresenius J. Anal. Chem. **355**, 343–344 (1996)

(10) McElroy, W.J., J. Phys. Chem. **94**, 2435–2441 (1990)

(11) Wang, T.X.,D. Margerum, Inorg. Chem. **33**, 1050–1055 (1994)

(12) Bielski, B.H.J., D.E. Cabelli, R.L. Arudi, A.B. Ross, J. Phys. Chem. Ref. Data **14**, 1041–1100 (1985)

(13) Fogelman, K.D., D.M. Walker, D.W. Margerum, Inorg. Chem. **28**, 986–993 (1989)

(14) Kläning, U.K.,T. Wolff, Ber. Bunsenges. Phys. Chem. **89**, 243–245 (1985)

(15) Herrmann, H., R. Zellner, in *N-Centered Radicals*, ed. by Z.B. Alfassi (Wiley, London, (1998), pp. 291–343

(16) Merényi, G., J. Lind, J. Am. Chem. Soc. **116**, 7872–7876 (1994)

(17) Wagner, I., H. Strehlow, Ber. Bunsenges. Phys. Chem. **91**, 1317–1321 (1987)

(18) Beckwith, R.C., T.X. Wang, D.W. Margerum, Inorg. Chem. **35**, 995–1000 (1996)

(19) Kumar, K., D.W. Margerum, Inorg. Chem. **26**, 2706–2711 (1987)

(20) Wang, T.X., M.D. Kelley, J.N. Cooper, R.C. Beckwith, D.W. Margerum, Inorg. Chem. **33**, 5872–5878 (1994)

(21) Buxton, G.V., C.L. Greenstock, W.P. Helman, A.B. Ross, J. Phys. Chem. Ref. Data **17**, 513–886 (1988)

(22) Ellison, D.H., G.A. Salmon, F. Wilkinson, Proc. Roy. Soc. (Lond.) **A 328**, 23–36 (1972)

(23) Fornier de Violet, P., R. Bonneau, S.R. Loan, J. Phys. Chem. **78**, 1698–1701 (1974); Barkatt, A., M. Ottolenghi, Mol. Photochem. **6**, 253–261 (1974)

(24) Troy, R.C., M.D. Kelley, J.C. Nagy, D.W. Margerum, Inorg. Chem. **30**, 4838–4845 (1991)

(25) Magi, L., F. Schweitzer, C. Pallares, S. Cherif, P. Mirabel, C. George, J. Phys. Chem. **A 101**, 4943–4949 (1997)

(26) Eigen, M., K. Kustin, J. Am. Chem. Soc. **84**, 1355–1361 (1962)

(27) Olsen, R.J., I.R. Epstein, J. Chem. Phys. **94**, 3083–3095 (1991)

(28) Wang, Y.L., J.C. Nagy, D.W. Margerum, J. Am. Chem. Soc. **111**, 7838–7844 (1989)

(29) Nagy, J.C., K. Kumar, D.W. Margerum, Inorg. Chem. **27**, 2773–2780 (1988)

(30) Troy, R.C., D.W. Margerum, Inorg. Chem. **30**, 3538–3543 (1991)

(31) Chinake, C.R., R.H. Simoyi, J. Phys. Chem. **100**, 1865–1871 (1996)

Table 8.37 Reactions with organic species

Reaction	k_{298}	B	Ref.
$OH + CH_4 \rightarrow CH_3 + H_2O$	1.1×10^8		(1)
$CH_3 + O_2 \rightarrow CH_3O_2$	4.1×10^9		(2)
$CH_3O_2 + CH_3O_2 \rightarrow HCHO + CH_3OH + O_2$	1.3×10^8		(3)
$\rightarrow 2CH_3O$	0.7×10^8		(3)
$CH_3O \rightarrow CH_2OH$	5.0×10^5		(3)
$CH_2OH + O_2 \rightarrow O_2CH_2OH$	4.5×10^9		(4)
$O_2CH_2OH \rightarrow HCHO + HO_2$	$<10\ (s^{-1})$		(3)
$OH + CH_3OH \rightarrow CH_2OH + H_2O$	$9.7 \times 10^{8\ a}$	580	(1, 5)
$SO_4^- + CH_3OH \rightarrow CH_2OH + HSO_4^-$	9.0×10^6	2,190	(6)
$NO_3 + CH_3OH \rightarrow CH_2OH + NO_3^- + H^+$	5.1×10^5	4,300	(7)
$OH + CH_2(OH)_2 \rightarrow CH(OH)_2 + H_2O$	1.1×10^9	1,020	(1, 8)
$SO_4^- + CH_2(OH)_2 \rightarrow CH(OH)_2 + HSO_4^-$	1.3×10^7	1,300	(9)
$NO_3 + CH_2(OH)_2 \rightarrow CH(OH)_2 + NO_3^- + H^+$	7.8×10^5	4,400	(7)
$CH(OH)_2 + O_2 \rightarrow OOCH(OH)_2$	3.5×10^9		(10)
$OOCH(OH)_2 \rightarrow HCOOH + HO_2$	$>1.0 \times 10^6$		(10)
$OH + HCOOH \rightarrow CO_2 + H + H_2O$	1.2×10^8	990	(1, 8)
$OH + HCOO^- \rightarrow CO_2^- + H_2O$	3.1×10^9	1,240	(1, 8)
$SO_4^- + HCOOH \rightarrow HSO_4^- + H + CO_2$	4.6×10^5		(11)
$SO_4^- + HCOO^- \rightarrow HSO_4^- + CO_2^-$	1.1×10^8		(11)
$NO_3 + HCOOH \rightarrow NO_3^- + H^+ + H + CO_2$	3.8×10^5	3,400	(12)
$NO_3 + HCOO^- \rightarrow NO_3^- + H + CO_2$	5.1×10^7	2,200	(12)
$OH + C_2H_6 \rightarrow C_2H_5 + H_2O$	1.8×10^9		(1)
$C_2H_5 + O_2 \rightarrow C_2H_5O_2$	2.1×10^9		(2)
$C_2H_5O_2 + C_2H_5O_2 \rightarrow products^b$	1.6×10^8		(12)
$OH + C_2H_5OH \rightarrow CH_3CHOH + H_2O$	$1.9 \times 10^{9\ c}$	1,200	(1, 13)
$SO_4^- + C_2H_5OH \rightarrow CH_3CHOH + HSO_4^-$	4.1×10^7	1,750	(6)
$NO_3 + C_2H_5OH \rightarrow CH_3CHOH + NO_3^- + H^+$	2.2×10^6	3,300	(7)
$CH_3CHOH + O_2 \rightarrow CH_3CH(O_2)OH$	4.6×10^9		(4)
$CH_3CH(O_2)OH \rightarrow CH_3CHO + HO_2$	$5.0 \times 10^1\ (s^{-1})$		(16)
$OH + CH_3CHO \rightarrow CH_3CO + H_2O$	3.6×10^9		(17)
$OH + CH_3CH(OH)_2 \rightarrow CH_3C(OH)_2 + H_2O$	1.2×10^9		(17)
$CH_3CO + O_2 \rightarrow CH_3CO_3$	$>1.0 \times 10^9$		d
$CH_3CO_3 + O_2^- (+ H^+) \rightarrow CH_3C(O)OOH + O_2$	$\sim 1.0 \times 10^9$		(17)
$CH_3C(OH)_2 + O_2 \rightarrow CH_3COOH + HO_2$	$>1.0 \times 10^9$		d, e
$OH + CH_3COOH \rightarrow CH_2COO^- + H^+ + H_2O$	1.7×10^7	1,330	(8)
$OH + CH_3COO^- \rightarrow CH_2COO^- + H_2O$	7.3×10^7	1,770	(8)
$CH_2COO^- + O_2 \rightarrow O_2CH_2COO^-$	1.7×10^9		(18)
$O_2CH_2COO^- + O_2^- (+ H^+) \rightarrow HOOCH_2COO^-$	1.7×10^7		(18)

(continued)

Table 8.37 (continued)

Reaction	k_{298}	B	Ref.
$CH_2COO^- + CH_2COO^- \rightarrow$ products^f	7.5×10^7		(18)
$SO_4^- + CH_3COOH \rightarrow SO_4^{2-} + CH_2COO^- + 2\ H+$	1.4×10^4		(8)
$SO_4^- + CH_3COO^- \rightarrow SO_4^{2-} + CO_2 + CH_3$	2.8×10^7	1,210	(19)
$NO_3 + CH_3COOH \rightarrow NO_3^- + CH_2COO^- + 2\ H^+$	1.4×10^4	3,800	(12)
$NO_3 + CH_3COO^- \rightarrow NO_3^- + CO_2 + CH_3$	2.9×10^6	3,800	(12)

^{a}This is the major reaction pathway, occurring with 93% probability; the remainder proceeds by abstraction of an H-atom from the hydroxyl group, with formaldehyde as the product (15)
^{b}The products are primarily ethanol and acetaldehyde
^{c}The reaction pathway shown proceeds with 84.3% probability. Hydrogen abstraction at the methyl group and at the hydroxyl group occur with 13.25 and 2.5% probability, respectively (Asmus et al. 1973)
^{d}The addition of oxygen to organic radicals generally occurs at nearly diffusion-controlled rates (20)
^{e}Similar to the oxidation of formaldehyde, the hydroxyalkylperoxy radical formed by the addition of oxygen to the hydrated acetyl radical decomposes rapidly toward the products shown
^{f}The major product is glyoxylic acid, other products are glycolic acid and formaldehyde

References

(1) Buxton, G.V., C.L. Greenstock, W.P. Helman, A.B. Ross, J. Phys. Chem. Ref. Data **17**, 513–886 (1988)
(2) Marchaj, A., D.G. Kelly, A. Bakac, J.H. Espenson, J. Phys. Chem. **95**, 4440–4441 (1991)
(3) Schuchmann, H.-P., C. von Sonntag, Z. Naturforsch. **40b**, 215–221 (1984)
(4) Adams, G.E., R.L. Wilson, Trans. Faraday, Soc. **63**, 2981–2987 (1969)
(5) Elliot, A.J., D.R. McCracken, Radiat. Phys. Chem. **33**, 69–74 (1989)
(6) Clifton, C.L., R.E. Huie, Int. J. Chem. Kinet. **21**, 677–687 (1989)
(7) Herrmann, H., R. Zellner, in *N-Centered Radicals*, ed. by Z.B. Alfassi (Wiley, London, 1998), pp. 291–343
(8) Chin, M., P.H. Wine, in *Aquatic and Surface Photochemistry*, ed. by G.R. Helz, R.G. Zepp, D.G. Crosby (Lewis Publication, Boca Raton, 1994), pp. 85–96
(9) Buxton, G.V., G.A. Salmon, N.D. Wood, in *Proc. 5th European Symp.*, ed. by G. Restelli, G. Angeletti. Physico-chemical Behaviour of Atmospheric Pollutants (Kluwer, Dordrecht, 1990), pp. 245–250
(10) McElroy, W.J., S.J. Waygood, J. Chem. Soc. Faraday Trans. **87**, 1513–1521 (1991); Bothe, E., D. Schulte-Frohlinde, Z. Naturforsch. **35b**, 1035–1039 (1980)
(11) Wine, P.H., Y. Tang, R.P. Thorn, J.R. Wells, J. Geophys, Res. **94**, 1085–1094 (1989)
(12) Exner, M., H. Herrmann, R. Zellner, J. Atmos. Chem. **18**, 359–378 (1994)
(13) Herrmann, H., A. Reese, B. Ervens, F. Wicktor, R. Zellner, Phys. Chem. Earth B **24**, 287–290 (1999)
(14) Ervens, B., S. Gligorowski, H. Herrmann, Chem. Phys. Phys. Chem. **5**, 1811–182 (2003)
(15) Asmus, K.-D., H. Möckel, A. Henglein, J. Phys. Chem. **77**, 1218 (1973)
(16) Bothe, E., M.N. Schuchmann, D. Schulte-Frohlinde, Z. Naturforsch. **38b**, 212–219 (1983)
(17) Schuchmann, M.N., C. von Sonntag, J. Am. Chem. Soc. **110**, 5698–5701 (1988)
(18) Schuchmann, M.N., H. Zegota, C. von Sonntag, Z. Naturforsch. **40b**, 215–221 (1985)
(19) Huie, R.E., C.L. Clifton, J. Phys. Chem. **94**, 8561–8567 (1990)
(20) Neta, P., R.E. Huie, A.B. Ross, J. Phys. Chem. Ref. Data **19**, 413–513 (1990)

Chapter 9
The Upper Atmosphere

9.1 Introduction

Conditions in the outermost region of the atmosphere differ radically from those in the lower atmosphere. The principal characteristic features are the following:

1. Because of the low density, eddy diffusion as a mechanism of vertical transport ceases to be important at altitudes above 100 km, and molecular diffusion becomes the main mode of transport. As a consequence, the components of air are no longer well mixed. Atoms and molecules develop individual scale heights determined by molecular mass, which leads to separation by diffusion in the Earth's gravitational field, such that the lighter species become more abundant with increasing altitude compared to heavier species, until ultimately hydrogen, helium and atomic oxygen are the dominant constituents at high altitudes. The transition region between turbulent mixing and molecular diffusion near 100 km altitude is called the *turbopause* or *homopause*. It can be made evident by the release of luminous substances from rockets. Below the turbopause the trails are confined to a narrow ribbon, usually strongly perturbed by horizontal winds, whereas above it they become diffuse and spread with increasing altitude like a smoke plume.
2. The absorption of solar radiation in the far ultraviolet region of the spectrum occurs at altitudes greater than 100 km. This radiation causes the photodissociation of oxygen, carbon dioxide and water vapor. Only molecular nitrogen remains present mainly in molecular form as no efficient photodissociation mechanism exists.
3. The lack of molecular species capable of emitting infrared radiation prevents an efficient dissipation of thermal energy by radiation. Heat must be conducted downward into regions of the mesosphere where carbon dioxide and water vapor are sufficiently abundant to allow the removal of excess energy by radiation to space. Consequently, temperature rises strongly with increasing altitude toward values between 700 and 2,000 K in the exosphere, depending on solar activity. This part of the atmosphere is called the *Thermosphere*. Solar radiation in the extreme ultraviolet varies considerably with solar activity. Temperatures in the

P. Warneck and J. Williams, *The Atmospheric Chemist's Companion: Numerical Data for Use in the Atmospheric Sciences*, DOI 10.1007/978-94-007-2275-0_9, © Springer Science+Business Media B.V. 2012

thermosphere vary accordingly, approaching low values at solar minimum and high values at solar maximum.

4. Extreme ultraviolet radiation from the sun also causes atoms and molecules to undergo photoionization, which produces electron densities up to some 10^6 cm^{-3} at altitudes near 300 km. The phenomenon is known as the *Ionosphere*. Its presence was originally inferred from the reflection of radio-waves back to earth, and prior to the advent of rockets and satellites, radio waves provided the only means of exploring electron densities in the ionosphere. Individual layers that were discerned with increasing altitude were named D, E, and F, a nomenclature still used today. At frequencies up to 4.4 MHz, radio waves are reflected from the E-layer at altitudes near 100 km. Wave penetration to the F-layer occurs at greater frequencies. During daytime the F-layer usually is split into a lower layer labeled F1 and a higher layer F2, but the F1 ledge disappears at night.

   The primary ions generated from nitrogen and oxygen in the region >100 km are N_2^+, O_2^+ and O^+. These ions enter into reactions with neutral species so that the stationary ion composition differs from that produced originally. For example, NO^+ is a prominent product appearing in the E-layer region. The loss of ions by ion-electron recombination occurs mainly via molecular ions, because only these allow excess energy to be released by dissociation, whereas the recombination of atomic ions requires energy release by radiation. Dissociative recombination is by several orders of magnitude more efficient than that of radiative recombination. The ionosphere extends downward to altitudes <100 km, with lower ion densities and different ion chemistry. Here, ionization occurs by solar X-rays, by ionization of NO via the solar Lyman a line, which penetrates deeper into the atmosphere due to a spectral window at 126 nm, and by cosmic rays. In this region, electrons attach to molecules whereby negative ions are formed that also enter into reactions with neutral molecules. These reactions give rise to a more complex ion chemistry.

5. The upper atmosphere also is the location of airglow phenomena; this includes a number of weak emission spectra only visible at night, although they are present much more strongly during the day, and quite spectacular displays of aurorae. Examples of airglow features in the mesosphere are emissions from long-lived molecular oxygen excited to the $^1\Delta_g$ state and vibrational excited OH molecules that emit the Meinel bands. Resonance emission from the sodium line at ~589 nm also occurs owing to a chemiluminescence reaction.

In the auroral region, energetic electrons, protons and heavier charged species precipitate into the atmosphere along magnetic field lines to produce sporadic ionization and excitation of atoms and molecules. Auroral lights are observed predominantly in circular belts between 15° and 30° from the geomagnetic poles in both hemispheres of the earth. The emissions derive from the excitation of oxygen and nitrogen at altitudes around 100 km, with the green and red lines of atomic oxygen being most prominent. Particles responsible for the excitation are projected from the sun. They originate from solar flares that liberate protons as well as electromagnetic radiation in the X-ray region.

9.2   Physical Conditions in the Thermosphere

The US Standard Atmosphere 1976 is an idealized steady state representation of Earth's atmosphere at 45° latitude (COESA 1976). Data in Tables 9.1 and 9.2 refer to conditions of mean solar activity. The subsequent Tables 9.3a–9.3f describe models of the thermosphere for different thermopause temperatures (Banks and Kockarts 1973).

Table 9.1 Temperature T, pressure p, density ρ, number concentration n, molar mass M and pressure scale height H_p as a function of geometric altitude in the thermosphere[a]

z (km)	T (K)	p (hPa)	ρ (kg m^{-3})	n (m^{-3})	M (g mol^{-1})	H_p (km)
86	186.87	3.734 (−3)	6.958 (−6)	1.447 (20)	28.95	5.621
90	186.87	1.836 (−3)	3.416 (−6)	7.116 (19)	28.91	5.636
95	188.42	7.597 (−4)	1.393 (−6)	2.920 (19)	28.73	5.727
100	195.08	3.201 (−4)	5.604 (−7)	1.189 (19)	28.40	6.009
105	208.84	1.448 (−4)	2.325 (−7)	5.021 (18)	27.88	6.561
110	240.00	7.104 (−5)	9.708 (−8)	2.144 (18)	27.27	7.723
115	300.00	4.010 (−5)	4.289 (−8)	9.681 (17)	26.68	9.882
120	360.00	2.538 (−5)	2.222 (−8)	5.107 (17)	26.20	12.091
125	417.23	1.735 (−5)	1.291 (−8)	3.013 (17)	25.80	14.254
130	469.27	1.251 (−5)	8.152 (−9)	1.930 (17)	25.44	16.288
135	516.59	9.357 (−6)	5.465 (−9)	1.312 (17)	25.09	18.208
140	559.63	7.203 (−6)	3.831 (−9)	9.322 (16)	24.75	20.025
145	598.78	5.669 (−6)	2.781 (−9)	6.858 (16)	24.42	21.746
150	634.39	4.542 (−6)	2.076 (−9)	5.186 (16)	24.10	23.380
155	666.80	3.693 (−6)	1.585 (−9)	4.012 (16)	23.79	24.934
160	696.29	3.040 (−6)	1.233 (−9)	3.162 (16)	23.49	26.414
165	723.18	2.528 (−6)	9.750 (−10)	2.532 (16)	23.19	27.826
170	747.57	2.121 (−6)	7.815 (−10)	2.055 (16)	22.90	29.175
175	769.81	1.794 (−6)	6.339 (−10)	1.688 (16)	22.62	30.466
180	790.07	1.527 (−6)	5.194 (−10)	1.400 (16)	22.34	31.703
185	808.51	1.308 (−6)	4.295 (−10)	1.172 (16)	22.07	32.890
190	825.31	1.127 (−6)	3.581 (−10)	9.887 (15)	21.81	34.030
195	840.62	9.749 (−7)	3.006 (−10)	8.400 (15)	21.55	35.127
200	854.56	8.474 (−7)	2.541 (−10)	7.182 (15)	21.30	36.183
210	878.84	6.476 (−7)	1.846 (−10)	5.337 (15)	20.83	38.182
220	899.01	5.015 (−7)	1.367 (−10)	4.040 (15)	20.37	40.043
230	915.78	3.928 (−7)	1.029 (−10)	3.106 (15)	19.95	41.781
240	929.73	3.106 (−7)	7.858 (−11)	2.420 (15)	19.56	43.405
250	941.33	2.477 (−7)	6.073 (−11)	1.906 (15)	19.19	44.924
260	950.99	1.989 (−7)	4.742 (−11)	1.515 (15)	18.85	46.346
270	959.04	1.608 (−7)	3.738 (−11)	1.215 (15)	18.53	47.678
280	965.75	1.308 (−7)	2.971 (−11)	9.807 (14)	18.24	48.925
290	971.34	1.069 (−7)	2.378 (−11)	7.967 (14)	17.97	50.095
300	976.01	8.770 (−8)	1.916 (−11)	6.509 (14)	17.73	51.193

(continued)

Table 9.1 (continued)

z (km)	T (K)	p (hPa)	ρ (kg m^{-3})	n (m^{-3})	M (g mol^{-1})	H_p (km)
320	983.16	5.980 (−8)	1.264 (−11)	4.405 (14)	17.29	53.199
340	988.15	4.132 (−8)	8.503 (−12)	3.029 (14)	16.91	54.996
360	991.65	2.888 (−8)	5.805 (−12)	2.109 (14)	16.57	56.637
380	994.10	2.038 (−8)	4.013 (−12)	1.485 (14)	16.27	58.178
400	995.83	1.452 (−8)	2.803 (−12)	1.056 (14)	15.98	59.678
450	998.22	6.447 (−9)	1.184 (−12)	4.678 (13)	15.25	63.644
500	999.24	3.024 (−9)	5.215 (−13)	2.192 (13)	14.33	68.785
600	999.85	8.213 (−10)	1.137 (−13)	5.950 (12)	11.51	88.244
700	999.97	3.191 (−10)	3.070 (−14)	2.311 (12)	8.00	130.630
800	999.99	1.704 (−10)	1.136 (−14)	1.234 (12)	5.54	193.862
900	1000.0	1.087 (−10)	5.759 (−15)	7.876 (11)	4.40	250.894
1,000	1000.0	7.514 (−11)	3.561 (−15)	5.442 (11)	3.94	286.203

[a]Powers of ten are shown in parentheses

Table 9.2 Atmospheric composition in the thermosphere (unit: m^{-3})[a]

Altitude	N_2	O_2	O	Ar	He	H
86	1.130 (20)	3.031 (19)	8.600 (16)	1.351 (18)	7.582 (14)	–
90	5.547 (19)	1.479 (19)	2.443 (17)	6.574 (17)	3.976 (14)	–
95	2.268 (19)	5.830 (18)	4.365 (17)	2.583 (17)	1,973 (14)	–
100	9.210 (18)	2.151 (18)	4.298 (17)	9.501 (16)	1.133 (14)	–
105	3.883 (18)	7.645 (17)	3.406 (17)	3.299 (16)	7.633 (13)	–
110	1.641 (18)	2.621 (17)	2.303 (17)	1.046 (16)	5.821 (13)	–
115	7.254 (17)	9.646 (16)	1.305 (17)	3.386 (15)	4.646 (13)	–
120	3.726 (17)	4.395 (16)	9.275 (16)	1.366 (15)	3.888 (13)	–
125	2.135 (17)	2.336 (16)	6.376 (16)	6.498 (14)	3.356 (13)	–
130	1.326 (17)	1.375 (16)	4.625 (16)	3.458 (14)	2.972 (13)	–
135	8.735 (16)	8.645 (15)	3.497 (16)	1.985 (14)	2.679 (13)	–
140	6.009 (16)	5.702 (15)	2.729 (16)	1.205 (14)	2.449 (13)	–
145	4.275 (16)	3.903 (15)	2.183 (16)	7.630 (13)	2.261 (13)	–
150	3.142 (16)	2.750 (15)	1.780 (16)	5.000 (13)	2.106 (13)	3.767 (11)
155	2.333 (16)	1.984 (15)	1.475 (16)	3.368 (13)	1.974 (13)	3.283 (11)
160	1.774 (16)	1.460 (15)	1.238 (16)	2.321 (13)	1.861 (13)	2.911 (11)
165	1.369 (16)	1.092 (15)	1.050 (16)	1.630 (13)	1.763 (13)	2.619 (11)
170	1.070 (16)	8.277 (14)	8.996 (15)	1.163 (13)	1.676 (13)	2.386 (11)
175	8.452 (15)	6.350 (14)	7.765 (15)	8.417 (12)	1.599 (13)	2.197 (11)
180	6.740 (15)	4.921 (14)	6.747 (15)	6.162 (12)	1.530 (13)	2.041 (11)
185	5.417 (15)	3.847 (14)	5.897 (15)	4.558 (12)	1.467 (13)	1.911 (11)
190	4.385 (15)	3.031 (14)	5.181 (15)	3.401 (12)	1.410 (13)	1.802 (11)
195	3.572 (15)	2.404 (14)	4.572 (15)	2.558 (12)	1.358 (13)	1.709 (11)
200	2.925 (15)	1.918 (14)	4.050 (15)	1.938 (12)	1.310 (13)	1.630 (11)
210	1.989 (15)	1.239 (14)	3.211 (15)	1.131 (12)	1.224 (13)	1.501 (11)
220	1.373 (15)	8.145 (13)	2.573 (15)	6.737 (11)	1.149 (13)	1.402 (11)
230	9.600 (14)	5.425 (13)	2.081 (15)	4.075 (11)	1.083 (13)	1.324 (11)
240	6.778 (14)	3.653 (13)	1.695 (15)	2.497 (11)	1.023 (13)	1.261 (11)

(continued)

Table 9.2 (continued)

Altitude	N_2	O_2	O	Ar	He	H
250	4.826 (14)	2.482 (13)	1.388 (15)	1.546 (11)	9.690 (12)	1.210 (11)
260	3.459 (14)	1.700 (13)	1.143 (15)	9.658 (10)	9.196 (12)	1.167 (11)
270	2.494 (14)	1.171 (13)	9.447 (14)	6.078 (10)	8.743 (12)	1.131 (11)
280	1.806 (14)	8.110 (12)	7.834 (14)	3.850 (10)	8.322 (12)	1.100 (11)
290	1.314 (14)	5.643 (12)	6.516 (14)	2.451 (10)	7.931 (12)	1.073 (11)
300	9.593 (13)	3.942 (12)	5.433 (14)	1.568 (10)	7.566 (12)	1.049 (11)
320	5.158 (13)	1.942 (12)	3.800 (14)	6.493 (9)	6.901 (12)	1.008 (11)
340	2.800 (13)	9.674 (11)	2.675 (14)	2.723 (9)	6.310 (12)	9.741 (10)
360	1.532 (13)	4.859 (11)	1.893 (14)	1.154 (9)	5.779 (12)	9.450 (10)
380	8.434 (12)	2.459 (11)	1.344 (14)	4.932 (8)	5.301 (12)	9.193 (10)
400	4.669 (12)	1.252 (11)	9.584 (13)	2.124 (8)	4.868 (12)	8.960 (10)
450	1.086 (12)	2.368 (10)	4.164 (13)	2.658 (7)	3.948 (12)	8.448 (10)
500	2.592 (11)	4.607 (9)	1.836 (13)	3.445 (6)	3.215 (12)	8.000 (10)
600	1.575 (10)	1.880 (8)	3.707 (12)	6.351 (4)	2.154 (12)	7.231 (10)
700	1.038 (9)	8.410 (6)	7.840 (11)	1.313 (3)	1.461 (12)	6.556 (10)
800	7.377 (7)	4.105 (5)	1.732 (11)	3.027 (1)	1.001 (12)	5.961 (10)
900	5.641 (6)	2.177 (4)	3.989 (10)	7.742 (−1)	6.933 (11)	5.434 (10)
1,000	4.626 (5)	1.251 (3)	9.562 (9)	2.188 (−2)	4.850 (11)	4.967 (10)

[a]Powers of ten are shown in parentheses

Comments: Temperature is expressed as a smooth mathematical function $T(z)$ with continuous first derivative. In the transition zone between homosphere and heterosphere (90–115 km) a combination of molecular and eddy diffusion coefficients is used to model vertical transport of individual species. Above 115 km eddy diffusion is negligible and molecular diffusion proceeds in a gravitational field $g = g_0/(1 + z/r_0)^2$, weakening with increasing altitude z, where r_0 is Earth's mean radius. Air composition is modeled to fit data from mass spectrometer, UV extinction, and satellite drag observations (N_2, O_2, O, Ar). As the species become separated by diffusion, the molar mass M changes with altitude and so does the pressure scale height $H_p = R_g T/gM$.

References

Banks, P.M., C. Kockarts, *Aeronomy (Part B, Appendix)* (Academic Press, New York, 1973)
Committee on Extension of US Standard Atmosphere (COESA), *US Standard Atmosphere 1976*, Washington, DC (1976)

Table 9.3a Thermosphere model for 750 K thermopause: Temperature T, pressure p, density ρ, total number concentration n, molar mass M, pressure scale height H_p and number concentrations of N_2, O_2, O, He, H as a function of altitude[a]

z (km)	T (K)	p (Pa)	ρ (kg m^{-3})	n (m^{-3})	M (g mol^{-1})	H_p (km)	$n(N_2)$ (m^{-3})	$n(O_2)$ (m^{-3})	$n(O)$ (m^{-3})	$n(He)$ (m^{-3})	$n(H)$ (m^{-3})
120	323.6	2.41 (−3)	2.34 (−8)	5.40 (17)	26.1	10.9	4.00 (17)	4.00 (16)	10.0 (16)	2.00 (13)	4.28 (12)
130	418.8	1.10 (−3)	7.95 (−9)	1.90 (17)	25.3	14.6	1.31 (17)	1.16 (16)	4.73 (16)	1.51 (13)	2.88 (12)
140	504.6	5.96 (−4)	3.48 (−9)	8.56 (16)	24.5	18.3	5.47 (16)	4.38 (15)	2.65 (16)	1.22 (13)	2.27 (12)
150	573.5	3.60 (−4)	1.79 (−9)	4.55 (16)	23.8	21.4	2.68 (16)	1.98 (15)	1.67 (16)	1.03 (13)	1.94 (12)
160	624.2	2.32 (−4)	1.03 (−9)	2.70 (16)	23.1	24.1	1.46 (16)	9.98 (14)	1.14 (16)	9.11 (12)	1.75 (12)
170	660.2	1.56 (−4)	6.38 (−10)	1.71 (16)	22.4	26.3	8.46 (15)	5.40 (14)	8.13 (15)	8.20 (12)	1.62 (12)
180	685.6	1.08 (−4)	4.14 (−10)	1.14 (16)	21.8	28.2	5.12 (15)	3.06 (14)	6.01 (15)	7.50 (12)	1.54 (12)
190	703.5	7.68 (−5)	2.78 (−10)	7.91 (15)	21.2	29.9	3.19 (15)	1.79 (14)	4.53 (15)	6.92 (12)	1.47 (12)
200	716.2	5.53 (−5)	1.92 (−10)	5.60 (15)	20.6	31.3	2.02 (15)	1.06 (14)	3.47 (15)	6.43 (12)	1.42 (12)
210	725.2	4.05 (−5)	1.35 (−10)	4.05 (15)	20.1	32.7	1.30 (15)	6.43 (13)	2.68 (15)	6.00 (12)	1.38 (12)
220	731.6	3.00 (−5)	9.66 (−11)	2.97 (15)	19.6	33.9	8.43 (14)	3.93 (13)	2.08 (15)	5.62 (12)	1.35 (12)
230	736.3	2.24 (−5)	7.02 (−11)	2.21 (15)	19.1	35.0	5.50 (14)	2.41 (13)	1.63 (15)	5.27 (12)	1.32 (12)
240	739.7	1.69 (−5)	5.16 (−11)	1.66 (15)	18.7	36.1	3.61 (14)	1.49 (13)	1.28 (15)	4.95 (12)	1.29 (12)
250	742.2	1.29 (−5)	3.83 (−11)	1.26 (15)	18.3	37.1	2.38 (14)	9.29 (12)	1.01 (15)	4.66 (12)	1.27 (12)
260	744.1	9.87 (−6)	2.87 (−11)	9.63 (14)	18.0	38.0	1.58 (14)	5.79 (12)	7.94 (14)	4.38 (12)	1.25 (12)
270	745.4	7.63 (−6)	2.17 (−11)	7.41 (14)	17.7	38.8	1.05 (14)	3.63 (12)	6.27 (14)	4.13 (12)	1.23 (12)
280	746.5	5.91 (−6)	1.66 (−11)	5.73 (14)	17.4	39.6	6.95 (13)	2.27 (12)	4.97 (14)	3.89 (12)	1.21 (12)
290	747.3	4.60 (−6)	1.27 (−11)	4.46 (14)	17.2	40.4	4.63 (13)	1.43 (12)	3.94 (14)	3.67 (12)	1.19 (12)
300	747.9	3.60 (−6)	9.81 (−12)	3.49 (14)	16.9	41.1	3.09 (13)	9.02 (11)	3.12 (14)	3.46 (12)	1.17 (12)
320	748.7	2.23 (−6)	5.92 (−12)	2.16 (14)	16.5	42.3	1.39 (13)	3.60 (11)	1.97 (14)	3.09 (12)	1.13 (12)
340	749.2	1.40 (−6)	3.64 (−12)	1.35 (14)	16.2	43.5	6.24 (12)	1.45 (11)	1.25 (14)	2.75 (12)	1.10 (12)
360	749.5	8.88 (−7)	2.26 (−12)	8.59 (13)	15.9	44.7	2.83 (12)	5.85 (10)	7.95 (13)	2.46 (12)	1.07 (12)
380	749.7	5.72 (−7)	1.42 (−12)	5.52 (13)	15.5	46.0	1.29 (12)	2.38 (10)	5.07 (13)	2.20 (12)	1.04 (12)

400	749.8	3.72 (−7)	9.04 (−13)	3.60 (13)	15.1	5.89 (11)	47.5	9.75 (9)	3.24 (13)	1.96 (12)	1.01 (12)
450	749.9	1.37 (−7)	3.01 (−13)	1.33 (13)	13.7	8.51 (10)	53.3	1.07 (9)	1.07 (13)	1.49 (12)	9.42 (11)
500	749.9	5.84 (−8)	1.06 (−13)	5.64 (12)	11.3	1.27 (10)	65.6	1.21 (7)	3.61 (12)	1.13 (12)	8.80 (11)
550	750.0	3.03 (−8)	4.00 (−14)	2.93 (12)	8.2	1.93 (9)	91.0	1.41 (7)	1.24 (12)	8.67 (11)	8.22 (11)
600	750.0	1.93 (−8)	1.71 (−14)	1.86 (12)	5.5	3.04 (8)	137.5	1.71 (6)	4.29 (11)	6.65 (11)	7.69 (11)
650	750.0	1.42 (−8)	8.63 (−15)	1.38 (12)	3.8	4.90 (7)	205.4	2.12 (5)	1.51 (11)	5.13 (11)	7.20 (11)
700	750.0	1.17 (−8)	5.20 (−15)	1.13 (12)	2.8	8.10 (6)	281.1		5.41 (10)	3.96 (11)	6.75 (11)
750	750.0	9.95 (−9)	3.63 (−15)	9.61 (11)	2.3	1.37 (6)	348.9		1.96 (4)	3.08 (11)	6.33 (11)
800	750.0	8.71 (−9)	2.78 (−15)	8.42 (11)	2.0	2.39 (5)	404.4		7.23 (10)	2.40 (11)	5.95 (11)
900	750.0	6.97 (−9)	1.88 (−15)	6.74 (11)	1.7		491.4		1.02 (9)	1.47 (11)	5.26 (11)
1,000	750.0	5.77 (−9)	1.39 (−15)	5.58 (11)	1.5		566.3		1.52 (8)	9.12 (10)	4.66 (11)

aPowers of ten are shown in parentheses; Source: Banks and Kockarts (1973), slightly condensed

Table 9.3b Thermosphere model for 1,000 K thermopause: Temperature T, pressure p, density ρ, total number concentration n, molar mass M, pressure scale height H_p and number concentrations of N_2, O_2, O, He, H as a function of altitude[a]

z (km)	T (K)	p^* (Pa)	ρ (kg m^{-3})	n (m^{-3})	M (g mol^{-1})	H_p (km)	$n(N_2)$ (m^{-3})	$n(O_2)$ (m^{-3})	$n(O)$ (m^{-3})	$n(He)$ (m^{-3})	$n(H)$ (m^{-3})
120	323.6	2.41 (−3)	2.34 (−8)	5.40 (17)	26.1	10.9	4.00 (17)	4.00 (16)	10.0 (16)	2.00 (13)	2.61 (12)
130	472.7	1.15 (−3)	7.42 (−9)	1.77 (17)	25.3	16.4	1.23 (17)	1.09 (16)	4.32 (16)	1.41 (13)	1.22 (12)
140	596.4	6.80 (−4)	3.38 (−9)	8.25 (16)	24.7	21.4	5.37 (16)	4.40 (15)	2.44 (16)	1.12 (13)	7.53 (11)
150	692.5	4.44 (−4)	1.86 (−9)	4.64 (16)	24.1	25.5	2.84 (16)	2.17 (15)	1.59 (16)	9.53 (12)	5.41 (11)
160	765.2	3.08 (−4)	1.14 (−9)	2.91 (16)	23.5	29.0	1.67 (16)	1.20 (15)	1.13 (16)	8.42 (12)	4.25 (11)
170	819.8	2.21 (−4)	7.48 (−10)	1.96 (16)	23.0	31.8	1.05 (16)	7.12 (14)	8.38 (15)	7.63 (12)	3.54 (11)
180	861.0	1.64 (−4)	5.15 (−10)	1.38 (16)	22.5	34.3	6.88 (15)	4.43 (14)	6.45 (15)	7.02 (12)	3.07 (11)
190	892.2	1.23 (−4)	3.67 (−10)	1.00 (16)	22.0	36.4	4.66 (15)	2.85 (14)	5.08 (15)	6.52 (12)	2.74 (11)
200	915.9	9.45 (−5)	2.68 (−10)	7.48 (15)	21.6	38.3	3.22 (15)	1.87 (14)	4.07 (15)	6.11 (12)	2.50 (11)
210	934.0	7.32 (−5)	1.99 (−10)	5.68 (15)	21.1	40.0	2.26 (15)	1.25 (14)	3.29 (15)	5.75 (12)	2.32 (11)
220	948.0	5.73 (−5)	1.50 (−10)	4.38 (15)	20.7	41.6	1.60 (15)	8.48 (13)	2.69 (15)	5.44 (12)	2.18 (11)
230	958.8	4.53 (−5)	1.15 (−10)	3.42 (15)	20.3	43.0	1.15 (15)	5.80 (13)	2.21 (15)	5.16 (12)	2.07 (11)
240	967.2	3.60 (−5)	8.91 (−11)	2.70 (15)	19.9	44.4	8.26 (14)	3.99 (13)	1.83 (15)	4.90 (12)	1.98 (11)
250	973.7	2.88 (−5)	6.96 (−11)	2.15 (15)	19.5	45.7	5.99 (14)	2.77 (13)	1.52 (15)	4.67 (12)	1.90 (11)
260	978.9	2.32 (−5)	5.48 (−11)	1.72 (15)	19.2	46.8	4.36 (14)	1.93 (13)	1.26 (15)	4.45 (12)	1.84 (11)
270	983.0	1.88 (−5)	4.35 (−11)	1.39 (15)	18.9	48.0	3.18 (14)	1.35 (13)	1.05 (15)	4.24 (12)	1.79 (11)
280	986.2	1.53 (−5)	3.47 (−11)	1.13 (15)	18.6	49.0	2.33 (14)	9.43 (12)	8.79 (14)	4.05 (12)	1.74 (11)
290	988.8	1.25 (−5)	2.79 (−11)	9.18 (14)	18.3	50.0	1.71 (14)	6.63 (12)	7.36 (14)	3.87 (12)	1.70 (11)
300	990.8	1.03 (−5)	2.25 (−11)	7.52 (14)	18.0	51.0	1.26 (14)	4.67 (12)	6.17 (14)	3.70 (12)	1.67 (11)
320	993.9	7.00 (−6)	1.22 (−11)	4.22 (14)	17.6	52.8	6.85 (13)	2.33 (12)	4.36 (14)	3.39 (12)	1.61 (11)
340	995.9	4.81 (−6)	1.00 (−11)	3.51 (14)	17.2	54.3	3.76 (13)	1.17 (12)	3.09 (14)	3.11 (12)	1.56 (11)
360	997.2	3.35 (−6)	6.84 (−12)	2.44 (14)	16.9	55.8	2.07 (13)	5.93 (11)	2.19 (14)	2.85 (12)	1.51 (11)
380	998.1	2.35 (−6)	4.72 (−12)	1.71 (14)	16.6	57.1	1.14 (13)	3.01 (11)	1.56 (14)	2.62 (12)	1.47 (11)
400	998.7	1.66 (−6)	3.29 (−12)	1.21 (14)	16.4	58.3	6.36 (12)	1.54 (11)	1.12 (14)	2.41 (12)	1.44 (11)

450	7.21 (−7)	1.38 (−12)	5.23 (13)	15.9	61.2	1.49 (12)	2.93 (10)	4.87 (13)	1.95 (12)	1.36 (11)
500	3.25 (−7)	5.99 (−13)	2.36 (13)	15.3	64.4	3.56 (11)	5.72 (9)	2.15 (13)	1.59 (12)	1.29 (11)
550	1.53 (−7)	2.68 (−13)	1.11 (13)	14.5	68.9	8.70 (10)	1.14 (9)	9.61 (12)	1.30 (12)	1.22 (11)
600	7.67 (−8)	1.24 (−13)	5.56 (12)	13.4	75.6	2.17 (10)	2.34 (8)	4.35 (12)	1.07 (12)	1.16 (11)
650	4.12 (−8)	5.91 (−14)	2.98 (12)	11.9	86.3	5.53 (9)	4.89 (7)	1.99 (12)	8.78 (11)	1.11 (11)
700	2.41 (−8)	2.95 (−14)	1.75 (12)	10.2	102.9	1.43 (9)	1.05 (7)	9.21 (11)	7.24 (11)	1.05 (11)
750	1.56 (−8)	1.56 (−14)	1.13 (12)	8.3	127.4	3.79 (8)	2.29 (6)	4.31 (11)	5.99 (11)	1.00 (11)
800	1.10 (−8)	8.87 (−15)	7.96 (11)	6.7	160.0	1.02 (8)	5.12 (5)	2.04 (11)	4.96 (11)	9.57 (10)
900	6.60 (−9)	3.68 (−15)	4.78 (11)	4.6	238.4	7.84 (6)		4.69 (10)	3.44 (11)	8.73 (10)
1,000	4.57 (−9)	2.03 (−15)	3.32 (11)	3.7	307.7	6.44 (5)		1.13 (10)	2.41 (11)	7.98 (10)

[a] Powers of ten are shown in parentheses; Source: Banks and Kockarts (1973), slightly condensed

Table 9.3c Thermosphere model for 1,250 K thermopause: Temperature T, pressure p, density ρ, total number concentration n, molar mass M, pressure scale height H_p and number concentrations of N_2, O_2, O, He, H as a function of altitude[a]

z (km)	T (K)	p (Pa)	ρ (kg m⁻³)	n (m⁻³)	M (g mol⁻¹)	H_p (km)	$n(N_2)$ (m⁻³)	$n(O_2)$ (m⁻³)	$n(O)$ (m⁻³)	$n(He)$ (m⁻³)	$n(H)$ (m⁻³)
120	323.6	2.41 (−3)	2.34 (−8)	5.40 (17)	26.1	10.9	4.00 (17)	4.00 (16)	10.0 (16)	2.00 (13)	2.61 (12)
130	523.3	1.20 (−3)	7.01 (−9)	1.66 (17)	25.4	18.2	1.16 (17)	1.04 (16)	4.01 (16)	1.33 (13)	1.01 (12)
140	677.5	7.48 (−4)	3.30 (−9)	8.01 (16)	24.8	24.2	5.28 (16)	4.40 (15)	2.29 (16)	1.05 (13)	5.90 (11)
150	796.5	5.14 (−4)	1.89 (−9)	4.69 (16)	24.3	29.1	2.93 (16)	2.29 (15)	1.53 (16)	8.96 (12)	3.99 (11)
160	888.4	3.73 (−4)	1.21 (−9)	3.05 (16)	23.8	33.2	1.81 (16)	1.34 (15)	1.11 (16)	7.94 (12)	2.94 (11)
170	959.9	2.81 (−4)	8.23 (−10)	2.12 (16)	23.4	36.7	1.19 (16)	8.43 (14)	8.43 (15)	7.21 (12)	2.28 (11)
180	1015.9	2.16 (−4)	5.88 (−10)	1.54 (16)	23.0	39.7	8.21 (15)	5.55 (14)	6.65 (15)	6.65 (12)	1.85 (11)
190	1060.2	1.69 (−4)	4.34 (−10)	1.16 (16)	22.6	42.2	5.83 (15)	3.77 (14)	5.37 (15)	6.20 (12)	1.54 (11)
200	1095.5	1.35 (−4)	3.28 (−10)	8.90 (15)	22.2	44.6	4.23 (15)	2.63 (14)	4.41 (15)	5.83 (12)	1.31 (11)
210	1123.6	1.08 (−4)	2.52 (−10)	6.97 (15)	21.8	46.6	3.12 (15)	1.86 (14)	3.66 (15)	5.52 (12)	1.14 (11)
220	1146.3	8.76 (−5)	1.97 (−10)	5.54 (15)	21.4	48.5	2.33 (15)	1.34 (14)	3.07 (15)	5.24 (12)	1.01 (11)
230	1164.6	7.16 (−5)	1.56 (−10)	4.46 (15)	21.1	50.3	1.76 (15)	9.71 (13)	2.60 (15)	5.00 (12)	9.05 (10)
240	1179.4	5.89 (−5)	1.24 (−10)	3.62 (15)	20.7	52.0	1.33 (15)	7.11 (13)	2.21 (15)	4.78 (12)	8.22 (10)
250	1191.5	4.87 (−5)	1.00 (−10)	2.96 (15)	20.4	53.4	1.02 (15)	5.24 (13)	1.89 (15)	4.57 (12)	7.54 (10)
260	1201.3	4.05 (−5)	8.15 (−11)	2.44 (15)	20.1	54.9	7.84 (14)	3.88 (13)	1.62 (15)	4.39 (12)	6.99 (10)
270	1209.4	3.39 (−5)	6.66 (−11)	2.03 (15)	19.8	56.3	6.05 (14)	2.89 (13)	1.39 (15)	4.21 (12)	6.53 (10)
280	1216.1	2.84 (−5)	5.48 (−11)	1.69 (15)	19.5	57.6	4.69 (14)	2.16 (13)	1.20 (15)	4.05 (12)	6.14 (10)
290	1221.6	2.39 (−5)	4.53 (−11)	1.42 (15)	19.2	58.8	3.64 (14)	1.62 (13)	1.04 (15)	3.90 (12)	5.82 (10)
300	1226.2	2.03 (−5)	3.76 (−11)	1.20 (15)	19.0	60.1	2.83 (14)	1.22 (13)	8.96 (14)	3.76 (12)	5.54 (10)
320	1233.1	1.72 (−5)	2.63 (−11)	8.58 (14)	18.5	62.3	1.73 (14)	6.93 (12)	6.74 (14)	3.49 (12)	5.10 (10)
340	1238.0	1.19 (−5)	1.87 (−11)	6.23 (14)	18.1	64.4	1.06 (14)	3.97 (12)	5.09 (14)	3.25 (12)	4.77 (10)
360	1241.5	8.76 (−6)	1.35 (−11)	4.58 (14)	17.7	66.3	6.57 (13)	2.29 (12)	3.87 (14)	3.03 (12)	4.51 (10)
380	1244.0	6.51 (−6)	9.80 (−12)	3.39 (14)	17.4	68.0	4.08 (13)	1.33 (12)	2.96 (14)	2.83 (12)	4.30 (10)
400	1245.9	4.87 (−6)	7.20 (−12)	2.53 (14)	17.1	69.6	2.54 (13)	7.76 (11)	2.24 (14)	2.64 (12)	4.14 (10)

450	2.41 (−6)	3.46 (−12)	1.26 (14)	16.6	73.3	7.94 (12)	2.05 (11)	1.15 (14)	2.23 (12)	3.82 (10)
500	1.24 (−6)	1.72 (−12)	6.44 (13)	16.1	76.4	2.53 (12)	5.54 (10)	5.99 (13)	1.90 (12)	3.60 (10)
550	5.85 (−7)	8.84 (−13)	3.39 (13)	15.7	79.6	8.18 (11)	1.53 (10)	3.14 (13)	1.61 (12)	3.42 (10)
600	3.17 (−7)	4.65 (−13)	1.83 (13)	15.3	83.3	2.70 (11)	4.29 (9)	1.67 (13)	1.38 (12)	3.27 (10)
650	1.76 (−7)	2.49 (−13)	1.02 (13)	14.7	87.8	9.03 (10)	1.23 (9)	8.92 (12)	1.18 (12)	3.13 (10)
700	1.01 (−7)	1.36 (−13)	5.89 (12)	13.9	93.8	3.07 (10)	3.59 (8)	4.82 (12)	1.01 (12)	3.00 (10)
750	6.09 (−8)	7.60 (−14)	3.53 (12)	13.0	102.2	1.06 (10)	1.06 (8)	2.62 (12)	8.66 (11)	2.89 (10)
800	3.83 (−8)	4.35 (−14)	2.22 (12)	11.8	129.9	3.72 (9)	3.21 (7)	1.44 (12)	7.46 (11)	2.78 (10)
900	1.77 (−8)	1.56 (−14)	1.03 (12)	9.1	151.1	4.77 (8)	3.07 (6)	4.46 (11)	5.56 (11)	2.48 (10)
1,000	1.01 (−8)	6.60 (−15)	5.84 (11)	6.8	208.5	6.47 (7)	3.14 (5)	1.42 (11)	4.18 (11)	2.40 (10)

[a]Powers of ten are shown in parentheses; Source: Banks and Kockarts (1973), slightly condensed

Table 9.3d Thermosphere model for 1,500 K thermopause: Temperature T, pressure p, density ρ, total number concentration n, molar mass M, pressure scale height H_p and number concentrations of N_2, O_2, O, He, H as a function of altitude[a]

z (km)	T (K)	p (Pa)	ρ (kg m⁻³)	n (m⁻³)	M (g mol⁻¹)	H_p (km)	$n(N_2)$ (m⁻³)	$n(O_2)$ (m⁻³)	n(O) (m⁻³)	n(He) (m⁻³)	n(H) (m⁻³)
120	323.6	2.41 (−3)	2.34 (−8)	5.40 (17)	26.1	10.9	4.00 (17)	4.00 (16)	10.0 (16)	2.00 (13)	2.37 (12)
130	527.9	1.20 (−3)	6.95 (−9)	1.65 (17)	25.4	18.4	1.15 (17)	1.03 (16)	3.98 (16)	1.33 (13)	9.73 (11)
140	702.1	7.57 (−4)	3.22 (−9)	7.81 (16)	24.8	25.1	5.16 (16)	4.30 (15)	2.23 (16)	1.03 (13)	5.53 (11)
150	844.9	5.29 (−4)	1.83 (−9)	4.54 (16)	24.3	30.9	2.85 (16)	2.24 (15)	1.47 (16)	8.67 (12)	3.66 (11)
160	959.8	3.92 (−4)	1.17 (−9)	2.96 (16)	23.9	35.8	1.77 (16)	1.33 (15)	1.06 (16)	7.62 (12)	2.64 (11)
170	1052.3	3.01 (−4)	8.09 (−10)	2.08 (16)	23.5	40.0	1.18 (16)	8.47 (14)	8.07 (15)	6.89 (12)	2.01 (11)
180	1127.2	2.37 (−4)	5.85 (−10)	1.53 (16)	23.1	43.7	8.29 (15)	5.70 (14)	6.39 (15)	6.34 (12)	1.60 (11)
190	1188.3	1.91 (−4)	4.39 (−10)	1.16 (16)	22.8	47.0	6.01 (15)	3.97 (14)	5.20 (15)	5.90 (12)	1.30 (11)
200	1238.3	1.55 (−4)	3.37 (−10)	9.06 (15)	22.4	49.9	4.46 (15)	2.84 (14)	4.31 (15)	5.54 (12)	1.09 (11)
210	1279.6	1.27 (−4)	2.64 (−10)	7.22 (15)	22.1	52.5	3.38 (15)	2.08 (14)	3.62 (15)	5.25 (12)	9.24 (10)
220	1313.8	1.06 (−4)	2.11 (−10)	5.83 (15)	21.7	54.8	2.59 (15)	1.54 (14)	3.08 (15)	4.99 (12)	7.97 (10)
230	1342.2	8.84 (−5)	1.70 (−10)	4.77 (15)	21.4	57.0	2.01 (15)	1.16 (14)	2.64 (15)	4.76 (12)	6.95 (10)
240	1366.0	7.44 (−5)	1.39 (−10)	3.95 (15)	21.1	59.0	1.58 (15)	8.78 (13)	2.28 (15)	4.56 (12)	6.13 (10)
250	1386.0	6.31 (−5)	1.14 (−10)	3.29 (15)	20.8	60.9	1.24 (15)	6.71 (13)	1.98 (15)	4.38 (12)	5.46 (10)
260	1402.8	5.36 (−5)	9.45 (−11)	2.77 (15)	20.6	62.7	9.87 (14)	5.16 (13)	1.73 (15)	4.21 (12)	4.91 (10)
270	1416.9	4.57 (−5)	7.89 (−11)	2.34 (15)	20.3	64.3	7.88 (14)	3.99 (13)	1.51 (15)	4.06 (12)	4.45 (10)
280	1428.9	3.93 (−5)	6.62 (−11)	1.99 (15)	20.0	65.9	6.31 (14)	3.10 (13)	1.33 (15)	3.92 (12)	4.07 (10)
290	1439.1	3.39 (−5)	5.59 (−11)	1.70 (15)	19.8	67.4	5.08 (14)	2.42 (13)	1.17 (15)	3.78 (12)	3.74 (10)
300	1447.8	2.92 (−5)	4.74 (−11)	1.46 (15)	19.5	68.9	4.09 (14)	1.90 (13)	1.03 (15)	3.66 (12)	3.46 (10)
320	1461.6	2.20 (−5)	3.45 (−11)	1.09 (15)	19.1	71.6	2.68 (14)	1.17 (13)	8.06 (14)	3.43 (12)	3.01 (10)
340	1471.7	1.67 (−5)	2.55 (−11)	8.22 (14)	18.7	74.1	1.77 (14)	7.31 (12)	6.34 (14)	3.22 (12)	2.68 (10)
360	1479.3	1.28 (−5)	1.91 (−11)	6.27 (14)	18.3	76.4	1.18 (14)	4.60 (12)	5.01 (14)	3.03 (12)	2.42 (10)
380	1485.0	9.89 (−6)	1.44 (−11)	4.83 (14)	18.0	78.6	7.91 (13)	2.91 (12)	3.98 (14)	2.86 (12)	2.22 (10)
400	1489.3	7.69 (−6)	1.10 (−11)	3.74 (14)	17.7	80.6	5.31 (13)	1.85 (12)	3.17 (14)	2.69 (12)	2.06 (10)

450	1496.2	4.21 (−6)	5.78 (−12)	2.04 (14)	17.1	85.1	2.00 (13)	6.06 (11)	1.81 (14)	2.34 (12)	1.79 (10)
500	1499.9	2.37 (−6)	3.16 (−12)	1.15 (14)	16.6	89.0	7.69 (12)	2.03 (11)	1.05 (14)	2.04 (12)	1.62 (10)
550	1501.9	1.38 (−6)	1.78 (−12)	6.60 (13)	16.2	92.6	3.01 (12)	6.94 (10)	6.11 (13)	1.78 (12)	1.51 (10)
600	1503.1	8.05 (−7)	1.02 (−12)	3.88 (13)	15.9	96.0	1.19 (12)	2.41 (10)	3.60 (13)	1.56 (12)	1.42 (10)
650	1503.7	4.83 (−7)	6.00 (−13)	2.33 (13)	15.5	99.7	4.80 (11)	8.52 (9)	2.14 (13)	1.37 (12)	1.36 (10)
700	1504.1	2.96 (−7)	3.58 (−13)	1.42 (13)	15.1	103.8	1.96 (11)	3.05 (9)	1.28 (13)	1.20 (12)	1.30 (10)
750	1504.4	1.84 (−7)	2.16 (−13)	8.89 (12)	14.7	108.7	8.07 (10)	1.11 (9)	7.73 (12)	1.06 (12)	1.25 (10)
800	1504.5	1.18 (−7)	1.33 (−13)	5.68 (12)	14.1	114.9	3.38 (10)	4.11 (8)	4.70 (12)	9.36 (11)	1.21 (10)
900	1504.7	5.24 (−8)	5.22 (−14)	2.52 (12)	12.5	133.2	6.12 (9)	5.83 (7)	1.77 (12)	7.33 (11)	1.13 (10)
1,000	1504.7	2.65 (−8)	2.21 (−14)	1.28 (12)	10.5	163.5	1.16 (9)	8.74 (6)	6.86 (11)	5.78 (11)	1.07 (10)

[a]Powers of ten are shown in parentheses; Source: Banks and Kockarts (1973), slightly condensed

Table 9.3e Thermosphere model for 1,750 K thermopause: Temperature T, pressure p, density ρ, total number concentration n, molar mass M, pressure scale height H_p and number concentrations of N_2, O_2, O, He, H as a function of altitude[a]

z (km)	T (K)	p (Pa)	ρ (kg m⁻³)	n (m⁻³)	M (g mol⁻¹)	H_p (km)	$n(N_2)$ (m⁻³)	$n(O_2)$ (m⁻³)	n(O) (m⁻³)	n(He) (m⁻³)	n(H) (m⁻³)
120	323.6	2.41 (−3)	2.34 (−8)	5.40 (17)	26.1	10.9	4.00 (17)	4.00 (16)	10.0 (16)	2.00 (13)	2.37 (12)
130	571.3	1.24 (−3)	6.64 (−9)	1.57 (17)	25.4	19.9	1.10 (17)	9.92 (15)	3.75 (16)	1.27 (13)	9.52 (11)
140	769.8	8.12 (−4)	3.16 (−9)	7.64 (16)	24.9	27.3	5.09 (16)	4.29 (15)	2.12 (16)	9.85 (12)	5.50 (11)
150	930.0	5.85 (−4)	1.85 (−9)	4.56 (16)	24.5	33.8	2.90 (16)	2.32 (15)	1.42 (16)	8.31 (12)	3.70 (11)
160	1059.6	4.45 (−4)	1.22 (−9)	3.04 (16)	24.1	39.2	1.86 (16)	1.42 (15)	1.04 (16)	7.32 (12)	2.70 (11)
170	1165.5	3.49 (−4)	8.56 (−10)	2.17 (16)	23.7	43.9	1.27 (16)	9.36 (14)	8.06 (15)	6.63 (12)	2.07 (11)
180	1252.8	2.81 (−4)	6.32 (−10)	1.63 (16)	23.4	48.0	9.16 (15)	6.48 (14)	6.47 (15)	6.11 (12)	1.65 (11)
190	1325.4	2.30 (−4)	4.82 (−10)	1.26 (16)	23.1	51.7	6.80 (15)	4.64 (14)	5.32 (15)	5.70 (12)	1.35 (11)
200	1386.1	1.91 (−4)	3.77 (−10)	9.98 (15)	22.8	54.9	5.17 (15)	3.42 (14)	4.47 (15)	5.37 (12)	1.12 (11)
210	1437.3	1.60 (−4)	3.01 (−10)	8.06 (15)	22.5	57.9	4.00 (15)	2.57 (14)	3.80 (15)	5.08 (12)	9.49 (10)
220	1480.6	1.35 (−4)	2.11 (−10)	5.83 (15)	22.2	60.6	3.14 (15)	1.96 (14)	3.27 (15)	4.84 (12)	8.14 (10)
230	1517.4	1.15 (−4)	1.99 (−10)	5.49 (15)	21.9	63.1	2.50 (15)	1.51 (14)	2.84 (15)	4.63 (12)	7.05 (10)
240	1548.8	9.84 (−5)	1.65 (−10)	4.60 (15)	21.6	65.4	2.00 (15)	1.18 (14)	2.18 (15)	4.28 (12)	5.43 (10)
250	1575.7	8.47 (−5)	1.38 (−10)	3.89 (15)	21.4	67.6	1.62 (15)	9.24 (13)	1.98 (15)	4.38 (12)	5.46 (10)
260	1598.8	7.32 (−5)	1.16 (−10)	3.32 (15)	21.1	69.6	1.32 (15)	7.31 (13)	1.92 (15)	4.12 (12)	4.83 (10)
270	1618.8	6.35 (−5)	9.84 (−11)	2.84 (15)	20.9	71.5	1.08 (15)	5.82 (13)	1.70 (15)	3.98 (12)	4.32 (10)
280	1635.9	5.53 (−5)	8.39 (−11)	2.45 (15)	20.6	73.3	8.84 (14)	4.65 (13)	1.52 (15)	3.85 (12)	3.88 (10)
290	1650.8	4.84 (−5)	7.18 (−11)	2.12 (15)	20.4	75.0	7.29 (14)	3.73 (13)	1.35 (15)	3.73 (12)	3.51 (10)
300	1663.7	4.24 (−5)	6.18 (−11)	1.85 (15)	20.2	76.7	6.03 (14)	3.01 (13)	1.21 (15)	3.62 (12)	3.20 (10)
320	1684.7	3.28 (−5)	4.63 (−11)	1.41 (15)	19.7	79.8	4.16 (14)	1.97 (13)	9.73 (14)	3.41 (12)	2.68 (10)
340	1700.8	2.57 (−5)	3.51 (−11)	1.09 (15)	19.3	82.7	2.90 (14)	1.31 (13)	7.88 (14)	3.22 (12)	2.29 (10)
360	1713.1	2.03 (−5)	2.70 (−11)	8.57 (14)	19.0	85.5	2.03 (14)	8.72 (12)	6.42 (14)	3.05 (12)	1.99 (10)
380	1722.7	1.61 (−5)	2.09 (−11)	6.77 (14)	18.6	88.0	1.43 (14)	5.86 (12)	5.24 (14)	2.90 (12)	1.76 (10)
400	1730.1	1.29 (−5)	1.64 (−11)	5.38 (14)	18.3	90.5	1.02 (14)	3.96 (12)	4.30 (14)	2.75 (12)	1.57 (10)

450	1742.6	7.51 (−6)	9.17 (−12)	3.13 (14)	17.7	95.9	4.38 (13)	1.51 (12)	2.65 (14)	2.43 (12)	1.25 (10)
500	1749.8	4.52 (−6)	5.33 (−12)	1.87 (14)	17.1	100.7	1.92 (13)	5.91 (11)	1.65 (14)	2.16 (12)	1.05 (10)
550	1754.0	2.79 (−6)	3.19 (−12)	1.15 (14)	16.7	104.9	8.59 (12)	2.35 (11)	1.04 (14)	1.92 (12)	9.24 (9)
600	1756.5	1.75 (−6)	1.96 (−12)	7.19 (13)	16.4	108.9	3.89 (12)	9.52 (10)	6.62 (13)	1.71 (12)	8.42 (9)
650	1758.1	1.11 (−6)	1.22 (−12)	4.58 (13)	16.1	112.6	1.78 (12)	3.90 (10)	4.24 (13)	1.53 (12)	7.82 (9)
700	1759.1	7.17 (−7)	7.74 (−13)	2.96 (13)	15.8	116.4	8.72 (11)	1.62 (10)	2.73 (13)	1.37 (12)	7.38 (9)
750	1759.8	4.71 (−7)	4.98 (−13)	1.94 (13)	15.5	120.4	3.88 (11)	6.84 (9)	1.77 (13)	1.23 (12)	7.04 (9)
800	1760.2	3.13 (−7)	3.24 (−13)	1.29 (13)	15.1	125.0	1.84 (11)	2.92 (9)	1.16 (13)	1.11 (12)	6.76 (9)
900	1760.7	1.45 (−7)	1.42 (−13)	5.98 (12)	14.3	136.3	4.28 (10)	5.52 (8)	5.03 (12)	8.99 (11)	6.31 (9)
1,000	1760.9	7.25 (−8)	6.47 (−14)	2.99 (12)	13.1	153.0	1.04 (10)	1.09 (8)	2.24 (12)	7.34 (11)	5.95 (9)

[a] Powers of ten are shown in parentheses; Source: Banks and Kockarts (1973), slightly condensed

Table 9.3f Thermosphere model for 2,000 K thermopause: Temperature T, pressure p, density ρ, total number concentration n, molar mass M, pressure scale height H_p and number concentrations of N_2, O_2, O, He, H as a function of altitude[a]

z (km)	T (K)	p (Pa)	ρ (kg m⁻³)	n (m⁻³)	M (g mol⁻¹)	H_p (km)	$n(N_2)$ (m⁻³)	$n(O_2)$ (m⁻³)	$n(O)$ (m⁻³)	$n(He)$ (m⁻³)	$n(H)$ (m⁻³)
120	323.6	2.41 (−3)	2.34 (−8)	5.40 (17)	26.1	10.9	4.00 (17)	4.00 (16)	10.0 (16)	2.00 (13)	2.37 (12)
130	590.9	1.26 (−3)	6.50 (−9)	1.54 (17)	25.4	20.5	1.08 (17)	9.74 (15)	3.65 (16)	1.24 (13)	9.44 (11)
140	807.9	8.35 (−4)	3.10 (−9)	7.49 (16)	24.9	28.7	5.01 (16)	4.24 (15)	2.06 (16)	9.60 (12)	5.48 (11)
150	985.5	6.12 (−4)	1.83 (−9)	4.50 (16)	24.5	35.7	2.89 (16)	2.32 (15)	1.38 (16)	8.07 (12)	3.70 (11)
160	1131.4	4.73 (−4)	1.21 (−9)	3.03 (16)	24.2	41.7	1.87 (16)	1.44 (15)	1.02 (16)	7.10 (12)	2.72 (11)
170	1252.4	3.77 (−4)	8.64 (−10)	2.18 (16)	23.8	47.0	1.30 (16)	9.64 (14)	7.89 (15)	6.42 (12)	2.10 (11)
180	1353.7	3.08 (−4)	6.44 (−10)	1.65 (16)	23.5	51.6	9.44 (15)	6.78 (14)	6.37 (15)	5.91 (12)	1.68 (11)
190	1439.2	2.56 (−4)	4.96 (−10)	1.29 (16)	23.2	55.7	7.10 (15)	4.94 (14)	5.27 (15)	5.52 (12)	1.38 (11)
200	1512.0	2.15 (−4)	3.92 (−10)	1.03 (16)	22.9	59.4	5.48 (15)	3.70 (14)	4.45 (15)	5.19 (12)	1.15 (11
210	1574.2	1.83 (−4)	3.16 (−10)	8.40 (15)	22.7	62.8	4.30 (15)	2.82 (14)	3.81 (15)	4.92 (12)	9.77 (10)
220	1627.6	1.56 (−4)	2.59 (−10)	6.96 (15)	22.4	65.9	3.43 (15)	2.19 (14)	3.30 (15)	4.69 (12)	8.40 (10)
230	1673.8	1.35 (−4)	2.15 (−10)	5.83 (15)	22.2	68.7	2.77 (15)	1.72 (14)	2.88 (15)	4.49 (12)	7.29 (10)
240	1713.8	1.17 (−4)	1.80 (−10)	4.94 (15)	21.9	71.4	2.26 (15)	1.37 (14)	2.54 (15)	4.31 (12)	6.38 (10)
250	1748.7	1.02 (−4)	1.52 (−10)	4.22 (15)	21.7	73.9	1.85 (15)	1.09 (14)	2.25 (15)	4.15 (12)	5.62 (10)
260	1779.0	8.91 (−5)	1.29 (−10)	3.63 (15)	21.4	76.2	1.53 (15)	8.83 (13)	2.00 (15)	4.00 (12)	4.99 (10)
270	1805.6	7.83 (−5)	1.11 (−10)	3.14 (15)	21.2	78.4	1.27 (15)	7.16 (13)	1.79 (15)	3.87 (12)	4.46 (10)
280	1828.9	6.91 (−5)	9.53 (−11)	2.73 (15)	21.0	80.5	1.06 (15)	5.84 (13)	1.61 (15)	3.75 (12)	4.00 (10)
290	1849.4	6.11 (−5)	8.25 (−11)	2.39 (15)	20.8	82.5	8.93 (14)	4.79 (13)	1.45 (15)	3.64 (12)	3.61 (10)
300	1867.5	5.41 (−5)	7.18 (−11)	2.10 (15)	20.6	84.4	7.52 (14)	3.94 (13)	1.31 (15)	3.53 (12)	3.27 (10)
320	1897.5	4.29 (−5)	5.49 (−11)	1.64 (15)	20.2	88.0	5.38 (14)	2.69 (13)	1.07 (15)	3.34 (12)	2.71 (10)
340	1921.1	3.44 (−5)	4.26 (−11)	1.30 (15)	19.8	91.3	3.89 (14)	1.86 (13)	8.85 (14)	3.17 (12)	2.29 (10)
360	1939.7	2.77 (−5)	3.34 (−11)	1.03 (15)	19.4	94.4	2.83 (14)	1.30 (13)	7.36 (14)	3.02 (12)	1.95 (10)
380	1954.6	2.25 (−5)	2.65 (−11)	8.34 (14)	19.1	97.3	2.08 (14)	9.11 (12)	6.14 (14)	2.88 (12)	1.69 (10)

400	1966.5	1.84 (−5)	2.11 (−11)	6.77 (14)	18.8	100.1	1.53 (14)	6.44 (12)	5.15 (14)	2.75 (12)	1.47 (10)
450	1987.1	1.13 (−5)	1.24 (−11)	4.13 (14)	18.1	106.4	7.27 (13)	2.75 (12)	3.35 (14)	2.46 (12)	1.10 (10)
500	1999.6	7.16 (−6)	7.59 (−12)	2.60 (14)	17.6	112.0	3.53 (13)	1.21 (12)	2.21 (14)	2.21 (12)	8.71 (9)
550	2007.4	4.63 (−6)	4.76 (−12)	1.67 (14)	17.2	117.1	1.74 (13)	5.38 (11)	1.47 (14)	1.99 (12)	7.23 (9)
600	2012.3	3.05 (−6)	3.06 (−12)	1.10 (14)	16.8	121.7	8.70 (12)	2.44 (11)	9.90 (13)	1.80 (12)	6.26 (9)
650	2015.6	2.04 (−6)	2.00 (−12)	7.32 (13)	16.5	126.0	4.40 (12)	1.12 (11)	6.70 (13)	1.63 (12)	5.58 (9)
700	2017.7	1.33 (−6)	1.33 (−12)	4.95 (13)	16.2	130.1	2.25 (12)	5.21 (10)	4.57 (13)	1.48 (12)	5.10 (9)
750	2019.2	9.44 (−7)	8.97 (−13)	3.39 (13)	15.9	134.2	1.17 (12)	2.45 (10)	3.13 (13)	1.35 (12)	4.75 (9)
800	2020.2	6.55 (−7)	6.11 (−13)	2.35 (13)	15.7	138.4	6.08 (11)	1.17 (10)	2.16 (13)	1.23 (12)	4.49 (9)
900	2021.4	3.25 (−7)	2.92 (−13)	1.16 (13)	15.1	147.7	1.71 (11)	2.73 (9)	1.04 (13)	1.03 (12)	4.11 (9)
1,000	2022.0	1.69 (−7)	1.45 (−13)	6.07 (12)	14.4	159.5	4.95 (10)	6.64 (8)	5.15 (12)	8.59 (11)	3.84 (9)

[a]Powers of ten are shown in parentheses; Source of data: Banks and Kockarts (1973), slightly condensed

9.3 Solar Radiation at Wavelengths Below 200 nm

Solar radiation in the far and extreme ultraviolet spectral regions is subject to variations with solar activity that are caused, both by the Sun's rotation (25 day at the solar equator, 35 day at the solar poles, 27 day average), and by the ~11 year sun spot cycle. Solar activity arises from the interaction of magnetic fields with the Sun's outer atmosphere. Thereby sites of activity are generated where photon emissions are either enhanced or depleted in comparison to the unperturbed solar atmosphere. Further sporadic perturbations are caused by solar flares, which affect mainly emissions in the X-ray region (~1,000 events per solar cycle). At wavelengths below about 200 nm the general character of the solar spectrum changes from a continuum with superimposed absorption features (such as Fraunhofer lines) to a much weaker continuum superimposed by emission lines; and the intensity decreases with decreasing wavelengths by more than four orders of magnitude (Tables 9.6 and 9.7 for solar cycles 21 and 22 below).

In the atmosphere of Earth, molecular oxygen absorbs radiation in the 120–175 nm wavelength range at altitudes above 100 km, and radiation between 175 and 200 nm in the mesosphere. Both, absorption at high altitudes and the dependence on solar activity make studies of solar radiation in the far ultraviolet spectral region quite demanding. Observations avoiding effects due to atmospheric absorption requires vehicles that can reach high altitudes. Since about 1960, both rockets and satellites have served as carriers of instruments designed to determine the spectral intensity distribution of solar radiation at short wavelengths. A further requirement arose to develop empirical models for estimating daily flux variations. Such models are based on a background at quiet solar conditions augmented by a wavelength-dependent contribution varying linearly with a proxy indicator such as the $F_{10.7}$ cm radio emission (see Table 9.5). For overviews on existing measurements and models see Geophysical Monograph 141 (Pap et al. 2004). In the following, the wavelength unit Ångström (Å) has been retained when it was used to describe the original data (1 nm = 10 Å) (Table 9.4).

Table 9.4 Overview on solar EUV irradiance measurements from spacecraft[a]

Spacecraft/Instrument	λ Range (nm)	Δλ Resolution (nm)	Time period
SOLRAD-1-11	1–10	1	1960–1976
OSO-3-6/SES	2–40	0.1	1967–1970
OSO-3-6/EUVS	27–131	0.2	1967–1970
AEROS-A-B	20–104	1	1973–1975
AE-C-E/EUVS	14–185	0.2–1	1974–1981
AE-C-E/ESUM	22–122	1–30	1974–1981
GOES/XRS	0.1–0.8	1	1974–present
Nimbus-7/SBUV	160–400	1.1	1978–1987
SME	115–300	0.75	1981–1989
NOAA-9/SBUV[b]	160–405	1.1	1984–1998
NOAA-16/SBUV[b]	160–405	1.1	2000–present
San Marco/ASSI	30–400	1	1988
Yohkoh/SXT	0.2–3	3	1992–present
SOHO CELIAS/SEM	26–34	8	1996–present

(continued)

Table 9.4 (continued)

Spacecraft/Instrument	λ Range (nm)	Δλ Resolution (nm)	Time period
SNOE/SXP	0.2–20	4–7	1998–present
UARS/SOLSTICE	119–420	0.1–0.3	1991–present
UARS/SUSIM	115–410	0.15, 1.1, 5	1991–present

[a]Source: Pap et al. (2004)
[b]Since 1984 SBUV instruments were flown on NOAA operational weather satellites, launched in 1984, 1989, 1995, 2000

Acronyms:

Satellites: AE = Atmospheric Explorer; AEROS = Aeronomy Satellites; OSO = Orbiting Solar Observatory; GOES = Geostationary Operational Environmental Satellite; SME = Solar Mesospheric Explorer; SNOE = Student Nitric Oxide Explorer; SOHO = Solar and Heliospheric Observatory; SOLRAD = Solar Radiation Spacecraft; UARS = Upper Atmosphere Research Satellite.

Instrumentation: ASSI = Airglow Solar Spectrometer Instrument; CELIAS = Charge, Element and Isotope Analysis System; ESUM = Extreme Solar Ultraviolet Monitor; EUVS = Extreme UV Spectrometer; SBUV = Solar Backscatter Ultraviolet; SEM = Solar EUV Monitor (EUV spectrometer); SOLSTICE = Solar Stellar Irradiance Comparison Experiment; SUSIM = Solar Ultraviolet Spectral Irradiance Monitor. SXP = Solar X-ray Photometer (Si photodiodes); SXT = Soft X-ray Telescope; XRS = X-ray sensor (ionization chamber);

Table 9.5 Observational indicators of solar activity[a]

Indicator	Description	Unit
Sunspot Number, R Since 1610	Number of dark regions of concentrated magnetic flux, best seen in white light spectroheliograms.	Number of sunspots or groups of sunspots
Plage Index, PI Since 1950	Sum of projected area a weighted by the observed intensity I of contiguous bright plage seen in Ca II K spectroheliograms at disc location $\mu = \cos\theta \cos\varphi$ for heliocentric latitude θ and longitude φ.	$10^3 \Sigma I a \mu$
10.7 cm Radio Flux, $F_{10.7}$ Since 1947	Full disc emission of 10.7 cm radiation from the upper chromosphere and corona, includes contributions from above both sunspots and plages.	$(10^{22}\ \mathrm{W\ m^{-2}\ Hz^{-1}})$
Ca II K 1A Index, $K1A$ Since 1976	Full-disc emission in a 1 (Å) core of the chromospheric Ca II K Fraunhofer line at 393.4 (nm) relative to emission in the nearby continuum.	
Helium I 1083-nm equivalent width since 1974	Full-disc measure of the equivalent width of the He 1083 (nm) line, primarily from the chromosphere, obtained by summing over digital spectroheliograms.	$10^{-3}(\text{Å})$
Fe XIV 530.3-nm limb flux since 1973	Latitude-averaged intensity of coronal Fe XIV emission from an annulus of width 1.1 arc min centred at 1.15 solar radii above the limb.	$10^{-6} \times$ disc center brightness

[a]Source: Lean (1991); Owing to their common 11 year cycle, time series of all indicators are highly correlated with each other, but none of the time series is identical to another, because each may originate in different solar layers and reflect different emission mechanisms

Table 9.6 Solar UV irradiances and prominent solar emission lines (solar cycle 21)[a]

λ (Å)	Emitter	I(low) 1974	I(high) 1979	λ (Å)	Emitter	I(low) 1974	I(high) 1979
50–100		4.00 (8)	1.15 (9)	1031.91	O VI	2.10 (9)	9.04 (9)
100–150		1.50 (8)	3.43 (8)	1,000–1,050		2.47 (9)	8.67 (9)
150–200		2.37 (9)	4.85 (9)	1,050–1,100		2.80 (9)	5.45 (9)
200–250		1.56 (8)	3.70 (9)	1,100–1,150		9.10 (8)	1.65 (9)
256.30	He II, Si X	4.60 (8)	5.95 (8)	1,150–1,200		4.40 (9)	1.06 (10)
284.15	Fe XV	2.10 (8)	3.17 (9)	1215.67	H I	2.51 (11)	8.64 (11)
250–300		1.68 (8)	4.14 (9)	1,200–1,250		4.00 (9)	1.06 (10)
303.31	Si XI	8.00 (8)	2.50 (9)	1,250–1,300		4.10 (9)	8.56 (9)
303.78	He II	6.90 (9)	1.13 (10)	1302.17	O I	1.10 (9)	2.11 (9)
300–350		9.65 (8)	5.63 (9)	1304.86	O I	1.13 (9)	2.17 (9)
368.07	Mg IX	6.50 (8)	1.39 (9)	1306.03	O I	1.23 (9)	2.36 (9)
350–400		3.14 (8)	2.20 (9)	1334.53	C II	1.84 (9)	3.90 (9)
400–450		3.83 (8)	9.93 (8)	1335.71	C II	2.52 (9)	5.34 (9)
465.22	Ne VII	2.90 (8)	3.62 (8)	1,300–1,350		4.58 (9)	1.19 (10)
450–500		2.85 (8)	1.67 (9)	1393.76	Si IV	1.30 (9)	3.02 (9)
500–550		4.52 (8)	1.55 (9)	1,350–1,400		6.10 (9)	1.56 (10)
554.37	O IV[b]	7.20 (8)	1.59 (9)	1402.77	Si IV	9.10 (8)	2.11 (9)
584.33	He I	1.27 (9)	4.87 (9)	1,400–1,450		9.49 (9)	2.34 (10)
550–600		3.57 (8)	1.02 (9)	1,450–1,500		1.62 (10)	3.04 (10)
609.76	Mg X	5.30 (8)	1.46 (9)	1548.20	C IV	3.80 (9)	9.17 (9)
629.73	O V	1.59 (9)	3.02 (9)	1,500–1,550		2.52 (10)	4.47 (10)
600–650		3.42 (8)	4.82 (8)	1550.77	C IV	1.90 (9)	4.74 (9)
650–700		2.30 (8)	4.55 (8)	1561.0	C I[b]	2.50 (9)	3.37 (9)
703.31	O III[b]	3.60 (8)	7.17 (8)	1,550–1,600		3.56 (10)	4.66 (10)
700–750		1.41 (8)	4.26 (8)	1,600–1,650		5.60 (10)	9.52 (10)
765.15	N IV	1.70 (8)	4.32 (8)	1657.2	C I[b]	8.50 (9)	8.16 (9)
770.41	Ne VIII	2.60 (8)	6.71 (8)	1,650–1,700		1.22 (11)	1.87 (11)
789.36	O IV	7.02 (8)	1.59 (9)	1,700–1,750		2.25 (11)	2.96 (11)
750–800		7.58 (8)	2.18 (9)	1,750–1,800		3.57 (11)	4.33 (11)
800–850		1.63 (9)	5.01 (9)	1808.01	Si II	9.2 (9)	1.77 (10)
850–900		3.54 (9)	1.33 (10)	1816.93	Si II	1.52 (10)	2.73 (10)
900–950		3.00 (9)	1.20 (10)	1,800–1,850		5.81 (11)	6.77 (11)
977.02	C III	4.40 (9)	1.32 (10)	1,850–1,900		7.77 (11)	9.06 (11)
950–1,000		1.48 (9)	4.42 (9)	1,900–1,940		8.29 (11)	9.66 (11)
1025.72	H I	3.50 (9)	1.31 (10)				

[a]Powers of ten in parentheses; intensity units are (photon cm^{-2} s^{-1}). Intensities of emission lines are not included in the integrated intensities for the 50 (Å) intervals; they should be added to obtain the total intensity in the wavelength range. Roman numbers behind the emitting elements indicate the degree of ionization (I neutral, II singly ionized, III doubly ionized, etc.). Data assembled from Heroux and Hinteregger (1978), Hinteregger et al. (1981), Torr et al. (1979) and Torr and Torr (1985)
[b]Wavelengths for multiplets are approximate, integrated intensities are shown

Comments: The data were obtained with spectrometers onboard the Atmospheric Explorer E satellite. Rocket experiments during the period 1974–1981 were used to calibrate the spectra. The solar cycle 21 lasted from about 1974 to 1986 with a maximum of solar activity in 1980–1981. Data for minimum conditions were derived from observations in April 1974 when $F_{10.7} = 74 \times 10^{-22}$ (W m^{-2} Hz^{-1}), data for maximum conditions are from observations in February 1979 ($F_{10.7} = 243 \times 10^{-22}$ (W m^{-2} Hz^{-1})). The evaluation of these data went through several revisions until the final reference spectra were obtained (designated F74113 and F79050N, respectively). A detailed listing of irradiances as a function of wavelength (about 1,600 values) has been archived at the US National Space Science Data Center.

The failure to reproduce photoelectron fluxes observed with instruments onboard the Atmospheric Explorer E satellite by calculations based on the F74113 spectrum has led to the recognition that intensities at wavelengths below 250 (Å) are too low and should be raised by a factor of 2–3 (see Richards et al. (1994) for a review). More recent X-ray measurements on the Student Nitric Oxide Explorer Satellite by Bailey et al. (2000) have confirmed this conclusion for the region 2–30 nm.

Table 9.7 UV Solar irradiances: reference spectrum for solar cycle 22[a]

λ (Å)	I(min)	R_{27d}	R_{11a}	λ (Å)	I(min)	R_{27d}	R_{11a}	λ (Å)	I(min)	R_{27d}	R_{11a}
5	5.01 (3)	3.0	100	265	2.40 (8)	1.64	4.10	525	1.24 (8)	1.56	3.19
15	1.00 (6)	6.0	20.0	275	5.28 (8)	1.57	3.27	535	1.55 /8)	1.21	1.59
25	1.00 (7)	3.0	10.0	285	3.89 (8)	1.72	6.21	545	1.12 (8)	1.16	1.38
35	2.85 (7)	1.66	4.70	295	2.47 (8)	1.49	2.47	555	5.01 (8)	1.17	1.42
45	6.82 (7)	1.56	3.29	305	7.46 (9)	1.24	1.57	565	1.49 (8)	1.19	1.52
55	1.91 (8)	1.42	2.26	315	1.11 (9)	1.48	2.41	575	1.43 (8)	1.22	1.64
65	1.94 (8)	1.44	2.37	325	6.30 (8)	1.74	7.28	585	8.00 (8)	1.24	1.71
75	2.56 (8)	1.37	2.11	335	6.41 (8)	1.74	7.23	595	1.93 (8)	1.16	1.41
85	1.77 (8)	1.41	2.12	345	7.06 (8)	1.48	2.41	605	5.42 (8)	1.49	2.49
95	2.48 (8)	1.31	1.80	355	5.62 (8)	1.53	2.77	615	2.53 (8)	1.16	1.38
105	1.00 (8)	1.27	1.68	365	7.80 (8)	1.53	2.79	625	1.06 (9)	1.21	1.51
115	7.98 (7)	1.22	1.64	375	2.40 (8)	1.10	1.17	635	2.57 (8)	1.22	1.64
125	5.48 (7)	1.22	1.61	385	1.06 (8)	1.10	1.17	645	6.65 (7)	1.16	1.38
135	5.69 (7)	1.22	1.63	395	9.55 (7)	1.17	1.43	655	4.91 (7)	1.21	1.58
145	1.65 (8)	1.40	2.05	405	1.64 (8)	1.17	1.43	665	4.88 (7)	1.16	1.38
155	1.61 (8)	1.54	2.98	415	1.01 (8)	1.73	6.50	675	4.87 (7)	1.17	1.44
165	3.48 (8)	1.28	1.72	425	9.46 (7)	1.10	1.17	685	1.67 (8)	1.20	1.48
175	1.56 (9)	1.36	1.86	435	1.85 (8)	1.30	1.60	695	1.19 (8)	1.25	1.61
185	6.71 (8)	1.49	2.45	445	9.06 /(7)	1.10	1.17	705	3.48 (8)	1.16	1.39
195	6.05 (8)	1.57	3.11	455	8.38 (7)	1.22	1.64	715	1.49 (8)	1.18	1.47
205	2.64 (8)	1.61	3.77	465	2.96 (8)	1.33	1.75	725	1.03 (8)	1.24	1.72
215	1.93 (8)	1.65	4.28	475	1.26 (8)	1.22	1.64	735	1.36 (8)	1.24	1.72
225	4.42 (8)	1.46	2.46	485	2.17 (8)	1.21	1.59	745	1.68 (8)	1.24	1.72
235	2.47 (8)	1.42	2.26	495	4.25 (8)	1.59	3.65	755	2.77 (8)	1.21	1.58
245	6.00 (8)	1.48	2.49	505	3.40 (8)	1.19	1.51	765	7.33 (8)	1.19	1.50
255	9.68 (8)	1.51	2.71	515	1.15 (8)	1.21	1.59	775	5.43 (8)	1.40	2.10

(continued)

Table 9.7 (continued)

λ (Å)	I(min)	R_{27d}	R_{11a}	λ (Å)	I(min)	R_{27d}	R_{11a}	λ (Å)	I(min)	R_{27d}	R_{11a}
785	9.20 (8)	1.25	1.61	1195	2.80 (9)	1.17	1.39	1605	1.53 (10)	1.06	1.15
795	4.60 (8)	1.18	1.48	1205	7.50 (9)	1.27	1.73	1615	1.81 (10)	1.05	1.14
805	4.38 (8)	1.24	1.72	1215	3.76 (11)	1.22	1.64	1625	2.09 (10)	1.06	1.14
815	5.04 (8)	1.24	1.72	1225	3.16 (9)	1.17	1.33	1635	2.29 (10)	1.06	1.16
825	7.52 (8)	1.24	1.72	1235	2.22 (9)	1.24	1.37	1645	2.55 (10)	1.10	1.17
835	1.41 (9)	1.19	1.51	1245	1.70 (9)	1.21	1.37	1655	4.10 (10)	1.09	1.12
845	9.57 (8)	1.24	1.72	1255	1.66 (9)	1.16	1.32	1665	2.93 (10)	1.04	1.07
855	1.08 (9)	1.24	1.72	1265	2.30 (9)	1.21	1.50	1675	3.40 (10)	1.07	1.12
865	1.16 (9)	1.24	1.72	1275	1.33 (9)	1.17	1.29	1685	3.38 (10)	1.04	1.06
875	1.39 (9)	1.24	1.72	1285	1.11 (9)	1.16	1.25	1695	5.13 (10)	1.05	1.07
885	1.61 (9)	1.24	1.72	1295	1.25 (9)	1.18	1.38	1705	5.92 (10)	1.04	1.08
895	1.98 (9)	1.24	1.72	1305	1.05 (10)	1.15	1.29	1715	5.87 (10)	1.05	1.11
905	2.37 (9)	1.24	1.71	1315	1.80 (9)	1.12	1.19	1725	6.43 (10)	1.05	1.09
915	1.86 (9)	1.23	1.68	1325	1.36 (9)	1.14	1.25	1735	6.53 (10)	1.05	1.08
925	6.62 (8)	1.23	1.70	1335	1.14 (10)	1.22	1.57	1745	8.07 (10)	1.05	1.07
935	5.98 (8)	1.22	1.65	1345	1.22 (9)	1.12	1.25	1755	9.98 (10)	1.05	1.08
945	4.29 (8)	1.23	1.67	1355	2.84 (9)	1.12	1.23	1765	1.09 (11)	1.04	1.07
955	2.74 (8)	1.17	1.44	1365	1.83 (9)	1.14	1.26	1775	1.31 (11)	1.04	1.08
965	2.88 (8)	1.17	1.44	1375	1.98 (9)	1.12	1.24	1785	1.47 (11)	1.05	1.08
975	5.00 (9)	1.20	1.54	1385	2.00 (9)	1.11	1.21	1795	1.44 (11)	1.05	1.08
985	5.96 (8)	1.18	1.45	1395	5.01 (9)	1.22	1.60	1805	1.76 (11)	1.07	1.12
995	9.29 (8)	1.18	1.45	1405	4.49 (9)	1.17	1.43	1815	2.11 (11)	1.08	1.15
1005	3.82 (8)	1.17	1.44	1415	2.91 (9)	1.10	1.22	1825	2.05 (11)	1.05	1.09
1015	4.46 (8)	1.17	1.44	1425	3.20 (9)	1.10	1.20	1835	2.21 (11)	1.05	1.09
1025	3.70 (9)	1.24	1.71	1435	3.73 (9)	1.10	1.21	1845	1.92 (11)	1.04	1.08
1035	3.88 (9)	1.23	1.69	1445	3.66 (9)	1.09	1.20	1855	2.20 (11)	1.04	1.08
1045	6.34 (8)	1.17	1.44	1455	3.97 (9)	1.09	1.20	1865	2.53 (11)	1.05	1.09
1055	5.42 (8)	1.17	1.44	1465	4.95 (9)	1.11	1.20	1875	2.94 (11)	1.04	1.09
1065	5.22 (8)	1.17	1.43	1475	6.29 (9)	1.07	1.16	1885	3.13 (11)	1.04	1.09
1075	5.60 (8)	1.17	1.43	1485	6.44 (9)	1.09	1.17	1895	3.53 (11)	1.04	1.10
1085	9.44 (8)	1.17	1.44	1495	5.83 (9)	1.08	1.17	1905	3.69 (11)	1.04	1.08
1095	6.41 (8)	1.17	1.44	1505	6.56 (9)	1.08	1.16	1915	4.06 (11)	1.04	1.09
1105	6.90 (8)	1.17	1.44	1515	7.13 (9)	1.08	1.16	1925	4.35 (11)	1.04	1.09
1115	7.18 (8)	1.10	1.17	1525	8.83 (9)	1.10	1.21	1935	3.32 (11)	1.03	1.08
1125	6.87 (8)	1.19	1.50	1535	9.95 (9)	1.10	1.20	1945	5.55 (11)	1.04	1.08
1135	6.21 (8)	1.10	1.17	1545	1.69 (10)	1.16	1.31	1955	5.43 (11)	1.04	1.08
1145	4.25 (8)	1.10	1.17	1555	1.45 (10)	1.12	1.24	1965	6.18 (11)	1.04	1.08
1155	7.03 (8)	1.10	1.17	1565	1.52 (10)	1.08	1.16	1975	6.30 (11)	1.04	1.07
1165	8.02 (8)	1.10	1.17	1575	1.38 (10)	1.06	1.16	1985	6.38 (11)	1.03	1.06
1175	2.97 (9)	1.19	1.51	1585	1.34 (10)	1.07	1.15	1995	6.97 (11)	1.03	1.06
1185	9.18 (8)	1.10	1.17	1595	1.36 (10)	1.06	1.13				

[a]Powers of ten are shown in parentheses, intensity units are (photon cm^{-2} s^{-1}); Source of data: Woods and Rottman (2002). Intensities outside Earth for solar minimum conditions are listed in 10 (Å) intervals (x ± 5 (Å)); R_{27d} and R_{11a} are solar cycle 22 variability ratios for the 27 day solar rotational period and the 11 year solar cycle. Variability ratios are maximum/minimum values for the time period of observation (1992–1997). Solar cycle 22 lasted from about 1986 to 1997

References

Bailey, S.M., T.N. Woods, C.A. Barth, S.C. Solomon, L.R. Canfield, R. Korde, J. Geophys. Res. **105**, 27179–27193 (2000)

Heroux, L., H.E. Hinteregger, J. Geophys. Res. **83**, 5305–5308 (1978)

Hinteregger, H.E., K. Fukui, B.G. Gilson, Geophys. Res. Lett. **8**, 1147–1150 (1981)

Lean, J., Rev. Geophys. **29**, 505–535 (1991)

Pap, J.M., P. Fox, C. Fröhlich, H.S. Hudson, J. Kuhn, J. McCormack, G. North, W. Sprigg, S.T. Wu, *Solar variability and Its Effects on Climate*. Geophysics Monograph, vol. 141 (American Geophysical Union, Washington, DC, 2004)

Richards, P.G., J.A. Fennelly, D.G. Torr, J. Geophys. Res. **99**, 8981–8992 (1994)

Torr, M.R., D.G. Torr, J. Geophys. Res. **90**, 6675–6678 (1985)

Torr, M.R., D.G. Torr, R.A. Ong, H.E. Hinteregger, Geophys. Res. Lett. **6**, 771–774 (1979)

Woods, T.N., G.J. Rottman, in *Atmospheres in the Solar System: Comparative Aeronomy*, ed. by M. Mendillo, A. Nagy, J.H. Waite. Geophysics Monograph, vol. 130 (American Geophysical Union, Washington, DC, 2002), pp. 221–233

9.4 Absorption and Photoionization Coefficients

Table 9.8 Ionization potentials of some important atmospheric constituents and the corresponding upper wavelength limits for photoionization by ultraviolet radiation[a]

Species	IP (ev)	λ_{IP} (nm)	Species	IP (ev)	λ_{IP} (nm)	Species	IP (ev)	λ_{IP} (nm)
Atoms								
H	13.5984	91.175	Mg	7.6462	162.151	Fe	7.9024	156.894
He	24.5874	50.426	Al	5.9858	207.132	Br	11.8138	104.949
Li	5.3917	229.953	Si	8.1517	152.097	Kr	13.9996	88.563
Be	9.3227	132.992	P	10.4867	118.230	Rb	4.1771	296.817
C	11.2603	110.107	S	10.3600	119.676	Sr	5.6949	217.713
N	14.5341	85.306	Cl	12.9676	95.613	I	10.4513	118.631
O	13.6181	91.044	Ar	15.7596	78.672	Xe	12.1298	102.214
F	17.4228	71.162	K	4.3407	285.634	Cs	3.8939	318.406
Ne	21.5645	57.495	Ca	6.1132	202.815	Ba	5.2117	237.898
Na	5.1391	241.258	Ti	6.8281	181.579	Hg	10.4375	118.787
Molecules								
H_2	15.426	80.374	NO	9.264	133.834	CH_3	9.843	125.962
OH	13.017	95.248	NO_2	9.586	129.339	CH_4	12.61	98.322
HO_2	11.350	109.237	CO	14.014	88.472	HF	16.044	77.278
H_2O	12.621	98.236	CO_2	13.773	90.020	HCl	12.749	97.250
H_2O_2	10.58	117.187	HCN	13.60	91.165	SO_2	12.349	100.400
O_2	12.070	102.721	CH	10.64	116.527	SO	10.294	120.443
N_2	15.581	79.574	CH_2	10.396	119.261	SF_6	15.32	80.930

[a]Data selected from Sansonetti and Martin (2005); see also National Institute of Standard Technology, http://webbook.nist.gov

Table 9.9 Ionization thresholds for excited states of oxygen and nitrogen ions[a]

	Term	v' (cm^{-1})	λ_{IP} (Å)	IP (eV)
Atomic oxygen[b]				
O^+ ($2s^2 2p^3$) (ground state)	$^4S^0_{3/2}$	109837.39	910.44	13.618
O^+ ($2s^2 2p^3$)	$^2D^0_{5/2}$	136647.94	731.81	16.942
	$^2D^0_{3/2}$	136667.96	731.70	16.945
O^+ ($2s^2 2p^3$)	$^2P^0_{3/2}$	150305.40	665.31	18.635
	$^2P^0_{1/2}$	150307.39	665.30	18.636
O^+ ($2s2p^4$)	$^4P_{5/2}$	229674.60	435.40	28.476
	$^4P_{3/2}$	229837.82	435.09	28.496
	$^4P_{1/2}$	229920.25	434.93	28.506
O^+ ($2s2p^4$)	$^2D_{5/2}$	275825.85	362.55	34.198
O^+ ($2s2p^4$)	$^2D_{3/2}$	275833.89	362.54	34.199
O^+ ($2s2p^4$)	$^2S_{1/2}$	305547.86	327.28	37.883
O^+ ($2s2p^4$)	$^2P_{1/2}$	324007.31	308.64	40.172
	$^2P_{3/2}$	324067.06	308.58	40.179
K shell ionization		4288450.	23.32	531.7
Molecular oxygen[c]				
O_2^+ (ground state)	X $^2\Pi_g$	97359.2	1027.1	12.071
O_2^+	a $^4\Pi_u$	129890.1	769.9	16.1
O_2^+	A $^2\Pi_u$	137525.9	727.1	17.1
O_2^+ (dissociating)[c]	b $^4\Sigma_g^-$	146557.0	682.3	18.2
O_2^+ (dissociating)	B $^2\Sigma_g^-$	163702.4	610.9	20.3
O_2^+ (dissociating repulsive)	$^2\Pi_u$	193573.4	516.6	24.0
Molecular nitrogen[d]				
N_2^+ (ground state)	X $^2\Sigma_g^+$	125667.64	795.8	15.581
N_2^+	A $^2\Pi_u$	134683.2	742.5	16.7
N_2^+	B $^2\Sigma_u^+$	151233.7	661.2	18.8
N_2^+ (dissociating)	C $^2\Sigma_u^+$	190209.7	525.7	23.6
N_2^+ (dissociating)	F $^2\Sigma_g^+$	233900.8	427.5	29.0

[a]Term symbol, wave number $v' = 1/\lambda$, threshold wavelength λ_{IP}, ionization potential, IP
[b]Source: Sansonetti and Martin (2005)
[c]The threshold for dissociation occurs at 661.87 (Å) (18.732 eV) which coincides with the energy of the $N = 9$ rotational level, $\upsilon = 4$, of the b $^4\Sigma_g^-$ state. Sources: Huber and Herzberg (1979), Samson et al. (1982)
[d]Predissociation occurs with $\upsilon = 3$ and higher vibrational levels of the C $^2\Sigma_u^+$ state, coinciding with the threshold for dissociation at 509.66 (Å) (24.327 eV). Sources: Huber and Herzberg (1979), Samson et al. (1987)

Table 9.10a Photoionization cross sections of atomic oxygen (multiple ionization)[a]

λ (Å)	σ(O$^+$) (10^{-20} cm^2 atom^{-1})	σ(O^{2+})	σ(O^{3+})	λ (Å)	σ(O$^+$) (10^{-20} cm^2 atom^{-1})	σ(O^{2+})	σ(O^{3+})
44.3	15.0	1.05	0.027	82.2	62.0	10.3	0.220
45.9	15.4	1.20	0.032	88.6	72.0	12.1	0.202
47.7	16.1	1.30	0.037	95.4	86.0	14.2	0.165
49.6	16.8	1.50	0.044	103.3	101.0	16.4	0.101
51.7	18.0	1.75	0.054	112.7	120.0	18.5	0.028
53.9	19.7	2.05	0.069	124.0	144.0	20.5	–
56.4	22.0	2.55	0.087	137.8	184.0	22.1	–
59.0	25.0	3.20	0.107	155.0	237.0	23.0	–
62.0	28.6	3.90	0.128	177.1	320.0	22.5	–
65.3	33.0	4.80	0.151	206.6	425.0	18.7	–
68.9	39.0	5.80	0.175	248.0	575.0	2.0	–
72.9	45.0	6.90	0.195	254.4	600.0	0.0	–
77.5	52.0	8.40	0.212	260.0	615.0	0.0	–

[a]The threshold for formation of O^{2+} occurs at 254.38 (Å) (48.739 eV), that for O^{3+} at 119.59 (Å) (103.675 eV). Source of data: Angel and Samson (1988)

Table 9.10b Photoionization cross sections of atomic oxygen (260 – 910 Å)[a]

λ (Å)	10^{18} σ (cm^2 atom^{-1})	λ (Å)	10^{18} σ (cm^2 atom^{-1})	λ (Å)	10^{18} σ (cm^2 atom^{-1})	λ (Å)	10^{18} σ (cm^2 atom^{-1})
270.0	6.50	479.43	15.60	690.0	9.80	760.0	3.90
280.0	6.85	486.6	12.65	690.6	9.47	760.22	4.06
290.0	7.40	490.0	12.0	692.2	9.51	765.0	4.00
300.0	7.55	500.0	11.9	692.5	9.60	780.0	3.70
303.8	7.70	510.0	11.9	693.8	9.00	785.0	3.70
310.0	7.90	520.0	12.0	695.3	9.06	825.0	3.30
320.0	8.25	530.0	12.1	705.0	9.60	830.0	3.30
330.0	8.60	540.0	12.3	705.9	9.71	835.0	3.30
340.0	8.92	550.0	12.5	707.6	9.88	840.0	3.30
350.0	9.25	560.0	12.7	709.2	9.71	845.0	3.25
360.0	9.57	570.0	12.9	710.0	9.15	850.0	3.20
370.0	9.90	580.0	13.0	710.8	9.30	855.0	3.10
380.0	10.20	590.0	13.2	712.5	9.26	860.0	3.10
390.0	10.50	600.0	13.3	714.1	9.18	865.0	3.10
400.0	11.80	610.0	13.4	715.7	9.24	870.0	3.10
410.0	11.00	620.0	13.4	717.6	8.79	885.0	2.85
416.0	11.17	630.0	13.4	719.2	9.00	890.0	2.70
420.0	11.30	640.0	13.3	720.0	9.00	895.0	2.75
430.0	11.50	650.0	13.0	720.9	8.91	900.0	2.85
436.67	11.38	660.0	12.6	722.5	9.02	905.0	2.75
449.0	11.42	665.3	12.0	727.5	8.90	907.5	2.75
457.5	12.26	677.5	10.0	729.0	9.17	909.8	2.45
462.0	12.22	682.8	9.69	730.6	8.40	910.0	1.15
464.3	11.17	683.0	9.65	732.2	8.03	910.5	0.70
471.53	11.25	683.9	9.25	752.3	4.15	911.0	0.15

[a]The threshold for formation of O$^+$ lies at 910.44 (Å) (13.618 eV). Source of data: Angel and Samson (1988), Samson and Pareek (1985). Smoothed experimental data are presented except in the region above 677.5 (Å) and between 430 and 490 (Å) where significant structure occurs

Table 9.11 State specific photoionization cross sections (10^{-18} cm^2 atom^{-1}) of atomic oxygen[a]

λ (Å)	$^4S^0$	$^2D^0$	$^2P^0$	^4P	^2P	O^{2+}	Total
50–100	0.190	0.206	0.134	0.062	0.049	0.088	0.730
100–150	0.486	0.529	0.345	0.163	0.130	0.186	1.839
150–200	0.952	1.171	0.768	0.348	0.278	0.215	3.732
200–250	1.311	1.762	1.144	0.508	0.366	0.110	5.202
250–300	1.628	2.325	1.488	0.637	0.383	0.00	6.461
300–350	2.259	3.446	2.173	0.815	0.00	0.00	8.693
350–400	2.523	3.883	2.422	0.859	0.00	0.00	9.687
400–450	3.073	4.896	2.986	0.541	0.00	0.00	11.496
450–500	3.394	5.459	3.274	0.00	0.00	0.00	12.127
500–550	3.421	5.427	3.211	0.00	0.00	0.00	12.059
550–600	3.620	5.910	3.494	0.00	0.00	0.00	13.024
600–650	4.250	6.159	2.956	0.00	0.00	0.00	13.365
650–700	5.128	11.453	0.664	0.00	0.00	0.00	17.245
700–750	6.739	3.997	0.00	0.00	0.00	0.00	10.736
750–800	5.091	0.00	0.00	0.00	0.00	0.00	5.091
800–850	3.498	0.00	000	0.00	0.00	0.00	3.498
850–900	4.554	0.00	0.00	0.00	0.00	0.00	4.554
900–950	1.315	0.00	0.00	0.00	0.00	0.00	1.315

[a]Weighted by EUV solar minimum reference spectrum F74113 (April 1974). Source of data: Richards et al. (1994). For energy thresholds see Tables 9.9 and 9.10a

Table 9.12 Photoionization cross sections of atomic hydrogen[a]

λ (Å)	10^{18} σ (cm^2 atom^{-1})	λ (Å)	10^{18} σ (cm^2 atom^{-1})	λ (Å)	10^{18} σ (cm^2 atom^{-1})
10	5.55 (−6)	200	8.33 (−2)	600	2.02
20	5.56 (−5)	250	1.62 (−1)	650	2.53
30	2.10 (−4)	300	2.78 (−1)	700	3.10
40	5.39 (−4)	350	4.36 (−1)	750	3.74
50	1.10 (−3)	400	6.43 (−1)	800	4.46
75	4.02 (−3)	450	9.01 (−1)	850	5.26
100	9.92 (−3)	500	1.22	900	6.12
150	3.47 (−2)	550	1.59	911.753	6.31

[a]Powers of ten are shown in parentheses. Calculated from theory; Source: Samson (1966)

Comments: The theoretical formula for the photoionization of atomic hydrogen is expected to be precise. The cross section is given by

$$\sigma = g \, Ry \left(e^2 / \varepsilon_0 h v\right)^3 / \left(2\pi \, 3^{3/2} \, n^5\right) = 1.045 \, g \, \lambda^3 \left[m^2\right]$$

where Ry is the Rydberg constant, e is the elementary charge, ε_0 is the permittivity of vacuum, h is the Planck constant, v is the frequency, n is the principal quantum number (here, n = 1), and g is the Gaunt factor, which is a function of frequency. The Gaunt factor varies from 0.8 at threshold to a maximum of unity at approximately 200 (Å), then falls rapidly to very low values in the X-ray region. Gaunt factors have been calculated and tabulated by Karzas and Latter (1961), and these data were used in the calculations.

Table 9.13 Photoionization cross sections of helium[a]

λ (Å)	$10^{18}\,\sigma$ (cm² atom⁻¹)	λ (Å)	$10^{18}\,\sigma$ (cm² atom⁻¹)	λ (Å)	$10^{18}\,\sigma$ (cm² atom⁻¹)	λ (Å)	$10^{18}\,\sigma$ (cm² atom⁻¹)
30.0	0.008	160.0	0.85	290.0	2.79	420.0	5.50
40.0	0.016	170.0	0.93	300.0	2.96	430.0	5.72
50.0	0.029	180.0	1.00	310.0	3.17	440.0	5.95
60.0	0.050	190.0	1.24	320.0	3.35	450.0	6.17
70.0	0.083	200.0	1.45	330.0	3.56	460.0	6.40
80.0	0.12	210.0	1.45	340.0	3.79	470.0	6.63
90.0	0.19	220.0	1.63	350.0	3.97	480.0	6.85
100.0	0.27	230.0	1.74	360.0	4.20	490.0	7.08
110.0	0.37	240.0	1.90	370.0	4.35	500.0	7.32
120.0	0.47	250.0	2.06	380.0	4.61	502.0	7.36
130.0	0.56	260.0	2.22	390.0	4.84	503.0	7.38
140.0	0.66	270.0	2.41	400.0	5.06	504.0	7.42
150.0	0.76	280.0	2.59	410.0	5.30	505.0	0.00

[a] The photoionization cross section is a continuous function of wavelength; the ionization threshold occurs at 504.26 (Å) (24 587 eV), where the photoionization cross section rises sharply to its peak value. Source: Kirby et al. (1979). Theoretical and experimental data differ by less than 5%

Table 9.14a Photoionization cross sections of molecular nitrogen (30 – 520 Å)[a]

λ (Å)	$\sigma(N_2^+)$ (10⁻¹⁸ cm² molecule⁻¹)	$\sigma(N^+)$ (10⁻¹⁸ cm² molecule⁻¹)	λ (Å)	$\sigma(N_2^+)$ (10⁻¹⁸ cm² molecule⁻¹)	$\sigma(N^+)$ (10⁻¹⁸ cm² molecule⁻¹)	λ (Å)	$\sigma(N_2^+)$ (10⁻¹⁸ cm² molecule⁻¹)	$\sigma(N^+)$ (10⁻¹⁸ cm² molecule⁻¹)
30.0	0.05	0.02	135.0	1.69	0.87	240.0	5.91	3.59
35.0	0.07	0.03	140.0	1.84	0.93	245.0	6.17	3.63
40.0	0.09	0.05	145.0	2.00	0.98	250.0	6.45	3.57
45.0	0.12	0.06	150.0	2.14	1.07	255.0	6.71	3.47
50.0	0.15	0.08	155.0	2.28	1.18	260.0	6.99	3.29
55.0	0.19	0.10	160.0	2.41	1.30	265.0	7.20	3.20
60.0	0.23	0.12	165.0	2.57	1.40	270.0	6.44	3.06
65.0	0.28	0.15	170.0	2.74	1.50	260.0	6.99	3.29
70.0	0.34	0.18	175.0	2.91	1.60	265.0	7.20	3.20
75.0	0.39	0.21	180.0	3.08	1.73	270.0	6.44	3.06
80.0	0.46	0.24	185.0	3.26	1.90	275.0	7.68	2.93
85.0	0.54	0.29	190.0	3.45	2.05	280.0	7.88	2.90
90.0	0.66	0.33	195.0	3.65	2.18	285.0	8.14	2.78
95.0	0.72	0.38	200.0	3.87	2.33	290.0	8.40	2.70
100.0	0.82	0.43	205.0	4.10	2.48	295.0	8.65	2.62
105.0	0.94	0.50	210.0	4.35	2.64	300.0	8.91	2.59
110.0	1.03	0.55	215.0	4.58	2.82	303.8	9.16	2.54
115.8	1.15	0.63	220.0	4.94	3.01	310.0	9.57	2.45
120.0	1.95	0.67	225.0	5.10	3.18	315.0	9.90	2.45
125.0	1.41	0.74	230.0	5.36	3.35	320.0	10.30	2.43
130.0	1.54	0.83	235.0	5.64	3.48	325.0	10.68	2.45

(continued)

Table 9.14a (continued)

λ (Å)	σ(N₂⁺)	σ(N⁺)	λ (Å)	σ(N₂⁺)	σ(N⁺)	λ (Å)	σ(N₂⁺)	σ(N⁺)
	$(10^{-18}\ cm^2\ molecule^{-1})$			$(10^{-18}\ cm^2\ molecule^{-1})$			$(10^{-18}\ cm^2\ molecule^{-1})$	
330.0	11.11	2.44	390.0	17.90	1.15	470.0	22.15	1.12
335.0	11.60	2.35	400.0	19.00	1.07	480.0	22.35	1.25
340.0	12.20	2.18	410.0	20.00	1.02	490.0	22.60	0.90
345.0	12.70	2.12	420.0	21.00	0.85	495.0	22.85	0.65
350.0	13.30	1.98	430.0	21.72	0.92	500.0	23.15	0.35
360.0	14.40	1.78	440.0	22.15	0.95	505.0	23.50	0.08
370.0	15.58	1.50	450.0	22.20	0.90	510.0	23.80	0.03
380.0	16.80	1.23	460.0	22.05	1.00	520.0	24.58	0.00

[a] The cross sections are smooth functions of λ in the wavelength region covered; the total ionization cross section is $\sigma(ion) = \sigma(abs) = \sigma(N_2^+) + \sigma(N^+)$. The energy threshold for the formation of N⁺ occurs at 509.66 (Å) (24.327 eV). Source of data: Samson et al. (1987) for the wavelength range >115 (Å). At lower wavelengths, the total ionization cross sections reported by Denne (1970) were interpolated; it was assumed that the nearly constant ratio $\sigma(N^+)/\sigma(ion)$ observed between 115 and 140 (Å) extends into the region below 100 (Å)

Table 9.14b Absorption and photoionization cross sections of N₂ (530 – 800 Å)[a]

λ (Å)	σ(abs)	σ(N₂⁺)	λ (Å)	σ(abs)	σ(N₂⁺)	λ (Å)	σ(abs)	σ(N₂⁺)
	$(10^{-18}\ cm^2\ molecule^{-1})$			$(10^{-18}\ cm^2\ molecule^{-1})$			$(10^{-18}\ cm^2\ molecule^{-1})$	
530.0	25.07	25.07	668.0	22.40	21.20	679.2	30.29	19.68
540.0	25.30	25.30	669.0	48.40	46.10	679.9	34.08	8.76
550.0	24.70	24.70	669.6	32.32	30.86	680.2	35.70	32.80
560.0	23.40	23.40	670.0	21.60	20.70	680.4	34.67	31.95
570.0	22.50	22.50	670.4	22.00	21.40	680.7	33.14	30.67
580.0	22.40	22.40	671.0	28.71	26.40	681.0	31.60	29.40
590.0	22.40	22.40	671.5	34.30	33.30	681.3	25.30	23.50
600.0	22.58	22.58	671.9	28.40	27.50	681.4	32.80	30.49
610.0	22.80	22.80	672.9	23.00	21.40	681.6	47.80	44.60
620.0	23.10	23.10	673.6	23.00	20.83	681.7	55.30	51.44
630.0	23.38	23.38	673.8	23.00	20.50	682.0	77.80	72.40
640.0	23.66	23.66	674.0	30.41	14.80	682.3	65.57	59.38
650.0	23.95	23.95	674.4	45.24	27.60	682.8	45.18	37.68
660.0	24.20	24.20	675.0	67.49	28.70	683.0	37.03	29.00
661.0	25.22	23.00	675.2	74.90	61.40	683.3	24.80	23.30
661.4	25.74	23.36	675.7	65.77	38.71	683.9	24.36	22.90
661.9	26.40	23.80	676.0	60.30	25.10	684.8	23.70	22.30
663.0	26.69	21.00	676.2	56.65	24.08	685.5	23.70	22.96
664.0	26.94	24.30	676.6	49.35	22.04	686.0	23.70	22.50
664.6	27.10	25.20	677.0	42.05	20.00	686.6	23.69	21.71
664.9	25.60	23.80	677.5	32.93	20.85	687.0	23.67	20.67
665.3	27.47	25.60	677.9	25.62	21.53	687.9	23.62	18.31
666.0	27.40	25.60	678.3	25.42	22.03	688.4	23.60	17.00
667.0	30.00	28.20	678.8	28.13	22.58	689.0	23.60	15.80
667.3	40.40	38.10	679.0	29.21	22.80	690.0	23.33	13.70

(continued)

Table 9.14b (continued)

λ (Å)	σ (abs)	σ (N$_2$$^+$)	λ (Å)	σ (abs)	σ (N$_2$$^+$)	λ (Å)	σ (abs)	σ (N$_2$$^+$)
	(10^{-18} cm^2 molecule^{-1})			(10^{-18} cm^2 molecule^{-1})			(10^{-18} cm^2 molecule^{-1})	
690.6	23.16	9.08	715.0	18.55	4.50	736.5	31.80	14.80
691.0	23.05	6.00	715.6	14.50	3.60	737.0	26.40	9.20
691.2	23.00	12.20	716.0	16.26	3.00	737.2	24.04	11.96
691.4	23.00	17.90	716.5	18.46	6.75	737.5	20.50	16.10
692.0	23.80	18.10	717.0	20.66	10.50	738.0	18.20	23.70
692.4	25.13	20.03	717.2	21.54	10.60	738.4	21.84	18.18
692.7	27.91	24.31	718.0	25.06	11.00	739.0	27.30	9.90
693.0	30.69	28.60	718.5	23.20	17.55	739.2	26.56	9.92
693.8	38.10	35.72	719.0	27.40	24.10	740.0	23.60	10.00
694.0	39.96	37.50	719.4	26.71	20.06	740.2	22.34	9.50
694.3	42.74	40.59	720.0	25.66	14.00	741.0	17.30	7.50
694.9	48.30	44.77	720.9	24.10	12.83	741.2	17.48	8.02
695.2	76.10	62.98	721.4	27.48	24.00	742.0	18.20	10.10
696.0	61.30	43.70	721.8	30.19	18.53	742.4	16.20	8.38
696.5	52.05	32.90	722.0	31.54	15.80	743.0	13.20	5.80
697.0	42.80	22.10	722.5	34.92	25.70	743.2	17.92	8.72
697.3	37.25	22.25	722.9	37.62	33.62	743.5	25.00	13.10
697.5	33.55	22.35	723.4	69.30	34.60	743.7	27.72	11.42
697.7	29.85	22.45	723.9	59.51	33.35	744.0	31.80	8.90
698.0	24.30	22.60	724.2	53.63	31.00	744.5	46.50	17.60
698.9	22.89	22.33	724.8	41.88	24.70	744.9	34.72	12.56
699.4	22.10	20.80	724.9	39.92	23.65	746.0	15.90	6.30
700.0	23.80	22.40	725.5	30.60	19.15	746.4	18.62	6.14
700.4	22.59	21.27	726.0	29.24	15.70	746.7	20.66	6.02
701.0	25.50	24.10	726.4	28.16	16.10	747.0	22.70	5.90
701.6	28.26	26.80	727.0	26.53	16.70	747.5	51.12	21.90
702.0	30.10	28.60	727.3	25.71	13.82	748.0	38.60	11.30
703.0	28.30	27.10	727.5	25.17	11.90	748.5	28.60	8.85
703.36	26.54	25.48	728.0	24.17	7.10	749.0	18.60	6.40
704.0	23.40	22.60	728.3	23.63	6.86	750.0	33.20	10.00
704.5	25.80	25.00	729.0	22.94	6.30	750.7	25.85	5.87
705.0	25.79	22.30	729.4	23.00	10.54	751.0	22.70	4.10
705.9	25.78	21.13	729.8	23.05	14.78	751.6	20.00	4.76
707.0	25.77	19.60	730.0	23.08	16.90	752.0	18.20	5.20
707.9	25.76	19.06	730.9	23.20	12.58	752.3	16.82	4.99
708.9	25.75	18.19	731.5	22.68	9.04	752.9	14.06	4.57
710.0	25.74	17.80	732.0	22.24	13.10	754.0	27.30	13.80
711.0	25.73	18.00	732.5	21.81	10.95	754.4	61.94	46.12
711.9	25.72	18.00	733.0	21.37	8.80	755.0	30.00	28.50
712.5	25.71	17.30	733.3	21.11	9.22	755.2	44.00	24.74
712.9	25.71	16.74	734.0	20.50	10.20	755.5	25.00	19.10
713.5	25.70	15.40	734.5	30.50	15.10	755.8	22.00	16.22
713.9	25.70	14.44	735.0	25.00	11.30	756.0	20.00	14.30
714.7	20.57	7.41	735.9	20.95	9.05	756.2	17.00	12.22

(continued)

Table 9.14b (continued)

λ (Å)	σ (abs) (10⁻¹⁸ cm² molecule⁻¹)	σ (N₂⁺)	λ (Å)	σ (abs) (10⁻¹⁸ cm² molecule⁻¹)	σ (N₂⁺)	λ (Å)	σ (abs) (10⁻¹⁸ cm² molecule⁻¹)	σ (N₂⁺)
756.5	12.50	9.10	768.7	42.30	28.50	783.5	150.00	63.20
757.0	50.00	24.20	769.0	25.00	16.00	783.8	234.10	125.30
757.2	69.20	20.24	769.2	32.70	14.72	784.0	224.36	48.40
757.5	42.68	14.30	769.5	42.30	12.80	784.4	204.87	18.80
757.7	25.00	12.70	770.0	26.85	9.40	784.8	155.00	8.40
758.0	16.80	10.30	770.2	20.67	8.96	785.0	125.00	6.40
758.3	66.70	37.60	770.4	14.49	8.52	785.5	50.00	9.70
758.4	62.85	32.20	770.8	19.56	10.16	786.0	45.50	22.80
758.68	27.86	19.15	771.0	25.00	11.40	786.2	56.80	18.04
759.0	19.04	14.50	771.5	141.80	51.30	786.4	35.60	13.28
759.4	11.90	10.20	772.0	50.00	26.00	787.0	10.50	7.80
760.0	23.82	12.30	772.4	30.00	14.80	787.5	8.48	7.30
760.2	27.80	15.80	773.0	21.50	10.30	787.7	7.67	6.94
760.4	27.80	13.27	773.5	18.00	8.30	788.0	9.04	6.40
760.7	36.90	17.00	774.0	14.50	8.00	788.5	11.32	5.80
761.0	46.00	24.50	774.5	33.10	13.20	789.0	13.60	7.20
761.5	48.00	18.80	775.0	40.90	12.50	789.5	45.50	11.30
762.0	36.40	11.70	775.7	143.81	126.90	790.0	25.58	11.00
762.2	25.00	10.46	776.0	125.00	63.30	790.5	18.60	9.60
762.5	17.30	8.60	776.5	59.10	20.30	790.8	125.00	101.90
763.0	16.00	10.60	777.0	24.50	14.05	791.0	58.0	41.00
763.5	25.00	21.40	777.5	20.20	7.80	791.3	53.24	37.94
763.7	68.10	38.30	778.0	15.90	7.20	791.4	51.62	36.92
764.0	18.20	12.20	778.5	38.61	9.50	791.8	125.00	44.90
764.4	13.30	9.20	778.7	47.70	9.14	792.4	71.80	38.90
764.7	18.00	11.40	779.0	40.54	8.60	792.8	43.60	21.62
765.0	150.00	106.90	779.5	28.60	8.10	792.92	35.14	15.91
765.3	91.00	52.90	779.8	14.90	7.70	793.14	25.69	10.56
765.4	70.59	44.30	779.9	14.70	7.35	793.5	15.90	6.60
765.7	31.80	27.02	780.3	25.15	8.50	794.0	25.00	6.20
766.0	14.50	14.00	780.5	32.25	9.50	794.5	39.75	6.30
766.5	13.95	11.40	781.0	50.00	41.70	795.0	54.50	12.30
767.0	13.41	9.70	781.2	118.20	70.70	795.2	46.22	5.92
767.3	13.30	10.06	781.5	75.00	32.30	796.0	13.10	4.50
767.7	13.30	10.86	782.0	34.00	13.40	797.0	8.00	1.00
767.9	40.90	11.42	782.5	15.00	6.70	797.7	8.00	0.30
768.3	22.30	14.58	782.9	17.56	7.58	798.0	8.00	0.00
768.4	27.30	15.54	783.2	70.92	29.96	800.5	16.70	0.00

[a]The ionization threshold for the formation of N₂⁺ occurs at 795.74 (Å) (15.581 eV); Source of data: Fennelly and Torr (1992), slightly condensed

Table 9.14c   Absorption cross sections of molecular nitrogen (801 – 987 Å)[a]

λ (Å)	$10^{18}\,\sigma$ (cm^2 molecule^{-1})	λ (Å)	$10^{18}\,\sigma$ (cm^2 molecule^{-1})	λ (Å)	$10^{18}\,\sigma$ (cm^2 molecule^{-1})
801.0	40.00	815.5	6.70	838.0	13.30
801.5	104.00	816.0	14.00	838.6	41.32
802.0	40.00	816.42	24.92	838.9	55.33
802.6	16.60	816.77	33.92	840.0	3.00
803.0	1.00	817.0	40.00	840.5	1.00
803.5	3.85	817.19	44.94	840.7	50.00
804.0	6.70	817.5	53.00	841.0	26.70
804.27	7.82	817.78	60.20	841.5	20.00
804.38	8.30	818.0	66.00	842.0	96.70
804.5	8.80	818.2	71.20	842.5	13.30
804.78	9.96	818.34	74.84	843.0	6.70
805.0	10.90	818.5	79.00	843.5	3.00
805.29	12.14	819.0	92.00	843.8	1.80
805.44	12.75	819.5	12.00	844.0	1.00
805.74	56.86	819.8	6.60	844.5	26.70
806.0	18.00	820.0	3.00	845.0	13.30
806.23	95.95	820.5	7.30	845.5	3.00
806.42	159.70	821.0	26.70	845.9	1.40
807.0	46.70	821.3	18.66	847.0	1.00
808.0	6.70	821.5	13.30	848.0	1.00
808.2	4.42	822.0	3.00	849.0	1.00
808.5	1.00	823.0	3.00	849.2	7.28
808.8	33.30	824.0	3.00	849.5	16.70
809.0	33.30	824.5	8.00	850.0	46.70
809.3	13.92	824.9	12.24	850.6	26.66
809.5	1.00	825.3	8.92	851.0	13.30
810.0	8.00	825.5	6.00	851.5	6.70
810.5	4.50	826.0	33.00	851.8	4.48
810.66	3.35	826.5	25.30	852.0	3.00
810.85	2.05	826.8	18.10	853.0	3.00
811.0	1.00	827.0	13.30	853.2	36.48
811.26	1.96	827.5	3.00	853.5	86.70
811.49	2.81			854.0	66.70
811.61	3.26	832.9	3.00	854.5	46.70
811.8	3.96	833.5	5.30	855.0	20.00
812.0	4.70	833.7	8.50	855.5	13.30
812.27	15.50	834.0	13.30	856.0	3.00
812.5	24.70	834.5	13.30	856.2	173.30
812.8	100.00	835.0	83.30	856.5	20.00
813.0	33.40	835.2	180.00	857.0	8.70
813.5	50.00	835.4	135.75	857.3	12.13
813.7	62.00	836.0	3.00	857.7	16.70
814.0	80.00	837.0	3.00	858.0	13.30
814.5	14.70	837.5	20.00	858.5	6.70
814.9	11.50	837.8	15.98	859.0	3.00

(continued)

Table 9.14c (continued)

λ (Å)	$10^{18}\,\sigma$ (cm² molecule⁻¹)	λ (Å)	$10^{18}\,\sigma$ (cm² molecule⁻¹)	λ (Å)	$10^{18}\,\sigma$ (cm² molecule⁻¹)
860.0	3.00	881.5	67.00	901.5	27.00
860.4	1.00	882.0	51.00	902.0	6.00
861.0	6.00	882.5	1.00	903.0	10.00
861.5	1.00	882.7	54.00	903.5	1.00
862.0	1.00	883.0	27.00	903.8	5.20
863.0	1.00	883.3	16.80	904.0	8.00
863.2	33.40	883.5	10.00	904.5	14.00
864.0	6.00	884.0	1.00	905.0	12.00
864.6	3.00	884.5	3.00	905.5	10.00
865.0	1.00	885.0	1.00	906.0	3.00
865.2	86.70	885.5	3.00	906.4	1.40
865.4	106.70	885.8	6.60	907.0	1.00
866.0	3.00	886.0	9.00	907.4	1.00
866.5	1.00	886.5	15.00	908.0	15.00
867.0	65.00	886.9	80.00	908.5	15.00
867.5	32.00	887.5	18.00	909.0	9.00
868.0	14.00	888.0	30.00	909.5	6.00
868.5	11.00	888.5	12.00	909.8	4.20
869.0	10.00	889.0	20.00	910.0	3.00
869.5	6.00	889.5	14.00	910.4	1.00
870.0	3.00	890.0	12.00	911.0	14.00
870.8	120.00	890.5	3.00	911.5	6.00
871.0	25.00	891.0	67.00	911.7	3.40
871.4	26.00	891.5	25.00	912.0	1.00
872.0	80.00	892.0	15.00	912.5	1.00
872.5	33.00	892.5	15.00	913.0	82.00
873.0	12.00	893.0	1.00	913.3	53.80
873.5	10.00	893.5	14.00	913.5	35.00
874.0	6.00	894.0	10.00	914.0	15.00
874.5	3.00	894.5	3.00	914.5	3.00
875.0	3.00	895.0	3.00	915.0	3.00
875.5	3.00	896.0	3.00	915.5	1.00
876.0	107.00	896.5	25.00	916.0	1.00
876.2	60.00	897.0	11.00	916.5	42.00
876.5	54.00	897.2	104.00	917.0	1.00
877.0	30.00	897.5	64.00	917.2	1.80
877.5	24.00	898.0	1.00	917.5	3.00
877.72	23.12	898.5	3.00	918.0	6.00
878.5	20.00	898.7	4.20	918.5	3.00
878.92	13.28	899.0	6.00	918.9	3.00
879.2	9.60	899.5	10.00	919.5	1.00
879.5	6.00	900.0	12.00	919.9	97.00
879.8	4.20	900.2	11.20	920.4	50.75
880.0	3.00	900.5	10.00	920.96	27.64
880.5	5.00	901.0	3.00	921.5	17.00
881.0	110.00	901.3	31.00	922.0	11.00

(continued)

Table 9.14c (continued)

λ (Å)	$10^{18} \sigma$ (cm² molecule⁻¹)	λ (Å)	$10^{18} \sigma$ (cm² molecule⁻¹)	λ (Å)	$10^{18} \sigma$ (cm² molecule⁻¹)
922.5	3.00	942.0	22.00	967.0	31.00
922.8	58.00	942.4	1.00	967.5	20.00
923.0	17.00	943.0	27.00	968.0	27.00
923.5	18.50	943.3	22.80	969.0	10.00
923.7	19.10	943.5	20.00	969.5	3.00
924.0	20.00	944.0	11.00	970.0	6.00
924.3	15.20	944.5	11.00	970.4	2.00
924.5	12.00	945.0	1.00	971.0	1.00
925.0	22.00	946.0	1.00	971.5	6.00
925.5	3.00	947.0	3.00	972.0	27.00
926.0	14.00	948.0	3.00	972.5	120.00
926.2	9.60	948.5	1.00	972.9	94.40
926.4	5.20	949.0	1.00	973.5	31.00
927.0	18.00	949.5	25.00	974.0	20.00
927.6	13.20	949.74	13.48	974.5	3.00
928.0	10.00	950.0	1.00	975.0	5.00
928.5	2.00	950.3	17.20	975.3	3.80
929.0	2.00	950.5	28.00	975.5	3.00
930.0	29.00	951.0	2.00	976.0	3.00
930.5	3.00	952.0	2.00	976.5	1.00
930.75	4.50	953.0	12.00	977.0	2.00
931.0	6.00	954.0	3.00	977.5	8.00
931.5	6.00	955.0	6.00	978.0	5.00
931.9	15.00	955.9	18.60	978.5	2.00
932.4	8.20	956.5	12.00	979.0	88.00
933.0	7.00	956.7	8.40	979.5	33.00
933.38	4.72	957.0	3.00	980.0	27.00
933.5	4.72	957.5	6.00	980.5	22.00
934.0	10.00	958.0	18.00	981.0	26.00
934.5	11.30	958.2	134.00	981.5	11.00
935.0	5.00	958.5	53.00	982.0	14.00
935.3	12.00	958.8	62.00	982.5	25.00
935.5	7.00	959.0	47.00	983.0	23.00
936.0	6.00	959.5	21.00	983.3	18.20
936.5	6.00	960.0	1.00	983.5	15.00
937.0	1.00	960.5	98.00	984.0	17.00
937.5	7.00	961.0	51.00	984.5	15.00
937.8	20.50	961.5	30.00	985.0	15.00
937.9	25.00	961.9	20.40	985.2	13.00
938.5	1.00	962.5	10.00	985.5	10.00
939.0	148.00	962.8	5.80	985.9	22.00
939.3	106.00	963.0	3.00	986.3	0.00
939.5	78.00	964.0	3.00	987.0	0.00
940.0	27.00	965.0	3.00	988.0	0.00
940.5	29.00	965.5	1.00	989.0	0.00
941.0	18.00	966.0	67.00		
941.5	15.00	966.5	60.00		

[a]The data were derived from absorption cross sections reported by Carter (1972) for a spectral resolution of 0.04 (Å); Vertical bars indicate no change in the cross section values. Source of data: Fennelly and Torr (1992)

Comments: In the wavelength region above 660 (Å) the spectrum of nitrogen features multiple absorption band systems superimposed on an ionization continuum, which sets in at 796 (Å). At longer wavelength the absorption of solar radiation by nitrogen in the upper atmosphere must still be taken into account because it partially shields the ionization of oxygen. For this reason, the table is extended to include the strongly varying absorption features up to 987 (Å). At wavelengths below 660 (Å), absorption and ionization cross sections are identical.

Table 9.15a Photoionization cross sections of molecular oxygen (30 – 600 Å)[a]

λ (Å)	$\sigma(O_2^+)$ $(10^{-18}$ cm^2 molecule$^{-1})$	$\sigma(O^+)$	λ (Å)	$\sigma(O_2^+)$ $(10^{-18}$ cm^2 molecule$^{-1})$	$\sigma(O^+)$	λ (Å)	$\sigma(O_2^+)$ $(10^{-18}$ cm^2 molecule$^{-1})$	$\sigma(O^+)$
30.0	0.01	0.12	250.0	8.10	4.70	470.0	17.1	4.95
40.0	0.02	0.24	260.0	8.67	5.03	480.0	17.3	5.26
50.0	0.04	0.39	270.0	9.12	5.59	490.0	17.7	5.29
60.0	0.08	0.58	280.0	9.65	5.85	500.0	18.1	5.51
70.0	0.14	0.82	290.0	10.2	5.94	510.0	18.5	5.53
80.0	0.24	1.06	300.0	10.7	6.03	520.0	19.1	5.43
90.0	0.38	1.38	310.0	11.2	5.77	530.0	20.0	4.93
100.0	0.57	1.64	320.0	11.7	5.52	540.0	20.7	4.67
110.0	0.82	1.90	330.0	12.1	5.20	544.7	21.1	4.15
120.0	1.13	2.11	340.0	12.6	4.95	548.9	21.3	4.55
130.0	1.58	2.37	350.0	12.9	4.91	550.0	21.5	4.42
140.0	2.11	2.51	360.0	13.3	4.70	551.4	21.7	4.24
150.0	2.56	2.76	370.0	13.9	4.54	555.6	21.6	4.53
160.0	3.02	3.02	380.0	14.1	4.74	560.0	21.6	4.51
170.0	3.51	3.20	390.0	14.4	4.80	570.0	21.4	4.44
180.0	4.07	3.26	400.0	14.8	4.79	580.0	20.6	4.24
190.0	4.60	3.55	410.0	15.2	4.81	585.8	18.1	3.72
200.0	5.09	3.77	420.0	15.5	4.77	588.9	16.1	3.18
210.0	5.65	3.92	430.0	15.8	4.87	590.0	16.1	3.28
220.0	6.14	4.16	440.0	16.2	4.79	594.7	24.6	3.14
230.0	6.78	4.32	450.0	16.6	4.82	596.7	24.6	3.14
240.0	7.35	4.55	460.0	16.8	4.93	600.0	25.7	2.93

[a] The cross sections are smooth functions of λ in the wavelength region covered; the total ionization cross section is $\sigma(\text{ion}) = \sigma(\text{abs}) = \sigma(O_2^+) + \sigma(O^+)$. Source of data: Samson et al. (1982) for the wavelength range >120 nm. At lower wavelengths, the total ionization cross sections reported by Denne (1970) were interpolated, and the trend in the decrease of $\sigma(O_2^+)/\sigma(\text{ion})$ below 150 nm was extrapolated by using a quadratic function fitted to the data of Samson et al. (1982)

Table 9.15b Absorption and photoionization cross sections of O_2 (600 – 665 Å)[a]

λ (Å)	σ(abs)	σ(O_2^+)	σ(O^+)	λ (Å)	σ(abs)	σ(O_2^+)	σ(O^+)
	(10^{-18} cm^2 molecule^{-1})				(10^{-18} cm^2 molecule^{-1})		
604.3	30.1	26.9	2.50	636.3	27.9	26.1	1.01
608.9	26.4	23.7	1.91	637.3	25.3	23.5	1.01
610.8	30.1	27.1	1.82	638.1	27.9	26.1	1.01
612.7	23.1	20.7	1.60	638.7	25.3	23.5	1.01
613.1	28.3	25.6	1.50	642.5	29.0	27.1	1.05
615.2	28.6	26.5	1.40	644.2	20.8	19.4	1.10
616.3	24.2	22.1	1.30	645.0	25.3	23.4	1.10
617.7	34.6	32.2	1.20	645.9	24.9	23.0	1.12
618.2	28.3	26.1	1.15	646.7	30.1	28.3	1.12
618.7	29.8	27.5	1.12	649.4	25.3	23.4	1.15
620.5	23.1	21.4	1.05	651.8	29.7	27.5	1.16
621.9	32.7	30.6	1.00	654.0	28.8	26.8	1.04
624.5	25.3	23.6	0.95	656.0	28.0	26.3	0.98
626.6	29.0	30.2	0.95	658.0	27.1	25.7	0.89
627.1	24.9	29.2	0.96	660.0	26.3	25.1	0.75
629.6	32.4	30.2	0.97	661.0	25.9	24.8	0.25
633.0	23.4	21.7	0.98	661.4	25.7	24.6	0.10
634.4	31.2	29.4	0.99	661.9	25.5	24.5	0.00
635.8	25.3	23.5	1.00	665.8	23.8	23.0	0.00

[a]The total ionization cross section is σ(ion) $= \sigma(O_2^+) + \sigma(O^+)$. The threshold for the formation of O^+ is 661.87 (Å) (18.732 eV). Source of data: Matsunaga and Watanabe (1967), Samson et al. (1982)

Table 9.15c Absorption and photoionization cross sections of O_2 (670 – 1077 Å)[a]

λ (Å)	σ(abs)	σ(O_2^+)	λ (Å)	σ(abs)	σ(O_2^+)	λ (Å)	σ(abs)	σ(O_2^+)
	(10^{-18} cm^2 molecule^{-1})			(10^{-18} cm^2 molecule^{-1})			(10^{-18} cm^2 molecule^{-1})	
669.6	22.7	21.9	684.9	27.1	26.4	703.1	29.0	25.3
670.0	20.8	20.1	686.1	16.7	16.4	703.6	26.0	22.7
670.5	21.6	20.8	686.6	23.8	23.0	705.3	63.2	55.0
673.6	19.3	18.6	687.9	16.7	16.4	706.6	24.9	21.2
675.0	21.6	20.8	688.7	20.8	20.8	707.9	26.4	22.7
675.7	20.1	19.3	689.1	16.4	15.6	708.9	25.7	21.9
676.2	21.6	20.8	690.1	26.0	25.7	709.3	26.8	22.7
676.6	20.1	19.3	691.3	17.8	16.0	709.7	25.7	21.2
677.0	21.2	20.4	692.4	34.6	29.7	711.0	41.6	35.7
679.2	18.2	17.5	694.0	18.2	17.1	711.9	32.3	25.7
679.9	23.8	23.0	695.2	30.5	27.9	712.9	37.9	31.2
680.7	19.3	18.6	696.0	19.3	17.5	714.0	33.1	27.1
681.1	22.3	21.6	697.3	30.5	27.5	714.7	34.9	28.6
681.4	21.9	21.2	697.7	31.6	28.6	716.0	29.0	23.8
681.6	22.3	21.6	698.3	34.6	31.2	716.5	30.5	25.3
682.3	18.6	18.6	699.6	23.4	21.6	717.2	27.5	20.0
682.8	21.9	21.2	700.8	34.9	30.1	717.6	29.7	24.5
683.8	17.5	16.4	701.6	19.0	16.4	719.4	24.9	20.1

(continued)

Table 9.15c (continued)

λ (Å)	σ(abs) $(10^{-18}\,cm^2\,molecule^{-1})$	$\sigma(O_2^+)$	λ (Å)	σ(abs) $(10^{-18}\,cm^2\,molecule^{-1})$	$\sigma(O_2^+)$	λ (Å)	σ(abs) $(10^{-18}\,cm^2\,molecule^{-1})$	$\sigma(O_2^+)$
720.4	34.2	24.9	764.6	17.5	9.67	809.3	22.3	9.29
721.3	26.8	21.9	765.4	22.3	11.9	811.8	55.0	15.2
721.8	27.5	23.8	766.7	19.0	10.0	812.5	29.4	8.18
722.6	25.7	20.4	768.4	17.1	11.5	813.7	34.9	9.67
722.9	26.0	20.8	768.8	20.8	13.0	814.9	18.6	9.29
723.4	25.3	20.1	769.2	17.8	11.2	817.2	45.4	21.9
724.2	27.5	22.7	769.6	21.2	13.0	818.2	20.1	7.06
725.0	25.3	21.6	770.2	15.2	9.29	819.8	31.6	11.2
726.4	49.1	36.4	770.5	20.4	11.5	821.3	16.0	6.55
727.3	30.1	25.3	770.8	19.7	9.67	823.2	28.3	11.2
728.3	30.5	25.7	771.6	24.2	10.4	824.1	16.0	6.32
729.0	27.9	23.4	772.4	23.8	10.0	924.9	24.2	8.92
729.4	29.8	25.03	773.1	26.8	14.1	825.3	20.1	7.06
729.8	29.0	24.5	775.1	13.8	10.8	826.0	29.0	10.4
731.1	35.3	29.7	778.1	29.4	17.8	826.8	11.9	5.58
731.8	31.6	27.1	778.8	24.5	13.0	827.8	12.3	5.95
732.5	51.3	44.2	780.0	27.9	10.8	828.3	11.2	5.95
733.3	32.7	29.7	781.5	14.5	11.2	829.4	22.7	8.18
735.3	35.3	31.6	782.9	20.8	9.29	829.6	22.3	7.81
737.2	32.3	29.7	783.2	20.4	8.92	829.8	23.0	8.55
737.5	34.2	31.2	784.4	25.7	11.2	831.0	10.8	4.09
739.3	24.9	22.7	784.8	23.8	10.0	832.5	32.7	9.67
740.0	25.7	21.9	786.0	26.8	10.8	834.1	10.4	4.09
741.2	20.1	17.5	786.4	19.7	10.0	834.5	10.8	4.46
742.2	21.6	18.6	787.6	25.7	15.2	835.4	10.0	4.09
743.2	17.8	13.4	788.0	24.2	16.4	836.3	17.1	5.58
745.0	20.4	16.7	788.6	27.5	11.2	837.8	11.2	4.09
746.4	18.6	14.9	789.0	26.8	10.8	838.6	24.5	10.8
747.0	21.2	16.7	790.0	28.3	11.5	838.9	23.4	10.4
748.0	15.6	11.5	791.3	21.2	9.29	839.1	24.9	10.8
750.0	23.8	14.9	792.4	27.9	11.8	842.1	8.18	3.05
751.0	17.5	11.5	792.8	24.2	11.8	843.8	12.3	4.09
751.6	19.0	13.8	794.1	33.8	12.3	844.6	10.0	3.72
752.9	14.9	10.0	795.0	23.0	7.43	845.9	18.6	5.20
755.0	23.4	14.5	795.2	23.4	8.18	847.6	7.43	2.45
756.0	19.0	11.2	796.0	21.9	7.43	848.5	7.81	2.97
756.2	19.3	13.4	797.7	31.6	16.0	849.2	7.43	3.23
758.0	17.5	10.8	798.1	26.8	12.3	850.6	9.67	4.46
758.4	19.3	12.3	799.5	39.8	14.5	851.8	8.55	3.42
759.4	17.1	10.4	801.6	27.1	7.43	853.2	12.6	5.58
760.7	21.2	11.2	802.6	34.2	11.2	955.0	7.06	3.01
761.5	19.3	10.4	803.5	26.0	13.0	857.3	10.0	4.09
762.1	19.7	11.2	805.1	49.1	12.3	859.2	6.69	2.86
762.5	19.3	10.8	807.0	27.9	10.0	861.0	7.43	3.05
763.2	21.9	13.0	808.2	36.4	13.4	864.6	9.29	4.46

(continued)

Table 9.15c (continued)

λ (Å)	σ(abs) (10⁻¹⁸ cm² molecule⁻¹)	σ(O₂⁺) (10⁻¹⁸ cm² molecule⁻¹)	λ (Å)	σ(abs) (10⁻¹⁸ cm² molecule⁻¹)	σ(O₂⁺) (10⁻¹⁸ cm² molecule⁻¹)	λ (Å)	σ(abs) (10⁻¹⁸ cm² molecule⁻¹)	σ(O₂⁺) (10⁻¹⁸ cm² molecule⁻¹)
867.6	5.58	2.90	960.0	2.49	1.71	1022.4	1.30	1.12
870.0	8.18	5.20	961.9	15.6	12.6	1023.4	1.86	1.12
870.6	7.81	4.83	962.8	6.69	5.20	1024.6	1.00	1.00
871.4	10.0	5.95	965.5	50.9	37.9	1025.3	1.78	0.97
875.2	5.58	3.31	970.4	2.41	2.04	1025.7	1.64	0.95
878.1	1.34	8.55	972.5	31.6	22.7	1027.6	1.16	0.20
883.3	4.83	3.31	972.9	42.8	29.7	1028.2	0.67	0.00
885.8	1.78	1.15	974.5	5.57	3.42	1029.3	1.90	0.00
889.1	1.34	3.05	975.3	26.8	9.67	1030.2	0.82	
891.6	10.8	7.06	980.5	2.42	1.67	1031.0	1.49	
893.1	5.58	4.46	983.3	46.1	31.2	1032.3	1.00	
894.0	11.2	7.43	985.2	4.83	3.72	1033.1	1.08	
895.8	5.58	3.35	985.9	7.43	5.20	1033.6	0.89	
897.3	7.43	5.20	988.0	2.97	2.42	1034.9	1.67	
898.7	4.83	4.09	988.5	4.83	4.46	1036.5	0.52	
900.2	8.92	5.20	989.6	1.49	1.30	1036.9	0.97	
901.1	13.4	8.18	992.9	16.9	20.1	1037.2	0.67	
902.0	9.67	5.95	993.2	21.6	18.2	1038.0	1.90	
903.8	10.8	6.32	993.5	24.5	20.8	1038.8	0.71	
906.4	4.09	3.27	997.2	1.45	1.26	1039.4	1.78	
909.6	17.1	11.2	1000.0	1.49	1.26	1041.1	0.89	
910.0	15.2	10.0	1004.0	6.32	5.20	1041.9	1.23	
910.5	17.5	11.2	1004.3	5.58	4.46	1043.8	0.12	
913.5	4.83	3.35	1004.6	6.32	5.20	1047.1	2.57	
914.7	7.43	5.58	1006.8	1.38	1.19	1047.8	1.41	
915.6	4.09	2.71	1007.9	1.82	1.51	1049.5	2.49	
917.3	23.4	14.5	1009.1	1.41	1.20	1050.1	0.41	
920.4	3.01	2.12	1009.4	1.52	1.07	1051.1	1.12	
923.1	9.67	8.92	1010.0	1.45	0.90	1051.9	0.82	
923.5	8.92	7.81	1011.4	1.30	0.86	1052.4	2.16	
924.5	23.4	19.0	1012.3	1.08	0.80	1054.2	0.59	
926.4	3.72	3.16	1013.5	1.34	0.72	1054.8	1.97	
927.6	4.09	3.12	1013.9	1.12	0.75	1055.6	0.32	
928.1	3.35	2.71	1015.8	1.75	0.93	1058.3	1.04	
929.1	4.09	3.42	1016.0	1.19	0.90	1058.7	0.63	
930.0	3.72	3.20	1016.4	1.38	0.85	1059.6	1.15	
930.6	26.0	16.7	1016.9	1.08	0.79	1060.6	0.86	
931.5	12.3	8.18	1017.2	1.60	0.78	1060.9	1.15	
932.4	28.6	20.1	1017.8	0.97	0.76	1061.1	0.59	
935.6	2.83	2.19	1018.3	1.52	0.87	1061.7	1.15	
939.3	45.0	23.4	1018.8	1.08	0.97	1062.6	0.41	
944.6	2.64	2.30	1019.4	1.41	1.05	1064.2	3.05	
947.7	55.8	39.4	1020.0	1.28	1.08	1064.8	1.75	
950.3	2.23	1.97	1020.4	1.19	1.10	1066.4	3.42	
955.9	54.6	43.5	1020.8	1.60	1.13	1068.9	0.12	
956.7	29.0	22.3	1021.1	1.30	1.13	1073.9	2.01	
957.0	35.3	27.1	1021.6	1.64	1.20	1077.0	0.14	

[a] The threshold for the formation of O_2^+ lies at 1027.2 (Å) (12.07 eV). Source of data: Matsunaga and Watanabe (1967), Watanabe and Marmo (1956)

Comments: In the wavelength region above 540 (Å), the spectrum of oxygen exhibits considerable discrete structure in both ionization and absorption cross sections. Between 870 and 1,030 (Å) the absorption peaks are well separated and are superimposed on a weak continuum. The table of Matsunaga and Watanabe lists values for peaks and valleys.

Table 9.16 Absorption cross sections (cm^2 molecule^{-1}) of O_2 (1215.7 and 1250 – 1750 Å)[a]

λ (Å)	$10^{18} \sigma$	λ (Å)	$10^{18} \sigma$	λ (Å)	$10^{18} \sigma$	λ (Å)	$10^{18} \sigma$
1215.7	0.0104	1299.0	0.520	1378.0	12.70	1550.0	7.84
1252.0	1.04	1302.0	0.444	1384.0	13.20	1560.0	7.34
1254.5	0.855	1306.0	0.357	1391.5	13.40	1570.0	6.49
1256.0	0.706	1309.0	0.516	1394.0	13.80	1580.0	5.94
1257.0	0.594	1312.5	0.721	1400.0	14.28	1590.0	5.22
1259.5	0.465	1317.0	1.10	1405.0	14.40	1600.0	4.75
1260.5	0.401	1321.5	1.60	1410.0	14.65	1610.0	4.04
1262.0	0.442	1325.0	2.05	1420.0	14.78	1620.0	3.45
1263.5	0.342	1329.0	2.31	1430.0	14.83	1630.0	2.92
1264.5	0.242	1333.5	2.29	1440.0	14.66	1640.0	2.48
1266.0	0.182	1336.5	2.20	1450.0	14.45	1650.0	2.16
1269.0	0.119	1339.5	2.24	1455.0	13.90	1660.0	1.78
1271.0	0.067	1343.0	2.79	1460.0	13.60	1670.0	1.48
1274.0	0.093	1345.0	3.64	1470.0	13.20	1680.0	1.24
1277.0	0.153	1349.0	5.76	1480.0	12.70	1690.0	0.973
1279.5	0.249	1351.0	7.10	1490.0	12.00	1700.0	0.843
1283.5	0.364	1355.0	7.10	1500.0	11.20	1710.0	0.698
1287.0	0.472	1361.0	8.18	1510.0	10.62	1720.0	0.579
1290.5	0.542	1366.0	9.62	1520.0	10.03	1730.0	0.465
1293.0	0.584	1369.0	11.30	1530.0	9.39	1740.0	0.369
1296.5	0.550	1375.0	12.40	1540.0	8.52	1750.0	0.255

[a] The photodissociation process $O_2 + h\nu \rightarrow O$ (^{3}P) + O (^{1}D) associated with the Schumann absorption continuum between 1,350–1,750 (Å) occurs with a quantum yield of unity. Source of data: Watanabe (1958), Metzger and Cook (1964), Blake et al. (1966). Good agreement exists except in the region 1,390–1,500 (Å) where Watanabe's values are ~5% lower, but he alone provided comprehensive numerical values. Allowance has been made for the slight difference in the region of maximum absorption by taking appropriate averages

Table 9.17 Absorption cross sections (cm^2 molecule^{-1}) of carbon dioxide (1200 – 1950 [Å]) at 298 K[a]

λ (Å)	$10^{19}\,\sigma$	λ (Å)	$10^{19}\,\sigma$	λ (Å)	$10^{19}\,\sigma$	λ (Å)	$10^{19}\,\sigma$
1200	0.411	1313.1p	11.95	1424.2p	6.37	1620	1.21
1210	0.578	1316.0p	11.52	1429.0	4.91	1630	1.16
1215.6	0.742	1320.0	5.37	1435.2p	6.49	1640	0.801
1220	0.764	1324.2p	11.27	1440.0	5.40	1650	0.623
1230	1.07	1327.0p	9.09	1446.6p	6.67	1660	0.620
1236.0	1.20	1330.5	6.40	1454.0	5.51	1670	0.454
1238.8p	1.51	1334.8p	10.51	1457.2p	6.69	1680	0.362
1243.1p	1.81	1339.0p	8.66	1463.5	5.02	1690	0.268
1245.5	1.53	1342.5	5.92	1467.7p	6.61	1700	0.207
1248.9p	2.38	1347.6p	12.45	1471.5	5.22	1710	0.176
1252.7p	2.39	1354.5	5.49	1480.9p	6.16	1720	0.121
1255.0	2.07	1359.4p	9.60	1489.5	4.95	1730	0.0965
1259.7p	3.32	1365.5	5.62	1492.0p	5.66	1740	0.0753
1262.0p	3.47	1369.3p	6.70	1500	4.19	1750	0.0640
1265.5	2.30	1376.0	5.47	1510	4.13	1760	0.0455
1269.3p	4.43	1379.8p	6.85	1520	4.17	1770	0.0348
1272.2p	5.04	1385.1p	5.51	1530	4.21	1780	0.0274
1275.5	2.87	1388.0	4.90	1540	3.69	1790	0.0157
1280.1p	6.38	1393.4p	5.85	1550	2.99	1800	0.0145
1283.2p	6.01	1396.0	4.83	1560	2.82	1820	0.0100
1286.0	3.54	1402.7p	6.14	1570	2.87	1840	0.00576
1291.0p	8.95	1406.5	5.33	1580	2.09	1860	0.00325
1297.0	4.24	1411.7p	6.11	1590	1.85	1880	0.00166
1302.2p	11.53	1417.7p	5.79	1600	1.76	1900	0.00082
1309.0	5.01	1419.0	5.19	1610	1.52	1950	0.00021

[a]Cross sections at wavelengths of prominent bands (peaks, p) and deep valleys between peaks in the spectral region 1,236–1,500 (Å); in other regions the differences between peaks and valleys are slight. Sources of data: Lewis and Carver (1983), Yoshino et al. (1996a), above 1,700 (Å) from Shemansky (1972). Photodissociation leads to CO($^1\Sigma^+$)+O (^{3}P) ($\lambda \leq 2,275$ (Å)) and CO($^1\Sigma^+$)+ O (^{1}D) ($\lambda \leq 1,671$ (Å)) with quantum yields of essentially unity in both cases (Welge 1974)

Table 9.18 Absorption cross sections (cm^2 molecule^{-1}) of water vapor (1142 – 1950 Å) at 298 K[a]

λ (Å)	$10^{18}\,\sigma$	λ (Å)	$10^{18}\,\sigma$	λ (Å)	$10^{19}\,\sigma$	λ (Å)	$10^{19}\,\sigma$
1142	1.67	1247.5	5.69	1319.5	62.8	1420	5.87
1151	4.57	1257.5	6.88	1328.5	42.7	1428	4.83
1161	2.08	1260	6.80	1334.5	49.8	1430	5.10
1173	8.55	1270	7.77	1340	41.9	1440	5.10
1180.5	2.53	1274.5	7.17	1350	35.4	1450	5.28
1192.5	11.52	1280	7.81	1360	23.8	1460	6.09
1205	3.53	1288	7.03	1370	19.1	1470	7.38
1215.6	14.39	1294	7.51	1380	16.6	1480	8.39
1220	18.22	1303	6.36	1390	11.5	1490	9.92
1230	2.57	1308	6.95	1400	8.58	1500	11.6
1239	13.01	1315.5	5.32	1410	7.51	1510	13.9

(continued)

Table 9.18 (continued)

λ (Å)	$10^{18}\,\sigma$	λ (Å)	$10^{18}\,\sigma$	λ (Å)	$10^{19}\,\sigma$	λ (Å)	$10^{19}\,\sigma$
1520	15.9	1630	45.5	1740	33.5	1850	0.675
1530	18.7	1640	48.0	1750	30.4	1860	0.423
1540	20.1	1650	49.7	1760	27.4	1870	0.259
1550	23.3	1660	50.1	1770	23.1	1880	0.165
1560	26.1	1670	50.1	1780	18.3	1890	0.101
1570	29.0	1680	49.6	1790	13.3	1900	0.0630
1580	32.3	1690	48.6	1800	8.53	1910	0.0464
1590	35.1	1700	47.6	1810	5.14	1920	0.0258
1600	38.3	1710	45.5	1820	3.01	1930	0.0168
1610	40.1	1720	42.0	1830	1.78	1940	0.0095
1620	43.3	1730	38.2	1840	1.16	1950	0.0058

[a]Cross sections in the region ≤1,340 (Å) at wavelengths of peaks and valleys. Data sources: 1,142–1,820 (Å), Watanabe et al. (1953); in the region 1,340–1,450 (Å) averaged with data of Cheng et al. (2004); in the region 1,450–1,820 (Å) averaged with data of Yoshino et al. (1996b) and Chung et al. (2001); in the region beyond 1,830 (Å) averages of data from Cantrell et al. (1997) and Chung et al. (2001). The continuum above 1,450 (Å) leads to photodissociation: $H_2O + h\nu \rightarrow OH + O\,(^3P)$ ($\varphi \approx 1$). The second continuum at $\lambda < 1,450$ (Å) shows increasing band structure with strong diffuse bands below 1,250 (Å)

References

Angel, G.C., J.A.R. Samson, Phys. Rev. **A 38**, 5578–5585 (1988)

Blake, A.J., J.H. Carver, G.N. Haddad, J. Quant. Spectrosc. Radiat. Transfer **6**, 451–459 (1966)

Cantrell, C.A., A. Zimmer, G.S. Tyndall, Geophys. Res. Lett. **24**, 2195–2198 (1997)

Carter, V.L., J. Chem. Phys. **56**, 4195–4205 (1972)

Cheng, B.M., C. Chung, M. Bahou, Y.P. Lee, L.C. Lee, R. van Harrevelt, M.C. van Hemert, J. Chem. Phys. **120**, 224–229 (2004)

Chung, C.Y., E.P. Chew, B.M. Cheng, M. Bahou, Y.P. Lee, Nucl. Instrum. Methods Phys. Res. **A 467/468**, 1572–1576 (2001)

Denne, D.R., J. Phys. **D 3**, 1392–1398 (1970)

Fennelly, J.A., D.G. Torr, Atomic Data Nucl. Data Tables **51**, 321–363 (1992)

Huber, K.P., G. Herzberg, *Molecular Spectra and Molecular Structure, IV. Constants of Diatomic Molecules* (Van Nostrand Reinhold Co., New York, 1979)

Karzas, W.J., R. Latter, Astrophys. J. Suppl. Soc. **6**, 167–212 (1961)

Kirby, K., E.R. Constantinides, S. Babeu, M. Oppenheimer, G.A. Victor, Atomic Data Nucl. Data Tables **23**, 63–81 (1979)

Lewis, B.R., J.H. Carver, J. Quant. Spectrosc. Radiat. Transfer **30**, 297–309 (1983)

Matsunaga, F.M., K. Watanabe, Sci. Light **16**, 31–42, 191–196 (1967)

Metzger, P.H., G.R. Cook, J. Quant. Spectrosc. Radiat. Transfer **4**, 107–116 (1964)

Richards, P.G., J.A. Fennelly, D.G. Torr, J. Geophys. Res. **99**, 8981–8992 (1994)

Samson, J.A.R., Adv. Atom. Molec. Phys. **2**, 177–216 (1966)

Samson, J.A.R., P.N. Pareek, Phys. Rev. **A 31**, 5578–5585 (1985)

Samson, J.A.R., G.H. Rayborn, P.N. Pareek, J. Chem. Phys. **76**, 393–397 (1982)

Samson, J.A.R., T. Masuoka, P.N. Pareek, G.C. Angel, J. Chem. Phys. **86**, 6128–6132 (1987)

Sansonetti, J.E., W.C. Martin, J. Phys. Chem. Ref. Data **34**, 1559–2349 (2005)

Shemansky, D.E., J. Chem. Phys. **56**, 1582–1587 (1972)

Watanabe, K., Adv. Geophys. **5**, 153–221 (1958)

Watanabe, K., F.F. Marmo, J. Chem. Phys. **25**, 965–971 (1956)

Watanabe, K., M. Zelikoff, E.C.Y. Inn, in *Absorption Coefficients of Several Atmospheric Gases*, Geophysical Research Papers 21, AFCRL Technical Report No. 53–23 (Air Force Cambridge Research Center, Cambridge, 1953)

Welge, K.H., Can. J. Chem. **52**, 1424–1435 (1974)

Yoshino, K., J.R. Esmond, Y. Sun, W.H. Parkinson, K. Ito, T. Matsui, J. Quant. Spectrosc. Radiat. Transfer **55**, 53–60 (1996a)

Yoshino, K., J.R. Esmond, W.H. Parkinson, K. Ito, T. Matsui, Chem. Phys. **211**, 387–391 (1996b)

9.5 Chemistry of the Ionosphere

Table 9.19 Characterization of the Ionosphere [a]

Region	Altitude (km) Range	Peak	Electron density (m^{-3}) Day	Night	Principal primary ion(s)	Major ion(s) observed
Topside	350–1,000	–	10^{10}–10^{11}		O^+ diffusion	O^+, H^+
F2	200–400	300	5×10^{11}	1×10^{11}	O^+	O^+
F1	140–200	170	2×10^{11}	8×10^8	N_2^+, O^+	O^+, NO^+
E	90–140	110	1×10^{11}	2×10^9	O_2^+	O_2^+, NO^+
D	65–90	70–90	1×10^9	1×10^8	N_2^+, O_2^+, NO^+	Cluster ions[b]

[a] Numerical values are approximate. Peak altitudes and electron densities vary with solar activity and with geographic latitude. Source of data: Rees (1989)

[b] Major cluster ions in the D-region are $H_3O^+(H_2O)_n$, and negative ions such as HCO_3^- and NO_3^- (also partly hydrated)

Comments: The main region of ion production is the F1 region, where absorption of solar radiation in the EUV spectral range 25–90 nm maximizes. Atmospheric constituents in this altitude regime are mostly N_2 and atomic oxygen. Photoionization leads to N_2^+ and O^+ as primary ions, but N_2^+ is largely converted to O_2^+ and NO^+ by reactions with neutral species. In the F2 region atomic oxygen is the most abundant neutral species. Although the rate of ionization is lower, electron densities are higher, because ion-electron recombination of atomic ions is much less efficient than that of molecular ions; O^+ is preferentially destroyed in ion-molecule reactions. The peak F2 ion density occurs at an altitude where chemical and diffusive losses are nearly equal. The E layer is formed by absorption of solar radiation of wavelengths 1–25 nm and $\lambda > 90$ nm (primarily the Lyman β line at 102.6 nm) that reach deeper into the atmosphere. Whereas the strength of the F1 layer decreases strongly at night, the E-layer persists, despite rapid ion-electron recombination, owing to ionization by (resonance-scattered) radiation from high altitudes: the 58.4 and 30.4 nm lines scattered from helium are a source of O^+ in the F1 region, and Lyman β at 102.6 nm is a source of O_2^+ in the E region. The production of ions in the D region is due to X-rays <1 nm in addition to ionisation of NO by the Lyman α line, which penetrates to low altitudes through a window in the O_2 absorption spectrum.

Table 9.20 Rate coefficients of positive ion-neutral reactions relevant to the ionosphere [a]

Reaction	k	Reaction	k
$H^+ + O \rightarrow O^+ + H$	3.75 (−10)	$He^+ + CO \rightarrow C^+ + O + He$	1.6 (−9)
$H^+ + O_2 \rightarrow O_2^+ + H$	1.17 (−9)	$He^+ + H_2O \rightarrow H_2O^+ + He$	5.5 (−11)
$H^+ + NO \rightarrow NO^+ + H$	1.9 (−9)	$\rightarrow H^+ + OH + He$	1.85 (−10)
$H^+ + CO_2 \rightarrow CO_2^+ + H$	3.8 (−9)	$\rightarrow OH^+ + H + He$	2.6 (−10)
$H^+ + H_2O \rightarrow H_2O^+ + H$	8.2 (−9)	$He^+ + CH_4 \rightarrow CH_4^+ + He$	3.2 (−11)
$H^+ + CH_4 \rightarrow CH_3^+ + H_2$	3.4 (−9)	$\rightarrow CH_3^+ + H + He$	8.1 (−11)
$\rightarrow CH_4^+ + H$	7.47 (−10)	$\rightarrow CH_2^+ + H_2 + He$	8.4 (−10)
$H^+ + HCN \rightarrow HCN^+ + H$	1.1 (−8)	$\rightarrow CH^+ + H_2 + H + He$	2.4 (−10)
$H_2^+ + H \rightarrow H^+ + H_2$	6.4 (−10)	$\rightarrow H^+ + CH_3 + He$	4.4 (−10)
$H_2^+ + H_2 \rightarrow H_3^+ + H$	2.0 (−9)	$He^+ + HCN \rightarrow CH^+ + N + He$	6.9 (−10)
$H_2^+ + He \rightarrow HeH^+ + H$	1.35 (−10)	$\rightarrow CN^+ + H + He$	1.55 (−9)
$H_2^+ + CH_4 \rightarrow CH_3^+ + H_2 + H$	2.28 (−9)	$\rightarrow C^+ + NH + He$	8.3 (−10)
$\rightarrow CH_4^+ + H_2$	1.41 (−9)	$\rightarrow N^+ + CH + He$	2.3 (−10)
$H_2^+ + O_2 \rightarrow O_2^+ + H_2$	7.8 (−10)	$HeH^+ + H \rightarrow H_2^+ + He$	9.1 (−10)
$\rightarrow HO_2^+ + H$	1.92 (−10)	$HeH^+ + H_2 \rightarrow H_3^+ + He$	1.77 (−9)
$H_2^+ + N_2 \rightarrow N_2H^+ + H$	2.0 (−9)	$HeH^+ + N_2 \rightarrow N_2H^+ + He$	1.7 (−9)
$H_2^+ + CO_2 \rightarrow HCO_2^+ + H$	2.35 (−9)	$HeH^+ + O_2 \rightarrow HO_2^+ + He$	1.1 (−9)
$H_2^+ + CO \rightarrow HCO^+ + H$	2.9 (−9)	$N^+ + H_2 \rightarrow NH^+ + H$	5.0 (−10)
$H_3^+ + O \rightarrow OH^+ + H_2$	8.0 (−10)	$N^+ + H_2O \rightarrow H_2O^+ + N$	2.7 (−9)
$\rightarrow H_2O^+ + H$	overall	$N^+ + O_2 \rightarrow NO^+ + O$	2.3 (−10)
$H_3^+ + O_2 \rightarrow HO_2^+ + H_2$	6.7 (−10)	$\rightarrow O_2^+ + N$	3.1 (−10)
$H_3^+ + N \rightarrow NH^+ + H_2$	2.6 (−10)	$\rightarrow O^+ + NO$	4.6 (−11)
$\rightarrow NH_2^+ + H$	3.9 (−10)	$N^+ + NO \rightarrow N_2^+ + O$	8.3 (−11)
$H_3^+ + N_2 \rightarrow N_2H^+ + H_2$	1.86 (−9)	$\rightarrow NO^+ + N$	4.7 (−10)
$H_3^+ + NO \rightarrow HNO^+ + H_2$	1.25 (−9)	$N^+ + CO_2 \rightarrow CO^+ + NO$	2.0 (−10)
$H_3^+ + NO_2 \rightarrow NO^+ + OH + H_2$	7.0 (−10)	$\rightarrow CO_2^+ + N$	9.2 (−10)
$\rightarrow NO_2^+ + H + H_2$	7.0 (−12)	$N^+ + CO \rightarrow CO^+ + N$	4.93 (−10)
$H_3^+ + CO_2 \rightarrow HCO_2^+ + H_2$	2.5 (−9)	$\rightarrow NO^+ + C$	6.2 (−11)
$H_3^+ + CO \rightarrow HCO^+ + H_2$	1.74 (−9)	$\rightarrow C^+ + NO$	5.6 (−12)
$\rightarrow HOC^+ + H_2$	1.1 (−10)	$N^+ + CH_4 \rightarrow CH_4^+ + N$	5.75 (−11)
$H_3^+ + H_2O \rightarrow H_3O^+ + H_2$	5.3 (−9)	$\rightarrow CH_3^+ H + N$	5.75 (−10)
$H_3^+ + CH_4 \rightarrow CH_5^+ + H_2$	2.4 (−9)	$\rightarrow HCNH^+ + H_2$	4.14 (−10)
$H_3^+ + HCN \rightarrow HCNH^+ + H_2$	7.5 (−9)	$\rightarrow HCN^+ + H_2 + H$	1.15 (−10)
$He^+ + H_2 \rightarrow H^+ + H + He$	8.3 (−14)	$N^+ + HCN \rightarrow HCN^+ + N$	3.7 (−9)
$\rightarrow H_2^+ + He$	1.7 (−14)	$N_2^+ + H_2 \rightarrow N_2H^+ + H$	2.0 (−9)
$He^+ + O_2 \rightarrow O^+ + O + He$	9.7 (−10)	$N_2^+ + O \rightarrow O^+ + N_2$	1.0 (−11)
$\rightarrow O_2^+ + He$	3.0 (−11)	$\rightarrow NO^+ + N$	1.4 (−10)
$He^+ + N_2 \rightarrow N^+ + N + He$	7.8 (−10)	$N_2^+ + O_2 \rightarrow O_2^+ + N_2$	5.0 (−11)
$\rightarrow N_2^+ + He$	5.2 (−10)	$N_2^+ + NO \rightarrow products$	4.1 (−10)
$He^+ + NO \rightarrow N^+ + O + He$	1.35 (−9)	$N_2^+ + CO_2 \rightarrow CO_2^+ + N_2$	8.0 (−10)
$\rightarrow O^+ + N + He$	1.0 (−10)	$N_2^+ + CO \rightarrow CO^+ + N_2$	7.3 (−11)
$He^+ + CO_2 \rightarrow CO^+ + O + He$	7.8 (−10)	$N_2^+ + H_2O \rightarrow H_2O^+ + N_2$	1.9 (−9)
$\rightarrow CO_2^+ + He$	5.0 (−11)	$\rightarrow N_2H^+ + OH$	5.0 (−10)
$\rightarrow O^+ + CO + He$	1.4 (−10)	$N_2^+ + CH_4 \rightarrow CH_2^+ + H_2 + N_2$	1.0 (−10)
$\rightarrow C^+ + O_2^+ + He$	2.0 (−11)	$\rightarrow CH_3^+ + H + N_2$	1.04 (−9)

(continued)

Table 9.20 (continued)

Reaction	k	Reaction	k
$O^+ + H \rightarrow H^+ + O$	6.4 (−10)	$OH^+ + O_2 \rightarrow O_2^+ + OH$	3.8 (−10)
$O^+ + H_2 \rightarrow OH^+ + H$	1.62 (−9)	$OH^+ + NO \rightarrow NO^+ + OH$	8.15 (−10)
$O^+ + N_2 \rightarrow NO^+ + N$	1.2 (−12)	$OH^+ + NO_2 \rightarrow NO^+ + HO_2$	1.3 (−9)
$O^+ + O_2 \rightarrow O_2^+ + O$	2.1 (−11)	$\rightarrow NO_2^+ + OH$	overall
$O^+ + NO \rightarrow NO^+ + O$	8.0 (−13)	$OH^+ + CO_2 \rightarrow HCO_2^+ + O$	1.35 (−9)
$O^+ + NO_2 \rightarrow NO_2^+ + O$	1.6 (−9)	$OH^+ + CO \rightarrow HCO^+ + H$	8.4 (−10)
$O^+ + CO_2 \rightarrow O_2^+ + CO$	1.1 (−9)	$OH^+ + H_2O \rightarrow H_2O^+ + OH$	1.59 (−9)
$O^+ + H_2O \rightarrow H_2O^+ + O$	2.6 (−9)	$\rightarrow H_3O^+ + O$	1.3 (−9)
$O^+ + CH_4 \rightarrow CH_4^+ + O$	8.9 (−10)	$H_2O^+ + H_2 \rightarrow H_3O^+ + H$	7.6 (−10)
$\rightarrow CH_3^+ + OH$	1.1 (−10)	$H_2O^+ + O_2 \rightarrow O_2^+ + H_2O$	3.3 (−10)
$O^+ + HCN \rightarrow CO^+ + NH$	1.17 (−9)	$H_2O^+ + NO \rightarrow NO^+ + H_2O$	4.6 (−10)
$\rightarrow HCO^+ + N$	1.17 (−9)	$H_2O^+ + NO_2 \rightarrow NO_2^+ + H_2O$	1.3 (−9)
$\rightarrow NO^+ + CH$	1.17 (−9)	$H_2O^+ + CO \rightarrow HCO^+ + OH$	4.25 (−10)
$O_2^+ + N \rightarrow NO^+ + O$	1.5 (−10)	$H_2O^+ + H_2O \rightarrow H_3O^+ + OH$	1.85 (−9)
$O_2^+ + NO \rightarrow O_2 + NO^+$	4.6 (−10)	$H_2O^+ + CH_4 \rightarrow H_3O^+ + CH_3$	1.12 (−9)
$O_2^+ + NO_2 \rightarrow O_2 + NO_2^+$	6.6 (−10)	$H_2O^+ + HCN \rightarrow H_3O^+ + CN$	2.1 (−9)
$OH^+ + H_2 \rightarrow H_2O^+ + H$	9.7 (−10)	$\rightarrow HCNH^+ + OH$	overall

[a] Powers of ten in parentheses, units of rate coefficients (cm^3 molecule^{-1} s^{-1}), $T \approx 300$ K. Source of data: Anicich (1993)

Table 9.21 Rate coefficients of positive ion reactions used in models of the ionosphere in the E and F regions [a]

Reaction	k (cm^3 molecule^{-1} s^{-1})	Reaction	k (cm^3 molecule^{-1} s^{-1})
$O^+(^4S) + N_2 \rightarrow N + NO^+$	1.533 (−12)	$O^+(^2P) + e^- \rightarrow O(^4S) + e^-$	1.7 (−7) $(300/T_e)^{0.5}$
	−5.92 (−13)$(T_f/300)$	$O^+(^2P) + O \rightarrow O(^4S) + O$	5.2 (−11)
	+ 8.6 (−14)$(T_f/300)^2$	$O^+(^2P) \rightarrow O^+(^2D) + h\nu$	0.173 s^{-1}
	$(T_f < 1{,}700$ K)	$O^+(^2P) \rightarrow O^+(^4S) + h\nu$	0.047 s^{-1}
$O^+(^2D) + N_2 \rightarrow O + N_2^+$	8.0 (−10)	$N_2^+ + O \rightarrow N_2 + O^+$	2.0 (−11)
$O^+(^2P) + N_2 \rightarrow O + N_2^+$	4.8 (−10)	$N_2^+ + O \rightarrow N + NO^+$	1.4 (−10)$(T/300)^{−0.44}$
$O^+(^4S) + N(^2D) \rightarrow O + N^+$	1.3 (−10)	$N_2^+ + O_2 \rightarrow N_2 + O_2^+$	5.0 (−11)$(T/300)^{−0.8}$
$O^+(^4S) + O_2 \rightarrow O + O_2^+$	2.82 (−11)	$N^+ + O \rightarrow N + O^+$	1.0 (−12)
	−7.74 (−12)$(T_f/300)$	$N^+ + O_2 \rightarrow O + NO^+$	2.6 (−10)
	+ 1.073 (−12)$(T_f/300)^2$	$N^+ + O_2 \rightarrow N + O_2^+$	3.1 (−10)
	− 5.17(−14)$(T_f/300)^3$	$O_2^+ + NO \rightarrow O_2 + NO^+$	4.4 (−10)
	+ 9.65 (−16) $(T_f/300)^4$	$O_2^+ + N \rightarrow O + NO^+$	1.2 (−10)
$O^+(^2D) + O_2 \rightarrow O + O_2^+$	7.0 (−10)	$O^+(^4S) + e^- \rightarrow O + h\nu$	4.0 (−12)$(300/T_e)^{0.7}$
$O^+(^2P) + O_2 \rightarrow O + O_2^+$	4.8 (−10)	$N_2^+ + e^- \rightarrow N + N$	1.8 (−7) $(300/T_e)^{0.39}$
$O^+(^2D) + O \rightarrow O(^4S) + O$	1.0 (−11)	$O_2^+ + e^- \rightarrow O + O$	1.95 (−7) $(300/T_e)^{0.7}$
$O^+(^2D) + e^- \rightarrow O(^4S) + e^-$	6.6 (−8) $(300/T_e)^{0.5}$	$NO^+ + e^- \rightarrow N + O$	4.0 (−7) $(300/T_e)^{0.9}$

[a] Powers of ten in parentheses; $T_f = (m_i T_n + m_n T_i)/(m_n + m_i)$, where T is temperature, m is mass and the subscripts i and n refer to ions and neutrals, respectively; T_e is electron temperature. Source of data: Buonsanto et al. (1992), Torr (1985)

Table 9.22 Ion reactions involving sodium and iron species, $T \approx 300$ K[a]

Reaction	k	Reaction	k
$N_2^+ + Na \rightarrow Na^+ + N_2$	1.3 (−9)	$O_2^+ + Fe \rightarrow Fe^+ + O_2$	1.1 (−9)
$O_2^+ + Na \rightarrow Na^+ + O_2$	2.9 (−9)	$NO^+ + Fe \rightarrow Fe^+ + NO$	9.2 (−10)
$NO^+ + Na \rightarrow Na^+ + NO$	8.0 (−10)	$H_2O^+ + Fe \rightarrow Fe^+ + H_2O$	1.5 (−9)
$H_2O^+ + Na \rightarrow Na^+ + H_2O$	6.2 (−9)	$Fe^+ + O_3 \rightarrow FeO^+ + O_2$	3.4 (−10)[c]
$Na^+ + N_2 + M \rightarrow Na^+ \cdot N_2 + M$	4.8 (−30)	$Fe^+ + O_2 + M \rightarrow FeO_2^+ + M$	1.7 (−29)
$Na^+ \cdot N_2 + X \rightarrow Na^+ \cdot X + N_2$	8.0 (−10)[b]	$Fe^+ + N_2 + M \rightarrow Fe^+ \cdot N_2 + M$	8.0 (−30)
$Na^+ \cdot N_2 + O \rightarrow NaO^+ + N_2$	6.0 (−10)	$Fe^+ \cdot N_2 + O \rightarrow FeO^+ + N_2$	7.0 (−10)
$NaO^+ + O \rightarrow Na^+ + O_2$	8.0 (−10)	$FeO_2^+ + O \rightarrow FeO^+ + O_2$	6.0 (−10)
$N_2^+ + Fe \rightarrow Fe^+ + N_2$	4.3 (−10)	$FeO^+ + O \rightarrow Fe^+ + O_2$	2.0 (−11)

[a]Powers of ten in parentheses, units of rate coefficients (cm^3 molecule^{-1} s^{-1}) for bimolecular reactions and (cm^6 molecule^{-2} s^{-1}) for termolecular reactions. Sources of data: Plane et al. (1999, 2003). Metals in the upper atmosphere arise from meteorite ablation
[b]$X = CO_2$ or H_2O
[c]The temperature dependence of the rate coefficient is $k = 7.6$ (−10) exp(−241/T)

Table 9.23 D-region positive ion chemistry and rate coefficients[a]

Reaction	k^*	α	B
$N_2^+ + O_2 \rightarrow O_2^+ + N_2$	5.1 (−10)		
$O_2^+ + O_2 + M \rightarrow O_4^+ + M$	4.0 (−30)	2.93	
$O_2^+ + N_2 + M \rightarrow N_2 \cdot O_2^+ + M$	1.0 (−30)	3.2	
$N_2 \cdot O_2^+ + O_2 \rightarrow O_4^+ + N_2$	5.0 (−10)		
$O_4^+ + M \rightarrow O_2^+ + O_2 + M$	2.8 (−5)	3.93	5,400
$N_2 \cdot O_2^+ + M \rightarrow O_2^+ + N_2 + M$	1.7 (−7)	4.2	2,700
$O_4^+ + O \rightarrow O_2^+ + O_3$	3.0 (−10)		
$O_4^+ + H_2O \rightarrow O_2^+ \cdot H_2O + O_2$	1.7 (−9)		
$O_2^+ \cdot H_2O + H_2O \rightarrow H_3O^+ \cdot OH + O_2$	9.0 (−10)		
$H_3O^+ \cdot OH + H_2O \rightarrow H^+(H_2O)_2 + OH$	2.0 (−9)		
$O_2^+ + H_2O_2 \rightarrow H_2O_2^+ + O_2$	1.5 (−9)		
$H_2O_2^+ + H_2O \rightarrow H_3O^+ + HO_2$	1.7 (−9)		
$NO^+ + N_2 + M \rightarrow N_2 \cdot NO^+ + M$	3.0 (−31)	4.7	
$NO^+ + CO_2 + M \rightarrow CO_2 \cdot NO^+ + M$	1.4 (−29)	4.7	
$N_2 \cdot NO^+ + M \rightarrow NO^+ + N_2 + M$	1.5 (−8)	5.3	2,093
$CO_2 \cdot NO^+ + M \rightarrow NO^+ + CO_2 + M$	6.2 (−7)	5.0	4,590
$NO^+ + H_2O + M \rightarrow NO^+ \cdot H_2O + M$	1.6 (−28)	4.7	
$CO_2 \cdot NO^+ + H_2O \rightarrow NO^+ \cdot H_2O + CO_2$	1.0 (−9)		
$N_2 \cdot NO^+ + CO_2 \rightarrow CO_2 \cdot NO^+ + N_2$	1.0 (−9)		
$NO^+ \cdot H_2O + N_2 + M \rightarrow NO^+ \cdot H_2O \cdot N_2 + M$	2.0 (−31)	4.4	
$NO^+ \cdot H_2O \cdot N_2 + M \rightarrow NO^+ \cdot H_2O + N_2 + M$	6.3 (−8)	5.4	2,150
$NO^+ \cdot H_2O + CO_2 + M \rightarrow NO^+ \cdot H_2O \cdot CO_2 + M$	7.0 (−30)	4.0	
$NO^+ \cdot H_2O \cdot CO_2 + M \rightarrow NO^+ \cdot H_2O + CO_2 + M$	3.8 (−6)	5.0	4,025
$NO^+ \cdot H_2O \cdot N_2 + CO_2 \rightarrow NO^+ \cdot H_2O \cdot CO_2 + N_2$	1.0 (−9)		
$NO^+ \cdot H_2O \cdot CO_2 + H_2O \rightarrow NO^+(H_2O)_2 + CO_2$	1.0 (−9)		
$NO^+ \cdot H_2O + H_2O + M \rightarrow NO^+(H_2O)_2 + M$	1.0 (−27)	4.7	
$NO^+(H_2O)_2 + N_2 + M \rightarrow NO^+(H_2O)_2 \cdot N_2 + M$	2.0 (−31)	4.4	
$NO^+(H_2O)_2 \cdot N_2 + M \rightarrow NO^+(H_2O)_2 + N_2 + M$	6.3 (−8)	5.4	1,800

(continued)

Table 9.23 (continued)

Reaction	$k*$	α	B
$NO^+(H_2O)_2+CO_2+M \rightarrow NO^+(H_2O)_2 \cdot CO_2+M$	7.0 (−30)	3.0	
$NO^+(H_2O)_2 \cdot CO_2+M \rightarrow NO^+(H_2O)_2+CO_2+M$	3.8 (−6)	5.0	3,335
$NO^+(H_2O)_2 \cdot N_2+CO_2 \rightarrow NO^+(H_2O)_2 \cdot CO_2+N_2$	1.0 (−9)		
$NO^+(H_2O)_2 \cdot CO_2+H_2O \rightarrow NO^+(H_2O)_3+CO_2$	1.0 (−9)		
$NO^+(H_2O)_2+H_2O+M \rightarrow NO^+(H_2O)_3+M$	9.0 (−28)	4.7	
$NO^+(H_2O)_3+H_2O \rightarrow H^+(H_2O)_3+HNO_2$	1.0 (−9)		
$H_3O^+ + N_2 + M \rightarrow H_3O^+ \cdot N_2+M$	3.5 (−31)	4.0	
$H_3O^+ \cdot N_2+M \rightarrow H_3O^+ + N_2+M$	1.0 (−8)	5.4	2,800
$H_3O^+ + CO_2 + M \rightarrow H_3O^+ \cdot CO_2+M$	8.5 (−28)	4.0	
$H_3O^+ \cdot CO_2+M \rightarrow H_3O^+ + CO_2+M$	5.5 (−3)	5.0	7,700
$H_3O^+ \cdot N_2+CO_2 \rightarrow H_3O^+ \cdot CO_2+N_2$	1.0 (−9)		
$H_3O^+ \cdot CO_2+H_2O \rightarrow H^+(H_2O)_2+CO_2$	1.0 (−9)		
$H_3O^+ + H_2O+M \rightarrow H^+(H_2O)_2+M$	4.6 (−27)	4.0	
$H^+(H_2O)_2+M \rightarrow H_3O^+ + H_2O+M$	2.5 (−2)	5.0	15,900
$H^+(H_2O)_2+N_2+M \rightarrow H^+(H_2O)_2 \cdot N_2+M$	3.5 (−31)	4.0	
$H^+(H_2O)_2 \cdot N_2+M \rightarrow H^+(H_2O)_2+N_2+M$	1.2 (−8)	5.4	2,700
$H^+(H_2O)_2+CO_2+M \rightarrow H^+(H_2O)_2 \cdot CO_2+M$	8.5 (−28)	4.0	
$H^+(H_2O)_2 \cdot CO_2+M \rightarrow H^+(H_2O)_2+CO_2+M$	1.0 (−3)	5.0	6,200
$H^+(H_2O)_2 \cdot N_2+CO_2 \rightarrow H^+(H_2O)_2 \cdot CO_2+N_2$	1.0 (−9)		
$H^+(H_2O)_2 \cdot CO_2+H_2O \rightarrow H^+(H_2O)_3+CO_2$	1.0 (−9)		
$H^+(H_2O)_2+H_2O+M \rightarrow H^+(H_2O)_3+M$	8.6 (−27)	7.5	
$H^+(H_2O)_3+M \rightarrow H^+(H_2O)_2+H_2O+M$	1.2 (−2)	8.5	9,800
$H^+(H_2O)_3+H_2O+M \rightarrow H^+(H_2O)_4+M$	3.6 (−27)	8.1	
$H^+(H_2O)_4+M \rightarrow H^+(H_2O)_3+H_2O+M$	1.5 (−1)	9.1	9,000
$H^+(H_2O)_4+H_2O+M \rightarrow H^+(H_2O)_5+M$	4.6 (−28)	14.0	
$H^+(H_2O)_5+M \rightarrow H^+(H_2O)_4+H_2O+M$	1.7 (−3)	15.0	6,400
$H^+(H_2O)_5+H_2O+M \rightarrow H^+(H_2O)_6+M$	5.8 (−29)	15.3	
$H^+(H_2O)_6+M \rightarrow H^+(H_2O)_5+H_2O+M$	4.0 (−3)	16.3	5,800
$H^+(H_2O)_6+H_2O+M \rightarrow H^+(H_2O)_7+M$	9.0 (−28)	15.3	
$H^+(H_2O)_7+M \rightarrow H^+(H_2O)_6+H_2O+M$	1.3 (−2)	16.3	5,400

[a]Powers of ten in parentheses, $M=N_2$ or O_2, $k=k*(300/T)^\alpha \exp(-B/T)$ in units of (cm^3 molecule^{-1} s^{-1}) for bimolecular reactions and (cm^6 molecule^{-2} s^{-1}) for termolecular reactions.
Source: Kopp (1996)

Table 9.24 D-region negative ion chemistry and rate coefficients[a]

Reaction	k	Reaction	k
$O_2+O_2+e^- \rightarrow O_2^-+O_2$	4.0 (−30)	$O^-+O_2(^1\Delta_g) \rightarrow O_3+e^-$	3.0 (−10)
$O_2+N_2+e^- \rightarrow O_2^-+N_2$	1.0 (−31)	$O^-+M \rightarrow O+e^-+M$	1.2 (−12)
$O_3+e^- \rightarrow O^-+O_2$	9.1 (−12)	$O^-+CH_4 \rightarrow OH^-+CH_3$	1.0 (−10)
$O^-+O_3 \rightarrow O_3^-+O$	8.0 (−10)	$O^-+HCl \rightarrow Cl^-+OH$	2.0 (−9)
$O^-+CO_2+M \rightarrow CO_3^-+M$	2.8 (−28)	$O^-+H_2 \rightarrow OH^-+H$	3.2 (−11)
$O^-+NO_2 \rightarrow NO_2^-+O$	1.0 (−9)	$O^-+H_2 \rightarrow H_2O+e^-$	5.8 (−10)
$O^-+O \rightarrow O_2+e^-$	1.9 (−10)	$Cl^-+O_3 \rightarrow ClO^-+O_2$	5.0 (−13)
$O^-+NO \rightarrow NO_2+e^-$	3.1 (−10)	$Cl^-+NO_2 \rightarrow NO_2^-+Cl$	6.0 (−12)
$O^-+CO \rightarrow CO_2+e^-$	5.5 (−10)	$Cl^-+H \rightarrow HCl+e^-$	9.6 (−10)

(continued)

Table 9.24 (continued)

Reaction	k	Reaction	k
$Cl^-+HCl+M \rightarrow Cl^-(HCl)+M$	1.0 (−27)	$NO_2^-+HCl \rightarrow Cl^-+HNO_2$	1.4 (−9) ·
$Cl^-+H_2O+M \rightarrow Cl^-(H_2O)+M$	2.0 (−29)	$NO_3^-+O_3 \rightarrow NO_2^-+2O_2$	1.0 (−13)
$Cl^-+CO_2+M \rightarrow Cl^-(CO_2)+M$	6.0 (−29)	$NO_3^-+NO \rightarrow NO_2^-+NO_2$	1.0 (−12)
$Cl^-+HNO_3 \rightarrow NO_3^-+HCl$	1.6 (−9)	$NO_3^-+HCl \rightarrow Cl^-+HNO_3$	1.0 (−12)
$O_2^-+O \rightarrow O^-+O_2$	1.5 (−10)	$NO_3^-+H_2O+M \rightarrow NO_3^-(H_2O)+M$	1.6 (−28)
$O_2^-+O \rightarrow O_3+e^-$	1.5 (−10)	$O_2^-+NO_2 \rightarrow NO_2^-+O_2$	7.0 (−10)
$O_2^-+O_2(^1\Delta_g) \rightarrow e^-+2O_2$	2.0 (−10)	$NO_3^-(H_2O)+M \rightarrow NO_3^-+H_2O+M$	2.4 (−4)
$O_2^-+O_3 \rightarrow O_3^-+O_2$	7.8 (−10)	$O_4^-+O \rightarrow O_3^-+O_2$	4.0 (−10)
$O_2^-+HCl \rightarrow Cl^-+HO_2$	1.6 (−9)	$O_4^-+NO \rightarrow NO_3^-+O_2$	2.5 (−10)
$O_2^-+O_2+M \rightarrow O_4^-+M$	3.4 (−31)	$O_4^-+CO \rightarrow CO_3^-+O_2$	2.0 (−11)
$O_2^-+CO_2+M \rightarrow CO_4^-+M$	4.7 (−29)	$O_4^-+CO_2 \rightarrow CO_4^-+O_2$	4.3 (−10)
$O_2^-+H_2O+M \rightarrow O_2^-(H_2O)+M$	2.2 (−28)	$O_4^-+H_2O \rightarrow O_2^-(H_2O)+O_2$	1.0 (−10)
$OH^-+O \rightarrow HO_2+e^-$	2.0 (−10)	$CO_3^-+O \rightarrow O_2^-+CO_2$	1.1 (−10)
$OH^-+O_3 \rightarrow O_3^-+OH$	9.0 (−10)	$CO_3^-+H \rightarrow OH^-+CO_2$	1.7 (−10)
$OH^-+NO_2 \rightarrow NO_2^-+OH$	1.1 (−9)	$CO_3^-+NO \rightarrow NO_2^-+CO_2$	1.0 (−10)
$OH^-+H \rightarrow H_2O+e^-$	1.4 (−9)	$CO_3^-+NO_2 \rightarrow NO_3^-+CO_2$	2.0 (−10)
$OH^-+HCl \rightarrow Cl^-+H_2O$	1.0 (−10)	$CO_3^-+HNO3 \rightarrow NO_3^-+HCO_3$	8.0 (−10)
$OH^-+CH_4 \rightarrow CH_3^-+H_2O$	1.0 (−12)	$CO_3^-+HCl \rightarrow Cl^-+OH+CO_2$	3.0 (−11)
$OH^-+CO_2+M \rightarrow HCO_3^-+M$	7.6 (−28)	$CO_3^-+H_2O+M \rightarrow CO_3^-(H_2O)+M$	1.0 (−28)
$OH^-+H_2O+M \rightarrow OH^-(H_2O)+M$	2.5 (−28)	$CO_3^-(H_2O)+NO \rightarrow NO_2^-+CO_2+H_2O$	7.0 (−12)
$ClO^-+O_3 \rightarrow Cl^-+2O_2$	6.0 (−11)	$CO_3^-(H_2O)+NO_2 \rightarrow NO_3^-+CO_2+H_2O$	1.5 (−10)
$ClO^-+O_3 \rightarrow O_3^-+ClO$	1.0 (−11)	$CO_4^-+O \rightarrow CO_3^-+O_2$	1.4 (−10)
$O_3^-+O \rightarrow O_2^-+O_2$	2.5 (−10)	$CO_4^-+H \rightarrow CO_3^-+OH$	2.2 (−10)
$O_3^-+O \rightarrow 2O_2+e^-$	1.0 (−10)	$CO_4^-+O_3 \rightarrow O_3^-+CO_2+O_2$	1.3 (−10)
$O_3^-+O_3 \rightarrow 3O_2+e^-$	1.0 (−10)	$CO_4^-+NO \rightarrow NO_3^-+CO_2$	4.8 (−11)
$O_3^-+H \rightarrow OH^-+O_2$	8.4 (−10)	$CO_4^-+H_2O \rightarrow O_2^-(H_2O)+CO_2$	2.5 (−10)
$O_3^-+CO_2 \rightarrow CO_3^-+O_2$	5.5 (−10)	$CO_4^-+HCl \rightarrow ClHO_2^-+CO_2$	1.2 (−9)
$O_3^-+NO_2 \rightarrow NO_3^-+O_2$	2.8 (−10)	$O_2^-(H_2O)+CO_2 \rightarrow CO_4^-+H_2O$	5.8 (−10)
$O_3^-+NO \rightarrow NO_3^-+O$	2.6 (−12)	$O_2^-(H_2O)+NO \rightarrow NO_3^-+H_2O$	2.0 (−10)
$O_3^-+H_2O+M \rightarrow O_3^-(H_2O)+M$	2.7 (−28)	$O_2^-(H_2O)+O_3 \rightarrow O_3^-+O_2+H_2O$	8.0 (−10)
$NO_2^-+O_3 \rightarrow NO_3^-+O_2$	1.2 (−10)	$NO_3^-(H_2O)+M \rightarrow NO_3^-+H_2O+M$	6.5 (−15)ᵇ
$NO_2^-+H \rightarrow OH^-+NO$	3.0 (−10)		
$NO_2^-+HNO_3 \rightarrow NO_3^-+HNO_2$	1.6 (−9)		

[a]Powers of ten in parentheses, k at 300 K in unit of (cm^3 molecule^{-1} s^{-1}) for bimolecular reactions and (cm^6 molecule^{-2} s^{-1}) for termolecular reactions. Source: Kopp (1996)
[b]Temperature dependence: $k = 2.4$ (−10) exp (−7,300/T)

Table 9.25 Photodetachment and photodissociation coefficients (s^{-1}) of negative ions[a]

Reaction	k (s^{-1})	Reaction	k (s^{-1})
$O^-+h\nu \rightarrow O+e^-$	1.4	$NO_2^-+h\nu \rightarrow NO_2+e^-$	8.0 (−4)
$O_2^-+h\nu \rightarrow O_2+e^-$	3.8 (−1)	$NO_3^-+h\nu \rightarrow NO_2+e^-$	5.2 (−2)
$O_3^-+h\nu \rightarrow O_2+O^-$	4.7 (−1)	$CO_3^-+h\nu \rightarrow CO_2+O^-$	1.5 (−1)
$O_3^-+h\nu \rightarrow O_3+e^-$	4.7 (−2)	$CO_3^-+h\nu \rightarrow CO_3+e^-$	2.2 (−2)
$O_4^-+h\nu \rightarrow O_2+O_2^-$	2.4 (−1)	$CO_4^-+h\nu \rightarrow CO_2+O_2^-$	6.2 (−3)
$OH^-+h\nu \rightarrow OH+e^-$	1.1		

[a]Powers of ten in parentheses; Source of data: Kopp (1996)

Table 9.26 Electron-ion recombination rate coefficients[a]

Reaction	k^*	α	Reaction	k^*	α
$O_2^+ + e^- \rightarrow O+O$	1.9 (−7)	0.7	$NO^+(H_2O)_2CO_2+e^-$	2 (−6)	
$NO^+ + e^- \rightarrow N+O$	4.2 (−7)	0.9	$H_3O^+ \, OH+e^-$	1.5 (−6)	
$H_2O_2^+ + e^-$	6.0 (−7)	0.5	H_3O+e^-	6.3 (−7)	0.5
$O_4^+ + e^-$	4.2 (−6)	0.5	$H_3O^+(H_2O)+e^-$	2.5 (−6)	
$O_2^+(H_2O)+e^-$	2 (−6)		$H_3O^+(H_2O)_2+e^-$	3 (−6)	
$NO^+(N_2)+e^-$	1.4 (−6)	0.4	$H_3O^+(H_2O)_3+e^-$	3.6 (−6)	
$NO^+(CO_2)+e^-$	1.5 (−6)		$H_3O^+(H_2O)_4+e^-$	5 (−6)	
$NO^+(H_2O)+e^-$	1.5 (−6)		$H_3O^+(H_2O)_5+e^-$	5 (−6)	
$NO^+(H_2O)_2+e^-$	2 (−6)		$H_3O^+(N_2)+e^-$	1.5 (−6)	
$NO^+(H_2O)_3+e^-$	2 (−6)		$H_3O^+(CO_2)+e^-$	2 (−6)	
$NO^+(H_2O)N_2+e^-$	2 (−6)		$H_3O^+(H_2O)N_2+e^-$	1.5 (−6)	
$NO^+(H_2O)CO_2+e^-$	2 (−6)		$H_3O^+(H_2O)CO_2+e^-$	3 (−6)	

[a]Powers of ten in parentheses, $k=k^* \, (300/T_e)^\alpha$ in units of (cm^3 molecule^{-1} s^{-1}). Only measured temperature exponents are given. For other reactions the theoretical value $\alpha=0.5$ may be applied. Source of data: Kopp (1996)

References

Anicich, V.G., J. Phys. Chem. Ref. Data **22**, 1469–1569 (1993)

Buonsanto, M.J., S.C. Solomon, W.K. Tobiska, J. Geophys. Res. **97**, 10513–10524 (1992)

Kopp, E., in *The Upper Atmosphere*, ed. by W. Dieminger, G.K. Hartmann, R. Leitinger (Springer, Berlin, 1996), pp. 620–630

Plane, J.M.C., C.S. Gardner, J. Yu, C.Y. She, R.R. Garcia, H.C. Pumphrey, J. Geophys. Res. **104**, 3773–3788 (1999)

Plane, J.M.C., D.E. Self, T. Vondrak, K.R.I. Woodcock, Adv. Space Res. **32**, 699–708 (2003)

Rees, M.H., *Physics and Chemistry of the Upper Atmosphere* (Cambridge University Press, Cambridge/New York, 1989)

Torr, D.D., Chapter 5, in *The Photochemistry of Atmospheres*, ed. by J.S. Levine (Academic Press, New York, 1985), pp. 165–278

9.6   Airglow Phenomena and Spectroscopy

Spectral emissions in the upper atmosphere arise from fluorescence in the presence of solar radiation and from chemiluminescence at all times. Rayleigh scattering during the day precludes the observation of emissions from the ground, although individual lines may be detected by interferometers at very high resolution. The nightglow spectrum observable from the ground is limited to the near UV, visible and near infrared wavelength region. The dayglow spectrum and the ultraviolet part of the nightglow can be observed with instruments on rockets and satellites. Aurorae are produced by interaction with atmospheric constituents of energetic electrons and protons precipitating from the magnetosphere into the upper atmosphere. Auroral spectra resemble those of the airglow but are more intense and superimposed on the regular airglow emissions (day and night).

Table 9.27 Summary of spectra observed in the airglow and aurorae[a]

Species	Transition	Wavelength or range[a] (nm)	Altitude[a] (km)	Excitation processes[b]
Nightglow				
O_2	$a\,^1\Delta_g - X\,^3\Sigma_g^-$	1270.4, 1583.6	85–100	$2\,O+M\rightarrow O_2{}^*+M$
O_2	$b\,^1\Sigma_g^+ - X\,^3\Sigma_g^-$	761.9, 864.5	60–95	$2\,O+M\rightarrow O_2{}^*+M$
O_2	$A\,^3\Sigma_u^+ - X\,^3\Sigma_g^-$	255–390	85–105	$2\,O+M\rightarrow O_2{}^*+M$
O_2	$A'\,^3\Delta_u - a\,^1\Delta_g$	330–430	85–105	$2\,O+M\rightarrow O_2{}^*+M$
OH	$X\,^2\Pi_i\,(\upsilon'\leq9)$	550–2,000	56–100	$H+O_3\rightarrow OH^*+O_2$
NO_2	$\tilde{A}(^2B_1) - X(^2A_1)$	400–1,400	85–110	$O+NO\rightarrow NO_2{}^*$
NO	$A\,^2\Sigma^+ - X\,^2\Pi_r$	225–245	80–150	$N+O\rightarrow NO^*$
	$C\,^2\Pi_r - X\,^2\Pi_r$	190–230	80–150	$N+O\rightarrow NO^*$
O	$(^1D_2) - (^3P_{2,1})$	630.0, 636.4	> 110	$O_2{}^+ + e \rightarrow O^*+O$
	$(^1S_0) - (^1D_2)$	557.7	150–250	$O_2{}^+ + e \rightarrow O^*+O$
			85–110	$2\,O+M\rightarrow O_2\dagger+M$ $O_2\dagger+O\rightarrow O^*+O_2$
Na	$Na(^2P_{1/2,\,3/2})–Na(^2S)$	589.6, 589.0	85–110	$Na+O_3\rightarrow NaO+O_2$ $NaO+O\rightarrow Na^*+O_2$
Dayglow[c]				
O_2	$a\,^1\Delta_g - X\,^3\Sigma_g^-$	1270.4, 1583.6	50	$O_3+h\nu\rightarrow O\,(^1D)+O_2{}^*$
NO	$A\,^2\Sigma^+ - X\,^2\Pi_r$	200–280	>110	resonance fluorescence
	$C\,^2\Pi_r - X\,^2\Pi_r$	190–230	>110	resonance fluorescence
N_2	$a\,^1\Pi_g - X\,^1\Sigma_u^+$	127–170	> 110	$N_2+e_p\rightarrow N_2{}^*+e$
	$A\,^3\Sigma_u^+ - X\,^1\Sigma_u^+$	260–400	> 110	$N_2+e_p\rightarrow N_2{}^*+e$
	$C\,^3\Pi_u - B\,^3\Pi_g$	290–405	> 110	$N_2+e_p\rightarrow N_2{}^*+e$
$N_2{}^+$	$B\,^2\Sigma_u^+ - X\,^2\Sigma_g^+$	358–427	> 110	resonance fluorescence
Aurora[d]				
N_2	$B\,^3\Pi_g - A\,^3\Sigma_u^+$	530–950	> 85	$N_2+e\rightarrow N_2{}^*+e$
$N_2{}^+$	$A\,^2\Pi_u - X\,^2\Sigma_u^+$	550–1,100	> 85	$N_2+e\rightarrow N_2{}^{+*}+e$
$O_2{}^+$	$b\,^4\Sigma_g^- - a\,^4\Pi_u$	500–890	> 85	$O_2+e\rightarrow O_2{}^{+*}+e$

[a]Wavelength and altitude ranges are approximate. See the following tables for spectroscopic details. The data are assembled from the reviews of McEwan and Phillips (1975) and Rees (1989)

[b]An asterisk indicates the exited atom or molecule responsible for the observed spectroscopic transition, a dagger indicates an energy-rich intermediate species (see comments), e_p denotes a photoelectron produced with excess kinetic energy, e is a thermal electron

[c]Emissions observed at night continue during the day

[d]These emissions are observed in addition to the above spectra

Comments:

(a) The four band systems of O_2 observed in the nightglow arise from the association of oxygen atoms. The very weakly bound $^5\Pi_g$ state of oxygen has been implicated as an intermediate in populating the excited states involved in the emissions. The $^5\Pi_g$ state may also be responsible for the energy transfer to atomic oxygen to form $O\,(^1S_0)$ in the 85–110 km altitude region (Wraight 1982).

(b) Only the $\upsilon'=0$ progressions appear in the atmospheric ($a\,^1\Delta_g - X\,^3\Sigma_g^-$) and the infrared atmospheric ($b\,^1\Sigma_g^+ - X\,^3\Sigma_g^-$) band systems.

(c) The nightglow spectrum in the wavelength range 250–380 nm is dominated by the Herzberg I band (A $^3\Sigma_u^+$ – X $^3\Sigma_g^-$) system (except for the O ($^1S_0$)–($^3P_1$) line at 297.2 nm, which parallels the O ($^1S_0$)–($^1D_2$) transition at 557.7 nm). The relative population of the A $^3\Sigma_u^+$ state favors the $\upsilon'=4$, 5, 6 levels. Higher levels are mostly quenched by collisions.

(d) In the spectral region below 170 nm dayglow and aurora features are dominated by lines due to excited oxygen and nitrogen atoms as well as the strong Lyman alpha hydrogen line at 121.6 nm, in addition to the Lyman-Birge-Hopfield (a $^1\Pi_g \rightarrow X\,^1\Sigma_g^+$) band system of nitrogen.

(e) The N_2^+ (B $^2\Sigma_u^+$ – X $^2\Sigma_g^+$) transition shows primarily the 1–0, 0–0, and 0–1 bands.

Table 9.28 Spectral lines of oxygen and nitrogen atoms observed in the airglow and aurora (wavelengths and energy levels)[a]

Nominal λ (nm)	Fine structure	Term energies involved (cm⁻¹)		Nominal λ (nm)	Fine structure	Term energies involved (cm⁻¹)	
Oxygen (OI)				135.851		$^5S^0_2$	73768.200
92.2	92.201	$^1F^0_3$	124326.791			$^3P_1$	158.265
		$^1D_2$	15867.862	164.1	164.131	$^3S^0_1$	76794.978
102.7	102.576	$^3D^0_3$	97488.538			$^1D_2$	15867.862
		$^3P_2$	0.00	297.2	297.228	$^1S_0$	33792.583
	102.576	$^3D^0_2$	97488.448			$^3P_1$	158.265
		$^3P_2$	0.00	394.7	394.729	$^5P_3$	120809.848
	102.743	$^3D^0_2$	97488.448			$^5S^0_2$	95476.728
		$^3P_1$	158.265		394.748	$^5P_2$	120808.629
	102.743	$^3D^0_1$	97488.378			$^5S^0_2$	95476.728
		$^3P_1$	158.265		394.759	$^5P_2$	120807.923
	102.816	$^3D^0_1$	97488.378			$^5S^0_2$	95476.728
		$^3P_0$	226.977	557.7	557.735	$^1S_0$	33792.583
115.2	115.215	$^1D^0_2$	102661.966			$^1D_2$	15867.862
		$^1D_2$	15867.862	630.0	630.030	$^1D_2$	15867.862
130.4	130.217	$^3S^0_1$	76794.978			$^3P_2$	0.00
		$^3P_2$	0.00		636.378	$^1D_2$	15867.862
	130.486	$^3S^0_1$	76794.978			$^3P_1$	158.265
		$^3P_1$	158.265	777.4	777.194	$^5P_3$	86631.454
	130.603	$^3S^0_1$	76794.978			$^5S^0_2$	73768.200
		$^3P_0$	226.977		777.417	$^5P_2$	86627.778
135.6	135.559	$^5S^0_2$	73768.200			$^5S^0_2$	73768.200
		$^3P_2$	0.00		777.539	$^5P_1$	86625.757
135.6	135.851	$^5S^0_2$	73768.200			$^5S^0_2$	73768.200
		$^3P_1$	158.265	844.6	844.625	$^3P_0$	88631.303
	135.559	$^5S^0_2$	73768.200			$^3S^0_1$	76794.978
		$^3P_2$	0.00		844.636	$^3P_2$	88631.146
						$^3S^0_1$	76794.978

(continued)

Table 9.28 (continued)

Nominal λ (nm)	Fine structure	Term energies involved (cm^{-1})		Nominal λ (nm)	Fine structure	Term energies involved (cm^{-1})	
	844.676	$^3P_1$	88630.587		67.295	$^2P_{3/2}$	189068.514
		$^3S^0_1$	76794.978			$^2P^0_{1/2}$	40470.00
Singly ionized oxygen (OII)				71.8	71.846	$^2D_{3/2}$	165996.50
43.0	43.018	$^4P_{5/2}$	232462.724			$^2D^0_{5/2}$	26810.55
	43.004	$^4P_{3/2}$	232535.949		71.856	$^2D_{3/2}$	165996.50
		$^4S^0_{3/2}$	0.00			$^2D^0_{3/2}$	26830.57
	42.992	$^4P_{1/2}$	232602.492		71.861	$^2D_{3/2}$	165988.46
		$^4S_{3/2}$	0.00			$^2D^0_{3/2}$	26830.57
48.4	48.376	$^2P_{1/2}$	233544.59	79.7	79.668	$^2D_{5/2}$	165988.46
		$^2D^0_{3/2}$	26830.57			$^2P^0_{3/2}$	40468.01
	48.403	$^2P_{3/2}$	233430.53		79.664	$^2D_{3/2}$	165996.50
		$^2D^0_{3/2}$	26830.57			$^2P^0_{1/2}$	40470.00
53.9	53.985	$^4P_{1/2}$	185235.281		79.663	$^2D_{3/2}$	165996.50
	53.945	$^4P_{3/2}$	185340.577			$^2P^0_{1/2}$	40468.01
	53.909	$^4P_{5/2}$	185499.124	83.4	83.447	$^4P_{5/2}$	119837.21
		$^4S^0_{3/2}$	0.00			$^4S^0_{3/2}$	0.00
55.5	55.512	$^2D_{5/2}$	206971.68		83.333	$^4P_{3/2}$	120000.43
		$^2D^0_{3/2}$	26830.57			$^4S^0_{3/2}$	0.00
55.5	55.512	$^2D_{3/2}$	206972.72		83.276	$^4P_{1/2}$	120082.86
		$^2P^0_{3/2}$	26830.57			$^4S^0_{3/2}$	0.00
	55.505	$^2D_{3/2}$	206972.72	247.0	247.022	$^2P^0_{1/2}$	40470.00
		$^2D^0_{5/2}$	26810.55			$^4S^0_{3/2}$	0.00
60.1	60.059	$^2D_{5/2}$	206971.68		247.034	$^2P^0_{3/2}$	40468.01
		$^2P^0_{3/2}$	40468.01			$^4S^0_{3/2}$	0.00
	60.058	$^2D_{3/2}$	206972.72	372.6	372.604	$^2D^0_{3/2}$	26830.57
		$^2P^0_{3/2}$	40468.01			$^4S^0_{3/2}$	0.00
	60.059	$^2D_{3/2}$	206972.72		372.882	$^2D^0_{5/2}$	26810.55
		$^2P^0_{1/2}$	40470.00			$^4S^0_{3/2}$	0.00
61.6	61.706	$^2P_{1/2}$	188888.543	374.9	371.275	$^4S^0_{3/2}$	212161.881
		$^2D^0_{3/2}$	26830.57			$^4P_{1/2}$	185235.281
	61.638	$^2P_{3/2}$	189068.514		372.733	$^4S^0_{3/2}$	212161.881
		$^2D^0_{3/2}$	26830.57			$^4P_{3/2}$	185340.577
	61.630	$^2P_{3/2}$	189068.514		374.949	$^4S^0_{3/2}$	212161.881
		$^2D^0_{5/2}$	26810.58			$^4P_{5/2}$	185499.124
64.4	64.415	$^2S_{1/2}$	195710.47	434.9	431.715	$^4P^0_{3/2}$	208392.258
		$^2P^0_{3/2}$	40468.01			$^4P_{1/2}$	185235.281
	64.416	$^2S_{1/2}$	195710.47		431.963	$^4P^0_{5/2}$	208484.202
		$^2P^0_{1/2}$	40470.00			$^4P_{3/2}$	185340.577
67.3	67.376	$^2P_{1/2}$	188888.543		432.676	$^4P^0_{1/2}$	208346.104
		$^2P^0_{3/2}$	40468.01			$^4P_{1/2}$	185235.281
	67.377	$^2P_{1/2}$	188888.543		434.558	$^4P^0_{1/2}$	208346.104
		$^2P^0_{1/2}$	40470.00			$^4P_{3/2}$	185340.577
	67.295	$^2P_{3/2}$	189068.514		434.945	$^4P^0_{5/2}$	208484.202
		$^2P^0_{3/2}$	40468.01			$^4P_{5/2}$	185499.124

(continued)

Table 9.28   (continued)

Nominal λ (nm)	Fine structure	Term energies involved (cm^{-1})		Nominal λ (nm)	Fine structure	Term energies involved (cm^{-1})	
	436.692	$^4P^0_{3/2}$	208392.258		116.854	$^2F_{5/2}$	104810.360
		$^4P_{5/2}$	185499.124			$^2D^0_{5/2}$	19233.177
441.6	441.491	$^2D^0_{5/2}$	211712.732	120.0	119.955	$^4P_{5/2}$	83364.620
		$^2P_{3/2}$	189068.514			$^4S_{3/2}$	0.00
	441.600	$^2D^0_{3/2}$	211527.117	120.0	120.022	$^4P_{3/2}$	83317.830
		$^2P_{1/2}$	188888.543			$^4S_{3/2}$	0.00
	445.238	$^2D^0_{3/2}$	211527.117		120.710	$^4P_{1/2}$	83284.070
		$^2P_{3/2}$	189068.514			$^4S_{3/2}$	0.00
464.9	463.885	$^4D^0_{3/2}$	206786.286	124.3	124.318	$^2D_{5/2}$	99663.427
		$^4P_{1/2}$	185235.281			$^2D^0_{5/2}$	19224.464
464.9	464.181	$^4D^0_{5/2}$	206877.865		124.317	$^2D_{3/2}$	99663.912
		$^4P_{3/2}$	185340.577			$^2D^0_{5/2}$	19224.464
	464.914	$^4D^0_{7/2}$	207002.482		124.331	$^2D_{3/2}$	99663.912
		$^4P_{5/2}$	185499.124			$^2D^0_{3/2}$	19233.177
	465.084	$^4D^0_{1/2}$	206730.762		124.331	$^2D_{5/2}$	99663.427
		$^4P_{1/2}$	185235.281			$^2D^0_{3/2}$	19233.177
	466.164	$^4D^0_{3/2}$	206786.286	149.3	149.263	$^2P_{3/2}$	86220.510
		$^4P_{3/2}$	185340.577			$^2D^0_{5/2}$	19224.464
464.9	467.623	$^4D^0_{5/2}$	206877.865		149.282	$^2P_{3/2}$	86220.510
		$^4P_{5/2}$	185499.124			$^2D^0_{3/2}$	19233.177
732.0	732.969	$^2P^0_{1/2}$	40470.00		149.468	$^2P_{1/2}$	86137.350
		$^2D^0_{3/2}$	26830.57			$^2D^0_{3/2}$	19233.177
	733.074	$^2P^0_{3/2}$	40468.01	174.3	174.273	$^2P_{3/2}$	86220.510
		$^2D^0_{3/2}$	26830.57			$^2P^0_{3/2}$	28839.306
	732.000	$^2P^0_{3/2}$	40468.01	174.3	174.273	$^2P_{3/2}$	86220.510
		$^2D^0_{5/2}$	26810.55			$^2P^0_{3/2}$	28839.306
Nitrogen (NI)					174.526	$^2P_{1/2}$	86137.350
96.5	96.504	$^4P_{1/2}$	103622.51			$^2P^0_{3/2}$	28839.306
		$^4S_{3/2}$	0.00		174.524	$^2P_{1/2}$	86137.350
	96.463	$^4P_{3/2}$	103667.61			$^2P^0_{1/2}$	28838.920
		$^4S_{3/2}$	0.00	346.6	346.650	$^2P^0_{3/2}$	28839.306
	96.399	$^4P_{5/2}$	103735.48			$^4S_{3/2}$	0.00
		$^4S_{3/2}$	0.00		346.655	$^2P^0_{1/2}$	28838.920
113.4	113.417	$^4P_{1/2}$	88170.570			$^4S_{3/2}$	0.00
		$^4S_{3/2}$	0.00	520.0	519.790	$^2D^0_{3/2}$	19233.177
	113.442	$^4P_{3/2}$	88151.170			$^4S_{3/2}$	0.00
		$^4S_{3/2}$	0.00		520.026.	$^2D^0_{5/2}$	19224.464
	113.543	$^4P_{5/2}$	88107.260			$^4S_{3/2}$	0.00
		$^4S_{3/2}$	0.00	821.6	818.487	$^4P^0_{5/2}$	95532.150
116.8	116.745	$^2F_{7/2}$	104881.350			$^4P_{3/2}$	83317.830
		$^2D^0_{5/2}$	19224.464		818.802	$^4P^0_{3/2}$	95493.690
	116.842	$^2F_{5/2}$	104810.360			$^4P_{1/2}$	83284.070
		$^2D^0_{5/2}$	19224.464		820.036	$^4P^0_{1/2}$	95475.310
						$^4P_{1/2}$	83284.070

(continued)

Table 9.28 (continued)

Nominal λ (nm)	Fine structure	Term energies involved (cm⁻¹)	Nominal λ (nm)	Fine structure	Term energies involved (cm⁻¹)
	821.072	$^4P^0_{3/2}$ 95493.690	64.5	64.463	$^3S^0_1$ 155126.73
		$^4P_{3/2}$ 83317.830			$^3P_0$ 0.00
	822.314	$^4P^0_{1/2}$ 95475.310		64.484	$^3S^0_1$ 155126.73
		$^4P_{3/2}$ 83317.830			$^3P_1$ 48.67
	821.634	$^4P^0_{5/2}$ 95532.150		64.518	$^3S^0_1$ 155126.73
		$^4P_{5/2}$ 83364.620			$^3P_2$ 130.80
821.6	824.239	$^4P^0_{3/2}$ 95493.690	66.0	66.029	$^1P^0_1$ 166765.66
		$^4P_{5/2}$ 83364.620			$^1D_2$ 15316.17
868.0	868.028	$^4D^0_{7/2}$ 94881.820	67.1	67.102	$^3P^0_2$ 149076.52
		$^4P_{5/2}$ 83364.620			$^3P_1$ 48.67
	868.340	$^4D^0_{5/2}$ 94830.890		67.139	$^3P^0_2$ 149076.52
		$^4P_{3/2}$ 83317.830			$^3P_2$ 130.80
	868.615	$^4D^0_{3/2}$ 94793.490		67.141	$^3P^0_1$ 148940.17
		$^4P_{1/2}$ 83284.070			$^3P_0$ 0.00
	870.325	$^4D^0_{1/2}$ 94770.880		67.163	$^3P^0_1$ 148940.17
		$^4P_{1/2}$ 83284.070			$^3P_1$ 48.67
	871.170	$^4D^0_{3/2}$ 94793.490		67.177	$^3P^0_0$ 148908.59
		$^4P_{3/2}$ 83317.830			$^3P_1$ 48.67
	871.883	$^4D^0_{5/2}$ 94830.890		67.200	$^3P^0_1$ 148940.17
		$^4P_{5/2}$ 83364.620			$^3P_2$ 130.80
	872.889	$^4D^0_{1/2}$ 94770.880	74.6	74.584	$^1P^0_1$ 166765.66
		$^4P_{3/2}$ 83317.830			$^1S^0_0$ 32688.64
	874.736	$^4D^0_{3/2}$ 94793.490	74.7	74.698	$^1P^0_1$ 149187.80
		$^4P_{3/2}$ 83317.830			$^1D_2$ 15316.17
1040.0	1039.77	$^2P^0_{3/2}$ 28839.306	91.6	91.561	$^3P^0_1$ 109216.44
		$^2D^0_{5/2}$ 19224.464			$^3P_0$ 0.00
	1040.73	$^2P^0_{3/2}$ 28839.306	91.6	91.596	$^3P^0_0$ 109223.34
		$^2D^0_{3/2}$ 19233.177			$^3P_1$ 48.67
	1040.76	$^2P^0_{1/2}$ 28838.920		91.602	$^3P^0_2$ 109216.93
		$^2D^0_{3/2}$ 19233.177			$^3P_1$ 48.67
Singly ionized nitrogen (NII)				91.602	$^3P^0_1$ 109216.44
52.9	52,935	$^3P^0_1$ 188909.17			$^3P_1$ 48.67
		$^3P_0$ 0.00		91.671	$^3P^0_2$ 109216.93
	52,941	$^3P^0_0$ 188937.24			$^3P_2$ 130.80
		$^3P_1$ 48.67		91.671	$^3P^0_1$ 109216.44
52.9	52,949	$^3P^0_1$ 188909.17			$^3P_2$ 130.80
		$^3P_1$ 48.67	108.5	108.399	$^3D_1$ 92251.46
	52.964	$^3P^0_2$ 188857.37			$^3P_0$ 0.00
		$^3P_1$ 48.67		108.457	$^3D_1$ 92251.46
52.9	52.972	$^3P^0_1$ 188909.17			$^3P_1$ 48.67
		$^3P_2$ 130.80		108.458	$^3D_2$ 92249.91
	52.987	$^3P^0_2$ 188857.37			$^3P_1$ 48.67
		$^3P_2$ 130.80		108.553	$^3D_1$ 92251.46
					$^3P_2$ 130.80

(continued)

Table 9.28   (continued)

Nominal λ (nm)	Fine structure	Term energies involved (cm^{-1})		Nominal λ (nm)	Fine structure	Term energies involved (cm^{-1})	
	108.555	$^3D_2$	92249.91		504.072	$^3F^0_2$	186511.58
		$^3P_2$	130.80			$^3D_3$	166678.64
	108.571	$^3D_3$	92236.46	504.5	500.270	$^3S_1$	168892.21
		$^3P_2$	130.80			$^3P^0_0$	148908.59
306.0	306.284	$^1S^0_0$	32688.64		501.062	$^3S_1$	168892.21
		$^3P_1$	48.67			$^3P^0_1$	148940.17
463.0	460.148	$^3P_2$	170666.23	504.5	504.510	$^3S_1$	168892.21
		$^3P^0_1$	148940.17			$^3P^0_2$	149076.52
	460.716	$^3P_1$	170607.89	567.9	566.663	$^3D_2$	166582.45
		$^3P^0_0$	148908.59			$^3P^0_1$	148940.17
	461.387	$^3P_1$	170607.89		567.602	$^3D_1$	166521.69
		$^3P^0_1$	148940.17			$^3P^0_0$	148908.59
463.0	462.139	$^3P_0$	170572.61		567.956	$^3D_3$	166678.64
		$^3P^0_1$	148940.17			$^3P^0_2$	149076.52
	463.054	$^3P_2$	170666.23		568.621	$^3D_1$	166521.69
		$^3P^0_2$	149076.52			$^3P^0_1$	148940.17
	464.308	$^3P_1$	170607.89		571.077	$^3D_2$	166582.45
		$^3P^0_2$	149076.52			$^3P^0_2$	149076.52
500.1	500.113	$^3F^0_2$	186511.58		573.064	$^3D_1$	166521.69
		$^1D_1$	166521.69			$^3P^0_2$	149076.52
	500.148	$^3F^0_3$	186570.98	575.5	575.464	$^1S_0$	32688.64
		$^3D_2$	166582.45			$^1D_2$	15316.17
	500.515	$^3F^0_4$	186652.49	658.3	654.801	$^1D_2$	15316.17
		$^3D_3$	166678.64			$^3P_1$	48.67
	501.638	$^3F^0_2$	186511.58		658.344	$^1D_2$	15316.17
		$^3D_2$	166582.45			$^3P_2$	130.80
	502.567	$^3F^0_3$	186570.98				
		$^3D_3$	166678.64				

[a]Observed transitions according to Rees (1989), energy levels from Sansonetti and Martin (2005), fine structure wavelengths are calculated, with values greater than 200 nm corrected from vacuum to air: $\lambda_{air} = \lambda_{vac}/n$, where n is the (wavelength-dependent) refractive index of air. Additional transitions given by Rees (1989) are for OI: 81.1, 87.8, 99.9, 115.2, 121.8, 436.8, 496.8, 799.0; OII 44.2; NI: 95.3, 110, 106.9, 116.4, 117.7, 118.9, 493.5; NII: 62.9, 63.5, 214.3, 500.6 (nominal wavelengths (nm))

Table 9.29 Term energies, life times, vibrational constants, and band systems of O_2, N_2 and NO observed in airglow and aurora[a]

	Term	T_e (cm^{-1})	Life time	ω_e	$\omega_e x_e$	Transition
O_2	B $^3\Sigma_g^-$	49793.3		709.31	10.65	Schumann-Runge (B $^3\Sigma_g^- \leftarrow$ X $^3\Sigma_g^-$)
	A $^3\Sigma_u^+$	35397.8	0.18 s	799.7	12.16	Herzberg I (A $^3\Sigma_u^+ \rightarrow$ X $^3\Sigma_g^-$)
	A' $^3\Delta_{u3}$	34605.8	2 s	817.6	12.76	Chamberlain (A' $^3\Delta_u \rightarrow$ a $^1\Delta_g$)
	A' $^3\Delta_{u2}$	34756.2	1.5 s			Herzberg III (A' $^3\Delta_u \rightarrow$ X $^3\Sigma_g^-$)
	c $^1\Sigma_u^-$	33057.3	3 s	794.2	12.73	Herzberg II (c $^1\Sigma_u^- \rightarrow$ X $^3\Sigma_g^-$)
	b $^1\Sigma_g^+$	13195.1	12 s	1432.77	14.0	Atmospheric (b $^1\Sigma_g^+ \rightarrow$ X $^3\Sigma_g^-$)
	a $^1\Delta_g$	7918.1	4 ks	1510.23	13.37	IR Atmospheric (a $^1\Delta_g \rightarrow$ X $^3\Sigma_g^-$)
	X $^3\Sigma_g^-$	0.0		1580.19	11.98	
O_2^+	b $^4\Sigma_g^-$	49552	1.15 μs	1196.77	17.09	1st negative (b $^4\Sigma_g^- \rightarrow$ a $^4\Pi_u$)
	a $^4\Pi_u$	32964		1035.69	10.39	
	X $^2\Pi_g$	0.0		1904.77	16.26	
N_2	C $^3\Pi_u$	89136.9	40 ns	2047.18	28.445	2nd positive (C $^3\Pi_u \rightarrow$ B $^3\Pi_u$)
	a $^1\Pi_g$	69283.1	100 μs	1694.20	13.949	LBH (a $^1\Pi_g \rightarrow$ X $^1\Sigma_g^+$)
	B $^3\Pi_g$	59619.3	5 μs	1733.39	14.122	1st positive (B $^3\Pi_u \rightarrow$ A $^3\Sigma_g^+$)
	A $^3\Sigma_u^+$	50203.6	2.5 s	1460.64	13.87	Vegard-Kaplan (A $^3\Sigma_g^+ \rightarrow$ X $^1\Sigma_g^+$)
	X $^1\Sigma_g^+$	0.0		2358.57	14.324	
N_2^+	B $^2\Sigma_u^+$	25461.4	60 ns	2419.84	23.19	1st negative (B $^2\Sigma_g^+ \rightarrow$ X $^2\Sigma_g^+$)
	A $^2\Pi_u$	9166.9	14 μs	1903.70	15.02	Meinel bands (A $^2\Pi_u \rightarrow$ X $^2\Sigma_g^+$)
	X $^2\Sigma_g^+$	0.0				
NO	C $^2\Pi_r$	52126.0	20 ns	2395	15	δ-bands (C $^2\Pi_r \rightarrow$ X $^2\Pi_r$)
	B $^2\Pi_r$	45913.6	1.8 μs	1037.2	7.70	β-bands (B $^2\Pi_r \rightarrow$ X $^2\Pi_r$)
	A $^2\Sigma^+$	43965.7	210 ns	2374.31	16.106	γ-bands (A $^2\Sigma^+ \rightarrow$ X $^2\Pi_r$)
	X $^2\Pi_r$	0.0		1904.20	14.075	

[a]Energy units of term electronic energy T_e and vibrational constants ω_e, $\omega_e x_e$ are (cm^{-1}). Data sources: Huber and Herzberg (1979), Slanger and Huestis (1983), Bates (1988); lifetimes are approximate; they depend on the vibrational level and the occurrence of predissociation. Vibrational energy levels can be calculated from the expression $G(\upsilon) = T_e + \omega_e(\upsilon + \frac{1}{2}) - \omega_e x_e(\upsilon + \frac{1}{2})^2$

Table 9.30 Band origins of the O_2 Herzberg I (A $^3\Sigma_u^+ \rightarrow$ X $^3\Sigma_g^-$) system[a]

$\upsilon' \backslash \upsilon'' =$	0	1	2	3	4	5	6	7	8	9	10	11	12
0	285.6	298.9	313.2	328.7	345.7	364.1	**384.3**	**406.5**	**431.1**	458.3	488.8	523.1	562.0
1	279.4	292.1	305.8	320.6	336.7	**354.1**	**373.2**	**394.1**	417.2	442.6	471.0	502.8	538.6
2	273.7	285.9	299.0	313.1	328.4	**345.0**	**363.1**	**382.9**	404.6	428.5	455.1	484.6	517.8
3	268.5	280.2	292.8	**306.3**	320.9	336.8	**354.0**	372.7	393.3	415.8	440.8	468.5	499.4
4	263.7	**275.0**	**287.1**	**300.1**	314.1	329.3	**345.7**	363.6	383.1	404.5	428.0	454.1	483.1
5	**259.4**	**270.3**	**281.9**	**294.5**	308.0	322.5	338.3	355.4	374.0	394.3	416.7	441.3	468.7
6	**255.4**	**266.0**	**277.3**	**289.4**	302.4	316.5	**331.6**	348.0	365.9	385.3	406.6	430.1	456.0
7	**251.9**	**262.2**	**273.2**	**284.9**	297.5	311.1	**325.7**	341.5	358.7	377.3	397.8	420.2	444.9
8	248.9	**258.9**	**269.6**	**281.1**	293.3	306.5	320.6	335.9	352.4	370.3	389.8	422.2	434.3
9	246.3	**256.2**	**266.6**	277.8	289.8	302.6	316.4	331.3	347.3	364.7	383.6	404.2	426.7
10	244.3	254.0	264.3	275.2	287.0	299.6	313.1	327.6	343.3	360.3	378.7	398.7	420.6

[a]Observed transitions are accentuated, wavelengths (nm) in air; Data sources: Degen (1977), Rees (1989)

Table 9.31 Band origins in the O_2 Herzberg II ($c\ ^1\Sigma_u^- \rightarrow X\ ^3\Sigma_g^-$) system [a]

$v'\setminus v''$	0	1	2	3	4	5	6
0	306.1	321.4	338.0	356.2	376.1	398.1	422.3
1	299.0	313.6	329.5	346.7	365.6	386.3	409.1
2	292.5	306.5	321.6	338.0	355.9	375.5	397.0
3	286.6	300.0	314.4	330.1	347.2	**365.8**	386.1
4	281.1	284.0	307.8	**322.9**	**339.1**	**356.9**	376.2
5	276.1	288.5	301.8	**316.2**	**331.8**	**348.8**	367.3
6	271.5	283.4	296.3	**310.2**	325.2	341.5	**359.2**

[a]Observed transitions are accentuated, wavelengths (nm) in air; Data sources: Gadsden (1996), Slanger and Huestis (1981)

Table 9.32 Band origins in the O_2 Chamberlain ($A'\ ^3\Delta_u - a\ ^1\Delta_g$) system [a]

$v'\setminus v''$	0	1	2	3	4	5	6	7	8
0	379.5	402.1	427.2	455.0	486.0	520.9	560.3	605.2	656.7
1	368.4	389.8	413.2	439.3	**468.0**	500.3	536.5	577.5	624.3
2	358.4	378.5	400.6	**424.9**	451.9	**481.9**	515.4	553.1	595.9
3	349.2	**368.3**	**389.2**	**412.1**	**437.4**	**465.4**	496.6	531.6	570.9
4	340.9	**359.0**	**378.9**	**400.6**	**424.4**	450.8	480.0	512.6	549.1
5	333.4	**350.8**	**369.7**	**390.3**	**412.9**	437.8	465.3	495.9	429.9
6	326.8	**343.4**	**361.5**	**381.2**	402.8	426.4	452.5	481.9	513.3

[a]Observed transitions are accentuated, wavelengths (nm) in air; Data source: Slanger (1979)

Table 9.33 Origins of the OH rotation-vibration bands [a]

$v'\setminus v''$	0	1	2	3	4	5	6	7	8
1	2801.3								
2	1434.0	2937.8							
3	**979.0**	1505.0	3085.9						
4	**752.3**	1028.5	1582.6	3248.6					
5	**617.0**	**791.2**	1082.8	1668.2	3429.0				
6	**527.4**	**649.8**	**834.1**	1143.2	1763.8	3632.1			
7	464.1	**556.2**	**686.1**	**882.2**	1211.2	1872.5	3865.3		
8	417.3	490.3	**588.5**	**727.3**	**937.0**	1289.3	1998.8	4139.3	
9	381.7	441.9	**520.1**	**625.5**	**774.7**	**1000.8**	1381.4	2189.6	4472.0

[a]Observed transitions are accentuated, wavelengths (nm) in air; Data sources: Gadsden (1996), Broadfoot and Kendall (1968)

Table 9.34 Band origins in the O_2 infrared atmospheric systems[a]

	Atmospheric bands $(b\,^1\Sigma_g^+ - X\,^3\Sigma_g^-)$					Infrared atmospheric bands $(a\,^1\Delta_g - X\,^3\Sigma_g^-)$			
$v' \backslash v''$	0	1	2	3	$v' \backslash v''$	0	1	2	3
0	**761.9**	**864.5**	996.6	1173.1	0	**1270.4**	**1583.6**	2091.4	3056.8
1	688.2	770.8	874.2	1007.1	1	1071.8	1286.5	1602.7	2114.4
2	628.7	696.7	780.2	884.4	2	929.2	1086.3	1303.5	1622.9
3	579.5	637.0	705.9	790.2	3	821.8	942.4	1101.5	1321.3

[a]Observed transitions are accentuated, wavelengths (nm) in air; Data sources: Gadsden (1996), Rees (1989)

Table 9.35 Neutral iron chemistry in the mesosphere: Reactions and rate coefficients[a]

Reaction	A	B	α
$Fe + O_3 \rightarrow FeO + O_2$	3.44 (−10)	146	0
$Fe + O_2 + M \rightarrow FeO_2 + M$	8.7 (−30)	2,038	0
$FeO + O \rightarrow Fe + O_2$	4.6 (−10)	350	0
$FeO + O_3 \rightarrow FeO_2 + O_2$	2.94 (−10)	174	0
$FeO + O_2 + M \rightarrow FeO_3 + M$	3.86 (−30)	0	0.5
$FeO + H_2O + M \rightarrow Fe(OH)_2 + M$	8.23 (−29)	0	−1.16
$FeO + CO_2 + M \rightarrow FeCO_3 + M$	3.09 (−31)	0	−1.19
$FeO_2 + O \rightarrow FeO + O_2$	1.4 (−10)	580	0
$FeO_3 + O \rightarrow FeO_2 + O_2$	2.3 (−10)	2,310	0
$FeO_2 + O_3 \rightarrow FeO_3 + O_2$	4.4 (−10)	170	0
$FeO_3 + H \rightarrow FeOH + O_2$	3.0 (−10)	796	0
$FeO_3 + H_2O \rightarrow Fe(OH)_2 + O_2$	5.0 (−12)		
$Fe(OH)_2 + H \rightarrow FeOH + H_2O$	4.4 (−10)	302	0
$FeOH + H \rightarrow FeO + H_2$	2.5 (−10)	850	0
$FeOH + H \rightarrow Fe + H_2O$	2.0 (−12)	600	0

[a]Powers of ten in parentheses; $k = A\,(T/300)^\alpha \exp(-B/T)$ in units of (cm³ molecule⁻¹ s⁻¹) for bimolecular reactions and (cm⁶ molecule⁻² s⁻¹) for termolecular reactions, respectively. Source of data: Plane et al. (2003), Plane (2003)

Table 9.36 Sodium chemistry in the mesosphere: Reactions and rate coefficients[a]

Reaction	A	B	α	Notes[b]
$Na + O_3 \rightarrow NaO + O_2$	1.1 (−9)	116	0	1
$NaO + O \rightarrow Na + O_2$	2.2 (−10) (overall)	0	0.5	2
$\rightarrow Na^* + O_2$	~2.7 (−11)			2
$NaO + O_3 \rightarrow NaO_2 + O_2$	1.1 (−9)	568	0	
$NaO + O_3 \rightarrow Na + 2\,O_2$	3.2 (−10)	550	0	
$Na + O_2 + M \rightarrow NaO_2 + M$	5.0 (−30)	0	−1.22	
$NaO_2 + O \rightarrow NaO + O_2$	5.0 (−10)	940	0	
$NaO + O_2 + M \rightarrow NaO_3 + M$	5.3 (−30)	0	−1	
$NaO_3 + O \rightarrow Na + 2\,O_2$	2.5 (−10)	0	0.5	3
$NaO + H_2O \rightarrow NaOH + OH$	4.4 (−10)	507	0	

(continued)

Table 9.36 (continued)

Reaction	A	B	α	Notes[b]
$NaO + H_2 \rightarrow NaOH + H$	1.1 (−9)	1,100	0	
$NaO + H_2 \rightarrow Na + H_2O$	1.1 (−9)	1,400	0	
$NaO + H \rightarrow Na + OH$	1.0 (−10)	668	0	
$NaO_2 + H \rightarrow Na + HO_2$	1.0 (−9)	1,000	0	
$NaO + CO_2 + M \rightarrow NaCO_3 + M$	1.3 (−27)	0	−1	
$NaCO_3 + O \rightarrow NaO_2 + CO_2$	5.0 (−10)	1,200	0	
$NaCO_3 + H \rightarrow NaOH + CO_2$	1.0 (−9)	1,400	0	
$NaOH + H \rightarrow Na + H_2O$	4.0 (−11)	550	0	
$NaOH + CO_2 + M \rightarrow NaHCO_3 + M$	1.9 (−28)	0	−1	
$NaHCO_3 + H \rightarrow Na + H_2O + CO_2$	1.0 (−12)	590	0	3
Photodissociation Reactions				
$Na + h\nu \rightarrow Na^+ + e^-$	2.0 (−5) (s^{-1})			
$NaO_2 + h\nu \rightarrow Na + O_2$	4.0 (−3) (s^{-1})			
$NaOH + h\nu \rightarrow Na + OH$	1.0 (−3) (s^{-1})			3
$NaO_3 + h\nu \rightarrow NaO + O_2$	1.0 (−4) (s^{-1})			3

[a]Powers of ten in parentheses; $k = A\,(T/T_{ref})^\alpha \exp(-B/T)$ with reference temperature $T_{ref} = 200$ K; units: (cm^3 molecule^{-1} s^{-1}) or (cm^6 molecule^{-2} s^{-1}) for bimolecular and termolecular reactions, respectively. Source of data: Plane et al. (1999), Plane (2003)

[b]Notes: (1) NaO is produced almost entirely in the long-lived, low-lying (A^2 Σ^+) excited electronic state, which is not efficiently quenched; (2) Na* designates the Na (^{2}P$_j$) excited state, Na the Na (^{2}S$_{1/2}$) ground state; (3) These values are estimates

References

Bates, D.R., Planet. Space. Sci. **36**, 875–881 (1988)

Broadfoot, A.L., K.R. Kendall, J. Geophys. Res. **73**, 426–428 (1968)

Degen, V., J. Geophys. Res. **82**, 2437–2438 (1977)

Gadsden, M., in *The Upper Atmosphere*, ed. by W. Dieminger, G.K. Hartmann, R. Leitinger (Springer, Berlin, 1996), pp. 818–828

Huber, K.P., G. Herzberg, *Molecular Spectra and Molecular Structure, IV Constants of Diatomic Molecules* (Van Nostrand Reinhold Co., New York, 1979)

McEwan, M.J., L.F. Phillips, *Chemistry of the Atmosphere* (Edward Arnold, London, 1975)

Plane, J.M.C., Chem. Rev. **103**, 4963–4984 (2003)

Plane, J.M.C., C.S. Gardner, J. Yu, C.Y. She, R.R. Garcia, H.C. Pumphrey, J. Geophys. Res. **104**, 3773–3788 (1999)

Plane, J.M.C., D.E. Self, T. Vondrak, K.R.I. Woodcock, Adv. Space Res. **32**, 699–708 (2003)

Rees, M.H., *Physics and Chemistry of the Upper Atmosphere* (Cambridge University Press, Cambridge/New York, 1989)

Sansonetti, J.E., W.C. Martin, J. Phys. Chem. Ref. Data **34**, 1559–2349 (2005)

Slanger, T.G., Chem. Phys. Lett. **66**, 344–349 (1979)

Slanger, T.G., D.L. Huestis, J. Geophys. Res. **86**, 3551–3554 (1981)

Slanger, T.G., D.L. Huestis, J. Chem. Phys. **78**, 2274–2278 (1983)

Wraight, P.C., Planet. Space Sci. **30**, 251–259 (1982)

Chapter 10
Measurement Techniques for Atmospheric Trace Species

10.1 Overview on Established Instrumental Techniques

Table 10.1 Instrumental analytical techniques for atmospheric trace gases[a]

Classification	Specific technique	Species measured
Infrared spectroscopy	Absorption, TSSL, MPAC in-situ extraction sampling	CH_4, H_2O, NO, NO_2, HNO_3, NH_3, O_3, HCHO, CO, H_2O_2, N_2O, HCl
	Absorption, FTIR, MPAC in-situ extraction sampling	H_2O, CO_2, CO, CH_4, NO_2, NO, C_2H_2, C_2H_4, CH_3COOH, CH_3OH, HCHO, HCOOH, HCN, NH_3
	Absorption, FTIR, remote column abundance	O_3, HCl, NO, NO_2, $ClONO_2$, HCl, HF
Far IR and microwave spectroscopy	Emission, FTIR, remote vertical profiles	OH, HO_2 ClO, HCl, HF (stratosphere)
UV-visible spectroscopy	Absorption, DOAS, 1–5 km path-length	OH, O_3, NO, NO_2, NO_3, HONO, NH_3, SO_2, HCHO, CS_2, ClO. OClO, BrO, OBrO, IO, OIO, I_2,
	Fluorescence	O, OH, HO_2, Cl, ClO (stratosphere)
	Laser-induced fluorescence	OH, NO, NO_2, NO_3, SO_2, HCHO, after conversion: HO_2, H_2O, ClO, BrO, $ClONO_2$, ClOOCl, peroxy nitrates, alkyl nitrates, HNO_3, N_2O_5, Hg
Mass spectrometry	Chemical ionization MS	OH, HNO_3, H_2SO_4, hydrocarbons, aldehydes, ketones, halocarbons, HO_2NO_2, alkyl nitrates, DMS, PAN
Chromatographic techniques	Gas chromatography grab sampling	Permanent gases, CO, CO_2, CH_4, hydrocarbons, halocarbons, DMS, aldehydes, ketones, alcohols, N_2O, PAN
	Liquid chromatography, sampling in cartridges or in liquids	Organic acids, inorganic ions, after derivatization: aldehydes, ketones, PAH, peroxides

(continued)

P. Warneck and J. Williams, *The Atmospheric Chemist's Companion: Numerical Data for Use in the Atmospheric Sciences*, DOI 10.1007/978-94-007-2275-0_10, © Springer Science+Business Media B.V. 2012

Table 10.1 (continued)

Classification	Specific technique	Species measured
Chemical techniques	Peroxy radical chemical Amplifier	Sum $HO_2 + RO_2$
	Chemiluminescence	O_3, NO, NO_2, isoprene
	Electrochemical cell	O_3
Electron spin resonance	Matrix isolation at low temperatures	HO_2, sum of RO_2, NO_3, NO_2, CH_3CO_3
Measurement from space	Satellites, IR and UV-visible optical techniques	O_3, N_2O, NO, NO_2, HNO_3, N_2O_5, CFCs, $ClONO_2$, HOCl, BrO, OClO, HCl, HF, CO, CO_2, CH_4, H_2O, O_4, SO_2, HCHO

[a]Abbreviations: *IR* Infrared, *UV* ultraviolet, *TSSL* Tunable solid state laser, *MPAC* Multi-path absorption cell, *FTIR* Fourier transform infrared spectroscopy, *DOAS* Differential optical absorption spectroscopy, *MS* mass spectrometry

Table 10.2 Specifications for trace gas analysis by gas chromatography[a]

Gas to be analyzed	Column type	Sample size (cm³)	Detector	Sensitivity (pmol mol⁻¹)
Light hydrocarbons $C_1 - C_6$	Open tubular PLOT type	100–3,000	FID	1–5
Isoprene/terpenes	WCOT siloxane	100–3,000	FID MS	1–10 FID 10–100 QMS full-scan <1 QMS single-ion 1–10 TOF MS
Chlorofluoro-carbons	WCOT typically methyl-polysiloxane	100–3,000	ECD MS	1–10 ECD 10–100 QMS full scan <1 QMS single-ion 1–10 TOF MS
Reactive halocarbons, organic nitrates	WCOT typically methyl-polysiloxane	100–3,000	ECD MS	1–10 ECD <1 QMS single ion <0.01 CI MS <0.01 mag-sector MS
Aromatics	WCOT, methyl-polysiloxane	100–3,000	FID MS	1–10 FID 10–100, QMS full scan <1 QMS single ion 1–10 TOF MS
PAN	packed column	1–10	ECD	10
DMS	WCOT methyl-polysiloxane	100–3,000	FID FPD	1–10
Carbonyl compounds	WCOT PLOT	100–3,000	FID MS	20–100 FID 20–200 QMS full scan 1–10 QMS single ion 1–10 TOF MS

[a]Source: Heard (2006) Abbreviations: *FID* flame ionization detector, *ECD* electron capture detector, *MS* mass spectrometry, *QMS* quadrupole mass spectrometer, *TOF* time of flight, *PLOT* porous layer open tubular, *WCOT* wall coated open tubular

Table 10.3 Instrumental techniques for the chemical analysis of atmospheric aerosol particles[a]

Classification	Specific technique	Species measured
Chromatographic techniques	Ion chromatography, aqueous extraction	Inorganic ions (cations and anions) di-carboxylic acids (C_2-C_9)
	Capillary zone Electrophoresis	Organic and inorganic ions
	Gas chromatography sequential organic solvent extraction	n-alkanes (C_{10}-C_{36}), polycyclic aromatic hydrocarbons, sesqui- and diterpenoids, after derivatization: n-alkanoic acids (C_9-C_{30}), alkenoic acids, α-hydroxy alkanoic acids, n-alcohols (C_{10}-C_{34}), others
Pyrolysis – Combustion	Determination of CO_2 by IR, GC or coulometry	Total carbon (organic plus elemental plus carbonate carbon)
Atomic absorption spectroscopy (AAS)	Vaporization of a sample at high temperatures	About 70 elements, but not C, S, and the halogens
Plasma Emission spectroscopy (ICP)	Vaporization of a sample into an argon plasma	Simultaneous detection of up to 48 elements per sample
X-ray fluorescence	Excitation by X-rays, electron or proton beams	All elements with atomic numbers 12 and higher
Electron microscopy	Analysis by X-ray emission	Allows quantification of particle size and elemental composition
Laser microprobe mass spectroscopy (LAMMS)	Vaporization and partial ionization of sample, time of flight mass spectrometry	Qualitative analysis of inorganic and some organic species
Secondary ion mass spectroscopy (SIMS)	Ion beam sputtering into a mass spectrometer	Detects elements and molecular fragments with high sensitivity

[a]Samples are collected on filters or by inertial impactors

Table 10.4a Instrumental techniques for the on-line measurement of atmospheric aerosol particles

Parameter measured	Technique	Description
Particle number concentration	Optical counting after condensational growth	Particles larger than activation diameter (typically 2–15 nm), water, butanol are typical working fluids, detection is by light scattering
Particle size distribution	Optical size measurement and counting	Particle size is determined from light scattering intensity; counting of particle number per size bin
	Aerodynamic size measurement and counting	(1) A double-laser velocimeter is used to count and measure the size-dependent velocity of particles accelerated in a nozzle (2) particles are charged, collected in a cascade impactor, and counted on each stage by means of an electrometer
	Electrical mobility size measurement and counting	Selection of particles in an electric field according to their mobility; detection is by condensation counter or electrometer

(continued)

Table 10.4a (continued)

Parameter measured	Technique	Description
Particle mass concentration	Measurement of inertia of collected mass	The mass of particles collected on an oscillating microbalance (e.g. a quartz crystal) is determined from the decrease of the Eigenfrequency of the system
	Measurement of reduction of β-radiation through a filter	The absorption of β-radiation by aerosol collected on a filter is converted to aerosol mass density

[a]This and the following table was kindly provided by F. Drewnick (Max-Planck-Institut für Chemie, Mainz)

Table 10.4b Instrumental techniques for the chemical on-line analysis of atmospheric aerosol particles

Parameter measured	Technique	Description
Multiple species analysis by mass spectrometry	Thermal desorption aerosol mass spectrometry	Particles focused into an evacuated chamber by an aerodynamic lens impact on a 600–700°C hot surface resulting in the flash vaporization of non-refractory species followed by electron impact ionization for MS. Particle sizes: ca. 40 nm–1 μm
	Laser desorption and ionization aerosol mass spectrometry	Particles focused into a vacuum chamber by an aerodynamic lens. Laser detection of particles trigger a high-intensity second laser pulse to desorb and ionize particles for mass spectral analysis; particle size range: ca. 150 nm – few μm
Multiple species analysis using chromatographic techniques	Direct dissolution of particle components coupled with ion chromatography	After removal of gas-phase species growth of particles in steam with subsequent inertial collection in impactor or cyclone. Direct transport of the solution to ion chromatograph for analysis of soluble aerosol components. Size range: >100 nm
Black carbon measurement	Light absorption measurement	The amount of light absorbed by aerosol collected on a filter is used for an estimate
Organic and elemental carbon measurement	Sample pyrolysis and CO_2 gas analysis	Particulate matter collected on quartz fiber filter is pyrolyzed in the presence and absence of O_2 using temperature ramps; OC and EC are operationally determined
Particulate sulfate measurement	Thermal conversion into SO_2 and gas analysis	Particulate matter is collected on metal strip and flash vaporized. Sulfate is converted into SO_2 and quantified using gas analyzer
Particulate nitrate measurement	Thermal conversion into NO_x and gas analysis	Particulate matter is collected on metal strip and flash vaporized. Nitrate is converted into NO_x and quantified using gas analyzer
Particulate PAH measurement	UV ionization of PAH and charge measurement	UV light from discharge lamp ionizes PAH near particle surface. Particle charge is determined using filter with electrometer

Table 10.5 Salient features of filters commonly used to collect aerosol particles[a]

Filter type	Material	Pore/fiber size (µm)	Porosity (%)	Filter thickness (mm)	Comments
Fibrous mats or weave	Cellulose, glass, quartz, polymer	0.1–100	60–99	0.15–0.5	High collection efficiency requires low air flow velocity
Micro-porous membrane	Polymer, sintered metal, ceramic	0.02–10	<85	0.05–0.2	High collection efficiency, large pressure drop
Straight-through cylindrical pores	polycarbonate films	0.1–8	5–10	0.01	Lower collection efficiency, large pressure drop
Granular bed, stationary	beads or granules special chemicals	≥200	40–60	Arbitrary	Low flow required to enhance diffusion

[a]Source of data: Willeke and Baron (1997)

Table 10.6 Origin of vibrational bands for some atmospheric species in the near and middle infrared spectral region and their integrated strength (cm^{-2} atm^{-1})[a]

Species		Band origin (cm^{-1})	(µm)	Strength integrated	Species		Band origin (cm^{-1})	(µm)	Strength integrated
H_2O	v_1	3,657	2.734	10	NO_2	v_1	1,320	7.576	2
	v_2	1,595	6.270	254		v_2	750	13.333	13
	v_3	3,756	2.662	192		v_3	1,617	6.184	1,400
CO	1-0	2,143	4.666	235		v_1+v_3	2,906	3.441	70
CO_2	v_2^1	667	15.0	200	N_2O	v_1	1,285	7.782	277
	v_3	2,349	4.257	2,230		v_2^1	589	16.978	29
CH_4	v_1	2,917	3.428	0.03		v_3	2,224	4.496	1,173
	v_2	1,534	6.519	2	HCN	v_1	2,097	4.769	0.1
	v_3	3,019	3.312	276		v_2	712	14.045	234
	v_4	1,306	7.657	136		v_3	3,312	3.019	219
O_3	v_1	1,103	9.066	10	SO_2	v_1	1,151	8.688	
	v_2	701	14.265	18		v_2	519	19.268	
	v_3	1,042	9.597	361		v_3	1,361	7.348	
HCHO	v_1	2,782	3.495	194	OCS	v_1	859	11.641	27
	v_2	1,746	5.727	301		v_2	521	19.194	12
	v_3	1,500	6.667	45		v_3	2,062	4.850	2,184
	v_4	1,167	8.569	26	HF	1-0	3,962	2.524	408
	v_5	2,843	3.517	356	HCl-35	1-0	2,886	3.465	121
NO	1.0	1,876	5.330	113	HBr	1-0	2,559	3.908	55

[a]Source of data: Heard (2006)

Table 10.7 Species measured by differential optical absorption spectroscopy (DOAS): wavelength range, $\Delta\lambda$ (nm), detection limit, DL (pmol mol^{-1}), path length required, L (km)[a]

Species	$\Delta\lambda$ (nm)	DL (pmol mol^{-1})	L (km)	Species	$\Delta\lambda$ (nm)	DL (pmol mol^{-1})	L (km)
OH	308	0.06	2	OClO	300–450	0.8	[c]
O$_3$	300–335	1,900	5	BrO	300–370	2	5
NO	200–230	50	1[b]	OBrO	450–550	1.5	[b]
NO$_2$	330–500	50	5	IO	415–450	1	5
NO$_3$	623–662	0.4	5	OIO	535–575	4	5
HONO	330–380	30	5	I$_2$	535–575	9	5
NH$_3$	200–220	150	1[b]	Benzene	230–280	200	1[b]
SO$_2$	290–310	10	5	Toluene	260–280	250	1[b]
HCHO	260–360	50	5	Naphthalene	310–320	100	5
CS$_2$	290–310	900	5	Phenol	260–280	20	1[b]
ClO	260–320	5	5	p-Cresol	260–290	50	1[b]

[a]Source of data: Heard (2006)
[b]A short path length is required, because absorption by O$_2$ and O$_3$ causes significant attenuation of light in the 200–300 nm wavelength region
[c]This compound is a stratospheric constituent, whose column abundance has been measured from the ground, by balloon-borne instruments and from satellites

References

Heard, D.E. (ed.), *Analytical Techniques for Atmospheric Measurement* (Blackwell Publishing, Oxford, 2006)

Willeke, K., P.A. Baron, (eds.), *Aerosol Measurement: Principles, Techniques and Applications* (Van Nostrand Reinhold, New York, 1997)

10.2   Description of Analytical Methods for Trace Gases

10.2.1   Chromatographic Techniques

Gas chromatography: This is a widely used technique for the separation and analysis of different species present in a gas sample, based on the principle of selective adsorption. An inert carrier gas, taken from a storage tank, flows through (a) a valve or port for the injection of an aliquot of the sample; (b) a column packed or coated with a suitable sorbent, with which the individual compounds present in the sample interact more or less strongly so that they are separated owing to different retention times; (c) a detector producing an electric signal when a compound elutes from the column and passes through the detector.

Analysis of air for trace gases generally requires sample pre-concentration and the removal of water vapor prior to the injection into a gas chromatograph. A wide variety of columns exist that allow good separation of compounds for most purposes. Common detectors are: (a) The *flame ionization detector* which generates ions when hydrocarbons and other carbon-containing compounds are burnt in a hydrogen-oxygen flame. It features not only a high sensitivity but also a linear response over several orders of magnitude; (b) The *flame photometric detector* which is a modification of the flame ionization detector designed for the detection of sulfur compounds. It utilizes the chemiluminescence at 294 nm of diatomic sulfur, which is produced when sulfur compounds are burnt in a hydrogen-oxygen flame. The emission is measured with a photomultiplier viewing the flame through an optical filter; (c) The *electron capture detector*, which is based on the variation of an electron current produced by a beta-emitter such as Ni-63 when some of the electrons are removed by the attachment (or dissociative attachment) to the compound being introduced; and (d) The *mass spectrometric detector*, which allows a mass spectrum to be obtained of the sample following electron impact ionization. (For applications see Tables 10.2 and 10.3). (e) Several other more specialized detectors are commercially available. An example is the nitrogen-selective detector, which is similar to the flame ionization detector, but instead of the ions produced in the flame it records the ionic emission from a heated ceramic bead. Another example is a detector for CO and H_2 based on the reaction with fine-grained mercury oxide, heated to about 200°C; the Hg released into the gas phase is detected by atomic absorption with light of the Hg 254 nm resonance line.

Liquid chromatography: The technique is similar to gas chromatography in that it separates compounds by the temporary adsorption on solid material in a column, but the medium is a liquid (frequently a mixture of water and another solvent); a high pressure pump (up to 20 MPa) is needed to push the mobile phase from the storage vessel through the (packed) column. Two variants of the technique are in use: *ion chromatography* and *high performance liquid chromatography*. They differ in the type of columns and detectors.

Ion chromatography separates either anions or cations present in an aqueous solution and detects them by a change in electric conductivity. An auxiliary ion exchange (suppressor) column in conjunction with the detector serves to reduce the conductivity of the mobile phase to low values so that good sensitivity is obtained. By careful selection of column materials and suitable electrolytes in the mobile phase all inorganic and many organic acids and bases are accessible to analysis. For example, the ionic constituents of the atmospheric aerosol are routinely determined by ion chromatography. The particles are collected by means of filters or inertial impactors and the ionic components are extracted from the deposits into an aqueous solution for analysis.

High precision liquid chromatography (HPLC) works largely with neutral solutions and utilizes primarily optical absorption or fluorescence for detection. Frequently, the compound to be analyzed is modified by a suitable chemical reaction to form a derivative containing a chromophore, thereby making optical detection possible. The reaction may be carried out prior to analysis or as a post-column reaction. For example, a widely used technique for the analysis of atmospheric carbonyl compounds is to draw ambient air through a cartridge impregnated with 2,4-dinitrophenyl hydrazine, which converts aldehydes and ketones to the corresponding hydrazones. These are extracted with a suitable solvent for analysis by HPLC with detection by optical absorption in the near UV spectral range. Another example is the analysis of hydrogen peroxide and organic peroxides, which are first separated by HPLC and subsequently reacted with horseradish peroxidase. One of the products is the *para*-hydroxyphenyl ethanoic acid dimer, which is detected by fluorescence at 414 nm wavelength.

Capillary zone electrophoresis: In contrast to using a sorption column to separate individual ions present in an aqueous solution, capillary zone electrophoresis exploits differences of the natural mobility of ions in an electrolyte when moving in an electric field. Optical techniques are used for detection, e.g. the decrease of attenuation in an optically absorbing medium, or the fluorescence of an ion.

10.2.2   Mass Spectrometry

This technique exploits the possibility of separating ions of different mass (actually different mass-to-charge ratio) by a suitable combination of electric and/or magnetic fields. As the controlled motion of ions requires a high vacuum, the challenge is to overcome the pressure difference between that of the sample (10^5 Pa) and that of the analyzer ($\sim 10^{-4}$ Pa). The basic setup of a mass spectrometer includes an ionization chamber, to which the gas sample is introduced, a series of ion focusing elements, the mass analyzer proper, and a detector, which usually is a secondary electron multiplier. The details of the setup may differ greatly depending on instrument design.

Ionization: The classical and most common method of ionizing a sample is by electron impact, with electrons being emitted from a hot filament and accelerated to the required energy in an electric field. The highest yield of ions is obtained with energies of 75–100 eV, which leads to fragment ion formation from molecular species. The extent of fragmentation can be greatly reduced by a technique called chemical ionization. Here, electron impact is used to form a dominant primary ion, which subsequently interacts with and ionizes the sample molecule(s) by charge transfer, proton transfer, or a similar process. For atmospheric samples, proton transfer has the advantage of avoiding ionization of the principle components of air (nitrogen, oxygen argon). Ion sources operated near atmospheric pressure, sampling air directly, are burdened with ion cluster formation due to the presence of moisture. This results in complicated mass spectra of hydrated ions. The extent of clustering can be reduced and the spectra simplified by breaking the clusters up by collision with rest gas molecules in the accelerating field adjacent to the ion source.

Magnetic field analyzer: The classic method for separating ions of different mass is the magnetic sector field. It provides directional focusing of diverging ions emerging from an aperture. Additional energy focusing is achieved by adding an electric field, which the ion traverses before entering or after leaving the magnetic field. The combination provides the high resolution necessary for isotope measurements.

Because magnets made of iron are heavy, the application of these instruments is restricted to the laboratory.

Quadrupole mass filter: This device consists of four parallel cylindrical rods (10–20 mm in radius, 20 cm long). Opposing rods are connected, a fixed DC potential is applied and an AC voltage in the radio-frequency range is impressed on the DC potential. The rod's potentials and/or the frequency can be adjusted so that an ion of selected mass-to-charge ratio passes through the quadrupole filter. To generate a mass spectrum, the potentials are changed in a suitable way as a function of time. Quadrupole filters are light, reasonably cheap, easy to operate, and provide adequate resolution so that they have become the favored instrument for many applications.

Time of flight mass spectrometers: This device makes use of the fact that ions having the same energy but different mass will travel at different speeds. A bunch of ions withdrawn from the ion source by the application of a pulsed electric field is injected into a field-free drift space of suitable length. Here, the ions undergo separation in accordance with individual velocities so that each group arrives at the detector at a different time. Ions with the lower masses arrive first, ions with larger mass correspondingly later. The arrival times are measured. Time of flight mass filters feature high transmission efficiencies, particularly at higher masses. The mass resolution is usually much better than that of a quadrupole, but it requires a sufficiently narrow pulse for the extraction of ions from the ion source.

10.2.3   Absorption Spectrometry

The basic arrangement consists of a light source and focusing elements to produce a suitable optical beam, which is passed through the space, where the optical absorption takes place, and is subsequently viewed by a detector. The degree of attenuation is determined and evaluated on the basis of the Beer-Lambert law. The details of the instrumental setup depend on application. Suitable spectral regions are the visible and ultraviolet, where absorption bands occur due to electronic transitions of molecules, and the infrared, where the absorption bands are due to the vibrational-rotational motion of molecules. The strength of infrared bands is weaker than that of the electronic bands in the visible/ultraviolet, and special attention must be devoted to the pressure dependence of rotational line shapes.

Double-beam Optical Absorption Spectroscopy: Instruments based on this principle make use of two identical optical cells, one for the sample, the other serving as reference. The optical train derives two beams from the same light source, usually by means of mirrors. One beam traverses the sample cell, the other the reference cell. The signals derived from two identical detectors positioned behind the absorption cells are electronically normalized to the intensity emitted from the light source and the difference is recorded. In a variant of this technique, a chopper is used to send a single light beam alternately through the absorption cells to converge on one detector. The amplitude of the alternating current is measured, or phase-sensitive detection is employed. Two-beam analyzers require appropriate electronics to stabilize the output from the light source and, if photomultiplier tubes are used as detectors, their amplification factor.

Ozone Analyzer: A commercial instrument making use of this principle measures the amount of ozone present in ambient air by using as a reference the same ambient air after it has passed through an ozone scrubber. Light from a mercury lamp is filtered to transmit primarily the strong Hg 254 nm resonance line. A detection limit of 1 nmol mol^{-1} is reached, which is sufficient for most practical applications.

Non-dispersive infrared absorption analyzer: This type of instrument has been exploited commercially to measure carbon monoxide in urban air. The reference cell contains a non-absorbing gas, while ambient air flows through the sample cell. The wide-band light source usually is a hot filament. A chopper sends the light beam alternately through the sample and the reference cell. The detector consists of two compartments separated by a thin metal diaphragm and filled with carbon monoxide. Heat generated by the absorption of infrared radiation raises the pressure in the detector compartments. The pressure difference resulting from the weakening of radiation transmitted by the sample cell moves the diaphragm with the frequency determined by the chopper, the corresponding change in capacitance is converted into an electric signal which is proportional to the concentration of carbon monoxide in the sample cell. The detector is also applicable to the measurement of carbon dioxide and sulfur dioxide by replacing the gas in the detector. A recent application based on a laser source has enabled the detection of ammonia.

Differential Optical Absorption Spectroscopy (DOAS): This technique is used to detect and measure the concentration of atmospheric constituents endowed with clearly distinguishable absorption features in the visible and near ultraviolet spectral region. Measurements in the lower troposphere are made with the help of broadband light sources such as a Xenon arc lamp. In this case, the collimated light beam is directed at a distant retro-reflector, which returns it to the transmitter-receiver station, where it is collected by a telescope, dispersed by a spectrometer and recorded by a detector. Path lengths on the order of 1–10 km are common. For measurements of stratospheric constituents light from the sun or the moon provides the background source, either directly or as zenith-scattered light. Generally, a greater number of constituents contribute to the total observed spectrum, so that the individual spectra must be de-convoluted with the help of computer-assisted differentiation techniques. In principle, individual spectra stored in the computer are subtracted one by one from the observed total spectrum after adjustment of amplitude in each case. Procedures use either polynomials for the generation of differential optical spectra combined with reference cross sections, or Fourier transforms in combination with high or low band path filters. These operations are applied after normalizing the raw spectrum in order to remove the contribution from scattered light. In addition to following the trend of important tropospheric constituents such as NO_2, HONO, SO_2, HCHO, all of which have absorption features in the near ultraviolet spectral region, the DOAS technique has found a major application in the detection of NO_3, which builds up at night, but is readily photolysed during the day. Passive DOAS utilizing scattered sunlight has been used to measure several halogen oxides and the associated radicals in the stratosphere.

Hydroxyl radical: Absorption in the (0,0) vibrational band of OH near 308 nm is utilized. Because the rotational constant of OH is large, the rotational lines can be resolved even at atmospheric pressure. Provided the spectral range covers more than one rotational line, their characteristic intensity distribution allows an unambiguous identification of the OH radical. A frequency-doubled laser and a high resolution echelle spectrograph are used and the spectrum is recorded with a photodiode-array detector. The spectral range 307.96–308.19 nm displays five rotational lines: $Q_1(2)$, $Q_{21}(2)$, $R_2(2)$, $Q_1(3)$, and $P_1(1)$, of which the first two cannot be fully resolved. Whereas earlier setups employed simple two way reflection light paths, more recent applications use multiple reflections in an open multi-path cell to achieve an absorption path length of 3 km. The detection limit is approximately 1×10^6 molecule cm^{-3}.

Tunable Infrared Diode Laser Absorption: The narrow line width of infrared diode lasers and the dependence of the output frequency on the laser injection current make it possible to sweep the laser frequency back and forth across a rotational absorption line so as to obtain a modulated signal, which can be detected with a lock-in amplifier. A variety of other techniques have also been used. Because of the weakness of absorption intensities it is still necessary to use a path length of 100 m by means of a multi-path cell. The pressure in the cell usually is lowered in order to reduce collisional line broadening. This requires that air samples enter through an inlet line, which requires the usual precautions when wall-sensitive gases are sampled. A great variety of trace gases have been studied with this technique, including N_2O, NO_2, HNO_3, HCHO, H_2O_2 by ground based as well as airborne measurements.

10.2.4   Fluorescence Analyzers

Instruments that make use of the fluorescence of a molecule for the measurement of trace gases include an optical cell, in which a beam of strong light (frequently from a laser source) of suitable wavelengths traverses a carrier gas containing the molecule to be analyzed. A detector views the excitation region at a right angle against a dark background. The fluorescence emitted from the center of the cell is collimated by lenses and generally passes a filter before entering the photomultiplier detector. In most applications the gas pressure in the cell is reduced in order to lower the degree of collisional quenching of fluorescence and to increase its yield. Pulsed laser sources with pulses shorter than the lifetime of the excited state allow the application of time-gated detectors, which greatly reduces the background noise resulting from scattered light. Fluorescence can be resonant as for OH detection at 308 nm; red-shifted from the excitation wavelength as for NO_2 detection; or blue-shifted as for NO detection by two-photon excitation. To date, only the detection of

OH radicals has justified the effort required in applying the laser-induced fluorescence technique for atmospheric measurements.

Detection of OH radicals: Fluorescence assay by gas expansion (FAGE) operates with a critical orifice through which air is sampled and expanded into a low-pressure detection cell; OH radicals are excited in a single path by means of a 1–20 mW high-repetition (1–10 kHz) laser tuned to 308 nm, the resonance fluorescence is detected with a gated photomultiplier. Detection limits reported are $(2.8-10) \times 10^5$ molecule cm^{-3}. Faloona et al. (2004) have used a multi-path cell with a micro-channel plate detector with a detection limit 1.4×10^5 molecule cm^{-3} in 30 s integration with signal to noise ratio of two. Instruments are calibrated on-line with OH radicals generated by photolysis of water using the radiation of a mercury lamp at 185 nm in oxygen to produce OH and HO_2 in a 1:1 ratio. HO_2 is detected after conversion to OH by reaction with NO, which is injected directly behind the critical orifice.

Detection of nitrogen dioxide: The absorption spectrum of NO_2 extends from about 300–600 nm peaking near 400 nm. It is highly congested due to many overlapping ro-vibronic transitions involving several electronic states overlying a broad continuum. Photodissociation to $NO + O$ with near unity quantum yield occurs at wavelength shorter than 398 nm, where fluorescence ceases. The excited state lifetime is fairly long, about 100 µs, so that collisional quenching is rapid. Fluorescence is observed in the near infrared spectral region, mandating the use of infrared-sensitive photomultiplier tubes. A cut-off filter removing light shorter than about 650 nm is applied. Lasers operating at wavelengths of either 532 or 585 nm and various techniques to achieve specificity to NO_2 are applied. A sensitivity of 6 pmol mol^{-1} with 1 min integration and signal to noise ratio of two can be reached.

Detection of nitric oxide: Laser-induced fluorescence of NO has primarily relied on two-photon excitation. A laser tuned to 226 nm excites NO to the first excited state, while a second laser at 1.1 µm brings the excited molecule to a higher state from which fluorescence at 187 nm occurs. The two light beams are synchronized and enter the optical cell together. By necessity, this technique requires instruments of much greater complexity than single photon fluorescence instruments. Furthermore, discrimination of 187 nm radiation against 226 nm requires either solar-blind photomultiplier tubes or dielectric coated mirrors. A detection limit of 3.5 pmol mol^{-1} with 2 min integration and signal to noise ratio of two has been achieved. Single photon excitation at 226 nm with fluorescence at wavelengths between 240 and 390 nm has also been used with detection of NO in the nmol mol^{-1} range of mixing ratios.

Detection of sulfur dioxide: A commercial instrument makes use of pulses of ultraviolet light generated by a lamp emitting light in the 190–230 nm wavelength region, combined with a gated photomultiplier, to detect sulfur dioxide by fluorescence at mixing ratios above about 2 nmol mol^{-1}.

10.2.5   Ozone Electrochemical Detector

The technique is based on the reaction of ozone with potassium iodide in aqueous solution as air is bubbled through the electrochemical cell (Komhyr et al. 1995):

$$2\,KI + O_3 + H_2O \rightarrow I_2 + O_2 + 2KOH$$

The Brewer-Mast cell works with a silver anode and a platinum cathode with a polarizing voltage of 0.42 V being applied. The free iodine generated in the reaction is converted back to iodide ions at the cathode ($I_2 + 2\,e^- \rightarrow 2I^-$), whereupon two electrons are released to the circuit at the anode owing to the ionization of two silver atoms ($2\,Ag \rightarrow 2\,Ag^+ + 2\,e^-$). The resulting current is continuously measured. The electrochemical cell of Komhyr et al. (1995) consists of two chambers, an anode chamber and a separate cathode chamber with solutions having different KI concentrations; they are connected by an ion bridge, which prevents mixing of the two solutions and provides a pathway for the ion current. As in the Brewer-Mast cell, a current is measured which is proportional to the ozone concentration.

Electrochemical detectors are routinely used by global network stations to record vertical ozone profiles up to 35 km altitude. Comparison studies have shown that in the stratosphere, between the tropopause and 28 km, precision is better than 3%. In the troposphere errors can be larger due to smaller concentrations of ozone and some interference by SO_2.

10.2.6   Chemiluminescence Analyzers

The technique is based on measuring the intensity of the luminescence produced in a chemical reaction of the molecule to be analyzed with a second reactant. The reaction may occur in a liquid solution, with a reagent adsorbed on a solid, or in the gas phase. In the first two cases a number of dyes have been used to detect ozone by chemiluminescence; and an instrument for measuring nitrogen dioxide has been designed based on the chemiluminescence from the reaction with luminol. The major gas phase process being utilized is the chemiluminescent reaction of ozone with nitric oxide.

Ozone Chemiluminescence Detector: In the majority of applications an organic dye is adsorbed onto a silica gel to produce a solid-phase reagent. The chemiluminescent disk is viewed by a photomultiplier across a narrow gap through which ambient air is drawn with a small pump, generating a signal proportional to the partial pressure of ozone. This type of device has been flown on balloons and rockets for measuring the ozone distribution in the stratosphere (Hilsenrath and Kirschner 1980). Calibration was performed by comparing the integrated altitude profile with total ozone determined simultaneously by ground-based measurement. The preferred dyes that were used are luminol, which emits in the blue region of the spectrum, and rhodamine-B, which emits in the red. Coumarin-47 also has been found suitable. All of them are sensitive to water vapor. In the case of rhodamine-B, the effect could be minimized by adding a hydrophobic agent on the silica gel surface. The advantages of this type of ozone detector are light weight, fast response, and high sensitivity. The disadvantage is the decrease of sensitivity with time, requiring frequent calibration and replacement of the chemiluminescent disk.

NO$_2$-Luminol Chemiluminescence Analyzer: The reaction of NO$_2$ with luminol (5-amino-2,3-dihydro-1,4-phthalazinedione) in aqueous solution leads to chemiluminescence at wavelengths near 425 nm. Since ozone reacts also with luminol generating a chemiluminescence 30 times stronger, the method requires discrimination against ozone. The interference of ozone can be reduced by adding sodium sulfite to the solution. Alternatively, an ozone scrubber may be placed in the inlet line. The design of the instrument makes use of a wetted strip of cellulose filter paper viewed by a photomultiplier at close distance. The luminol solution is cycled over the filter paper by means of a peristaltic pump. The gas stream is fed through the narrow gap between paper strip and photomultiplier window. Detection limits on the order of 30 pmol mol^{-1} have been achieved, although non-linearity occurring at low concentrations require corrections. For mixing ratios greater than 1 nmol mol^{-1} the device gives results comparable to that of other instruments. Advantages of the luminol detector are its light weight, rapid response and low power consumption. The instrument is sensitive to pressure and temperature changes and appropriate precautions are needed. Interference due to peroxyacetyl nitrate must be tolerated. A commercial instrument based on this technique has been marketed.

NO$_x$-Chemiluminescence Analyzer: The instrument utilizes the reaction sequence

$$NO + O_3 \quad \rightarrow NO_2 + O_2$$
$$\rightarrow NO_2^* + O_2$$
$$NO_2^* \quad \rightarrow NO_2 + h\nu$$
$$NO_2^* + M \rightarrow NO_2 + M$$

where the asterisk indicates an electronically excited NO$_2$ molecule. The radiation emitted extends from 590 to 2,600 nm with a peak near 1,200 nm, so that detection requires a photomultiplier sensitive to the near-infrared spectral region. The instrument consists of a gold-plated cylindrical reaction chamber viewed end-on through a 590 nm cut-off filter by a photomultiplier cooled to −20°C to reduce the thermal dark current. The reagent flow is high purity oxygen containing 3–4% O$_3$, produced in a silent discharge, at about 100 cm^3 min^{-1}. The air inlet flow of about 1,000 cm^3 min^{-1} is intimately mixed with the reagent flow directly in front of the photomultiplier. The pressure in the chamber is reduced to 5–10 hPa by means of a vacuum pump and a control valve. The photomultiplier signal is processed by photon counting electronics. A sensitivity of 7 counts per pmol mol^{-1} NO, and a detection limit of 2 pmol mol^{-1} at 10s have been achieved in research instruments.

The instrument has also been applied to measure the sum of NO and NO_2 by using converters that dissociate NO_2 to form NO. Heated metal-catalytic converters, for example the molybdenum converter operated at 350°C, have been found to dissociate compounds such as peroxyacetyl nitrate, HONO, organic nitrates and some nitric acid in addition to NO_2, so that their use leads to signals that represent essentially the total oxides of nitrogen (often termed NOy). More efficient gold tube converters have been used to measure NOy, in particular in the stratosphere. Ferrous sulfate crystals at ambient temperatures also convert NO_2 to NO at nearly 100% yield, but again peroxyacetyl nitrate interferes, leading to signals in excess of those due to NO_2. The only converter reasonably free from interference is the photodissociation of NO_2, which is accomplished by passing the inlet air through a photolytic cell where the sample is irradiated with light from a high-pressure arc lamp. Conversion efficiencies of up to 70% can be achieved (Kley and McFarland 1980).

Ozone determination by gas-phase chemiluminescence: Although the above described system may also be used to measure the concentration of ozone by adding NO as the second reagent, the instrument is rarely used in this way. However, an ozone analyzer based on the same principle, except that NO is replaced by ethene as chemiluminescent reagent, provides a favorable alternative to other techniques, as it shows excellent sensitivity, linearity, and absence of interference from other components in the air.

10.2.7 Collection Followed by Liquid Phase Chemical Analysis

In contrast to instruments based on physico-chemical processes that are designed to produce an electric signal, liquid chemical analysis generally involves a series of steps including (a) the collection of a sample from ambient air, (b) the extraction of collected material or its soluble fraction into solution, and finally (c) the chemical analysis of the extract. Collection devices include filters, spray samplers, denuder tubes for water-soluble components and adsorption tubes for organic compounds. By drawing air through a filter, a spray sampler, or an absorption tube, certain trace components that occur in the gas phase as well as on aerosol particles are collected together and the subsequent analysis provides the sum of both. Important examples are NH_3/NH_4^+ and HNO_3/NO_3^-. The differentiation between gaseous and particulate concentrations requires special arrangements. In all cases care must be exercised to avoid contamination of samples by impurities. Both nitric acid and ammonia are difficult to measure with instruments due to interferences occurring in or with the inlet lines.

Collection on filters: Two filters in series are common. A pump is used to draw air through the filter pack at constant speed, with a flow meter indicating the amount per unit time. The dose is determined by the collection period. The first filter, which is exposed to ambient air and collects essentially all the aerosol particles, should be chemically inert to the gas-phase species, which after having passed the first filter is then absorbed on the second. Teflon and quartz fibers have been found essentially immune to interaction with atmospheric trace constituents, so that they provide suitable materials for the first filter. The material for the second filter depends on the application. Following exposure, the material absorbed on each filter is extracted with a small volume of deionized water (or a suitable aqueous solution) in an ultrasonic bath. The solutions obtained are analyzed by ion chromatography or capillary zone electrophoresis.

Nitric acid/Nitrate: The standard double filter pack consists of a Teflon membrane filter and a Nylon filter in series. Extraction is made with a millimolar NaOH solution. Cellulose fibers impregnated with tetrabutyl ammonium hydroxide have been found equivalent to Nylon as material for the second filter. The retention of nitric acid on both is close to 100%.

Ammonia/Ammonium: The Teflon membrane filter is followed by an acid-coated cellulose fiber filter. Although several types of acids have been used, oxalic acid appears to have been most favorable for impregnation of the filter. The system tends to overestimate gaseous ammonia owing to the volatilization of ammonia from ammonium nitrate deposited on the first filter. Following exposure, the filters must be stored and handled in an environment free of ammonia in order to prevent subsequent contamination (particularly in laboratory air).

Collection in a spray sampler: This technique combines steps (a) and (b) by absorbing the gaseous species of interest directly into water. Ambient air is pumped through a glass chamber, containing a finite amount of liquid water. A nebulizer, activated by the air current, produces a fine mist of water drops so that the equilibrium between gas and aqueous phase is rapidly established for water-soluble gases. The loss of water at the exit of the chamber is prevented by a hydrophobic filter. The method has proven successful for ammonia and organic acids.

Diffusion denuders: This technique of differentiation is based on the different mobility of gas molecules and aerosol particles in the gas phase. A laminar flow of air is drawn through a narrow tube, or an assembly of such tubes, to collect gases by diffusion to the coated walls, while the less mobile aerosol particles pass through the tube and are collected on a back-up filter. To increase the air flow so-called annular denuders are used that consist of two concentric tubes with an annular spacing of a few millimeters, through which the air is drawn. Artifacts may occur due to interference of ambient trace gases with the denuder coating. Such effects can be overcome by using two or more denuder tubes in series with different coatings so as to remove interfering gases before collecting the species to be determined.

Nitric acid, Ammonia, Ammonium nitrate: Two tubes in series are used: the first, coated with sodium fluoride, collects nitric acid, and the second, coated with phosphoric acid, collects ammonia. Ammonium nitrate is collected on the back-up filter. Losses of ammonium nitrate are avoided by a filter combination consisting of a PTFE filter followed by cellulose filters coated with NaF and H_3PO_4, respectively.

Following sampling the denuders and filters are treated with water and the extracts are analyzed for NO_3^-, NH_4^+ as well as other ions. Collection periods are about 20 min for a detection limit of 0.1 μg m^{-3}.

A different technique has explored the use of denuders for the detection of HNO_3 and NH_3 by chemiluminescence. The denuder coating is prepared from tungstic acid. Following collection, the tube is heated to 350°C, whereupon NH_3 is released while HNO_3 desorbs as NO and NO_2. The desorbed gases are separated by passing them through an intermediate tube, where NH_3 is recollected, whereas NO and NO_2 continue and are determined in an NO_x analyzer. Subsequently, the intermediate tube is heated to release NH_3, which is also analyzed in the NO_x analyzer.

Nitrous Acid: Denuders coated with sodium chloride or sodium carbonate may be used if precautions are taken for artifacts resulting from the absorption of sulfur dioxide, nitrogen dioxide, and/or oxidation of nitrous acid to form nitric acid. Good results have been obtained with an assembly of three tubes in series, the first coated with tetrachloromercurate for the removal of SO_2, and the other two coated with Na_2CO_3. The second tube absorbs all the HNO_2, whereas the artifact caused by NO_2 occurs equally on the second and third tube. The difference of nitrite found on the second and third tube is taken as a measure of nitrous acid. This assembly does not allow the determination of nitric acid, because it is absorbed on the first tube together with SO_2.

Automated wet annular denuder: This ingenious device keeps the walls of the tubes wetted with an aqueous solution in which the species to be determined accumulate (Genfa et al. 2003). The inner tube of the annular denuder has an outer diameter of 42 mm, and the outer tube an inner diameter of 45 mm, leaving an annulus of 1.5 mm. The denuder is kept in a nearly horizontal position. About 20 cm^3 of solution is placed in the annulus. During operation the denuder is rotated around its axis so that a solution film is maintained on the denuder walls. After 30 min sampling time the solution is pumped into a storage unit for analysis by ion chromatography and the denuder is reloaded. Ammonia is determined as NH_4^+, nitric acid as NO_3^-, nitrous acid as NO_2^-, HCl as Cl^-, and SO_2 as SO_3^{2-}. H_2O_2 can also be determined.

References

Faloona, I.C., D. Tan, R.L. Lesher, N.L. Hazen, C.L. Frame, J.B. Simpas, H. Harder, M. Martinez, P. Di Carlo, X.R. Ren, W.H. Brune, J. Atmos. Chem. **47**, 139–167 (2004)

Genfa, Z., S. Slanina, C. Brad Boring, P.A.C. Jongejan, P.K. Dasgupta, Atmos. Environ. **37**, 1352–1364 (2003)

Hilsenrath, E., P.T. Kirschner, Rev. Sci. Instrum. **51**, 1381–1389 (1980)

Kley, D., M. McFarland, Atmos. Technol. **12**, 63–69 (1980)

Komhyr, W.D., R.A. Barnes, G.B. Brothers, J.A. Lathrop, D.P. Opperman, J. Geophys. Res. **100**, 9231–9244 (1995)

10.3   Description of Aerosol Measurement Techniques

The principal parameters needed to characterize the atmospheric aerosol are: number density, mass concentration and chemical composition of the particles. Each quantity is a function of the particle size. The complete characterization of all aerosol parameters involves a substantial effort requiring the simultaneous application of a larger number of measurement techniques.

10.3.1   Number Concentration

Condensation nucleus counter (CNC): This type of counter determines the total number concentration of particles larger than a minimum detectable size. The basic principle is the growth of particles by the condensation of a vapor (usually water or *n*-butyl alcohol) under conditions of supersaturation. Diameter growth factors of 100–1,000 are common, whereby the particles reach a size range accessible to optical detection. Supersaturation can be achieved by volume expansion, as in the original version developed by Aitken, or by heat transfer from the warm aerosol to the walls of a condenser maintained at typically ~10°C. The accuracy of determining particle concentration depends on the optical detection scheme. Single-particle-counting instruments determine particle concentration by counting individual drops formed by condensation, an indirect measurement is based on the attenuation of light scattered by the droplet cloud. The latter method requires calibration with an independent concentration standard. CNCs are able to detect particles as small as 3 nm.

Electrical mobility analyzers (Differential mobility analyzer, DMA): This technique exploits the behavior of charged particles moving in an electric field. Their drift velocity is inversely proportional to particle size and increases with the field strength. Before entering the analyzer chamber, the initially neutral aerosol particles are exposed to small ions, which can attach to the particles so that a fraction of them acquires an electric charge. Both unipolar and bipolar charging methods are in use. Electrical mobility analyzers work with an electric field perpendicular to a viscous flow of air, so that particles having mobilities larger than a value predetermined by flow rate and electric field strength are removed from the air stream and are precipitated, whereas particles with lower mobilities are carried along with the air flow and emerge from the exit aperture of the analyzer, where particles in a narrow mobility range are delivered to the detector. Detection may be made by the collection of charged particles in a Faraday cup, but the condensation nucleus counter is more sensitive, and its use is common today. A complete size distribution is obtained by scanning the classifying voltage applied to the electrodes. The raw data require deconvolution to determine the contribution of each size to measurements at a given classifying voltage. Multiple charging, which affects mainly the larger particles, limits the technique at the high end of the size spectrum. Thus, the use of differential mobility analyzers is limited to the aerosol size range of approximately 3–500 nm.

Optical particle counter: Optical counters measure the amount of light scattered by individual particles as they traverse a tightly focused light beam. A fraction of the light is collected and directed onto a photo-multiplier detector, which converts it to a proportional voltage pulse. The magnitude of the pulse is an indicator of particle size. The instrument is calibrated with spherical particles of known size. Lasers, which provide illuminating intensities several orders of magnitude higher than incandescent sources, enable the detection of smaller particles. Optical particle counters allowing the detection of ~0.05 μm sized particles are available, whereas the detection limit of optical counters operating with white light is ~0.3 μm. A basic problem with optical counters is that the properties of atmospheric particles, such as shape, morphology, refractive index, generally are unknown, and that the atmospheric aerosol represents a mixture of different types of particles. Difficulties arise for strongly absorbing particles such as soot. Ideally, particles having the same size but different shape, refractive index, etc. may produce distinctly different pulse heights. Valuable information about shape and/or refractive index can be obtained from the angular distribution of the scattered light. These so-called differential light scattering instruments can distinguish between spherical and non-spherical particles in the 0.2–2 μm size range. Hygroscopic salt particles usually acquire sufficient water to appear nearly spherical in shape, whereas particles of crustal origin are less hygroscopic and appear non-spherical in shape.

10.3.2   Mass Concentration

Manual method: The most commonly used technique for measuring the mass concentration of particles is deposition on filters and weighing. Filter samplers often are equipped with a critical orifice as inlet to eliminate particles above a specific size cut. Fiber and Nucleopore filters are used. Larger particles undergo interception and impaction, particles smaller than ~0.1 μm are collected by diffusion. Particles in the size range 0.1–0.5 μm are most likely to escape collection by the filter. Nevertheless, most filters in use collect all particles with better than 99% efficiency. The sensitivity for gravimetric analyses is near ±1 μg. Thus, with typical flow rates of 1 m^3 h^{-1} it is necessary to apply longer sampling periods. A sampling time of 24 h is standard.

Piezoelectric crystal: A quartz plate cut along specific crystallographic planes will undergo mechanical vibration when a periodic electric potential is applied across the crystal plane. Piezoelectric crystals used for particulate mass monitoring have natural resonance frequencies of 5–10 MHz. The change of resonant vibrational frequency with incremental mass is of the order of kHz per μg, and this frequency change is measured by electronically mixing with the frequency of an identical crystal, which serves as standard. Piezoelectric crystals can measure particulate mass in the low ng range, but the response becomes non-linear when the mass loading exceeds 5–10 μg. The need for periodic cleaning is a detriment for using this technique for routine monitoring of atmospheric aerosol concentration. The response also is sensitive to pressure fluctuations.

Inertial impactor: This device collects particles by drawing ambient air through a nozzle and directing the air stream at normal incidence onto a collecting surface. Larger particles are deposited due to inertial forces, while smaller particles are carried along with the air flow around the collector. Size and shape of the nozzle determines the cut-off size of the particles that are collected. The dimensionless parameter that determines whether a particle is collected is the Stokes number, defined by $St = (\rho\, D^2 CU)/(9\mu D_n)$, where ρ and D are density and diameter of the particle, U is the mean velocity through the accelerating nozzle, D_n is the nozzle diameter (or an equivalent width for rectangular slots), C is the slip correction factor, and μ is the viscosity of air. Owing to the dependence on particle density, the size classification occurs in accordance with the aerodynamic diameter of particles, which is defined as the diameter of a unit density sphere having the same settling speed as the particle. Impactors collect particles that have Stokes numbers larger than a critical value of approximately 0.22. A single impactor should collect all particles larger than a predetermined cut-off size. In practice, however, the collection probability near the cut-off limit is not an ideal a step function but resembles an S-shaped curve. Two identical impactors in series consequently will on the second plate collect particles in the narrow size range given by the spread in the collection probability. A set of such double stage impactors can be used to cover the whole size range accessible by impactors, approximately 0.1–10 μm, to determine the size distribution of mass or any chemical substance being part of the aerosol. The size range covered contains virtually all of the particulate mass. The mass distribution can be determined by weighing if thin foils are used for the collection of particles. Commonly, however, impactors are used to determine the size distribution of chemical composition. Particle bounce, which is an inherent problem in using impactors, can be overcome by coating the collection plates with liquid oil or sampling at higher relative humidity.

Cascade impactor: A series of impactors with nozzles of decreasing size arranged in sequence such that the air flow passes successively through each impactor stage. In modern cascade impactors the nozzles are arranged equidistantly on a circle and the number of nozzles is increased at each subsequent impactor stage in order to keep the air flow approximately constant. Ring–shaped foils are used to collect particulate material for chemical analysis.

10.3.3 Individual Particle Sizing and Morphology

Visual inspection by microscopy provides information on the appearance of selected particles, which leads to further conclusions about the nature of the particle.

Light microscope: The resolving power of a normal light microscope is limited by the wavelength of light (diffraction limit). Direct examination and sizing with a light microscope is straight forward for particles >1 μm, but particles at the 0.5 μm size level can still be observed and counted. Glass slides or cellulose filters are commonly used as support. Particles that have an index of refraction different from that of the substrate mounting material are most easily viewed because of enhanced contrast. The contrast can be improved by suitable techniques such as coating or chemical treatment of the support. For identification of a particle the observed appearance may be compared to known structures.

Electron microscope: Electron imaging also provides information on particles size and morphology. The transmission electron microscope probes the sample with a beam of electrons and the image is shown on a phosphorescent screen. Magnification up to a million times is possible, which allows very small particles (<0.5 μm diameter) to be detected. The scanning electron microscope uses a finely focused electron beam to raster over the sample area. Scattered electrons resulting from elastic or inelastic collisions of beam electrons with the sample are detected, the signal is amplified and displayed on the screen of a cathode ray tube with its scanning pattern matched to that of the electron beam. The image provides morphological as well as topographical information with magnification up to 100 thousand times.

10.3.4 Chemical Composition

A complete chemical analysis is generally not possible, as no analytical procedure exists that can provide a full characterization of the complex chemical composition of atmospheric aerosol particles. As a consequence, the focus of study and the applied analytical technique are always closely related. The principal collection devices are filters and impactors. Water-soluble compounds can be extracted by leaching and the resulting aqueous solution analyzed by wet chemical methods. Organic compounds are extracted by organic solvents and analyzed by chromatographic techniques (see Rogge et al. 1993). These techniques have been described above under *Collection Followed by Liquid Chemical Analysis*. A description of instrumental methods for elemental analysis follows below. An important consideration is whether the species measured are on the surface of the aerosol, where they are available for reaction with the gas phase, or in the bulk.

X-ray microanalysis: The technique has developed from electron microscopy. The X-ray energy spectrum emitted from the sample when being irradiated with fast electrons is detected. Windowless or thin-window detectors allow the detection of X-rays from elements with atomic number 11 (sodium) and greater. The location of elements on or within the particles can be determined by using electron beams that are small relative to particle size. X-ray microanalysis provides a non-destructive micro-analytical technique, but the sample is kept under vacuum so that volatile components will be lost. Instruments can be engineered to scan particles for size and elemental composition simultaneously, so that, in principle, the distribution of the elements with particle size can be determined.

X-ray fluorescence analysis (XRFA): Both energy-dispersive and wavelength-dispersive techniques may be used. Wavelength-dispersive instruments subject the sample (deposited on a filter) to radiation from an X-ray tube, a crystal disperses the emitted X-rays in accordance with Bragg's law, and a detector placed at an appropriate angle measures the intensity of characteristic X-rays at a particular wavelength. Energy-dispersive instruments use a solid-state detector, such as a silicon diode doped with lithium, and a multichannel pulse height analyzer to detect and record the full X-ray fluorescence spectrum simultaneously. The method allows quantitative analysis of elements with atomic numbers of about 12 and higher. Detection limits for samples collected on Teflon filters range from 20 to 200 ng cm^{-2} for many elements.

Proton-induced X-ray emission (PIXE): This technique differs from that of X-ray fluorescence analysis only in the excitation source. Although an expensive accelerator is required to generate high energy protons, the technique excels by large excitation cross section even at low atomic number and low X-ray background. A lithium-doped silicon semiconductor detector is again used in combination with a pulse height analyzer to determine the X-ray fluorescence energy spectrum. The method provides good precision for samples of up to 50 μm thickness.

Neutron activation: When a sample is exposed to an intense thermal neutron flux, nuclear reactions result in the emission of γ-rays characteristic for an element. The amount of each radio-nuclide produced is proportional to the mass of that element. The radiation is detected with a lithium-drifted germanium detector with high resolution for separating the energies of the γ-rays. Advantages of this method include the ability to analyze thick and inhomogeneous samples. Disadvantages are the need for a nuclear reactor and special techniques for the handling of radioactive samples. In addition, several elements (Ni, P, among others) are not accessible to measurement.

Laser microprobe mass spectroscopy (LAMMS): A particle from an assembly collected on a substrate is selected by a microscope and pulse-irradiated with a high-power laser focused on the sample. The sample is vaporized, the ion fragments ejected are accelerated to a uniform translational energy and analyzed with a time-of-flight mass spectrometer. An advantage of this technique is that all elements usually are ionized in a single laser pulse and detected in a single mass scan. LAMMS can detect trace levels of metals in individual particles at the ppm level; anion molecular species common in atmospheric particles show up in the spectra as NO_3^-, SO_4^-, PO_x^- and CO_3^-. Organic compounds often appear as molecular fragments similar to those occurring in electron impact ionization, which makes identification difficult, because aerosol particles contain complicated mixtures of organic compounds.

Secondary ion mass spectrometry (SIMS): When a beam of inert ions, such as argon ions, is directed at a solid target, the target material undergoes sputtering. A small fraction of the ejected material consists of ions that can be collected and accelerated into a mass spectrometer. SIMS provides a large dynamic range and excellent detection limits, useful for trace analysis. High beam currents cause successive ablation of the material so that depth profiles of elements of interest can be obtained, whereas low beam intensities avoid excessive sputtering and can provide information on the presence of molecular compounds. A disadvantage is that the combined sputtering and ion production efficiencies of elements can vary over 5 orders of magnitude, and a corresponding need for calibration standards.

Online aerosol mass spectrometry: The principal feature of aerosol mass spectrometry is the removal of gaseous material from the sample while retaining the particulate fraction. This is accomplished by collimating the particles in an aerodynamic lens, which consists of a series of apertures with successively decreasing diameters, combined with differential pumping during expansion of the sample into the mass spectrometer. Vaporization and ionization may be carried out in different ways. One is by impacting the particles on a metallic surface heated to a high temperature to produce surface ionization; another is vaporization of the particles at a lower temperature followed by electron impact ionization of the vapor; and a third method is the use of lasers to vaporize and ionize the sample. The size of the particles to be analyzed can be determined prior to entering the ionization region. The currently favored technique of particle sizing is to measure the velocity after nozzle acceleration by means of a laser and two optical scattering detectors viewing different stages of the particle beam. The main advantage of using online aerosol mass spectrometers is the time resolution offered by such instruments. The information obtained on the composition of particles refers mainly to the non-refractory material and depends much on a correct interpretation of the mass spectra.

Atomic absorption spectroscopy (AAS): This is one of the most widely used techniques, applicable to about 70 elements. Unlike the X-ray emission techniques discussed above, AAS can be applied only for each element separately. The method is based on the principle of optical absorption by an atomic line. A hollow cathode lamp provides the atomic emission spectrum needed to observe the absorption. The sample is vaporized at sufficiently high temperatures to liberate the elements into the gas phase. One common procedure is to use a nebulizer to disperse the analyte solution into a high-temperature flame, another uses a graphite furnace to stepwise dry, char and atomize the analyte at temperatures up to 3,000°C.

Plasma emission spectroscopy: An argon discharge generates a plasma, which is induced by a high frequency electric field applied through inductive coils (inductively coupled plasma, ICP). The sample solution is injected into the plasma by a nebulizer and dissociates to form atoms of the elements contained in the sample. The elements are excited by electron collisions to emit their characteristic atomic emission lines, which are separated by a grating spectrometer and detected with a photomultiplier tube. The response remains linear over five orders of magnitude change in concentration. The method is suitable for most elements over a concentration range in solution of $10–10^6$ ng cm^{-3}.

Reviews of Atmospheric Measurement Techniques

Clemitshaw, K.C., Review of instrumentation and measurement techniques for ground-based and airborne field studies of gas-phase tropospheric chemistry. Crit. Rev. Environ. Sci. Technol. **34**, 1–108 (2004)

Heard, D.E. (ed.), *Analytical Techniques for Atmospheric Measurement* (Blackwell Publishing, Oxford, 2006)

Heard, D.E., M.J. Pilling, Measurement of OH and HO_2 in the troposphere. Chem. Rev. **103**, 5163–5198 (2003)

Lodge Jr., J.P. (ed.), *Methods for Air Sampling and Analysis*, 3rd edn. (Lewis, Chelsea, 1989)

McMurry, P.H., A review of atmospheric aerosol measurements, Atmos. Environ. **34**, 1959–1999 (2000)

Parrish, D.D., F.C. Fehsenfeld, Methods for gas-phase measurements of ozone, ozone precursors and aerosol parameters. Atmos. Environ. **34**, 1921–1957 (2000)

Rogge, W.F., M.A. Mazurek, L.M. Hildemann, G.R. Cass, B.R.T. Simoneit, Quantification of urban organic aerosols at a molecular level: identification, abundance and seasonal variation. Atmos. Environ. **27A**, 1309–1330 (1993)

Willeke, K., P.A. Baron (eds.), *Aerosol Measurement: Principles, Techniques and Applications* (Van Nostrand Reinhold, New York, 1997)

10.4   Satellite Sensors

Remote sensing instruments viewing the Earth from satellites offer the advantage of a nearly complete global coverage for any type of information that can be retrieved from the radiances leaving the top of the atmosphere in various spectral regions. The data must be subjected to mathematical inversion techniques in order to derive the global distribution of parameters such as the temperature or the concentrations of trace species. A variety of mathematical approaches have been developed to extract these quantities by taking into account the transfer of radiation within the atmosphere. The majority of instruments are passive sensors that observe either thermal emissions or solar backscattered radiation, although active measurement techniques such as LIDAR have also been developed. Three viewing modes are commonly employed: (a) nadir (downward) viewing observes the entire atmospheric column vertically, yielding the total atmospheric column density; (b) limb sounders view the atmosphere at various angles directed at the horizon, so that vertical profiles can be determined; (c) occultation measurements view the sun through the atmosphere at sunrise and sunset. Many satellites operate in near-polar sun-synchronous orbits at altitudes of 700–900 km, providing a global coverage within ~1 day. Occultation instruments in sun-asynchronous orbits typically observe 14 sunrise and sunset measurements around the globe per day. Geostationary satellites at

35,790 km altitude such as GOES, which carries instruments designed for the monitoring of surface and meteorological fields, provide an alternative platform for monitoring a specified region continuously.

Three wavelength regions are currently being exploited for observations: (a) the spectral region 250–2,500 nm encompassing ultraviolet, visible and near infrared radiation, in which backscattered solar radiation is dominant; (b) the thermal infrared region, 3–100 μm, which is characterized by thermal radiation from the Earth's surface and the atmosphere; and (c) the microwave region, 0.1–10 cm, where O_2 and H_2O vapor are important absorbers and RADAR techniques can be used to obtain information on the distribution of liquid water and ice particles. Measurements in the thermal infrared and microwave spectral regions provide information also at night, whereas solar backscatter is active only during the day.

Classic spectroscopic techniques for wavelength dispersion and detection are applied to observe the 250–2,500 nm spectral region. In the thermal infrared, diffraction gratings as well as filters are used for wavelength selection. Gas correlation sensors, which work with the specific gas to be monitored as an optical filter, provide an efficient method of measurement. In addition, Fourier transform spectrometers based on the Michelson interferometer have been adapted for remote sensing. The solid state detectors used are cooled to reduce the local thermal noise. Microwave radiometers work with dipole antennas and solid state devices to amplify and process the thermal radiation received. A frequent calibration against suitable black body emitters is required to minimize the influence of fluctuations and drift of amplifier gain.

Table 10.8 Overview on satellite sensors[a]

Satellite[b]	Revisit period	Instrument[c]	Spectral range[d]	Foot print (km²)[e]	Atmospheric constituents observed[f]
NIMBUS 7 1978–1992	6 days	LIMS	6.3, 7.3, 9.9, 11.4, 15.3 ± 0.4, 15.3 ± 2.0 μm	L scan	O_3, NO_2, HNO_3, CO_2, H_2O
		SAM II	1 ± 0.019 μm	30 arc-sec (O)	Stratospheric aerosol
		SAMS	4.3, 5.3, 7.7, 15, 100, 25–100 μm GCR	20–100 km L	T, CO_2, CO, N_2O, CH_4, H_2O
		SBUV	252–340 nm (12 λ, 1 nm res.)	200×200	O_3
		TOMS	310–360 nm (6 λ, 1 nm res.)	50×50	Total O_3, O_3 (Tr low latitudes), SO_2 (volcanoes)
		LRIR	8.8–10.1, 14.6–15.9, 14.2–17.3, 20–25 μm	L	O_3, CO_2 (St profiles)
UARS 1991–2005	36 days	CLAES	3.52, 5.27, 6.23, 7.96, 10.8, 11.4, 11.9, 12.6, 12.8 μm	L	T, P, O_3, H_2O, CH_4, N_2O, NO, NO_2, N_2O_5, HNO_3, $ClONO_2$, HCl, CFC-11, CFC-12, aerosol (St)
		HALOE	2.45–10.04 μm (several channels) GCR	O	O_3, NO_2, NO, CO_2, H_2O, HF, HCl, CH_4 (St)
		ISAMS	4.43–16.5 μm (14 λ)	2.6×13 L	O_3, NO_2, NO, CO_2, CO, H_2O, N_2O, HNO_3, CH_4 (St)
		MLS	MW 63, 183, 205 GHz	L	T, RH, ice, O_3, SO_2, HNO_3, H_2O, ClO, CH_3CN, (St)
ERS-2 1995–2003	5 days	GOME	230–790 nm (cont.)	40×320	O_3, NO_2, HCHO, CHOCHO, BrO, OClO, SO_2, H_2O
		ATSR	555, 659, 865 nm, 1.6, 3.7, 11, 12 μm	1×1	SST, cloud properties, aerosol properties
TERRA 1999–	1–2 days	MODIS	0.41–14.2 μm (36 λ)	10×10	Cloud properties, aerosol properties
		MISR	450–870 nm (4 λ)	18×18 (9 <)	Aerosol properties
		MOPITT	4.7 μm GCR	22×22	CO
		AMSR	MW 19, 22, 37, 85 GHz	4×6	SST, H_2O, clouds, rain,
AQUA 2002–	1 day	AIRS	3.7–4.6, 6.2–8.2, 8.8–15.4 μm	13.5×13.5	O_3, CO_2, CO, H_2O, CH_4
		CERES	0.3–5 μm, 8–12 μm, 0.3–100 μm	20×20	Cloud's and Earth's radiation budget
ENVISAT 2002–	6 days	AATSR	555, 659, 865 nm, 1.6, 3.7, 11, 12 μm	1×1	SST, cloud properties, aerosol properties
		GOMOS	248–693, 750–776, 915–956 nm	O	O_3, NO_2, NO_3, H_2O, O_2
		SCIAMACHY	0.24–1.75 (cont.), 1.94–2.04 (cont.) 2.265–2.38 μm (continuous)	30×60 N 240×400 L	O_3, NO_2, NO, N_2O, HCHO, BrO, OClO, SO_2, CO, CO_2, CH_4, H_2O, O_2, O_4, aerosol, cloud properties

	Instrument[c,d]	Wavelength/channels[d]	Pixel (km²)	Measured quantities[f]
	MERIS	412–900 nm (15 λ, 1.8 nm res.)	0.3×0.3	H_2O, cloud properties, aerosol properties
	MIPAS	4.15–14.6 μm (continuous) FTS	3×30 L	O_3, NO_2, N_2O, HNO_3, H_2O, CH_4 (standard)
AURA 2004– 1 day	MLS	118, 190, 240, 640, 2,250 GHz	5×500 L	T, RH, ice, O_3, N_2O, HNO_3, CO, H_2O, HCl, HOCl, ClO, BrO, OH, HO_2, HCN, CH_3CN, SO_2 (St)
	OMI	270–500 nm (continuous)	13×24	O_3, NO_2, SO_2, OClO, BrO, HCHO, CHOCHO, O_4 Cloud properties, aerosol properties
	TES	3.3–15.4 μm (continuous)	5×8 N, L	O_3, NO_2, NO, HNO_3, N_2O, CO, H_2O, CH_4
PARASOL 2004– 1 day	POLDER	443, 490, 565, 670, 763, 765, 865, 910, 1,020 nm (polarized)	16×18	Clouds properties, aerosol properties
CALIPSO 2006– 1 day	CALIOP	LIDAR 532 (polarized), 1,064 nm	100×330 30–60 vertical	Aerosol layers: height, thickness (for $\tau > 0.005$); Cloud layers: height ($\tau > 0.01$), thickness ($\tau < 5$)
METOP A 2006– 1.5 day	GOME-2	230–790 nm (cont., 0.2 nm res.)	40×80	O_3, NO_2, HCHO, CHOCHO, BrO, OClO, SO_2, H_2O, Cloud properties, aerosol properties
	IASI	3.6–15.5 μm (continuous) FTS	12×12	O_3, N_2O, HNO_3, CH_4, CO, CO_2, H_2O, CFC-11, CFC-12, HCFC-22, SO_2, C_2H_4, HCOOH, CH_3COOH
	AVHRR	0.58–0.68, 0.73–1.0, 1.58–1.64, 3.55–3.93, 10.3–11.3, 11.5–12.5 μm	5×5	Vegetation, cloud properties, aerosol properties
	HIRS	690 nm, 3.7–4.6 μm (7 λ), 6.5–15 μm (12 λ)	20×20	SST, T and P profiles, cloud height/cover, surf. albedo
	AMSU	MW 23.8–89 GHz (15 channels)	48×48	T profiles, cloud height/cover
	MHS	MW 89–190 GHz (5 channels)	15×15	SST, H_2O profiles, cloud cover, precipitation

[a] Assembled largely from data in Burrows et al. (2011)

[b] All satellites except UARS are in polar orbits; the orbit of UARS was inclined 57° toward the equator

[c] For acronyms see Table 10.9; TOMS is currently flown on Earth Probe satellite; MODIS and AMSU are flown also on AQUA

[d] Ultraviolet/visible/near infrared: wavelength (nm), thermal infrared: wavelength (μm), microwave: frequency (GHz); (n λ) number of discrete wavelengths, GCR gas correlation radiometer, FTS Fourier transform spectrometer

[e] Abbreviations: N nadir, L limb view, O solar occultation

[f] SST sea surface temperature, Tr troposphere, St stratosphere, cloud properties: generally includes cloud fraction, optical thickness, cloud top height and pressure, liquid water path, average droplet/crystal size; aerosol properties: generally include optical depth and Ångström exponent

Table 10.9 Abbreviations and acronyms of satellite sensors

AATSR	Advanced Along-Track Scanning Radiometer (ESA ENVISAT)
ACE-FTS	Atmospheric Chemistry Experiment – Fourier Transform Spectrometer (CSA SCISAT)
AIRS	Atmospheric Infrared Sounder (NASA AQUA)
ATMOS	Atmospheric Trace Molecule Spectroscopy Experiment (NASA Spacelab)
ATSR	Along-Track Scanning Radiometer (ESA ERS)
ATOVS	Advanced TIROS Operational Vertical Sounder
AVHRR	Advanced Very High Resolution Radiometer (NASA TIROS-N, Metop A)
BUV	Backscatter Ultraviolet Ozone Experiment (NASA NIMBUS 4)
CALIOP	Cloud-Aerosol Lidar with Orthogonal Polarization (NASA CALIPSO)
CERES	Cloud's and the Earth's Radiant Energy System (NASA AQUA)
CLAES	Cryogenic Limb Array Etalon Spectrometer (NASA UARS)
GLAS	Geoscience Laser Altimeter System (ICESat)
GOME	Global Ozone Monitoring Experiment (ESA ERS, EUMETSAT)
GOMOS	Global Ozone Monitoring by Occultation of Stars (ESA ENVISAT)
HALOE	Halogen Occultation Experiment (NASA UARS)
HIRS	High-resolution Infrared Radiation Sounder (EUMETSAT METOP A)
HIRDLS	High Resolution Dynamics Limb Sounder (NASA AURA)
IASI	Infrared Atmospheric Sounding Interferometer (EUMETSAT METOP)
ILAS	Improved Limb Atmospheric Spectrometer (NASA ADEOS)
IMG	Atmospheric Infrared Sounder (NASDA ADEOS)
ISAMS	Improved Stratospheric and Mesospheric Sounder (NASA UARS)
LIMS	Limb Infrared Monitor of the Stratosphere (NASA NIMBUS 7)
LITE	Lidar In-space Technology Experiment (Space Shuttle Discovery)
LRIR	Limb Radiance Inversion Radiometer (NASA Nimbus 7)
MAPS	Millimeter Wave Atmospheric Sounder (NASA Space Shuttle)
MAS	Microwave Atmospheric Sounder (Space Shuttle ATLAS)
MERIS	Medium Resolution Imaging Spectrometer for Passive Atmospheric Sounding (ESA ENVISAT)
MIPAS	Michelson Interferometer for Passive Atmospheric Sounding (ENVISAT)
MISR	Multiangle Imaging Spectro-Radiometer (NASA TERRA)
MLS	Microwave Limb Sounder (UARS, AURA)
MODIS	Moderate Resolution Imaging Spectrometer (NASA TERRA)
MOPITT	Measurement of Pollution in the Troposphere (NASA TERRA)
OMI	Ozone Monitoring Instrument (NASA AURA)
OSIRIS/IRI	Optical Spectrograph and Infrared Imaging System (SWEDISH ODIN)
POLDER	Polarization and Directionality of the Earth's Reflectances (NASA PARASOL)
POAM	Polar Ozone and Aerosol Measurement (CNES SPOT)
SAGE-2	Stratospheric Aerosol and Gas Experiment (NASA Earth Radiation budget)
SAM	Stratospheric Aerosol Measurement Instrument (NASA NIMBUS 7)
SAMS	Stratospheric and Mesospheric Sounder (NASA NIMBUS 7)
SBUV	Solar Backscatter Ultraviolet Ozone Experiment (NASA NIMBUS 7)
SCAMS	Scanning Microwave Sounder (NASA NIMBUS 4, 6)
SCIAMACHY	Scanning Imaging Absorption Spectrometer for Atmospheric Cartography (ESA ENVISAT)
SCR	Selective Chopper Radiometer (NASA NIMBUS 4, 5)
SEVIRI	Spinning Enhanced Visible and Infrared Imager (METEOSAT)
TANSO	Thermal and short wave infrared Sensor for observing greenhouse gases (JAXA GOSAT)
SeaWiFS	Sea-viewing Wide Field-of-View Sensor (ScaStar)
TES	Tropospheric Emission Spectrometer (NASA AURA)
TOMS	Total Ozone Monitoring Spectrometer (NASA Nimbus 7, Earth Probe)

Table 10.10 Further abbreviations and acronyms

ADEOS	Advanced Earth Observation Satellite
ATLAS	Atmospheric Laboratory for Application and Space
CALIPSO	Cloud-Aerosol Lidar and Infrared Pathfinder Satellite Observations
CFC	Chlorofluorocarbon
CNES	Centre National d'Etude Spatiales
CSA	Canadian Space Agency
ENVISAT	Environmental Satellite
EOS	Earth Observing System
EPS	EUMETSAT's Polar System
ESA	European Space Agency
EUMETSAT	European Organization for the Exploitation of Meteorological Satellites
GOMAS	Geostationary Observatory for Microwave Atmospheric Sounding
GOSAT	Greenhouse gases Observing Satellite
GSFC	Goddard Space Flight Center
ICESat	Ice, Cloud, and land Elevation Satellite
JAXA	Japanese Aerospace Space Agency
MAS	Microwave Atmospheric Sounder (Space Shuttle ATLAS)
METOP	A series of three satellites, forming a segment of EPS
METEOSAT	Meteorological Satellite
MIR	Millimeter–wave Imaging Radiometer
NASA	National Aeronautics and Space Agency
NASDA	National Space Development Agency in Japan
NOOA	National Oceanic and Atmospheric Administration
SME	Solar Mesospheric Experiment
SPOT	Système Pour l'Observation de la Terre
TIROS	Television Infrared Observation Satellite
TOVS	TIROS operational vertical sounder
TROPOSAT	Use and usability of satellite data for tropospheric research
UARS	Upper Atmospheric Research Satellite

Reference

Burrows, J.P., U. Platt, P. Borrell, P. (eds.), *The Remote Sensing of Tropospheric Composition from Space* (Springer, Heidelberg, 2011)

Glossary of Atmospheric Chemistry Terms

Absorber A device used for the collection or removal of material by absorption, which transfers gases or liquids contained in another gas or liquid into a different phase. The device may be a packed column, scrubber, impinger, or a spray chamber.

Absorption (material) The process by which one material (the absorbent) is retained by another (absorbate); the transfer of a component from one phase to another; in atmospheric chemistry usually the transfer of gas or vapor molecules to the liquid phase.

Absorption (optical) The loss of light intensity at characteristic wavelengths caused by the interaction of light with, and the transfer of radiative energy to, an absorbing substance. In contrast to scattering, absorption results in a real loss of light intensity, whereas scattering causes merely a redirection of light in its path.

Absorption coefficient Wavelength-dependent coefficient, $\varepsilon(\lambda)$, quantifying the extent of absorption of light by a pure compound determined by the Beer-Lambert law: $\log_{10}(I_0(\lambda)/I_t(\lambda)) = \varepsilon(\lambda)\, l\, [A]$, where I_0 and I_t are the incident and the transmitted radiances (intensities), respectively, for an optical path length l and uniform concentration of absorbing compound $[A]$. If the units for concentration are mol per volume, $\varepsilon(\lambda)$ is called the decadic molar absorption coefficient. This form of the Beer-Lambert law is commonly used in reporting absorption coefficients for liquids and aqueous solutions. When Naperian (natural) logarithms are employed, a symbol different from ε should be used to avoid ambiguity. In any case it is necessary to state the base of the logarithm, units of concentration and path length when reporting absorption coefficient data.

Absorption cross section A wavelength-dependent coefficient, $\sigma(\lambda)$, specifying the extent of absorption of light by the Beer-Lambert law in the form $I_t(\lambda)/I_0(\lambda) = \exp(\sigma(\lambda)\, l\, n)$. Here, I_0 and I_t are the incident and the transmitted radiances (intensities), respectively, l is the optical path length (unit: cm), and n is the number concentration (units: molecule cm^{-3}) of the absorbing molecules. The absorption cross section $\sigma(\lambda)$ (unit: cm^2 molecule^{-1}) is commonly used in reporting optical absorption data for gases. It is related to the corresponding molar cross section κ by $\sigma = \kappa/N_A$ where $N_A = 6.02214 \times 10^{23}$ is the Avogadro constant.

P. Warneck and J. Williams, *The Atmospheric Chemist's Companion: Numerical Data for Use in the Atmospheric Sciences*, DOI 10.1007/978-94-007-2275-0, © Springer Science+Business Media B.V. 2012

Absorption spectrum A plot of the absorption coefficient or absorption cross section versus wavelength (usual) or reciprocal wavelength. A spectrum may consist of a series of discrete narrow lines (line spectrum), a group of bands that with increasing spectral resolution may be shown to consist of a series of lines (band spectrum), or a true continuous spectrum that cannot be further resolved.

Accommodation coefficient (Also called sticking coefficient) A measure of the efficiency by which molecules (atoms, particles, etc.) impinging on aerosol particles, cloud drops, etc. are captured from the gas phase. The accommodation coefficient is the fraction of the total number of collisions that result in the capture of molecules (atoms, particles, etc.) by the particle, cloud drop, etc. The others are returned to the gas phase.

Accretion The external addition of new matter to particles leading to their growth in size. In the early development of the solar system, proto-planets are said to have grown by accretion, that is by the addition of material due to collisions with other small bodies being present.

Accumulation mode A maximum in the size distribution of atmospheric aerosol particles formed by coagulation of smaller particles and deposition of condensable material from the gas phase.

Achondrite Stony meteorite without chondrules, mainly igneous in origin.

Actinic flux The total spectral flux of photons per unit area and wavelength interval available to molecules at a particular point in the atmosphere where the term *actinic* refers to radiation capable of causing photochemical reactions. The actinic flux (actually a flux density) consists of three components: direct solar radiation, diffuse radiation originating from scattering in the atmosphere, and diffuse radiation originating from reflection at the earth's ground surface. Accordingly, the actinic flux at a particular point in the atmosphere is calculated by integrating the spectral radiance over all directions of space. The actinic flux must be distinguished from spectral irradiance, which is the hemispherically integrated radiance weighted by the cosine of the angle of incidence and represents the photon flux per unit area through a plane surface. Both quantities have the same units (m^{-2}, s^{-1} nm^{-1}).

Activated particle Said of an aerosol particle that with rising relative humidity grows by the attachment of water and, when r.h. >100%, overcomes the peak in the Köhler growth curve to grow further and form cloud drops.

Activation energy An energy barrier preventing a chemical reaction (even an exothermic one) unless the reaction partners are endowed with sufficient energy to overcome the energy barrier.

Activity (chemical) In thermodynamics, the activity a_B of a substance B in a mixture of substances is related to the chemical potential $\mu_B = R_g T \log(a_B)$, where R_g is the gas constant and T the absolute temperature. In dilute solutions the activity is a relative concentration defined by $a_B = \gamma m_B / m^{\ominus}$, where m_B is the concentration in mol kg^{-1} (molality), $m^{\ominus}$ is a standard value of molality, often chosen to be 1 mol kg^{-1}, and γ is the activity coefficient of B.

Activity (optical) The phenomenon displayed by many substances whereby plane-polarized light, when passing through the substance, suffers a rotation of the plane of polarization. In the case of molten or dissolved substances, the effect is due to the asymmetric molecular structure of the substance.

Activity (radioactive) The amount of radioactive material in a unit of air, or the associated number of disintegrations per second in the same unit.

Activity coefficient A factor with which to multiply the concentration of a component in solution; it is used to correct for the departure from ideal behavior. Thus, the true activity of a component A with concentration $[A]$ is $a = \gamma [A]$. The activity coefficient γ is usually determined empirically. It depends not only on the concentration of A but also on its properties and on the concentration and kind of other substances present.

Adsorption The process by which gas molecules, dissolved substances or liquids (adsorbate) adhere to the surface of solids (adsorbent) through either physical forces (physical adsorption) or stronger chemical forces (chemical adsorption).

Adsorption isotherm A function relating the volume of vapor adsorbed on a surface to the pressure of the vapor in the gas phase at fixed temperature.

Advection The transport of air, its properties and trace materials solely by mass motion of the atmosphere, usually in horizontal direction.

Aerodynamic diameter See *equivalent diameter*.

Aerosol Suspension of an assembly of small particles (liquid, solid or a mixed variety) in a carrier gas, usually air. In the atmosphere, the size of the particles ranges from less than 10 nm to more than 100 μm, although the largest particles occur in concentrations so small that they are difficult to detect. Atmospheric aerosol particles are an integral part of and undergo transport with the air, only particles larger than about 25 μm are subject to gravitational settling. The atmospheric aerosol may be characterized by size distribution, optical properties, electric charge, radioactivity or chemical composition.

Aethalometer An instrument used to measure the optical absorption of collected aerosol samples.

Agglomerate (a) Aerosol: a cluster of particles resulting from the adherence of individual particles following successive collisions. (b) Geology: a mass of volcanic rock fragments compacted by heat, typically occurring in volcanic vents.

Aggregate A heterogeneous particle in which the individual components are not easily separated.

Air mass A qualitative term used in meteorology (mainly) to describe a body of air in the atmosphere with essentially uniform physical (not necessarily chemical) characteristics that are maintained for a limited period of time. An air mass residing over a region for a longer time undergoes chemical changes due to the addition of substances by emissions at the earth surface or the removal of substances by wet and dry deposition, etc. Frontal systems separate different air masses. A cold front, for example, separates a warm air mass in its front from cold air occurring in the rear.

Air pollutant A substance, harmful to humans, animals, or materials, which has been brought into the air, usually by human activity, at concentrations high enough to cause harmful effects. Examples for *primary* pollutants, which are emitted directly into the atmosphere, are sulfur dioxide, nitrogen oxides, and carbon monoxide. Pollutants that derive from primary pollutants by reactions in the atmosphere are called *secondary* pollutants. Ozone is a secondary air pollutant. High concentrations of ozone occur with photochemical smog, which develops when automobile exhaust gases accumulate in the air under adverse meteorological conditions (presence of an inversion layer and intensive sunshine).

Air pollution Atmospheric conditions resulting from excessively high concentrations of substances in the air that impair comfort, health or welfare of people, animals or biota.

Air pollution control Measures taken to reduce the degree and effects of air pollution.

Air pollution control district A geographical region designated by law where emissions of individual specified air pollutants are controlled to a degree specified by law.

Aitken nuclei Atmospheric particles in the approximate size range 0.01–0.1 μm.

Albedo The fraction of energy of electromagnetic radiation reflected from the surface of the earth relative to the incident energy. Whereas the terms reflectivity and spectral albedo are used to describe the reflection of monochromatic radiation, the term albedo refers to a broad wavelength band of radiation in the visible, ultraviolet or infrared wavelength range.

Aliquot A representative portion of the whole.

Alkadiene A hydrocarbon containing two >C=C< double bonds.

Alkali Any chemical substance, which when dissolved in water, forms an alkaline solution, especially the hydroxides of sodium and potassium.

Alkali metals The (monovalent) elements of the first group in the periodic table, Li, Na, K, Rb, Cs, Fr, except hydrogen. Francium (Fr) is a radioactive element not occurring naturally.

Alkaline earth metals The (divalent) elements of the second group in the periodic table, Be, Mg, Ca, Sr, Ba, Ra.

Alkalinity The extent to which a solution is alkaline. The alkalinity of seawater is defined as the sum $[Alk] = [HCO_3^-] + 2[CO_3^{2-}] + [B(OH)_4^-] + [OH^-] - [H^+]$, where the square brackets are concentrations in mmol kg^{-1}.

Alkane A fully saturated hydrocarbon. Alkanes exist as straight chains and in branched form.

Alkene A hydrocarbon containing one >C=C< double bond.

Alkoxy(l) A radical of the type RO·, where R represents an alkyl radical, produced in the oxidation sequence of an alkane by the removal of an oxygen atom from an alkyl peroxyl radical, *Alkoxy* is common usage, although the correct form would be *alkoxyl*.

Alkyl A radical arising from an alkane when a hydrogen atom is removed.

Alkyl peroxy(l) A radical formed by the addition of an alkyl radical to molecular oxygen, *Alkyl peroxy* is common usage, although the correct form would be *alkyl peroxyl*, similar to hydrogen peroxyl (HOO·).

Ambient air The outdoor air in the particular location.

Ångstrom Unit of length equal to 0.1 nm = 1×10^{-10} m. Although still in use, especially in vacuum ultraviolet spectroscopy, it is not formally defined within the International System of Units (SI).

Anomalistic year The time between two successive perhelion passages of the earth in its orbit. The time (365.25964 days) is longer than the sidereal and tropical years owing to the advance of Earth's perhelion as a result of planetary perturbations. Perhelion is the closest orbital point of Earth from the sun. It occurs around January 3rd each year.

Anthropogenic Produced by human activities.

Apogee The point in the orbit of the moon or an artificial earth satellite that is farthest from Earth.

Appearance potential (energy) In mass spectrometry with electron impact ionization, the minimum accelerating voltage required to produce observable ions. For atomic ions and molecular ions not undergoing fragmentation the appearance potential corresponds closely to the ionization energy (in electron volt). In the case of fragmentation the energy corresponds to the ionization energy augmented by the dissociation energy required to produce the fragment ions. In the case of photo-ionization it is the minimum energy of the quantum of light required to produce ionization of the absorbing molecule.

Arctic Haze The phenomenon of a visible brown haze in the Arctic atmosphere caused by the advection of polluted air.

Aromatic compound An organic compound derived from benzene.

Arrhenius equation The equation $k = A\exp(-E_a/R_gT)$, which describes the temperature dependence of the rate coefficient k for a bimolecular reaction in terms of a pre-exponential factor A and an activation energy E_a, where R_g is the gas constant and T is the temperature.

Ash The solid residue remaining after the combustion of a fuel such as coal. It consists largely of heat-treated mineral matter, but may also contain products of incomplete combustion of the fuel.

a.s.l. Abbreviation for *above sea level*, used to indicate the elevation in mountainous terrain.

Aspirator An apparatus which leads to the movement of a fluid by suction, such as a squeeze bulb, a pump, or a Venturi.

Atmosphere (a) The entire mass of air surrounding the earth, which is largely composed of nitrogen oxygen and argon. (b) A pressure unit roughly equivalent to the pressure of air at sea level: 1 atm = 101,325 Pa (exactly) ≡ 1013.25 mbar ≡ 760 Torr ≡ 760 mmHg. Although the use of atmosphere as pressure unit is no longer recommended by IUPAC (it should be replaced by the corresponding value in Pascal), it is for convenience still used in atmospheric chemistry.

Atomic absorption spectroscopy A technique to determine the abundance of elements by vaporization in a furnace and measuring the optical absorption at the wavelengths characteristic of each element.

Atomic mass The atomic mass refers to the mass of an isotope of an element; the atomic weight is the abundance-weighted average of all isotopes of the same element in the same units relative to ^{12}C. Thus, in the case of ^{12}C the atomic mass is 12.000 g, whereas the atomic weight of carbon, 12.0107 g, takes into account the contribution of about 1.1% of ^{13}C (13.00335 g). See also under standard atomic weight.

Atomizer A device used to produce droplets by mechanical disruption of a bulk liquid.

Auger Process See predissociation.

Aurora The colored light resulting from the precipitation of electrons and protons from the magnetosphere into the Earth's atmosphere, frequently occurring as a cusp or oval surrounding the magnetic poles at altitudes above 100 km.

Auto-ionization The process by which an atom or molecule excited to an electronic state at an energy level higher than the ionization threshold undergoes ionization rather than emission of radiation.

Background concentration (Background level) The concentration of a given species in pristine air as opposed to the higher concentration in polluted air. The background concentration is presumed to be nearly constant at a low level. This applies only to short-lived anthropogenic impurities, however. The global background concentration of many longer lived atmospheric trace gases such as methane, carbon dioxide, the halocarbons, etc. are variable with time due to anthropogenic sources. The true background level would be of purely natural origin if anthropogenic sources were entirely absent.

Barometric law The barometric law states that the pressure $p(z)$ at any altitude z in the atmosphere equals the pressure exerted by the total column of air overhead. The pressure decreases with increasing altitude because of the decrease in air column density. For an atmosphere in hydrostatic equilibrium, the differential decrease of pressure is given by $dp(z) = -(gM/R_g T)p(z)dz$, where g is the gravitational acceleration, M is the molar mass of air, R_g is the gas constant and T is the temperature.

Basalt Dark, fine-grained, mafic igneous rock consisting mainly of plagioclase feldspar and pyroxene.

Beer-Lambert law (also Beer's law) Relates the absorption of light to the properties of the material through which the light is traveling; for details see absorption coefficient and absorption cross section.

Benthic A term applied to the sedentary animal and plant life occurring on the sea bottom.

Bergeron-Findeisen process The initiation of precipitation in mixed-phase clouds when the ice particles grow at the expense of liquid water drops.

Bimodal distribution The occurrence of two maxima in a frequency distribution.

Biomass The mass of parts of or the total biosphere existing on earth. On a global scale, the biomass is dominated by the terrestrial biomass occurring in plants, primarily forests, amounting to approximately 650 Pg of carbon.

Biomass burning The (usually incomplete) combustion of living and dead plant organic matter in the open air, such as occurring naturally by forest fires, by the burning of agricultural wastes, or in the clearing of forest and brush land by fire for agricultural purposes. Biomass burning leads to the emission of air pollutants that make a large impact on the global budgets of many trace gases in the troposphere.

Biosphere The portion of the globe that encompasses all forms of life on the earth. It includes life forms on the continents as well as in the oceans.

Boundary layer Tropospheric region closest to the earth surface where air motions are subject to friction forces. The boundary layer is characterized by strong diurnal variations in the degree of turbulence, resulting from convective buoyancy induced by solar heating of the ground surface during the day, and by temperature inversion layers arising from radiative cooling of the surface at night.

Brownian motion The random motion of particles resulting from collisions with molecules of the environment.

Bunsen coefficient A parameter used to describe the absorption of a gas in a liquid. The Bunsen coefficient α is defined as the volume of gas reduced to 273.15 K and 1 atm (101,325 Pa) pressure that is absorbed by a unit volume of solvent under a partial pressure of 1 atm. If ideal gas behavior and Henry's law is obeyed $\alpha = (273.15/T)(V_{gas}/V_{liquid})$.

Carbonaceous chondrites Primitive stony meteorites containing up to 4% carbonaceous material by mass. Most of the carbonaceous chondrites are highly oxidized and have a chemical composition similar to that of the solar photosphere, except for very volatile elements.

Carbonyl compounds Organic compounds featuring a C=O group (carbonyl group) as part of the carbon skeleton; *aldehydes* contain the carbonyl group at a terminal position, in *ketones* the carbonyl group is positioned between two carbon atoms.

Carbonyls Compounds arising from the association of carbon monoxide with a metal, such as nickel tetracarbonyl, $Ni(CO)_4$, a colorless liquid (melting point $-25°C$, boiling point $43°C$). In the laboratory jargon of atmospheric chemists, organic carbonyl compounds (aldehydes and ketones) are sometimes called carbonyls, but this practice is discouraged to avoid possible misunderstandings.

Cascade Impactor A device used for the classification of aerosol particles according to size. It consists of a series of impaction stages with decreasing particle cut-off size so that particles are separated stepwise by momentum differences into a number of relatively narrow intervals of aerodynamic diameter and are simultaneously collected for subsequent chemical analysis.

Catalysis The increase of the rate of a chemical reaction by the addition of a substance (the catalyst), which itself does not generally undergo a net chemical change. In most applications, the use of a suitable catalyst is essential to make exploitation of a reaction for practical purposes possible. If the substance reduces the reaction rate, it is called an inhibitor. The term "negative catalysis" for this type of process has been abandoned.

Catalyst Generally, a substance whose presence accelerates the rate of a chemical reaction. In automobile technology a device used to convert harmful exhaust gases such as nitric oxide or carbon monoxide to less harmful gases (nitrogen and carbon dioxide, respectively); short for catalytic converter.

CCN Abbreviation for cloud condensation nuclei.

Chain reaction A chemical reaction mechanism involving a series of steps that are called initiation, propagation and termination. If the propagation steps recur a number of times for each initiation and termination step, the reaction is called a chain reaction. This requires that a chemical intermediate consumed in the first step of propagation is regenerated in a second or subsequent step of the propagation mechanism. A pertinent example is the chlorine-hydrogen system, which when irradiated within the Cl_2 absorption band produces two chlorine atoms that react with hydrogen, thus $Cl + H_2 \rightarrow HCl + H$ followed by $H + Cl_2 \rightarrow HCl + Cl$, whereupon the reaction continues. Termination occurs by association of chlorine or hydrogen atoms assisted by a third body. Under suitable conditions (low light intensity) the system leads to chain lengths of up to a million, that is 10^6 HCl molecules are formed for each Cl_2 molecule being photo-dissociated. The oxidation of hydrocarbons in the atmosphere occurs by complex chain reactions involving OH, HO_2 and RO_2 radicals, the oxidation of NO to NO_2 and the photo-dissociation of the latter.

Chemiluminescence The emission of light from an electronically excited product resulting from a chemical reaction.

Chemiluminescence analyzer An instrument utilizing a chemiluminescent reaction for the detection and analysis of a species. The instrument consists of a reaction chamber with separate inlets for the sample and the reactant gas, an optical filter, a photo-multiplier and signal processing electronics. The reactive gas usually is introduced in excess. The quantity of light produced is proportional to the sample flow rate and the concentration of the measured substance in the sample under specified conditions of temperature and pressure. The filter limits the wavelength to the spectral region of interest and helps to eliminate interferences. The most widely used chemiluminescence analyzer makes use of the reaction between O_3 and NO and is used for the determination of NO_x in air.

Chondrites The most abundant class of stony meteorites, so-called because they contain chondrules. The term is also applied to all meteorites with bulk composition close to that of the solar photosphere, even if they do not contain chondrules (e.g. CI chondrites).

Chondrules Millimeter-sized, nearly spherical grains consisting mainly of olivine and/or low calcium pyroxene, found in chondritic meteorites.

Chromosphere (of the Sun) Transparent, intermediate region between photosphere and corona. Temperatures in the solar chromosphere range from ~4,000 K to 50,000 K at its top.

Cirrus Fibrous ice clouds.

Clastic rocks A term used in geology to describe rocks that consist of fragments of previous rocks. The term is broadly synonymous with sedimentary rocks.

Cloud Generally, an atmospheric aerosol which is dense enough to be perceptible to the eye. The term usually is applied to water clouds, that is, an assembly of small water drops or ice particles suspended in the atmosphere. However, dust clouds also may occur. In this case, a cloud consists of an assembly of particles that has a density about 1% higher than the density of air alone.

Cloud condensation nuclei The fraction of atmospheric aerosol particles that can be activated to form cloud drops at low supersaturation of water vapor, that is, at relative humidities less than about 102%. See also under Köhler diagram.

Cloud element A water drop or ice crystal in a cloud below the size characterizing a rain drop (usually <100 μm).

Coagulation A process by which collisions of aerosol particles with each other and their agglomeration shift the size distribution of an aerosol toward larger values.

Coarse particle mode In the size distribution of the atmospheric aerosol: the largest particles (ca. 2–20 μm) consisting primarily of particles generated by mechanical disintegration processes.

Column density Number of atoms or molecules per unit area. Often obtained from optical absorption measurements with the sun as background source and used to express the total abundance of a constituent in the atmosphere integrated with regard to altitude.

Concentration The chemical amount, number of molecules, or mass of a substance of concern in a given volume divided by that volume. Concentration is the basic quantity required to describe the rate of a chemical reaction. The preferred concentration unit used in gas phase kinetics in atmospheric chemistry is molecule cm^{-3} or molecule m^{-3}; the SI units are mol m^{-3}. Mass concentration is appropriate when the chemical composition of a sample is ill-defined or unknown. Number concentration and mass concentration are used to characterize the atmospheric aerosol (particle m^{-3}, mg m^{-3}).

Condensation The physical process of converting a material from the gas or vapor phase to a liquid or solid phase. Condensation occurs when the partial pressure of the substance exceeds the saturation vapor pressure at a given temperature. This is achieved either by raising the partial pressure or by lowering the temperature.

Condensation nuclei counter (Also known as Aitken nuclei counter) A device in which sub-micrometer-sized particles are grown by vapor supersaturation to a larger size so that they can be detected by light scattering.

Coordinated Universal Time (abbreviated UTC) is the time standard based on International Atomic Time (TAI) with leap seconds added at irregular intervals to compensate for the Earth's slowing rotation. Leap seconds are used to allow UTC to closely track the mean solar time at the Royal Observatory, Greenwich (UT1). Since the difference between UTC and UT1 is not allowed to exceed 0.9 s, the general term Universal Time (UT) may be used if high precision is not required. In casual use, when fractions of a second are not important, Greenwich Mean Time (GMT) can be considered equivalent to UTC or UT1.

Corona (a) Outermost extended region of the Sun, where temperatures reach up to 2×10^6 K. (b) An abbreviation for a corona discharge.

Corona discharge A (sometimes visible) electric discharge resulting from the ionization of a gas, such as air, surrounding a wire or other electric conductor maintained at high electric potential.

Correlation cell radiometer See under gas correlation radiometer.

Cosmic rays Highly energetic particles continuously bombarding the earth from all directions. The particles are composed of all atomic nuclei. Cosmic rays are produced by supernova explosions, pulsars, and perhaps other cosmic events. Cosmic rays interact with nitrogen, oxygen and argon in the atmosphere to produce radioactive nuclei, primarily via spallation reactions.

Coulomb force The magnitude of the electrostatics force of interaction between two point charges is directly proportional to the scalar multiplication of the magnitudes of charges and inversely proportional to the square of the distances between them.

Coulter counter An instrument that measures the volume of individual particles in a liquid by means of the change in resistance as the liquid passes through an orifice.

Critical orifice An orifice through which a constant airflow is maintained when the pressure drop across the orifice is sufficient to cause sonic flow.

Cumulus Clouds with significant vertical extent.

Cut-off particle diameter The diameter of a particle, which has 50% probability of being removed by a device or stage and 50% probability of passing through; a design characteristic especially of an inertial impactor.

Deactivation Loss of energy (total or partial) from an excited (energy-rich) atom or a molecule due to energy transfer by collisions with other atoms or molecules in the surroundings; also called quenching.

Deliquescence The process of dissolution of a salt particle by the absorption of moisture from the surrounding air. The process occurs when the vapor pressure of the saturated aqueous solution of a substance is less than the vapor pressure of water in the air. A crystalline salt aerosol particle will deliquesce in the atmosphere when the relative humidity surpasses a characteristic value, the so-called *deliquescence point.*

Denitrification The process by which bacteria reduce oxides of nitrogen (nitrate, nitrite) to molecular nitrogen (and nitrous oxide, a minor fraction). All denitrifying bacteria are aerobic species that turn to oxides of nitrogen as a source of oxygen when the level of oxygen becomes too low.

Deposition velocity A measure of the efficiency of dry deposition. The deposition velocity is defined by the ratio of flux density of a substance being deposited at the ground surface to the concentration of the substance at a reference height above ground, usually 1 m. The deposition velocity also represents the inverse of the sum of resistances exerted on the transfer of material by the gas phase and the uptake of material at the ground.

Diagenesis In geology, the process by which ocean sediments are modified to form sedimentary rocks. Diagenetic processes include cementation, compaction, diffusion, redox reactions, transformation of organic and inorganic material, and ion exchange phenomena.

Diffraction Change in direction and amplitude of radiation as it passes near an object or through an orifice.

Diffusion Net movement of particles or molecules from a region of higher to one of lower concentration.

Diffusion battery An aerosol spectrometer for sub-micrometer-sized aerosol particles, in which the size is measured by diffusive loss of particles in ducts of various lengths.

Diffusion denuder An apparatus for the separation of particles and gases based on the difference in diffusion velocity, often used for the purpose of separate chemical analysis. Usually a narrow flow tube is used, which removes gases by reaction with a wall coating, whereas particles pass through the tube and are collected on a filter.

Dispersion (In the atmosphere) The dilution of a pollutant or tracer substance by spreading due to diffusion or turbulent mixing (eddy diffusion); (In chemistry) A system consisting of particles suspended in a fluid; a colloid; (In spectroscopy) The spectral resolving power of a device producing a spectrum from white light; Specifically, the variation of the refractive index of material used in a prism to produce a spectrum.

DOAS (Differential optical absorption spectrometry) A technique for the detection and measurement of certain trace gases in the atmosphere that show a discrete absorption spectrum in an optically accessible spectral region. A long path length (km) is required and the spectrum must be differentiated against the background absorption.

Dobson unit Unit used to describe the total amount of ozone in a vertical column of air overhead (DU). It is given as the thickness (in units of 10^{-3} cm or 10^{-5} m) of the layer that would form if the total ozone in the column were reduced to 1 atm pressure (1 atm = 1013.25 hPa) and 0°C temperature. One DU is approximately equivalent to an ozone column density of 446.2 μmol m^{-2}. A typical column abundance of ozone of 300 DU corresponds to approximately 134 mmol m^{-2}.

Drag coefficient A parameter that relates the drag force experienced by an object to its velocity.

Drag force The resistance experienced by an object when moving in a fluid.

Dry deposition Removal of a trace substance from the troposphere by downward transport and absorption by, and/or chemical reaction with, materials at the earth surface.

Dust Air-borne solid, dry, small particles formed by wind erosion, volcanic eruption or other mechanical disintegration of parent material. Dust particles generally are of irregular shape and occur in the size range 1–100 μm in diameter. They settle slowly under the influence of gravity.

Dust collector A device for monitoring dust emissions. Also, the equipment used to remove and collect dust from exhaust gases. This may employ simple sedimentation (dustfall jars) inertial separation (cyclones), precipitation (electrostatic), or filtration.

Eddy In turbulent fluid motion a small vortex that maintains its character for some time before it blends with the rest of the fluid.

Eddy diffusion A synonym for the process by which individual air parcels are formed and mixed into the surroundings due to atmospheric turbulence, equivalent to the dispersal of an atmospheric tracer due to turbulent mixing. This process can be treated mathematically by a (three-dimensional) diffusion equation, in which the flux of the tracer is proportional to the gradient of the mixing ratio of the tracer. The coefficient involved is treated as an empirical parameter called the *eddy diffusion coefficient.*

Eddy diffusion coefficient A parameter used in the eddy diffusion equation to quantify the rate of turbulent mixing or eddy diffusion. Its magnitude increases with the degree of turbulence in the air, which, in turn, increases with wind force. The eddy diffusion coefficient can be determined empirically by measuring the rate of dispersion of a tracer in the atmosphere.

Efflorescence The reverse of deliquescence, that is, the drying of a salt solution when the vapor pressure of water in the saturated solution of a substance is greater than the partial pressure of water in the ambient air. The term is also used to describe the loss of crystal water from solid salts such as $Na_2CO_3.10H_2O$.

Electrical aerosol analyzer (also electrical aerosol classifier) An aerosol size spectrometer in which the particles are separated by their different mobilities in an electric field. Those particles occurring within a certain size range can be selected for analysis.

Electrochemical methods (of analysis) Methods in which the change of either current or potential resulting from an electrochemical reaction is measured. The gas or liquid containing the species to be analyzed is sent through an electrochemical cell, where the reaction with the species takes place.

Electrostatic precipitator A device in which airborne particles are charged in a unipolar field and precipitated in a high-voltage electric field.

Elemental carbon (EC) Carbon remaining after a filter sample is oxidized at high temperatures (340°C) in oxygen to convert organic carbon (OC) to carbon dioxide.

Emission General term to describe the discharge of material into the atmosphere. Frequently it is used more specifically for the rate at which a solid, liquid or gaseous pollutant is emitted from a given source. The rate is usually expressed as mass per unit time.

Emission flux The emission per unit area of an appropriate surface of an emitting source.

Emission inventory A systematic survey of air pollution emissions in a given area. The information usually includes the types of sources, such as electric power plants, refineries, traffic routes, etc., as well as regional source distributions, that is, composition and rate of discharge per area. The inventory may focus on a specific pollutant. The variation with time of day, month or year also is desirable information in inventories.

Enrichment factor This factor quantifies the enrichment often observed with elements of the tropospheric aerosol compared to their abundances in Earth's crust. The enrichment factor is defined by $EF=([X]/[Ref])_{aerosol}/([X]/[Ref])_{crust}$; [X] is the concentration of the element in either the aerosol sample or the average crust, and [Ref] is a suitably chosen reference element (usually aluminum), again in either the aerosol sample or the Earth's crust.

Entrainment The mixing of surrounding air into a cloud, frequently from the top.

Epiphanometer An instrument which measures the surface area of aerosol particles.

Equation of state A thermodynamic equation describing the state of matter under a given set of physical conditions. It is a constitutive equation which provides a mathematical relationship between two or more state functions associated with the matter, such as its temperature, pressure, volume, or internal energy. Equations of state are useful in describing the properties of fluids, mixtures of fluids, solids, and even the interior of stars.

Equinox The dates around March 21st and September 21st, when the sun crosses from south to north of the earth's equator in the first case (vernal equinox), and from north to south in the second case (autumnal equinox). Equinoxes are not fixed in position but are moving retrograde (westward) at 50.28 arc seconds per year (precession of equinoxes). See also Tropical year.

Equivalent A valence concentration of a solute in aqueous solution to express concentration of a multiply charged ion for the purpose of demonstrating charge balance. For example, the neutralization of NaOH by H_2SO_4 requires one half the amount of acid compared to that of alkali, because two protons are liberated from the acid, whereas one hydroxyl ion is liberated from NaOH. The customary procedure has been to multiply the concentration of each ionic species i by its valence z_i and to compare the totals. Nowadays, the use of terms such as equivalent, gram equivalent, equivalent concentration, etc. is considered obsolete. It has been replaced by the equivalent entity defined as the fraction $(1/z_i)$ of the molecular entity. Thus, the equivalent entity of the sulfate ion is $½SO_4^{2-}$. For a one molar solution of sulfuric acid $([SO_4^{2-}]=1\,M)$ the corresponding equivalent entity is $[½SO_4^{2-}]=2\,M$.

Equivalent diameter The (theoretical) diameter of an aerosol particle that corresponds to the diameter of a particle with known properties, which produces the same effect. For example, the *aerodynamic equivalent* diameter is that of a unit density sphere having the same gravitational settling velocity as the particle in question; an *optical equivalent* diameter is that of a calibration particle, which scatters as much light in a specific instrument as the particle being measured.

External mixture A term used to describe an aerosol composed of at least two groups of particles with distinctly different chemical composition.

Extinction coefficient A parameter used to specify the loss of intensity along the path of a light beam caused by the combined action of optical absorption and scattering by atmospheric molecules and particles. While absorption causes a real loss of light, scattering results only in a redirection of light out of the beam.

Feldspar Solid solution of calcium-, sodium-, and potassium-aluminosilicate minerals (anorthite, albite, orthoclase).

Ferrel cell In atmospheric meridional circulation the Ferrel cell is a weak secondary feature, which is induced by and occurs poleward of the main Hadley cell but rotates in opposite direction.

Filter A porous material on which particles present in the air (or in another fluid) are caught and retained.

Filter, ceramic A filter made of porous ceramic material.

Filter, fibrous A filter consisting of a mat of fibers, such as fabric or glass.

Filter, membrane A filter prepared with well-defined pore size in a sheet of suitable material (polyfluoroethene, polycarbonate, cellulose esters). The pores consist of 80–85% of filter volume; several pore sizes are available for air sampling (0.45 and 0.8 μm are commonly employed).

Filtration A process of separating different phases in a mixture of substances, such as the separation of suspended solids from a liquid or gas, by forcing the liquid or gas through a porous medium.

Fixed nitrogen This term is used to describe nitrogen contained in chemical compounds that all plants and microorganisms can use. Bacterial nitrogen fixation (see there) is an energetically costly process; the biosphere as a whole tends to preserve fixed nitrogen by extensively circulating it within an ecosystem.

Flash point The lowest temperature at which a substance such as fuel oil will give off a vapor that will flash or burn momentarily when ignited.

Flocculation The process of contact and adhesion of particles, whereby larger clusters are formed. The term is synonymous with *agglomeration* and *coagulation*.

Floccule A small loosely aggregated mass of material suspended in or precipitated from a liquid; a cluster of particles.

Fluid flow The movement of air or other fluid in the open or in a duct, pipe, or passage. Several types of flow are distinguished: *Uniform flow* is steady in time, or the same at all points of space; *steady flow* is one for which the velocity at a point in a fixed system of coordinates is independent of time; *viscous flow* (also called laminar flow or streamline flow), occurring at low Reynolds numbers, features a transverse velocity gradient due to viscous shear and zero velocity at the stationary boundary or boundaries, but the fluid velocity at a fixed point remains constant; *turbulent flow*, occurring at high Reynolds numbers, is characterized by a fluid velocity at a fixed point that fluctuates randomly in time.

Fluorescence The spontaneous re-emission of radiation from an excited state of a molecule or atom, following the absorption of light. If the re-emission connects the initially excited energy level and the ground state, the fluorescence light has the same frequency as the excitation radiation. In this case, the process is called *resonance fluorescence*. It is the only process available to atoms. In molecules the energy may be degraded by collisional quenching to a vibrational level lower than the excitation level. This, and transitions to higher vibrational levels of the ground state, result in a shift of fluorescence light to longer wavelengths. Fluorescence occurs essentially only during irradiation with light of specific wavelengths.

Fly ash Particles of ash entrained in the flue gas resulting from the combustion of fossil fuels.

Fog A general term applied to a suspension of droplets in a gas. In meteorology, it refers to a suspension of water drops resulting in a visibility of less than 1 km.

Fourier-transform-spectrometry A technique for the collection of infrared spectra using a Michelson interferometer. Incoming radiation is split into two beams by means of a half-transparent mirror. One of the two beams is reflected from a fixed mirror, the other beam is reflected from a moving mirror introducing an optical path difference. Both beams are brought to interference and the intensity variations are recorded as a function of the position of the moving mirror. To obtain the source spectrum, the interferogram is subjected to a mathematical procedure known as inverse Fourier transformation.

Fraunhofer lines Absorption lines in the spectrum emitted by the photosphere of the Sun. At visible wavelengths, the most prominent lines are due to singly ionized calcium, neutral (atomic) hydrogen, sodium and magnesium. Many weaker lines derive from iron.

Free troposphere The region of the troposphere outside the boundary layer, located between the top of the boundary layer and the tropopause.

Front The concept of a front refers to the phenomenon of a sharp boundary between geophysical fluids of different density, or temperature or salinity. In the atmosphere, fronts commonly are associated with low pressure systems (cyclones), but may occur also on a lager scale such as the polar front in the 30–40° latitude belt. A typical front in the ocean is the gulf stream front.

Fumarole(s) Small vent(s) on the flanks of a volcanic cone, or in the crater itself, from which gaseous products emanate.

Fume Fine solid particles, predominantly less than 1 μm, that are the result of condensation of a vapor produced in a chemical reaction (often from combustion). In popular usage, the term *fumes* is often used to describe unpleasant and malodorous airborne effluents that may arise from chemical processes.

Gas correlation radiometer An instrument used to measure the concentration of a trace gas in the air by means of infrared absorption spectrometry combined with a modulation technique. The radiation of a suitable light source (or solar radiation) is passed through a cell containing the same gas as that to be monitored. Within an absorption band, the response is greatest at the wavelengths of the absorption lines and smallest at wavelengths in between. Thereby, the instrument offers good selectivity and signal to noise ratio.

Gaussian distribution A distribution function describing the normal curve of error: $f(x) = [\sigma(2\pi)^{1/2}]^{-1}\exp[-(x - \bar{x})^2/2\sigma^2]$, where $\bar{x}$ is the arithmetic mean of x and σ is the standard deviation.

Geostrophic Implies balance between the Coriolis force and the horizontal pressure gradient force in the atmosphere.

Global circulation (atmosphere) The large-scale, global circulation of the atmosphere is characterized by circulation cells (Hadley cell and Ferrel cell), the intertropical convergence zone, the tropical easterly wind regime, the extra-tropical westerly wind regime, and meandering jet streams. (ocean) The circulation of the global ocean describes the transport of water masses through the large ocean basins. It includes wind-driven circulation, thermo-haline circulation controlled by surface sources of heat and salt inflow, which generates convective sinking of water masses at high latitudes and gradual up-welling at low latitudes, connected by the conveyor belt mechanism of heat and salt transport.

Global warming potential A factor, which compares the amount of heat trapped by a certain mass of the gas in question to the amount of heat trapped by a similar mass of carbon dioxide over a specific time interval.

GMT (Greenwich Mean Time) See Coordinated Universal Time (UTC).

Granite Igneous rock consisting mainly of alkali feldspar and quartz.

Graupel A larger ice particle formed by the attachment and freezing of water drops onto an ice crystal.

Gregorian calendar The calendar, now in use throughout most of the world, was instituted by Pope Gregory XIII in 1582, when it became necessary to correct the Julian calendar, used previously, for an accumulated error of 10.4 days. The correction involved removing 10 days from the running year, so that October 15th followed immediately October 4th. A year has 365 days, leap years with 366 days are inserted every year divisible by 4 (as in the Julian Calendar), but centesimal years are leap years only if divisible by 400 (unlike the Julian calendar). Thus, the year 2000 was a leap year, but not the year 1900.

Gross Primary Production *gross primary production* (GPP) is the rate at which primary producers within an ecosystem, that is, plants, algae, etc. capture and store a given amount of chemical energy as biomass in a given length of time. Some fraction of this fixed energy is used by primary producers for cellular respiration and maintenance of existing tissues (i.e., "growth respiration" and "maintenance respiration"). The remaining fixed energy (equivalent to the mass of photosynthate) is referred to as *net primary production* (NPP).

GWP Abbreviation for Global Warming Potential.

Hadley cell The zonal mean mass circulation of the atmosphere in the vertical-meridional plane exhibits an upward transport in equatorial latitudes and a downward transport in the sub-tropics. The pattern exists in both hemispheres but is most pronounced in the winter hemisphere.

Half-life time Time period in which a given species decays to one half of its initial concentration. Usually used for radio-active elements, rarely for chemical reactions. If the decay constant is λ [s^{-1}], the half-life time is $t_{\frac{1}{2}} = (\ln 2)/\lambda$.

Haze The condition of reduced visibility resulting from increased light scatter in the presence of aerosol particles. The effect may be produced both by an increased concentration of the particles or by the swelling of aerosol particles with the uptake of water with increased relative humidity.

Henry's law Originally the statement that in a closed system containing a gas phase and an aqueous phase at equilibrium the partial pressure p of a substance in the gas phase is proportional to its concentration c in the aqueous phase under ideal conditions: $p = k \times c$. The coefficient k depends on temperature. The condition of ideality requires that the substance does not interact with water, e.g. to form dissociated or ionized species, and it requires in addition that the solution is sufficiently dilute so that the interaction of solute molecules with each other becomes negligible. Henry's law partitioning of gases and vapors is important in clouds and other systems of aqueous particles. In atmospheric chemistry there is an increasing tendency to apply Henry's law in the form $c_a = K_H \times p$ with units mol dm^{-3} for aqueous concentration and atm (1 atm = 1.01325 Pa) for pressure.

Heterosphere The upper portion of the atmosphere where individual atmospheric constituents undergo diffusive separation so that the molar mass of air changes as a function of altitude.

Homopause (also termed turbopause) A border region occurring at ~100 km altitude, where the physical conditions of the atmosphere change from that of a mixed state characterized by an essentially constant molar mass to a state characterized by diffusive separation of individual constituents, causing each constituent to follow its own individual scale height.

Homosphere The bulk of the atmosphere below about 100 km altitude where transport and mixing of air parcels is dominated by turbulent motions. In the homosphere the relative proportions of the main constituents of air, nitrogen, oxygen and argon, remain constant.

Hydrometeor Any atmospheric particle with water as the dominant constituent.

Hydrosol A suspension of particles in water, an aqueous colloid.

Hydrostatic equilibrium This condition applies to fluids at rest. Hydrostatic equilibrium is said to exist when there is no net force in vertical direction. In this case, the vertical pressure gradient force balances the force of gravity: $dp/dz = \rho\, g$, where ρ is the density of the fluid and g is the acceleration due to gravity. For liquids ρ is constant and the pressure is $p = \rho\, g\, h$, where h is the column height of the liquid above the point of measurement. In the atmosphere, $\rho(z) = (M/R_g T(z)) p(z)$ is a function of altitude, which depends on pressure as well as on temperature; M is the molar mass of air and R_g is the gas constant. The simple equation for liquids remains nevertheless valid, provided the column height h is replaced by the scale height $H_z = R_g T_z / Mg$ at the altitude considered. The column density of air above the selected altitude is $\rho_z H_z$ so that $p_z = \rho_z\, g\, H_z$.

Hydroxyl The radical derived from water (H_2O) by the removal of a hydrogen atom. In the atmosphere, the OH radical is responsible for initiating most oxidation processes. In the gas phase it arises mainly from the reaction of water vapor with excited oxygen atoms $O(^1D)$, produced in the ultraviolet photolysis of ozone, and also by photolysis of nitrous acid, HONO.

Hygroscopicity The property of a substance such as salt indicating its tendency to absorb water from the surrounding air.

Hysteresis Generally, a memory effect of a material after the removal of a force or strain, best known in magnetism. Specifically for a salt particle in the air, the observation that when the relative humidity is increased marginally beyond the deliquescence point the salt crystal suddenly acquires sufficient water to form a solution, whereas upon reducing the relative humidity below the deliquescent point, the salt solution remains liquid until much lower values of relative humidity are attained.

Ideal gas law The equation of state of an ideal gas, which is based on the assumption that intermolecular attractive forces can be neglected. It may be expressed by $p = N\, k_B T$, where p denotes the pressure, N is the number concentration of atoms or molecules, k_B is Boltzmann's constant, and T is the absolute temperature; an alternative expression is $p = (n/V)R_g T$, where n [mol] is the chemical amount present in the volume V, and R_g is the universal gas constant, equal to the product of Boltzmann's constant and Avogadro's number. The ideal gas law is a good approximation for air at pressures encountered in the atmosphere and many trace gases.

Igneous Descriptive of the rock produced when hot molten material from the earth' interior cools either at depth or at the land surface.

Igneous rock Rock formed by melting and subsequent solidification.

Impaction A forcible contact of (aerosol) particles with a surface.

Impactor See inertial impactor.

Impingement Equivalent to impaction; usually refers to a liquid surface.

Impinger A sampling device for the collection of particulate matter in water or an aqueous solution contained in a bottle, with a concentric inlet tube ending in a nozzle of millimeter size immersed in the liquid. Common volumes are the (a) midget impinger, 1–10 cm^3 water, (b) the standard impinger, 75 cm^3 water. Impingers have also been used for the collection of trace gases and vapors in the air.

Inert gas A non-reactive gas under particular conditions, usually a carrier gas; examples are the noble gases, and nitrogen at ordinary temperatures.

Inertial impactor A device for the collection of aerosol particles by impaction on a surface. Ambient air is drawn through a nozzle and directed at normal incidence onto the collecting surface, where the larger particles are deposited due to inertial forces. Smaller particles are carried along with the air flow around the collector. Size and shape of the nozzle determines the cut-off size of the particles that are collected (See also Cascade Impactor).

Insolation Exposure to solar radiation.

Intensity (radiation) In a strict sense, the radiant power per unit solid angle (W sr^{-1}); the term spectral intensity implies differentiation with regard to wavelength (W sr^{-1} nm^{-1}). On the other hand, *intensity* is also used rather loosely to signify the *amount of light*; there is no objection to this usage if the appropriate units are provided.

Interception The process by which particles are caught and retained by an object when the particle carried along with the air flow meets the object's boundary within one particle radius.

Internal mixture A term used to describe an aerosol in which the chemical constituents present in the mixture occur in all the particles of the aerosol. A uniform chemical composition is not necessary. The tropospheric aerosol is – to a large extent – internally mixed because coagulation of particles and deposition of condensable vapors tend to favor mixing processes. Compare this with the term *external mixture*, which indicates the presence of groups of particles that are chemically different.

Intertropical Convergence Zone (ITCZ) is the region encircling the earth near the equator where winds originating in the northern and southern hemispheres come together.

Inversion (temperature) A departure from the normal decrease of temperature with increasing altitude in the troposphere. A temperature inversion may be produced, for example, by the advection of a warmer air mass over a cool one. The inversion layer impedes convection and hampers the vertical exchange of air, so that ground level emissions become trapped in the region below the inversion.

Ionic strength A measure of the total number of ions present per unit volume of an ionic solution. It is defined by $I = \frac{1}{2}\Sigma\, z_i^2 C_i$, where z_i is the charge of the ion of type i, C_i is its concentration, and the sum includes all types of positive and negative ions.

Ionization potential In mass spectrometry, the minimum electron energy necessary for the production of an atomic or molecular ion, given in units of electron volt (eV). The quantity corresponds to the voltage required for acceleration of the electron to reach the required energy for ionization.

Isoaxial sampling Sampling condition in which the air flowing into a sampling inlet has the same direction as the ambient air flow.

Isokinetic sampling Sampling condition in which the air flowing into a sampling inlet has the same velocity and direction as the ambient air flow.

Isotope(s) Atoms that contain the same number of protons but a different number of neutrons are called isotopes. The number of protons determines the atomic number, whereas the total number of nucleons (sum of protons and neutrons) in the nucleus, determines the mass number, as well as the atomic mass.

Jet stream (subtropical jet) A region of westerly maximum wind speed concentrated in mid-latitudes of both hemispheres at a pressure of about 200 hPa. It originates from the differential heating of the rotating planet and is a consequence of the Hadley circulation.

j-**Value** See photodissociation coefficient.

Kelvin effect The increase in partial vapor pressure over a curved surface such as that of a small water drop or a deliquesced aerosol particle compared to the vapor pressure in equilibrium with a flat surface of the same liquid.

Knudsen number Ratio of gas molecular mean free path to the physical dimension of a particle; indicator of free molecular flow versus continuum gas flow.

Köhler diagram A semi-logarithmic plot of the radius of an aerosol particle as function of relative humidity, usually for particles composed of a pure salt. The Köhler curve combines two effects: Raoult's law, which relates the vapor pressure of water above a salt solution with the concentration of dissolved salt, and Kelvin's law describing the influence of surface curvature upon the vapor pressure of water. The Köhler diagram shows a moderate increase of particle size with rising relative humidity at r.h. <100%. In the region of supersaturation (r.h. >100%) a critical size exists, which depends on the size of the dry particle and the degree of supersaturation. Particles that reach and overcome the critical size are able to grow further and form cloud drops.

Lapse rate The rate of change of temperature with altitude, dT/dz. In the troposphere the gradient is negative, that is, the temperature decreases with altitude due to adiabatic expansion. In dry air the gradient is (minus) 9.8 $K\,km^{-1}$ (Dry adiabatic lapse rate). The presence of moisture lowers the lapse rate to a tropospheric average of about 6.5 $K\,km^{-1}$.

Lee wave A wave-like cloud pattern in the rear of a mountain induced by adiabatic air flow over the mountain top.

Lidar (Light detection and ranging) A technique for the measurement of aerosol particles and trace gases in the atmosphere based on the emission of light pulses from a laser and the detection of the backscattered signal following the emission.

Life time Period of time (τ) during which a substance decays to $1/e$ of its initial concentration ($e = 2.718...$, base of the natural logarithm). For substances that decay by first order processes $\tau = 1/k$, where k is the rate coefficient for the reaction A → products. In atmospheric chemistry, second order reactions are often treated as pseudo-first order processes by assuming constant average concentrations of the second reactant(s). Thus, for a substance undergoing simultaneous reactions with other species and photolysis, the life time is $\tau = 1/(\Sigma k_i [X_i] + j)$, where $[X_i]$ designates the reactant concentration of the ith reactant, the k_i are the corresponding bimolecular (second order) rate coefficients and j is the photolysis rate coefficient (j-value) of the substance under consideration.

Life time (global average) The life time of a given species averaged over latitude, season, year, etc.

Limestone Limestone is a sedimentary rock composed largely of the minerals calcite and/or aragonite, which are different crystal forms of calcium carbonate ($CaCO_3$).

Liquid water content (LWC) the mass of liquid water in a volume of moist air.

Loess An aeolian sediment formed by the accumulation of wind-blown silt, typically in the 20–50 μm size range, and lesser and variable amounts of sand and clay that are loosely cemented by calcium carbonate.

Lognormal distribution A distribution function in which the logarithm of a quantity is normally distributed, i.e. $F(y) = F_{Gauss}(\ln y)$ where $F_{Gauss}(x)$ is a Gaussian distribution. The size distribution of aerosols is often described by a combination of lognormal distribution functions.

Luminescence A general term used for the emission of light by a molecule at temperatures below those required for incandescence. Luminescence generally originates from electronically excited states of molecules (or atoms captured in a solid matrix). If the state is short-lived as for an allowed electronic transition, the emission follows optical excitation almost immediately. In this case the emission is called *fluorescence*. If the emission of radiation occurs from a long-lived energy level by a spin-forbidden transition to the ground state, the emission is significantly delayed and the process is called *phosphorescence*. Optical absorption usually involves allowed transitions to an excited singlet state, from which long-lived lower-lying triplet states are populated by internal energy transfer or intersystem crossing.

Luminol (5-amino-2,3-dihydro-1,4-phthalazinedione, $C_8H_7N_3O_2$) is a versatile chemical that exhibits chemiluminescence, with a striking blue glow, when mixed to react with an appropriate oxidizing agent. It is a white to slightly yellow crystalline solid that is soluble in most polar organic solvents, but insoluble in water. It has been used for measurements of nitrogen dioxide.

Mafic A term used to describe one of the two classes of primary silicate minerals in igneous rocks, namely the ferromagnesian or mafic series, composed of iron and magnesium minerals (olivine, pyroxene). The other type is the felsic series, composed of aluminum silicates (feldspars).

Meridional Along a meridian, that is, in the north–south direction.

Metamorphic Refers to an earth material altered by intense heat or pressure.

Meteor The visible path of a meteoroid that enters Earth's atmosphere, colloquially called a shooting star. The incandescence generally ceases to be visible before the body falls to the earth. If it reaches the ground and survives impact, then it is called a meteorite.

Meteoroid An up to boulder-sized body of debris in the interplanetary space (smaller than an asteroid).

Mie Scattering In the theory of light scattering by dielectric spheres of radius r, Rayleigh scattering refers to the case $r \ll \lambda$ whereas Mie scattering refers to the case $r \geq \lambda$. In the atmosphere, the latter conditions are met for aerosol particles with radii greater than about 0.25 μm. Mie scattering is nearly independent of wavelength in the near ultraviolet and visible spectrum of light, in contrast to Rayleigh scattering.

Milankowitch Cycles Climatic patterns on Earth caused by orbital variations in eccentricity and oscillations of tilt and precession of the Earth's axis.

Mixing ratio This is a dimensionless fraction specifying the ratio of the amount, mass or volume of a substance in a homogeneous chemical system to the amount, mass or volume of all substances present in the system. In atmospheric chemistry the molar mixing ratio (amount of substance fraction) is often used to describe the abundance of a trace gas in an air parcel because it is independent of pressure and temperature changes associated with altitude or meteorological variability. Since in this case the mixing ratio refers to the total gas mixture, the presence of water vapor causes the mixing ratio to vary with humidity. For water vapor the mass mixing ratio is often used with reference to the mass of dry air. For solid materials, such as soils, minerals, or aerosol particles the mass mixing ratio is used to quantify the abundance of trace elements. The type of mixing ratio used must be specified by the appropriate SI unit, for example pmol mol^{-1}, or μg g^{-1}.

Mole fraction The fraction that an amount of substance contributes to the total amount of the sample (chemical amount fraction); the molar mixing ratio.

Net Primary Productivity The amount of organic matter that is produced by photosynthesis in plants (the difference between autotrophic photosynthesis and respiration).

Nimbus A raining cloud.

Nitrification The oxidation of ammonia to nitrous acid and nitrate by specialized bacteria in well-aerated soils (see also nitrogen fixation and denitrification).

Nitrogen fixation The process by which a small group of bacteria fitted with the necessary enzyme system reduce molecular nitrogen and incorporate it in their cells. The decay liberates fixed nitrogen in the form of ammonium and makes it available to plants and other bacteria.

NMHC Non-methane hydrocarbon.

Noble gases The inert gases He, Ne, Ar, Kr, Xe, and Rn. Radon is a radioactive element produced by the decay of radium.

Normal distribution A bell-shaped or Gaussian distribution (see there).

Non refractory (Aerosol) A particle not having the ability to retain its physical shape and chemical identity when subjected to high temperatures.

Nucleation The process by which particles are initially formed from a (supersaturated) vapor.

Nucleation, heterogeneous The formation of droplets by condensation onto pre-existing nuclei, so-called condensation nuclei. In the atmosphere, condensation nuclei are provided by aerosol particles that can grow to form cloud drops at a low degree of supersaturation (<102% relative humidity).

Nucleation, homogeneous The formation of droplets in the absence of condensation nuclei; also called self-nucleation. This process can be observed only in the laboratory. In the atmosphere the ubiquitous presence of aerosol particles favors heterogeneous nucleation.

ODP Abbreviation for Ozone Depletion Potential.

Ostwald coefficient A parameter used to describe the absorption of a gas in a liquid. The Ostwald coefficient is defined as the ratio of the volume of gas absorbed to the volume of the absorbing liquid, all measured at the same temperature.

Oxidant A qualitative term used for atmospheric species with an oxidative power stronger than that of oxygen. Examples include ozone, hydrogen peroxide, the hydroxyl radical, and the nitrate radical.

Ozone depletion potential A factor, which determines the effect of a trace gas on the density of the ozone layer. The ozone depletion potential (ODP) is the relative amount of degradation of the ozone layer, which a compound can cause relative to trichlorofluoro-methane (CFC-11), fixed at an ODP of 1.0.

Partial pressure The pressure exerted by a specific gaseous component occurring in a mixture of gases (and vapors).

Passive sampler A device for the collection of gaseous trace substances from the air by molecular diffusion without controlled conveyance of the gas to be investigated.

Pelagic A term applied to organisms living in the middle depth and surface waters of the ocean. Also applied to sediments formed in deep waters along the slopes of the continents.

Perigee The point in the orbit of the moon or an artificial earth satellite that is closest to Earth. At perigee, the orbital velocity reaches a minimum.

Photochemical reaction Strictly speaking, a reaction involving an electronically excited state of a molecule or atom, which has undergone excitation by the absorption of light. This requires sufficient energy per quantum (usually ultraviolet or visible radiation).

Photochemical smog The mixture of high concentrations of ozone and other oxidants as well as nitrogen oxides, hydrocarbon oxidation products such as aldehydes, and aerosols that forms in urban atmospheres exposed to pollution by automobile exhaust and high sun intensities under stagnant meteorological conditions. Photochemical smog leads to pulmonary and other health-related problems, eye irritation, plant damage, and reduction of visibility.

Photochemistry Chemistry induced by reactions of electronically excited atoms and molecules formed by the absorption of light providing sufficient energy for the excitation. Primary photochemical processes are the various reaction channels open to the excited atoms or molecules. These include molecular decomposition into stable atoms, molecules, or radicals, molecular isomerization, fluorescence, phosphorescence, electronic quenching by collisions, etc. The subsequent reactions of molecular fragments thus formed are correctly referred to as thermal reactions, in that they are analogous to reactions induced by thermal decomposition at high temperatures. Atmospheric scientists sometimes use the term photochemistry (incorrectly) more widely to describe the chemical system as a whole, so that it includes primary processes and thermal (dark) reactions as well.

Photodissociation The process in which a molecule absorbs light endowed with sufficient energy to cause the molecule to split into fragments.

Photodissociation coefficient (j-value) An effective first order rate coefficient for a primary photochemical reaction in the atmosphere. Consider a photochemical process allowing two channels leading to different products, $X + h\nu \rightarrow A + B$ (1), $X + h\nu \rightarrow C + D$ (2); the corresponding two photodissociation coefficients are given by the integrals $j_1 = \int \sigma(\lambda)\phi_1(\lambda)F(\lambda)d\lambda$ and $j_2 = \int \sigma(\lambda)\phi_2(\lambda)F(\lambda)d\lambda$, where $\sigma(\lambda)$ is the absorption cross section of the species X being photo-decomposed, $\phi_1(\lambda)$ and $\phi_2(\lambda)$ are the quantum yields for processes (1) and (2), respectively, $F(\lambda)$ is the actinic flux occurring at the location being considered (all quantities depend on the wavelength λ), and the integral is to be taken over the photochemically active relevant wavelength region $\Delta\lambda$.

Photolysis The term means literally the cleavage of a molecular bond by light. It has also been used (incorrectly, however) to indicate the act of irradiating a substance with light. Specific terms such as photodecomposition, photo-isomerization, photodissociation that describe the physical consequences of the irradiation are more precise and are recommended.

Photosphere The visible region of the Sun at temperatures ranging from 4,000 to 6,000 K.

Photosynthesis A metabolic process occurring in plants and some types of bacteria, whence the absorption of visible light by chlorophyll and other photosynthetic pigments results in the reduction of carbon dioxide followed by the production of organic compounds. In plants the overall process leads to the conversion of CO_2 and H_2O to carbohydrates (and other plant material) and the release of oxygen.

Plagioclase feldspar A solid solution of calcium- and sodium-aluminosilicates (anorthite, albite).

Potential temperature The temperature which a dry air parcel would assume if it were lowered or raised adiabatically to a standard pressure (usually 1,000 hPa).

Precipitation (In chemistry) The formation and sedimentation of solid material in a liquid solution. For a precipitate to form, the solid must be present in amounts exceeding its solubility. (In meteorology) The occurrence of rain, snow, hail, etc. (In power plant technology) The separation of particles from the flue gases by application of a high potential difference (12–30 kV) in an electrostatic precipitator. Charged particles are attracted to an electrode of opposite charge and collected.

Precipitation element A condensed water particle of sufficient size to reach the ground before evaporating completely (usually >100 μm)

Predissociation (Spectroscopy, Auger process) The process by which a molecule excited to an electronic state at an energy level higher than the dissociation threshold undergoes decomposition rather than emission of radiation. The phenomenon is due to the overlap of vibrational and/or rotational levels of one electronic state of a molecule by the dissociation continuum belonging to another electronic state of the same molecule. The probability of the process depends on the strength of the mutual interaction of the involved states in competition with the probability of radiation to a lower energy level. In the region of predissociation the spectrum becomes diffuse.

Quantum yield Broadly speaking, the efficiency of a physical or chemical conversion process induced by the absorption of light. Specifically, (a) for a primary process it is the fraction of electronically excited molecules that decompose, isomerise, fluoresce or react by some other specific pathway. The quantum yield for each individual process is less than unity, but the sum of quantum yields for all primary processes following the excitation of a molecule is unity. (b) For secondary products generated by thermal reactions that follow the primary step the quantum yield of any stable product C is defined as the number of molecules of C formed in the system during a given short period of time divided by the number of light quanta absorbed by the photochemically active molecules in the system during the same time period. This type of quantum yield may be greater than unity and can reach large values when chain reactions are involved.

Quenching (Usually collisional quenching, also called deactivation) Removal of energy by collisions with other gaseous molecules or atoms from a molecule excited to a higher energy level, which leads to fluorescence or photodissociation, so that the intensity of fluorescence or the extent of photodissociation is markedly lowered.

Radical (also free radical) A fragment of a stable molecule, generally containing an odd number of bonding electrons. Atoms are radicals by definition, but are not usually called radicals. Radicals occur as intermediates in many complex chemical processes, such as in hydrocarbon oxidation and photochemical reactions. They are very reactive species determining the rate of the reaction in many cases. Important radicals in the atmosphere, responsible for much of the chemical transformation there, are OH, HO_2, Cl, ClO, NO_3, CH_3O_2. The widespread practice of designating a radical by the chemical formula with a dot placed next to the atom carrying the odd electron (example: $CH_3CH_2O\cdot$) is usually not followed in atmospheric chemistry.

Rain-out The mechanism by which small particles occurring in cloud drops are removed by the formation of raindrops followed by precipitation. This process is to be distinguished from *wash-out*, which occurs below cloud level. Both terms are not always used in accordance with these definitions. For clarity, the terms in-cloud scavenging and below-cloud scavenging are recommended.

Raoult's law For dilute solutions, the relative lowering of the vapor pressure of a liquid by dissolved substances. For a binary solution, Raoult's law can be expressed by the ratio of the actual vapor pressure p to the saturation vapor pressure p_s in the absence of a solute $p/p_s = n_{solvent}/(n_{solvent} + n_{solute})$, where $n_{solvent}$ and n_{solute} are the chemical amounts of the two substances present in the solution.

Rayleigh A measure of airglow brightness; one Rayleigh is the brightness of a source emitting 10^6 photon cm^{-2} s^{-1} in all directions (this corresponds to a surface brightness of $10^6/4\pi$ photon cm^{-2} sr^{-1} s^{-1}).

Rayleigh scattering The scattering of light by particles much smaller than the wavelength λ, specifically by atmospheric molecules. The scattering coefficient depends on the square of isotropic polarizability and rises with the fourth power of the reciprocal of the wavelength. The blue appearance of the sky is due to the consequence that blue rays of light are scattered more strongly than red rays of light.

Refraction (Optics) Change in speed and direction of radiation when passing from one medium into another.

Refractive index Ratio of the speed of light in vacuum to that in a material considered.

Refractory elements Elements that condense or vaporize at high temperatures, such as Al, Ca, Os, Re, Ti, W.

Relative humidity A measure of the amount of water vapor in a mixture of air and water vapor. It is commonly defined by the ratio of the partial pressure of water vapor to the saturation vapor pressure at the prevailing temperature, expressed in percent.

Residence time (referring to a geochemical reservoir) The ratio G_s/Q_s, averaged over a suitably long time period, where G_s is the mass content of a substance in the geochemical reservoir and Q_s is the sum of all sources, internal as well as external. For a balanced budget the sum of sources equals the sum of all sinks S_s, internal as well as external. If S_s is proportional to G_s, so that $S_s = (\Sigma k_i)G_s$, where the k_i are coefficients determining the rates of individual loss processes contributing to the total sink strength, the residence time is $\tau = 1/(\Sigma k_i)$.

Residence time (atmospheric) In this case, residence time refers to the whole atmosphere or a part thereof, such as the troposphere, which is treated as a geochemical reservoir exchanging material with adjacent geochemical reservoirs. Adjacent reservoirs of the troposphere are the stratosphere, the oceans and the biosphere. Anthropogenic emissions and production in the troposphere add to the content of materials in the troposphere. The residence time of a substance in the troposphere is determined by the mass content divided by the sum of rates of all emissions, transfer from adjacent reservoirs and internal production processes in the troposphere.

Residence time (instrumentation) Time required for air or any other reagent to pass from the entrance of a duct to the detection unit. This may be approximated by the ratio of interior volume of the duct to the flow rate.

Resonance Fluorescence See under fluorescence.

Respirable fraction Fraction of aerosol that can reach the pulmonary region of the human respiratory system.

Resuspension The process by which dust particles deposited on the ground surface are reactivated by wind force to become airborne again.

Reticle A transparent disk with lines or marks place in the focal plane of optical systems for calibration or alignment.

Reynolds number A non-dimensional ratio used for assessing the similarity of motion in viscous flows: $R = \rho \upsilon L/\eta$, where ρ is the density, υ is the velocity, η the viscosity of the fluid, and L is a characteristic length of the system considered.

Riming The growth of ice crystals in a mixed cloud due to the collection and freezing of liquid water drops. The process leads to the formation of graupel, and eventually to hail stones.

Rotameter A device used to measure the flow rate as indicated by the height of a float centered in a vertical tapered tube.

Salinity The salinity of the oceans is defined as grams dissolved salt per kilogram seawater in parts per thousand, ‰. The average salinity of seawater is $S = 34.7‰$. Salinity is related to chlorinity by $S(‰) = 1.80655 \, Cl(‰)$.

Saturation vapor pressure The (partial) pressure exerted by a pure substance at a given temperature when vapor and condensed phase of the substance are at equilibrium.

Scale height A parameter H indicating the rate at which pressure and density decrease with altitude z in an isothermal atmosphere; for example $p = p_0 \exp(-z/H)$. The local scale height in Earth's atmosphere is given by $H = R_g T/Mg = 29.2 \, T$ [m]. Here R_g is the gas constant, T is the absolute temperature, M is the molar mass of air, and g is the acceleration due to gravity. The scale height has also been used to indicate the total height of the atmosphere when compressed to a constant density under standard conditions of temperature and pressure.

Scrubber A device for the removal of material by transfer from a gas to a liquid medium. The gas is passed through a space containing the liquid in bulk, as a spray, or as a wetted surface. In the sampling of trace constituents from ambient air, the term *scrubber* is used synonymously with *absorber* (see there).

Sea breeze A local circulation occurring at the shore-lines throughout the world. During the day the sea breeze consists of a shoreward air flow at the surface, rising currents inland, seaward return flow near 900 hPa and sinking currents several kilometers out to sea. At night the air flow may be reversed.

Secondary aerosol Atmospheric particulate matter that is formed within the atmosphere by the chemical conversion of gaseous precursors to less volatile species, which condense to generate either new particles or attach to already existing particles. The most prominent example is ammonium sulfate resulting from the association of ammonia with sulfuric acid produced by the oxidation of sulfur dioxide.

Sedimentary rocks In geology, rocks that are formed from the sediments formed by particles carried to the ocean (and continental basins) with the rivers and by winds. These rocks experience tectonic uplift, become part of the continental crust, and are partly exposed at the earth's surface.

Sidereal year The time for earth to complete one revolution around the sun with respect to fixed stars: 365.25636 days. The sidereal year is 20 min longer than the tropical year.

Sinclair-LaMer Generator A device that produces monodisperse aerosols by the condensation of a vapor onto nuclei

Slip correction factor A factor applied to correct for slip flow using continuum gas flow equations.

Slip flow regime Transition between free molecular flow and continuum gas flow.

Small ion In aerosol physics, a cluster of molecules (mostly water) usually carrying one charge; as opposed to large ions, which represent aerosol particles having acquired charges by the attachment of ions from the surrounding air.

Smog Originally the combination of smoke and fog, a term used in Great Britain to describe the heavy pollution, rich in sulfur dioxide, resulting from coal burning in winter in large cities such as London. More common today in cities is photochemical smog characterized by high concentrations of ozone and other oxidants.

Smog chamber A large reaction chamber in which sunlight or simulated sunlight is used to irradiate a mixture of selected hydrocarbons and nitrogen oxides in air for the purpose of studying the complex reactions occurring during irradiation. While the results of such studies have led to valuable knowledge regarding the nature of photochemical smog in the atmosphere, unwanted wall reactions and ill-defined spectral intensities inside the chamber require precautions to be exercised when extrapolating the results to atmospheric systems.

Smoke A visible aerosol made up of particles formed by the incomplete combustion of organic matter. Smoke consists largely of carbon and carbon-rich products but does not include steam (condensed water vapor).

Solstice The dates, around June 22nd and December 22nd each year, when the sun's declination is +23.45° and −23.45°, respectively. The sun passes directly overhead at noon at 23.45° latitude north (Tropic of Cancer) in the first case, and at 23.45° south (Tropic of Capricorn) in the second case.

Solubility (General) Maximum amount of material (pure solid, liquid, or gaseous) that will dissolve at equilibrium temperature in a given amount of solvent. The system is in equilibrium when, at a fixed temperature, the solution phase as well as the solid, liquid or gaseous phase remain in contact indefinitely without further net change in amount of either phase. (Specific) The mass of a dissolved substance that will saturate 100 g of a (liquid) solvent. The solubility of gases in water is often considered in terms of Henry's law.

Solute (General) A substance which is dissolved in another. (Specific) A solid, liquid or gaseous substance dissolved in a liquid, the solvent.

Solution A homogeneous mixture of two or more elements or compounds. The term may be applied to solids as well as liquids, but unless so stated, solution normally refers to a liquid medium.

Solvation The association of ions or molecules of a solute with those of a solvent. The interaction is due to electrostatic or van der Waals forces, as well as chemically more specific effects such as hydrogen bonding. The term is also applied to the attachment of water molecules to ions in the gas phase.

Solvent The component of a solution which is present in excess, or whose physical state is the same as that of the solution.

Spallation A nuclear reaction in which the collision of a high energy particle with an atom or a molecule generates a larger number of disintegration products.

Specific Humidity The ratio of water vapor to air (including water vapor and dry air) in a particular air mass expressed in kg of water vapor per kg of total moist air.

Spectrum In the field of optics, a display of the intensity of electromagnetic radiation versus the wavelength; more generally, the frequency distribution of a physical quantity, such as the mass, the velocity or the size of a particle.

Spores Dormant cells of microorganisms.

Spray sampler A device for the efficient collection of water-soluble gases and vapors from ambient air. The air is pumped through a small chamber, where intimate contact with a fine mist of water drops is established, which are generated by a nebulizer from a fixed volume of liquid water. The loss of water at the exit of the chamber is prevented by a hydrophobic filter.

Standard atomic weight The abundance-weighted sum of the atomic masses of the isotopes of an element in the environment of the Earth's crust and atmosphere as determined by the IUPAC Commission on Atomic Weights and Isotopic Abundances. These values are included in a standard periodic table and are used in most bulk calculations. For synthetic elements the isotope formed depends on the means of synthesis, so the concept of natural isotope abundance has no meaning. Therefore, for synthetic elements the total nucleon count of the most stable isotope (i.e., the isotope with the longest half-life) is listed in brackets in place of the standard atomic weight. Recently, the Commission has proposed a new notation for ten elements, for which published distributions of isotopic composition span a larger range due to variations caused by sampling location rather than the precision of measurement. The ten elements are H, Li, B, C, N, O, Si, S, Cl, Tl. The preference now is to list the range rather than the average. Thus, for hydrogen as example the atomic weight is listed as [1.00784; 1.00811], for carbon the corresponding values are [12.0096; 12.0116] (M.E. Wieser, T.B. Coplen, Pure Appl. Chem. **83**, 359–396 (2011)).

Stern-Volmer Equation An expression for the pressure dependence of the fluorescence yield resulting from an excited atom or molecule, or the product quantum yield associated with the photodecomposition of a molecule. The relation takes into account that an excited atom or molecule may undergo collisions with other atoms or molecules present in the surroundings, whereby excitation energy is transferred and lost (deactivation or quenching). If Φ is the quantum yield of fluorescence or photodecomposition, the Stern-Volmer relation is written $(1/\Phi) = (1/\Phi_0)(1 + \alpha p)$ where p is the pressure and α an appropriate constant; Φ_0 is the quantum yield approached at zero pressure for the process under consideration. If only one process occurs, e.g. fluorescence, $\Phi_0 = 1.0$, otherwise Φ_0 is smaller than unity and represents the maximum yield that the process can attain.

Stratosphere The atmospheric region above the troposphere, which is characterized by temperatures increasing with altitude. The stratosphere extends from the tropopause at 10–15 km altitude toward the stratopause at 50 km, where temperatures reach a relative maximum, signaling the beginning of the mesosphere.

Stratus A layer cloud.

Subduction In geology, the tectonic processes by which rocks are moved downward to lower strata of the earth's crust or mantle.

Subsidence In meteorology, the sinking of an air mass, usually caused by the horizontal outflow of air at lower altitudes such as in the center of a high pressure region.

Tectonic Refers to earth movements that typically lead to the faulting, folding, and/or vertical movement of rock.

Terminal settling velocity The downward constant velocity of raindrops reached when falling under the opposing influence of gravity and fluid drag.

Texture The relationship of mineral grains in a rock; the physical graininess of soil.

Thermophoresis Motion of particles in a temperature gradient, i.e. from a hotter to a cooler region.

Thermosphere The outer region of the atmosphere; here the temperature rises with increasing altitude, reaching 700–2,000 K depending on solar activity. This phenomenon is caused by the absence of infrared-active molecules capable of radiating the energy deposited by the absorption of solar ultraviolet radiation back toward space. Heat conduction serves to transport the excess energy to lower atmospheric regions, where infrared radiative emissions become possible.

Total carbon (TC) Sum of elemental carbon and organic carbon (see elemental carbon).

Tropical year The time between two successive passages of the sun through the vernal equinox (around March 21st): 365.24219 days. The tropical year is about 20 min shorter than the sidereal year because precession produces an annual net retrograde motion of 50.28 arc sec of the equinoxes relative to fixed stars.

Tropopause A narrow region in the atmosphere between the troposphere and the stratosphere. It is characterized as a change in the temperature gradient (lapse rate) from negative to positive. The WMO (World Meteorological Organization) definition is: the lowest level at which the lapse rate decreases to 2 K/km or less, and the average lapse rate from this level to any level within the next 2 km does not exceed 2 K/km.

Troposphere The lowest region of the atmosphere ranging from the earth surface to the tropopause, which locates the base of the stratosphere, at 10–15 km altitude depending on latitude and meteorological conditions. The troposphere harbors about 70% of the total mass of the atmosphere.

Turbulent flow See fluid flow.

Vapor pressure For a vapor derived from a liquid or solid, the partial pressure of the vapor in equilibrium with the condensed liquid or solid at the given temperature.

Venturi A convergent-divergent duct in which the pressure energy of the air stream is converted into kinetic energy by acceleration through the narrow part of the wasp-waist passage. It is a common method of accelerating the air flow in a wind tunnel. Small venturis are used in an aircraft to provide a suction source for vacuum operated instruments, which are connected to the low-pressure neck of the duct.

Visibility Defined as the greatest distance at which a black object of suitable dimension can be seen and recognized against the horizon sky. The criterion of recognizing the object, not just seeing the object without recognition, is applied.

VOC (Volatile organic compound) It is defined by the World Health Organization (WHO) as any organic compound having a saturation vapor pressure at 25°C greater than 102 kPa.

Volatile substances Elements or compounds that condense or vaporize at relatively low temperatures.

Wash-out The removal of small particles and gases from the atmosphere by the attachment to, or dissolution in, raindrops on their way from a raining cloud to the earth surface. Unfortunately, the term has also been used for the absorption of material by cloud drops followed by rain formation and precipitation. For clarity, the expression wash-out should be replaced by the term below-cloud scavenging.

Wet deposition The removal of a trace substance from the troposphere by incorporation into cloud, fog or rain drops, followed by their precipitation to the earth surface.

White cell A multi-path optical cell based on the design of White (J. Opt. Soc. Am. **32**, 285–288 (1942)). It consists of three spherical mirrors of identical radii of curvature. The larger front mirror is placed opposite two identical D-shaped rear mirrors, suitably adjusted so that the input beam is directed back and forth between the front mirror and the rear mirrors alternating between both. The re-imaged beam forms two rows of spots on the front mirror whereby interferences are avoided.

Zonal The term zonal means along a latitude circle, that is, in the west–east direction

Index

A

Absorption coefficients
  hydrogen peroxide, 302
  iron ions, 303
  nitrate anion, 302
  nitrite anion, 303
  nitrous acid, 303
Absorption spectra
  absorption cross sections, 204–224
  primary quantum yields, 204–224
Acetaldehyde, 72, 86, 89, 215, 246, 249, 281, 293, 314
Acetic acid, 72, 85, 281, 296
Acetone, 72, 86, 89–91, 215–216, 246, 250, 293
Acetonitrile, 71, 292
Air
  basic properties, 50–52
  bulk chemical composition, 49, 51
  rare gases, 51, 53
Air glow phenomena
  aurora, 363–367
  bands systems, 362, 363, 368
  day glow, 361–367
  nightglow, 361–363
Ammonia, 43, 71, 82–83, 127, 134, 290, 296, 382, 387–389
Association reactions, 227–229, 255, 268–269
Atmospheric aerosol
  biological particles, 134
  characterization, 128
  chemical composition, 128, 129, 147–148
  continental, 129, 132, 134
  desert, 132, 149
  dicarboxylic acids, 176, 177
  global production/emission rates, 133–138

  marine, 129
  mineral dust, 129, 133
  n-alkanals, 167
  n-alkanes, 164, 166, 168–172, 178
  n-alkanoic acids, 167, 168, 170–173, 178
  n-alkanols, 167, 170, 171, 178
  n-alkenoic acids, 168
  organic compounds, derived from automobile traffic, 164–166
  phenols, 168–169
  phytosterols, 169, 178
  polycyclic aromatic hydrocarbons, 163, 165–169, 174, 175
  polycyclic aromatic ketones, 166
  resins, 134, 166, 168
  sea salt, 127–129, 132–134, 136, 185
  size distribution, 128, 129, 131
  terpenoids, 167
  trace metals in, 135–137

B

Benzene, 72, 90, 245, 250, 378
Biosphere
  carbon content, 39, 42
  global net production, 41
  terrestrial biomass, 40

C

Carbon disulfide, 292
Carbon monoxide, 30, 70, 79, 238, 291, 292, 382
Carbon tetrachloride, 75, 217–219
Carbonyl sulfide, 71, 129, 217, 292
Chlorofluorocarbons, 374, 401

P. Warneck and J. Williams, *The Atmospheric Chemist's Companion: Numerical Data for Use in the Atmospheric Sciences*, DOI 10.1007/978-94-007-2275-0,
© Springer Science+Business Media B.V. 2012